Ludwik Hirszfeld

Rochester Studies in Medical History

Senior Editor: Theodore M. Brown
Professor of History and Preventive Medicine
University of Rochester
ISSN 1526–2715

Ludwik Hirszfeld

The Story of One Life

Translated and Edited by MARTA A. BALIŃSKA
Edited by WILLIAM H. SCHNEIDER

R UNIVERSITY OF ROCHESTER PRESS

First published 2010
Reprinted in paperback 2013

University of Rochester Press
668 Mt. Hope Avenue, Rochester, NY 14620, USA
www.urpress.com
and Boydell & Brewer Limited
PO Box 9, Woodbridge, Suffolk IP12 3DF, UK
www.boydellandbrewer.com

ISSN: 1526-2715
ISBN: 978-1-58046-338-6

Library of Congress Cataloging-in-Publication Data
Hirszfeld, Ludwik, 1884-1954.
    [Historia jednego życia. English]
    Ludwik Hirszfeld : the story of one life / translated and edited by Marta A.
Balińska and William H. Schneider.
        p. cm. – (Rochester studies in medical history, ISSN: 1526-2715 ; v.16)
    Includes bibliographical references and index.
    ISBN 978-1-58046-338-6 (hardcover : alk. paper) 1. Hirszfeld, Ludwik, 1884-
1954. 2. Microbiologists–Poland–Biography. 3. Serology. 4. Blood groups–ABO
system. I. Balińska, Marta A. II. Schneider, William H. (William Howard), 1945-
III. Title.
    QR31.H57A3 2010
    579.092–dc22
    [B]

2010010851

A catalogue record for this title is available from the British Library.

This publication is printed on acid-free paper.
Printed in the United States of America

Contents

Illustrations

Foreword to
The Story of One Life

Next to that of Karl Landsteiner, the name of Ludwik Hirszfeld is unquestionably the greatest in the early history of blood grouping and its applications. He was at the same time an outstanding personality, inspiring the work of others, and much loved by generations of students and colleagues. And his life encompassed times of fruitful fundamental discovery, and some of the most terrifying and painful experiences that a man can live through and yet survive, events which will have a permanent place in history. With all this he had the gift of writing—of clear description of scientific research and the human interactions which bring it about, of poetic description of nature and of human love, and of writing movingly and without bitterness of the almost unbelievable tragedies which he witnessed in the beleaguered Warsaw Ghetto. Thus we welcome as a great literary event the appearance of an English translation of his autobiography.

It is mainly a great human document, which will appeal to people of all backgrounds, whether scientific or not. However, it is being issued as one of a series of historic scientific works, and it is therefore appropriate to try to assess the position which Hirszfeld's researches occupy in the field of blood-group investigation.

His first blood-group studies were carried out on dogs and were inspired not by the work of Landsteiner on human beings but by the independent and simultaneous work of Ehrlich and Morgenroth on goats. In the course of it he discovered that dogs had blood groups and that these were hereditary characters. It was only when he proposed to extend his researches to man that he found out about the work of Landsteiner.

Subsequently the greater part of his research work was on human blood groups, a field in which he made numerous discoveries, the most important of which were the demonstration that human blood groups too behaved as Mendelian characters, and the finding of the subgroups of A, both in collaboration with von Dungern; and some years later, in collaboration with his

The following foreword was prepared by Arthur E. Mourant for an earlier attempt at an English translation of Hirszfeld's autobiography. It remains entirely fitting and serves as a tribute to the work of both men.

wife, the discovery that the frequencies of the blood groups varied widely in different populations.

He was deeply conscious of the probable biological importance of the blood groups and was perhaps unfortunate not to be the first to identify materno-fetal isoimmunization as the cause of haemolytic disease of the newborn.

One of his many other important contributions to blood-group research was his work, following the initial findings of Schiff, in the elucidation of the relationships between the A_1, A_2, B, and H antigens—work recently brilliantly crowned by the genetic work of Race and Sanger and the biochemical explanations of Morgan and Watkins.

To revert, however, to his personal character, perhaps the true greatness of himself and his wife was most clearly shown when, as an aging couple worn out by the unspeakable privations and horrors of war and the tragic death of their only daughter, they might easily have chosen a life of honor and ease in America, but they chose instead to stay and help to rebuild the edifice of medicine and science in war-worn Poland.

A. E. Mourant

Preface

The publication of this English translation of *The Story of One Life* has taken more than sixty years and the efforts of a number of people, beginning with Hirszfeld himself. He was not successful at reaching that goal nor were subsequent attempts by a number of prominent scientists such as Arthur F. Coca, A. E. Mourant, Tibor Greenwalt, and Frank Camp, the latter who headed a U.S. Army project at Fort Knox, Kentucky, to translate classic works in the field of serology. There were at least two English manuscripts prepared, one of which we have located. The present translation was entirely reworked according to the 2000 Polish edition of *The Story of One Life*, with all missing passages reinserted. The present editors and translator are grateful to be associated with such a distinguished group and humbled by the honor of seeing the project finally come to fruition.

To our knowledge, there exist to date only two biographies of Ludwik Hirszfeld, both in Polish. The first one, published in Warsaw in 1977 and translated into German (East Germany) some years later, is a short monograph that mostly summarizes Hirszfeld's own memoirs. It cites no sources, provides little information on Hirszfeld's background, and does not make an attempt to delve into the significance of his scientific contributions.[1] The author, Marek Jaworski, must have interviewed close relations (although no names are referenced) because he writes that Hirszfeld's dying words were: "But I believe that [in the future] people will be better."[2] The second biography appeared in Poland in 2005 and was written by the scientist Waldemar Kozuschek, a former student of Hirszfeld.[3] It too quotes extensively from Hirszfeld's *The Story of One Life*, but Kozuschek goes further by providing bibliographical and biographical sources. Concerning Hirszfeld's personal life, he provides no more (and rather less) information than is in the present edition.

This is not a definitive, exhaustive, scholarly edition of the autobiography. But because many names and events may be unfamiliar, the editors have attempted to provide at least birth and death dates for individuals mentioned. For more important individuals, where information exists, a brief description is given to show relevance to the autobiography and some references for further information where possible and useful. The goal is to provide background and make more accessible the richness of the work without inhibiting the flow of the narrative. Some people and references are better known than others, and in questionable cases, the authors have erred on the side of adding material when available. In many places, however, Hirszfeld provides names impossible to identify. It is a kind of Kaddish, a Jewish ritual of remembering the dead by repeating their names. The editors have left them as such in this autobiography.

An additional help to readers is a biographical annex in the appendix that offers a quick way to check for names repeated in subsequent chapters. A notation is given only at the first mention. The reader can thus check the biographical appendix if a name appears without reference in subsequent chapters to see if information was provided in an earlier chapter.

Given Hirszfeld's travels and the changing rulers of those lands, the names of places and individuals may have more than one spelling. Hirszfeld himself, for example, published under the Germanized name Ludwig Hirschfeld until he returned to Poland after World War I. In this case, we have used his Polish name throughout and in other cases we have attempted to use contemporary names, unless the historical and more common ones are familiar to the English-speaking reader (e.g., Warsaw instead of Warszawa).

Thank you to those who helped make this work possible, among whom are Craig Chosney, Kelly Canaday, Vanessa d'Amico, Kelly Gascoine, Judi Izuka-Campbell, the archivists at the Polish Academy of Sciences, Pauline Mazumdar, Anne-Marie Moulin, Andrzej Górski, Felix Milgrom, and Stanisław Dubiski. Special thanks go to Joanna Kiełbasińska-Belin. We are very grateful to the National Library of Medicine, which provided a Health Sciences Publication Award for support of this project and to Shifra Shvarts and Paweł Sztwiertnia who arranged additional support from the Gertner Institute in Israel and INFARMA in Poland.

Introduction

Marta A. Balińska and William H. Schneider

For most autobiographies, an introduction might seem superfluous, the authors having recorded what they wished to leave to posterity. But the memoirs of Ludwik Hirszfeld have a special story. To begin with, they were not intended as a literary work or as the reminiscences of a prominent scientist in the twilight of his career. In fact, Ludwik Hirszfeld would probably never have written this book had he not lost his only child and been forced to bury her under another name. He had thus been deprived not only of his continuity but of his identity. This above all—as well as the barbarity he experienced firsthand in Nazi-occupied Poland, and particularly the Warsaw Ghetto—was what prompted him to revisit his past. It was undoubtedly as much a question of maintaining his sanity as it was of bearing witness. "I wrote my memoirs in the greatest of pain," Hirszfeld explained to a Swiss acquaintance,

> directly following the death of my only child, in a house in the middle of a forest where I was living under a foreign name. My only wish was to tell the world of the unspeakable suffering war brings about and in what an appalling way the war had deformed the human spirit.[1]

The death of Hirszfeld's daughter was a personal motivation, and the book commemorates her as well as his own life, something his daughter could obviously no longer do. Hirszfeld's desire to bear witness to the horrors of the Nazis was similar to what other survivors felt, but Hirszfeld had an added responsibility given his status as a senior medical and scientific authority. Ludwik Hirszfeld was one of the pioneers of immunology and serology in the first half of the twentieth century. Having made a number of fundamental discoveries about human blood groups, as well as establishing Poland's first immunological laboratory between the wars, his career was abruptly interrupted by World War II and Nazi occupation. Hirszfeld was confined to the Warsaw Ghetto, and it was only after he escaped with his family that his daughter tragically died while in hiding.

Hirszfeld was not the only respected medical or scientific figure to witness the Nazi atrocities. Poland had a rich tradition of Jewish doctors and scientists, many of whom ended up in the Warsaw Ghetto. A number of them had experience and prestige comparable with Hirszfeld's, and a few of their observations survived in the form of memoirs and journals that were hidden or smuggled out. But Hirszfeld was undoubtedly the most prominent to live to tell the tale.

Hirszfeld wrote by compulsion, rising in the early morning hours and then dictating his recollections to his wife who typed them. In two months, the book was completed. At the time—1943—he was hiding in the Polish countryside, and as soon as the Germans were thought to be in the vicinity, the typewriter was quickly shut away in a back cupboard. When he left this hiding place in Miłosna to go to another, he buried the manuscript under the floorboards, where he retrieved it after the war.[2]

With a mixture of guilt and gratitude but above all responsibility, Hirszfeld undertook the writing of the autobiography. And he was well up to the task. Given his training and experience, he understood better than most around him what was happening, and his descriptions, especially of the typhus epidemic and starvation in the ghetto, contain a penetrating perspective without being cold or aloof. More surprising, given his lack of experience as a writer, is the humanity and compassion he showed in conveying the individual horror faced by those forced into the ghetto. The combination of these two qualities makes for a powerful account.

The Story of One Life is not, however, an account of only World War II and the Holocaust. Almost half of the book is devoted to the happy years that preceded it: his medical studies in Germany, his extraordinary adventures with the Serbian Army during World War I, his enthusiastic participation in the construction of an independent Poland . . . and throughout, his passion for science and teaching. In the course of reading, one becomes aware that *The Story of One Life* was written by a suffering individual trying to come to terms with the total annihilation of the values of his youth and a world he had once loved. Thus, in addition to describing the life of a man of science, Hirszfeld's autobiography depicts the destiny of one man against the background of the history of his time. His life was both exceptional and revealing of the major events that affected millions of human beings in the first half of the twentieth century.

Previous Attempts at Translation

The Story of One Life was first published in Warsaw in 1946 and met with immediate success. Ten thousand copies sold within the first two years, which was a large number considering postwar conditions in Poland.[3] The Polish press was unanimous in praising the book's depiction of the sufferings of millions, its convincing critique of German science, and its frank assessment of the Polish-Jewish question. Hirszfeld received hundreds of letters from enthusiastic readers as well as a certain number of virulent verbal attacks that accused him of being a traitor to the Jewish cause as a convert and an advocate of assimilation.[4]

One of Hirszfeld's ardent wishes was that his memoirs appear in the West, first and foremost in the United States where he had spent three months beginning in May 1946. In particular, he hoped that his accounts of German cruelty might have an impact on postwar negotiations about the occupation and treatment of Germany.[5] The book was accepted for publication by Little, Brown, and Company, a contract signed in December 1946, and a translator identified. After eleven months of translation work and revisions, much to Hirszfeld's dismay, the publisher decided the book would not sell and the contract was canceled.

The problem was one of timing. "The first half of the book [his life before the Second World War] is more interesting to an American audience now than the last part," wrote the publisher James Brown to Hirszfeld when he broke the bad news in November 1947.

> But even it does not have the dramatic qualities necessary to capture the imagination of the American reader. The latter part of the book, we feel, kills it completely as something for the American audience now. Unfortunately, the people of this country have reached the saturation point with accounts of aspects of the war and its attendant miseries.[6]

Hirszfeld was devastated and not convinced by the reasons stated for breaking the contract. He wondered whether the book had been purposefully blocked by opposition to his perspective,[7] which his family suspected existed among certain Americans who were unpleasantly surprised by Hirszfeld's largely (though far from wholly) positive portrayal of the Poles during the war. In France, the prominent pediatrician Robert Debré, a close colleague of Hirszfeld's wife, Hanna, also made fruitless attempts to have the book published.[8] Debré found a publisher who was interested but wanted to see the English translation first. A German translation, which Hirszfeld himself worked on with a Swiss woman in Wrocław, was found among the manuscripts in the apartment in which he died. Until the appearance of the present English edition, the book was published in translation only in Serbo-Croatian, in 1963, as a tribute to his work with Serbs during World War I.

Structure and Themes of the Autobiography

The first part of Hirszfeld's autobiography is about his life in science before World War II. It can best be understood as comprising three periods, the first of which covers the years before World War I and describes a "golden age" of international science in a number of countries with Germany at the center.[9] These chapters reveal much about scientific discovery, the real workings of research laboratories, and the master-apprentice model of training and collaboration. Despite Hirszfeld's inclination sometimes to avoid discussion of his research, it is a valuable contribution to understanding how German scientists worked in the most fruitful period of history of science in that country.

The chapters describing World War I mark an abrupt change in Hirszfeld's life. Residing in neutral Switzerland at the outbreak of war, Hirszfeld describes how he volunteered twice to join the Allied cause. First he left Zurich in response to the typhus epidemic in Serbia. Then, after being evacuated in the face of an Austrian offensive, he went to Salonica where he spent more than two years running a bacteriology laboratory with his wife. As bad as the war was, viewing it from the perspective of 1943 he presents an almost nostalgic view of the international cooperation of allies and a war fought under rules that had been observed for centuries. The Serbian evacuation, an overland march of 150 miles as winter set in, is described by Hirszfeld in noble as well as frightful terms but is nothing like his later experience of World War II. The Salonica front where he spent his second tour was a contrasting anomaly of lack of fighting that permitted Hirszfeld time for research. Still, the World War I episodes he describes are hardly unique, since Hirszfeld was but one of a number of individuals thrown into the melee who recounted their extraordinary experiences.[10]

The third phase of Hirszfeld's life described in this book falls between the two world wars, and it covers his work both building a public health service while conducting scientific research in Poland and participating in international science. Hirszfeld found public health service tedious, albeit necessary. He was frustrated by the politics and inefficiencies of scientific research institutions but obviously took great satisfaction in what he was able to accomplish.[11] Two chapters (7 and 8) are particularly instructive of the workings of internal politics, both scientific and professional, and governmental administration. His tone brightens considerably in the next chapter when he describes his scientific work and especially his assistants and collaborators at the Polish State Institute of Hygiene (Państwowy Zakład Higieny, often called the National Institute of Hygiene or just Institute of Hygiene). In light of the disastrous fate that befell most of them during World War II, Hirszfeld clearly means to recognize and thank them.

International science is presented mostly through Hirszfeld's descriptions of the congresses he attended between the wars. In hindsight at least, Hirszfeld was keenly aware of the corrosive influences of nationalism on science, especially at the meetings in Italy during the rule of Mussolini. Earlier glimmers are also seen, as when German scientists supported Kaiser Wilhelm during World War I.[12] Overall, Hirszfeld makes a series of observations about the similarities and differences in science in the various countries in which he works and visits between the wars. These include "new" countries such as Poland and Yugoslavia, which were just establishing their scientific infrastructures. Indeed, Hirszfeld has a particular sympathy for the Yugoslavs, stemming from his work in Serbia during World War I.[13] He also kept in close touch with colleagues in countries with long-established scientific institutions, such as France, Denmark, Switzerland, and Italy thanks to his participation in the League of Nations Biological Standardization Commission, as well as conferences he describes.

The second part of the book is about World War II, and it could not be in greater contrast to the hope and optimism of nation building and scientific collaboration in the 1920s and 1930s. Many accounts, both autobiographical and otherwise, have described the fate of Poland and Warsaw, and especially the Holocaust. Hirszfeld's descriptions of what he experienced are both powerful and alive, but his age (in his late 50s) and experience give them a reflective dimension, rarely found in other memoirs.[14] In particular, Hirszfeld sheds light on some of the more prominent ghetto leaders' actions that could appear "fatalistic" at first sight. But in the context of the times as described by Hirszfeld, they appear to be a more dignified and chosen "suicide" to support the cause they believed in.

And this leads to one of the central points of the book: Hirszfeld's story is especially compelling not only in itself but also because of his high position within the Jewish Health Council of the Warsaw Ghetto. He was thus in close contact with German officials responsible for administering the "district," as the ghetto was called, and the city of Warsaw. He participated in oversight of the makeshift clinics and what was left of the hospitals, which enabled him to appreciate the larger picture of what was happening. His chronicle of death in an unimaginable setting is both "clinical" and eerily detached, yet also sympathetic to the people involved and how they coped. In addition, the chapters about his escape and movements around occupied Poland show facets of Polish (non-Jewish) life not usually included in accounts by people with fewer resources and connections.

Ludwik Hirszfeld (1884–1954)

Rarely does autobiography tell all, and such is the case with Hirszfeld's book. For example, he does not comment on his scientific work or the early years of his life. In the case of his scientific work, he makes this explicit (chapter 2), but there are two important chronological periods that are noteworthy in their absence. To begin, Hirszfeld's boyhood years and family background are completely ignored in the first chapter, which, as the title indicates, starts with his years as a medical student in Germany. The second missing period is more obviously explained since the book ends in 1944 (two years before publication) and therefore does not include the last ten years of his life. These were fraught with personal suffering (the Hirszfelds never spoke of their daughter's death), a sense of duty and accomplishment (he provided scientific leadership as one of the few scientists of international distinction from the interwar period who survived in a war-devastated country deprived of its intellectual elite), and ideological tension owing to the communist regime (Hirszfeld did not live to see the downfall of Stalinism).

As to the first omission, Ludwik Hirszfeld was born in Warsaw on August 5, 1884, the son of Stanisław, a banker, and Jenny Ginsburg. Little is known about his mother's background apart from the fact that her family was highly cultivated and well-to-do. They owned factories in the textile center of Łódź as well as Warsaw. His father came from a prominent family of the Warsaw Jewish intelligentsia. Ludwik's paternal grandfather, also named Ludwik and well known as a benefactor of the arts, was reputed also to have "financed" the 1863 Polish insurrection against the Russians, which gives some indication of the extent of his Polish patriotism.[15] Although Stanisław and Jenny did not convert to Christianity, several of their children did. According to Felix Milgrom, who was Hirszfeld's last student and was designated to be his successor at his research laboratory, Hirszfeld was circumcised and attended Jewish religious classes for some time.[16] This was in contrast to his first cousin and close colleague Ludwik Rajchman who was baptized at birth, even though his parents were agnostic. Hirszfeld's sister Józia was baptized only after their mother's death in the mid-1930s. Hirszfeld's other sister, Róża, married Władysław Sterling, a prominent professor of neurology. Both sisters perished during World War II. Róza and her husband were shot and Józia died of heart failure.[17]

Ludwik Hirszfeld spent the first years of his life in Warsaw until his father went bankrupt. The family then moved to Łódź, where his mother's family employed Stanisław at their factory. Stanisław's business failure appears to have been a major event in the Hirszfelds' life. Several of Stanisław's siblings had invested in his bank and lost their fortunes as well, which created a long-lasting rift in the family. It explains, for example, why Ludwik Hirszfeld and Ludwik Rajchman, who worked together as adults, hardly knew each other in childhood. It was Rajchman, Medical Director of the League of Nations Health Organization and future founder of UNICEF, who persuaded Hirszfeld to work at the new State Polish Institute of Hygiene in 1920.[18]

The family member with whom Hirszfeld most identified was his paternal uncle, Bolesław Hirszfeld, a chemist by training, a teacher by vocation, a defender of political prisoners by conviction, and a socialist by ideology. He was a champion of "education for all" and a passionate Polish patriot.

Interestingly, Bolesław Hirszfeld was also a "role model" for two of Ludwik's cousins: the aforementioned Ludwik Rajchman, as well as his sister Helena Radlińska who became one of the pillars of Polish pedagogy.[19] The only mention of family history in the first few paragraphs of *The Story of One Life* is an account of Bolesław Hirszfeld's funeral in 1899, at which hundreds of mourners turned out despite the period of tremendous political repression. Little could Ludwik Hirszfeld have guessed that his own funeral in 1954 would be similar in both regards. He also does not mention that Bolesław Hirszfeld committed suicide. In fact there was more than one suicide among the Hirszfeld relatives, and the family had a history of being subject to severe depression.[20]

As a young boy in Łódź, Hirszfeld attended the Russian *gimnazium*, well known for its draconian methods and anti-Polish repression. Those times were marked by great patriotic and social fervor. "Education for all" and "a common Polish culture" were the rallying cries of an entire generation who believed that Polish independence and social equity would be attained through making instruction accessible to all social classes and uniting the many ethnic minorities around shared cultural values. In his *gimnazium*, Ludwik was the librarian of the clandestine self-learning group.[21]

At the age of sixteen, he fell in love with Hanna Kasman, a young girl he met while attending dancing lessons. The Kasmans were a well-to-do Jewish family who owned a large property near Modlin, a nineteenth-century fortress to the north of Warsaw. Hanna's father was a physician who had died, and her mother remarried giving birth to two other children: Izabela and Szymon Belin. By all accounts, Hanna was a very beautiful and serious girl who, like Ludwik, was taken with socialist and humanitarian ideals. Although neither of them belonged to the Polish Socialist Party (Polska Partia Socjalistyczna, PPS), they both sympathized with its progressive ideals and the struggle for Polish independence. The young couple liked to stroll together, she with Lenin's *Sparks* under her arm, and he with Marx's *Das Kapital*.[22] It is not clear what the Hirszfelds's attitude was toward Soviet Russia between the wars, but their reputation of being "to the left" must have provided them with at least some protection in communist Poland after World War II.

Ludwik and Hanna married as medical students while he was studying in Berlin, and he obtained his medical degree after completing his thesis in pathology in 1907.[23] Faced with the problem of obtaining positions for both of them, they chose Heidelberg, where Ludwik worked at the Cancer Research Institute, eventually joining the laboratory of Emil von Dungern.

In 1911, Hirszfeld accepted the offer of a position in Zurich at the Institute of Hygiene where he had an opportunity to teach. Hanna, who had specialized in pediatrics, obtained an appointment at the Children's Clinic. In 1913, when Ludwik received his *Habilitation* at the Institute, his career seemed set. With the outbreak of war the following year, Hirszfeld hesitated to become involved, in part because being in Switzerland he did not need to, but also because it was not clear how to, since there was no belligerent with whom he strongly identified. Thus, it was not clear where he could be useful.

The report of a typhus epidemic in Serbia early in 1915 gave him the opportunity to use his skills in the service of a Slavic country. The disease broke out among Austrian prisoners and quickly spread throughout Serbia where approximately half a million cases were reported in the first six months of 1915.[24] Fortunately, there was a lull in the fighting and Serbia's allies and neutral countries still had access to the country, so Hirszfeld joined a number of foreign medical relief personnel who quickly responded. One reason for the speed was that the French bacteriologist Charles Nicolle had recently (1913) discovered the louse-borne nature of the disease, and countries had already established mechanisms to monitor and control typhus among refugees and immigrants. After a few months of service, Hirszfeld was joined, despite his objections, by Hanna, but they both were soon forced to retreat with the Serbian Army following a German offensive beginning in October 1915.[25] They were evacuated and returned to Zurich in December 1915. Once committed to the war, however, Hirszfeld found it difficult to remain on the sidelines. He therefore volunteered again, this time to serve with British, French, and other Entente Cordiale forces, and he and his wife were sent to Salonica to set up a hospital laboratory for the *Armée d'Orient*, which had retreated from the Dardanelles campaign.[26] This remained one of the few inactive fronts for the rest of the war, and it was there that he and Hanna conducted their groundbreaking population genetics study of blood-type distribution among soldiers from the many different countries stationed there.

Hirszfeld did not leave Greece until the spring of 1919, passing through Belgrade, Zagreb, and Vienna on his way to visit family and friends in Poland. He decided to settle in Warsaw, after his cousin Rajchman persuaded him to work at the Polish State Institute of Hygiene. He soon took over most administrative responsibilities when Rajchman left to work for the League of Nations in Geneva. Hirszfeld was also able to establish within the institute his own research laboratory where he and coworkers continued immunological research.[27]

The political situation in Poland became more perilous in the late 1930s, and along with it nationalism, anti-Semitism, and authoritarianism grew stronger. Hirszfeld describes how this affected his work: he was increasingly restricted in whom he could hire, and his responsibilities shifted as the institute was affected by changes in politics. Nevertheless, he remained a key figure in Polish public health until World War II and even after his dismissal from the institute shortly after the Germans appointed their own director.

Hirszfeld remained in Warsaw despite offers of assistance to leave. He and his family were forced into the ghetto during February 1941 where they remained until their escape in 1943. After being smuggled out, they spent a few weeks in Warsaw, then a year and a half in various locations in the Polish countryside, during which time Hirszfeld drafted his autobiography.

Hirszfeld provides great detail about his life in hiding but not about his family's escape from the ghetto. He originally devoted an entire chapter to the Army of the Polish Resistance (*Armia Krajowa*, or Home Army), which orchestrated the Hirszfelds's escape from the ghetto and thus saved their lives. Because the underground Home Army was seen as a potential threat, it was disbanded by the Soviets when they occupied Poland at the end of the war. Hence, it would have been far too politically dangerous for Hirszfeld to acknowledge it in 1946, when *The Story of One Life* was published. Entitled "A Window on the World" ("Okno na świat"), the chapter "described Hirszfeld's friendship with Zygmunt Zaremba who was one of the leaders of the Polish Socialist Party and one of the important figures of the [Polish] underground. . . . The Hirszfelds' escape from the ghetto, as well as their subsequent hiding places, was orchestrated by Hanna's brother-in-law, Stanisław Kiełbasiński, and they received their falsified papers from the Home Army."[28]

The afterword to *The Story of One Life* was not included in the 1946 edition.[29] It was written at the request of Little, Brown and Company, who convinced Hirszfeld that readers would be curious to find out how he survived the war and what became of him afterward. After the fighting ended in Poland, Hirszfeld was very active in the scientific reconstruction of the country. In 1944, he participated in the creation of the Maria Skłodowska Curie University in Lublin as "prorector" and was then asked to reestablish the Medical School in Wrocław (formerly Breslau) in German territory that had been ceded to Poland after the war. He was chair of the Department of Microbiology as well as dean of the school, and founded the country's first Institute of Immunology in Wrocław, which now bears his name.[30]

Although Hirszfeld had new opportunities to rebuild the framework for medical research after the destruction of the war, he also faced threats to his scientific and personal life in nascent communist Poland. Understandably, Hirszfeld says nothing in the afterword about the heavy political atmosphere in which Poland had been plunged. Nor could he mention the difficulties he experienced as an independent-minded scientist, while Lysenko reigned supreme in Soviet biology, according to two of his last students who were forced out of positions and emigrated to the United States and Canada after Hirszfeld's death.[31]

An example of the tensions and the bitter paradoxes Hirszfeld had to face can be seen in the following excerpt from an unpublished memoir written in 1962 by Hanna shortly before her own death. It concerns a hematology

conference in Gdańsk (former Danzig) in 1952, on the occasion of the fiftieth anniversary of Landsteiner's discovery of blood groups. Given Hirszfeld's
contribution to the field, he figured prominently at the conference. While
he and his wife were having breakfast, the director of the Warsaw Institute
of Hematology asked to join them, along with two leading Soviet hematolo-
gists, G. A. Alekseyev and A. S. Kiselov.

> We spent a long time over breakfast discussing the discoveries of
> [Hirszfeld's] Institute, research on serological conflicts, theories of
> allergy, spontaneous abortions and the like. [The Soviets] were very inter-
> ested by everything we said and invited Ludwik to go the Soviet Union
> and teach them about his findings. In return, Hirszfeld invited the Soviets
> to visit Wrocław after the Gdańsk meeting. But, the following day at the
> hematology conference, Ludwik was giving his talk entitled "50 years of
> research in blood groups," and behind him the wall was covered with
> graphs, amongst which was our table showing the anthropological
> research into blood groups of different nations and races. We had carried
> out that research during the First World War and it was confirmed by
> research on 5 million people . . . all over the world. I was sitting in the
> first row, near the two Soviet scientists. . . . At a certain point I overheard:
> "You see what a reactionary and antisemite: he put the Russians next to
> the Jews."[32]

The problem with blood groups in communist Poland, as one of
Hirszfeld's students put it, was that they were inherited! As supporters of
Lysenko, the Soviets could hardly approve of Hirszfeld, whose methods they
claimed to be antiquated.[33]

Hanna Hirszfeld did not reveal in this account the extent to which such
incidents may have affected her husband. It is known from other sources
that at the height of Stalinism, he feared for his life, particularly when a
former Polish friend and outstanding biochemist, Jakub Parnas, "disappeared." Hirszfeld had earlier tried to bring him back to Poland from
Moscow, but he suspected Parnas had been swept up in the "Doctors' Plot"
in 1952, when Stalin arrested physicians who were accused of causing the
death of a number of Bolshevik leaders because they were supposedly
"Jewish nationalist agents" working for the United States.[34]

Another example of Hirszfeld coping with the new communist regime
concerned Jacek Prentki, Hirszfeld's "adoptive" son who was imprisoned in
a sordid underground police station run by the secret services on Koszykowa
Street in Warsaw. This was at the time that Hirszfeld was helping women with
incompatible rhesus blood group factors give birth to healthy babies. One
day the head of the secret police turned up with his pregnant wife and asked
for help. Hirszfeld said to them, "You have an innocent child about to be

born. I have an innocent child in prison. We can get along very well together. I will save your child by giving him blood transfusions and I ask you to get Jacek out of prison and out of this country." Two days later, Jacek was in Paris.[35]

There were also numerous paradoxes and ironies as a result of Hirszfeld's life, personal identity, and the extraordinary events he witnessed. For example, as mentioned above, Hirszfeld's ancestry was Jewish but he was not raised as a practicing Jew. He can best be described as an agnostic intellectual until his return to Poland in 1920, when he converted to Catholicism, in part for patriotic but also personal reasons that are explained at length in the autobiography. Hirszfeld never renounced his Jewish origins and experienced firsthand the savagery of Nazism, choosing to remain in the ghetto, which he could have left, and deciding to flee only in a vain attempt to save his daughter's life. Nonetheless, given the political and religious atmosphere during his life, he was considered a Jew by anti-Semites and suffered the dire consequences as a result. Yet many Jews accused him and others like him who converted of being opportunist to avoid the rising anti-Semitism in Poland between the wars, or worse, of becoming a fascist or self-hating anti-Semite who had betrayed his own roots.

On the question of Hirszfeld's Jewishness, the book speaks for itself. Hirszfeld identified above all as a Pole, but he did not reject his Jewish origins. His conversion to Christianity has been explained by his wife's niece in the following terms: "Ludwik Hirszfeld was baptized when he returned to Poland after independence. He was a great Polish patriot and was passionately attached to Polish culture. He felt that Catholicism was an integral part of that culture and this was for him—I believe—the decisive factor."[36] According to acquaintances, the Hirszfelds never attended mass after World War II. Yet, in his personal papers there is a text, undoubtedly written by Hirszfeld, entitled "My Dialog with Christ" in which he seeks to come to terms with the contradiction between a belief in God and the horrors he had witnessed.[37]

It should be remembered that Hirszfeld's book was written for a Polish postwar audience, and he hoped as such that it could contribute to bridging the gap between Poles and Jews.[38] When his memoir was being translated for Little, Brown and Company, he realized that certain passages might shock the American reader unacquainted with the Polish situation and asked his translator to pass over any parts of the book that could involuntarily widen the gap. In particular, he suggested removal of chapter 27, "Race or Tradition?" which describes his observations about Jews based on his experiences in the ghetto.[39]

A final and crucial part of Hirszfeld's life that is missing from the autobiography concerns his daughter. The attentive reader of *The Story of One Life* will note that Hirszfeld never mentions the exact cause of his daughter's death. There is strong evidence elsewhere that it was from anorexia, and between the lines he hints at this, for example, when he suggests that he might have saved her life had he spared her the vision of the Warsaw Ghetto.[40] He also confesses his inability to have passed on to his daughter what he calls "the hunger for life." The expression "hunger for life" is found throughout the book. Hirszfeld criticizes those who have too much "hunger for life" as being basically egocentric, but recognizes that those who have not enough are restless, creative spirits who must suffer and sometimes even perish.

In the end, there is a tragic irony in the fact that the Hirszfelds' daughter saved both her parents' lives, in the sense that attempting to save her was the reason why they finally took advantage of offers to assist them in escaping from the ghetto. That she perished nonetheless, albeit not directly at the hands of the Nazis, was an added motivation for Hirszfeld to tell his story.

Hirszfeld's Scientific Work

Whether or not he intended it, Hirszfeld's book is part of the tradition of biographical memoirs that has long been an important source for historians of science.[41] Autobiographies often follow controversy, as individuals on one side or another of a debate seek to tell their own stories. An example in the history of twentieth-century science has been in the field of physics, especially among many of the figures who developed the atomic bomb.[42] Their memoirs have understandably served as apologies, while other autobiographies have served such purposes as "settling scores" about priority and motivation for research. In the process, they also reveal much about the nature of scientific discovery. Accounts of those in the life sciences have often served similar interests. For example, James Watson stimulated a great deal of controversy with his 1962 account of the competition for discovery of the structure of DNA. Likewise, disputes surrounding the research on HIV/ AIDS promise to produce much autobiographical literature.[43] Some disciplines have made a systematic effort to encourage the founders of their fields to record their experiences, such as psychology and neuroscience, but these are exceptions.[44]

Hirszfeld felt his memoirs bore a resemblance to the unconventional autobiography written by Hans Zinsser, *As I Remember Him*.[45] Zinsser was a renowned bacteriologist at Columbia who then chaired Harvard's Department of Bacteriology from 1923 to 1940 making a number of important scientific discoveries, especially about typhus. These included the discovery of the typhus strain he named Brill's Disease, which subsequently was renamed Brill-Zinsser Disease, and a vaccine against typhus. It was his very popular book *Rats, Lice and History* that revealed his gift as a writer. There is no indication that Hirszfeld and Zinsser ever met, although their careers followed close and parallel paths. For example, Zinsser, who died in 1940, worked in the same laboratory at Heidelberg as Hirszfeld, although just before him, and he served at the same time as Hirszfeld in Serbia in 1915. They likely never met during the war, however, since Zinsser only spent a few months in a Red Cross laboratory in the south (Macedonia) while Hirszfeld was in the northern town of Valjevo. Although Hirszfeld identified with Zinsser, it was not only because of his scientific work but also because Hirszfeld appreciated and aspired to his literary abilities. Zinsser's autobiography was translated into Polish before the war, and the Hirszfelds had a copy on their bookshelf.[46]

Hirszfeld's autobiography omits details of his scientific accomplishments, so it is worthwhile summarizing them here to provide a proper context. Ludwik Hirszfeld was one of the leading figures of immunology in the first half of the twentieth century, thanks to his discoveries about the inheritance of ABO blood types and human population genetics made before and during World War I.[47] Hirszfeld continued his research at the laboratory of experimental medicine he helped establish at the National Institute of Hygiene in the new Polish state between the wars. In addition to serving the country's needs in vaccines, testing, and production of other biological materials, this institute became—in large part thanks to Hirszfeld's work—an important European experimental research center.

An assessment of Hirszfeld's work by Felix Milgrom, his protégé and designated successor, lists the following as Hirszfeld's main accomplishments: proof of Mendelian inheritance of blood groups in human beings; discovery of Paratyphoid C bacillus; and discovery of the population genetics of human blood groups.[48] These are similar to what Arthur Mourant wrote in the foreword of this volume and which led him to conclude that "next to that of Karl Landsteiner, the name of Ludwik Hirszfeld is unquestionably the greatest in the early history of blood grouping and its applications." Less well-known or accepted were Hirszfeld's theory of "Konstitutionserologie" (individual immunological differences), links between blood groups and disease, and his research on maternal-fetal incompatibility of blood groups. Ernest Witebsky cited these same contributions in his nomination of Hirszfeld for the Nobel Prize in 1950, noting in particular his "entirely original concept of heterospecific pregnancy in which he pointed at the possible significance of the differences in blood groups between mother and child."[49]

The scientific research of Hirszfeld touched upon many fields, combining experimental work with research inspired by different disciplines and different methods—epidemiological, anthropological, and others. This is shown, above all, in his research on blood groups, an area of human biology that did not exist before 1900, yet which has grown exponentially since its discovery. Thanks to the work of Hirszfeld and others, blood groups have been of critical importance in the theory and practice of medicine through most of the twentieth century, from their practical value in transfusions, to the evidence they have provided about the workings of heredity and immunology.[50]

Hirszfeld belongs to the handful of pioneers who suspected the biological individuality of human beings,[51] all of whom built on the first landmark discovery of Landsteiner. He recognized that when he mixed the sera and blood cells of individual coworkers with that of others, their properties of coagulation could be described by placing them in only three groups, which he called A, B, and C. Although Landsteiner immediately saw the applications of this discovery for criminology and transfusion, he had no proof of whether the new categories were stable, nor any idea that they might be inherited.[52] Researchers such as Hirszfeld quickly filled in the broader picture of what Landsteiner had sketched. For example, the first attempts at transfusion often used family members as donors, and because of the higher success in finding matches, the suspicion arose that blood groups were inherited. It was the result of research on dogs, described in chapter 2, that prompted Hirszfeld while working in von Dungern's Heidelberg laboratory to test systematically and prove that the human blood type was a permanent biological feature transmitted from parent to child, thus constituting a hereditary trait.[53] In the process, Von Dungern and Hirszfeld established the present-day nomenclature of ABO.

Hirszfeld's second major discovery about blood groups occurred while working with his wife, Hanna, and was based on large-scale blood testing of Allied troops stationed outside Salonica during World War I. They had at their disposal a cosmopolitan army made up of men from five continents who found themselves trapped on a front near Salonica in the last few years of the war. Thanks to the development of rapid blood group testing techniques for transfusion purposes on the Western battlefront, which permitted blood group determination in minutes instead of hours or days, the Hirszfelds tested more than eight thousand individuals from sixteen different ethnic groups and discovered a pattern of variation in the distribution of ABO groups in the distinct populations.[54] To put it simply, they found a higher proportion of blood type A in populations in northern and western Europe, while the proportion of type B rose in populations to the south and east, with the highest found in subcontinent India. This discovery inspired a wave of testing around the world to confirm and extend the discovery. Although some researchers attempted to use the results to prove a racial hierarchy, the more important result of this line of research was to lay the groundwork for the new field of human population genetics.[55]

After World War I, Hirszfeld expanded his research to the field of "constitutional medicine," a name he used for the older idea of inheritance of predisposition to disease. Hirszfeld published a summary of this work in 1928, adding his ideas and findings about possible links between blood types and inherited diseases. This broadly influenced biologists and served as a stimulus for speculation about immunity and serology, while also lending support to the idea that resistance to germs varied from one individual to another.[56] Another line of research grew out of Hirszfeld's own work on blood types and diphtheria using Schick tests, which led him to study fetal-maternal blood group compatibility.[57]

Although serology was at the heart of Hirszfeld's work, he also published on public health issues, primarily because of the circumstances he found himself in rather than a long-term research plan. For example, as indicated by his description of wartime service with the Serbian Army during World War I, he sought to limit the effects of the typhus epidemic. Later, in Salonica, he discovered the C strain of the paratyphoid bacillus that was ultimately named for him (*Salmonella hirszfeldii*). After the war, he became a prominent figure in Polish public health through his scientific director-ship of the National Institute of Hygiene in Warsaw. This was Poland's first and most important center for research in microbiology and public health, created by his cousin Rajchman who, as mentioned earlier, recruited Hirszfeld to work there. After Rajchman left Poland to set up a Health Organization at the League of Nations in 1921, Hirszfeld took a leading role in running the institute, which attracted a number of prominent scientists between the wars, including the discoverer of vitamins, Kazimierz Funk.[58] Funk, Hirszfeld, and others benefited from support from the Rockefeller Foundation, both directly and indirectly, through the League's Health Organization. After World War II, Hirszfeld's Department of Microbiology and Institute of Immunology in Wrocław gave him further opportunities to continue his research both in serology and public health, although most of his efforts were directed to institution building and train-ing of the next generation of researchers.[59] One key activity at the institute has been in bacteriophage research and therapy, which Hirszfeld first inves-tigated between the wars and introduced at the Wrocław institute in the late 1940s.[60]

The Warsaw Ghetto and German Scientists

There have been numerous accounts—both eyewitness and historical—of the Warsaw Ghetto established by the Germans in 1940. Each narrative adds in its own way to the larger picture of one of the most intense experiences of collective human suffering ever known.[61] Still, Hirszfeld's story stands out as exceptional for several reasons. First, as mentioned above, he was a con-vert to Christianity who had considered himself more Polish than Jewish before the war. The fact that he and his family lived within the walls of the Catholic parish in the ghetto and attended mass with other Jews who had adhered to the Christian faith symbolizes the separate identity of Hirszfeld within the Jewish Ghetto, which itself was deliberately separated from the rest of Poland. His status as a "convert" sometimes, though not always, put him in an awkward position. He was both loved and hated by his fellow captives, since some saw him as a "turncoat."[62]

Another exceptional feature of Hirszfeld's account of the ghetto is that it was written by a man who had spent probably the happiest time of his youth and early professional years in Germany, a country and culture he came to love before World War I. He spoke and wrote perfect German and published regularly in German periodicals throughout the interwar period.[63] Still more striking, he deliberately pointed this out to some of his former German colleagues who came to Poland in the position of the occupier. He mentions, for example, meeting a German scientist visiting the Polish State Institute of Hygiene shortly after the Polish surrender whom he had met a little over a year earlier in Paris and with whom his daughter had dined at a banquet at the Pasteur Institute.[64] According to Rudolf Wohlrab, who was deputy director of the Institute of Hygiene in Warsaw when it was taken over by the Germans, Hirszfeld was a scientist whose work was well known far beyond the borders of Poland: "everybody knew who Hirszfeld was."[65] In that regard, of the scientists in the Warsaw Ghetto, he was the best known internationally.[66]

Shortly after arriving in the ghetto, Hirszfeld joined the Jewish Health Council, a body set up with the permission of the German authorities to fight the typhus epidemic. Its main purpose was to isolate people infected with typhus, to oversee personal delousing, to disinfect the homes of the sick, and more generally to fight louse infection. Hirszfeld's account is, therefore, an extremely valuable eyewitness description of what was in many ways an untold story of World War II, at least until the publication of Charles Roland's and Paul J. Weindling's recent books.[67] Although there have been studies of the importance of typhus in medical history, Roland and especially Weindling show the central place it occupied in the Nazi vision of Eastern Europe and indeed in the Final Solution.[68]

Typhus was the rationale for confining the non-Aryan population to ghettos in occupied Poland. The fear of spreading typhus to the *Wehrmacht* and possibly all the way to the *Reich* was the pretext for shooting dead any Jew who ventured unauthorized out of a ghetto. In Warsaw, an artificially created typhus epidemic killed those ghetto residents who could not be shipped off fast enough to the death camps. Those being forced into the gas chambers were often told that these were disinfecting showers to protect them against typhus, such as those that had existed in the refugee camps at the end of World War I.

Many died of typhus in the camps up until and following liberation. Typhus vaccines were produced in both German and clandestine Polish laboratories, and they were smuggled, forged, and sold for fabulous sums on the black market.[69] Finally and paradoxically, the terror inspired by typhus was such that the Germans left the job of delousing and care largely to Jewish and Polish doctors. This had two advantages. First, it protected the doctors against deportation, and second, typhus provided a shield for conspiratorial activities, although it could also be a cause of suspicion of collaboration with the Germans.

The efforts (not always successful) of the Germans in identifying typhus with the Jews and the ghetto is seen in an anonymous memorandum entitled "The Fight against Typhus," written in 1942,

When I arrived in Warsaw in April 1940, I noticed in several places large signposts reading SEUCHENSPERRGEBIET [epidemic no-go area]—throughway forbidden. As I subsequently discovered, the signs had been posted in those places where the ghetto was to be, although its actual creation was being put off. As the time drew closer to the true opening of the ghetto, these signs were shifted from one place to another and yet—despite the signs—there was no epidemic. The Germans had almost immediately devised the thesis that exanthematous typhus—for that was the main disease in question—was obviously being spread by the Jews. However, even after the ghetto was opened and sealed, there was no typhus—or at least it did not spread in epidemic proportions in the Jewish district. It was only in 1941 that yellow signs began to appear above doorways marked FLECKFIEBER [typhus].[70]

The crucial discovery of body lice as vectors of the illness had been made at the beginning of the twentieth century, and while this opened the way for prevention measures, it also led to the image of typhus as a "filthy" disease. For any East European or deportee who lived through the World War II years, the word "typhus" conjured up the ominous flavor of the times, the remembrance of humiliation (typhus "patients" were stripped of their clothes and last belongings), and damnation. In 1944, the Allies feared a major epidemic in the aftermath of liberation, but this scenario was averted as a result of the massive use of DDT by the United Nations Relief and Rehabilitation Administration (UNRRA).[71]

Weindling also shows how the Germans invested in the science of typhus prevention as part of the overall plan for the *Drang nach Osten*. Between the wars, typhus came to be regarded as an East European disease, a Bolshevik disease, and *in fine* a Jewish disease.[72] In Poland, the disease had been brought under control with the exception of certain rural foci in the eastern part of the country,[73] but the mass deportations and the sealing off of the ghettos by the German authorities resulted in a considerable flare-up, thus supposedly justifying the Nazi thesis that typhus was a Slavic, and above all, a Jewish disease. The German "Sanitätsrat" (health council) rapidly began creating Jewish ghettos, supposedly to control typhus, after the German invasion of Poland. The first such ghetto was established in November 1939 in Piotrków, followed by Łódź in December of the same year, while ghettos in Warsaw and Cracow were respectively sealed off in November 1940 and March 1941. Later in 1941, Jost Walbaum, the chief medical officer of the Polish so-called *Generalgouvernement*, wrote a substantial and shocking book (more than two hundred pages and with photographs of "degenerate" Slavs and Jews) entitled *Kampf den Seuchen!* [Fight the Epidemics!], intended for Germans posted in occupied Poland.[74]

Hirszfeld was in a privileged position to observe and describe at length both the draconian and hypocritical measures enforced by the German health authorities to "contain" typhus. His account, as a single testimony, is undoubtedly one of the most complete and is confirmed by other Polish sources on the same subject.[75] In fact, as can be seen by Hirszfeld's account, the unofficial activities of the Jewish Health Council went further. As a member of the council, Hirszfeld was able to visit all parts of the ghetto, including hospitals and dispensaries. He was indeed particularly active in the struggle to contain typhus and even wrote a memorandum in German explaining that the typhus epidemic had quite simply been brought about by the occupation. As he pointed out in an undated text entitled "The Recollections of an Epidemiologist":

> Martini was right when he stated that the fight against louse infection is based on the principle that more lice should die than are born. And this can be obtained without delousing columns. It is enough to have a change of both day and night shirts. . . . Whereas the Germans take away the last bit of work, and the last shirt on a man's back, they persecute and chase people from one place to another, not even bothering to ask whether they come from an infected region. And yet at the same time they proudly announce that where there are Germans to be found, there is no typhus. *Typhus in this war is the work of the Germans.*[76]

Hirszfeld was an important teacher in the underground medical school set up in the ghetto, quixotic in hindsight, but which Charles Roland describes as an "inspirational reaction" of those in the ghetto "to fill the depleted ranks of the Jewish medical profession," if the war lasted. Or at least it demonstrated "some control over their lives and their future."[77] Although Roland corrects those who attributed the creation of the medical school to Hirszfeld (something Hirszfeld himself never claimed), he does state quite simply, "Hirszfeld was by far the most scientifically prominent member of the faculty . . . , remembered by the students from the clandestine school as a gifted teacher."[78] Hirszfeld was also allowed for a time to go back and forth between the ghetto and the "Aryan" side, a surreal experience unknown to most, with the exception of the little children who smuggled food back to the ghetto and were subsequently gassed, if they had not first died of starvation or simply been shot by a German soldier when climbing over the wall.

Amazingly, under the circumstances, the Hirszfelds kept up their research while shut up in the district. Hanna studied hunger states in children; Ludwik set up a laboratory and even employed some of his former associates from the Institute of Hygiene who did research on typhus and starvation in addition to the bacteriological work they carried out for the hospital.[79] For understandable reasons, Hirszfeld was particularly attached to the research conducted under such arduous conditions with the help of young scientists, none of whom survived. As a result, in 1947 he was bitterly disappointed when an American scientific journal refused to publish a paper he had written with his close colleague Tekla Epstein, because it had already been brought out in Polish. Hirszfeld argued that if it were only published in Polish, it would have no international impact and went on to say in his letter to the editor of the *Texas Reports*:

> I am old enough not to care [about] it. But this paper has for me a special significance. These experiments had been done in 1942 in the Warsaw ghetto during the terriblelest [*sic*] time of the German occupation, two months before the slaughter of 400 000 people. There was not one day when going to the laboratory I did not see on my way people killed by Germans or dead from hunger. In spite of the daily danger of life, we succeeded to continue our work which was presented 29th of May 1942 [at] the Medical Society in the ghetto. I succeeded [in sending] the [abstract] of our findings to my [present] successor [at] the State Institute of Hygiene, Dr. Przemyscki, asking him to publish this paper in Polish and English in case of our death which seemed inevitable. My friends succeeded [in saving] me, but my collaborators and assistants [have all] been murdered. Miss Epstein with whom I did this work [was] killed in the street on the way to the laboratory some weeks after the presentation of this paper.[80]

The other major theme of the autobiography, Hirszfeld's critique of the actions of German scientists during the war, is especially and understandably stinging. His status as a ghetto survivor and distinguished research scientist lent weight to the outrage he expressed at the complacency on the part of the international community with regard to German scientists after the war, and he spared no measure in voicing his protests, which had an impact in Germany.[81] For example, in a note to a Swiss colleague in Basel two years after the war, he commented:

> You rightly say that science must be international. After the Great War I was, as you know, of the same opinion. I considered the reconciliation of intellectuals as a way to bring the peoples of different nations together. . . . But today things are different. You know that 3 million Poles and 3 million Jews were killed not in battle, but simply murdered. This mass murder required technical but also mental preparation. Some of the German intellectuals took part in this preparation, others said nothing.[82]

Hirszfeld was even more concerned that the murdered scientists, both Jewish and Polish, should not be forgotten. In 1947 he wrote to the famous medical scientist, sinologist, and champion of peace, Joseph Needham:

> I believe it is our duty to write a volume upon the scientists murdered during this war. In Copenhagen [at an international meeting in microbiology], I was unpleasantly impressed by the fact that nobody spoke one word about [our] dead colleagues. The memory of them ought to be the starting point of all scientific international [meetings].[83]

And yet despite all that he experienced and lost, Hirszfeld knew how to distinguish between those German scientists who had collaborated with the system and those who had not. Most notably, he remained faithful to his teacher and friend Emil von Dungern, with whom he met one last time in Switzerland after World War II, and whom he nominated for the Nobel Prize in medicine in 1952.[84] Although some thirteen years older than Hirszfeld, Dungern outlived his friend by more than a decade. Unable to attend Hirszfeld's funeral in 1954 for political reasons—Dungern settled in West Germany—he paid tribute by writing the following poem:

> You wished to gaze into Nature's face,
> And you led with her a lifelong struggle,
> While others wandered in flowering fields.
>> Nor were you able to pierce through her armor,
>> The time is up and you are dead
>> So have you been defeated by eternal Nature.
> No rosy dawn shall shine for you again,
> For earth always covers over its disputes,
> And unrelieved remains your yearning need.
> How well I know—and yet we struggle on.[85]

***

Paweł Jasienica (pseudonym for Lech Beynar [1909–70]), a scientific journalist who wrote a popular book based largely on Hirszfeld's work after the war, recalled a conversation he had with the Hirszfelds while strolling along the Odra, in Wrocław, on a Sunday afternoon. Suddenly Ludwik Hirszfeld said to him, "More than once my wife and I have asked ourselves whether we did well to come back to Poland [in 1920]. After all, you know what we went through in the last war."[86]

> There was a moment's silence, necessary whenever there is thought of the events and shocks that occurred and which cannot be erased nor redressed. "In theory, if one disregards questions of patriotism and considers only the strictly personal ones," the professor continued in a purely matter-of-fact tone, was it wise to come back in [1920] to settle for good in Poland? We had been invited to Switzerland. We would have each had a chair and would be working to this day in the same branch of science, without those interruptions, humiliations and mortifications, without that terminal tragedy. [Instead,] we returned to Poland. . . .
>
> Mrs. Hirszfeld interrupted, "Our first apartment was on Marszałkowska street, in a courtyard, and the windows gave out onto a blank wall. I couldn't get used to such a view, because earlier both in Switzerland and the Balkans the surroundings were beautiful. But I found a practical way to deal with it—I tried always to look at the sky. And no other country has such a beautiful sky as Poland."
>
> "So the general conclusion," the professor continued, "is that we didn't regret our decision and we still don't. It is hard to explain. Our quality of life would have been quite different than it has been in Poland, had we lived in that glistening, peaceful, and bourgeois Switzerland. Yet here at home we are closer to the sense of history."

Hirszfeld's research and concept of the nascent science of immunology are largely inseparable from the dramatic circumstances that marked most of his life. Indeed, it is possible that adversity favored his scientific creativity because it challenged his profound belief in the beauty of science. That belief ultimately allowed him to overcome (or escape) the tragedies he witnessed and survived.

Ludwik Hirszfeld

Foreword to the original 1946 edition

Ludwik Hirszfeld

It is not my desire to tell the story of my life, nor have I striven to write a literary work. Rather, I wish to describe the suffering of a man and the lot of a scientist who believed that science could render man better. I wish to recount how he wandered through life and how ultimately destiny allowed him to peer into the greatest of suffering: the death of entire nations. A death surrounded by no halo—not even by the hope that it would give rise to a legend that would place on the brows of the sufferers the prickly wreath of martyrdom. Maybe my account will accomplish this task.

It is, however, a difficult task. Writing only about those who perished is not sufficient to expunge the guilt of those who live on and who should have been the gardeners of human souls and not the arsonists of the world. But I am not a historian nor yet a man of letters and I am unable to describe an entire epoch point by point, date by date. That is why I shall endeavor, despite all my reluctance, to link the history of that epoch to that of my own life.

Chapter One

University Years

I grew up in the tradition of Polish underground struggle in which my father's family—and especially my uncle—took an active part. My first memories of this underground struggle are associated with the burial of my uncle, Bolesław Hirszfeld.[1] The university and the polytechnic closed on the day of his burial. I can still see the students in their navy blue uniforms and the police attacking at the cemetery. I was the librarian of our illegal student circle where we secretly studied Polish history. I had had my first encounter with the Russian gendarmes and been in danger of being arrested already as a high school student. And it was at a workers' meeting in a forest near Łódź, when I was sixteen years old, that I heard the battle song of socialism for the first time.[2] To those who still remember those years, these few words will suffice to bring back the breath of that era; that is why I shall not dwell on the topic nor recount the period of decadence that enveloped me in its wings. The few verses I wrote as a young boy show how far I was from Dionysian joy of life. Was it the hopelessness of Polish life, or was it the exhaustion of mankind?

> Around, a battle is boiling and seething. Whistle of swords,
> Wild calls of victors, and roars of the defeated resound.
> Mankind will perish, it seems. Nothing will save it.
> Man's breast is possessed by the wild fury of destruction.

> Like a symbol of suffering, a sublime man stands there.
> His powerful arm will not twitch with the threat of a blow.
> Gods have lifted him above the pleasure of a victorious fight
> But put on him the burden of the consciousness of existence.

> To him, the fight is vain; there is no point to it.
> He will not and cannot crush these worms of people.
> While throngs are fighting and perishing around him,
> He feels only strange sadness and boredom.

Or another verse:

> Why so wounded? Will I lack the strength
> To draw the bloody thread of death around me?

Does my heart still thirst for life?
An extinct volcano can gush no lava.

This "extinct volcano" was then sixteen years old. Two years later, with a feeling of internal dissatisfaction and disdain for the two forms of life which surrounded me—I knew of no others—I left for Wurzburg to study medicine.[3]

The small university towns in southern Germany had a remarkable charm. Most were located on quietly flowing rivers and surrounded by hills with steep slopes covered with vineyards and beautiful leafy forests. The people in those days were gentle and simple. Town life embraced the university: the students were everyone's children, and everyone took a most understanding attitude toward their little follies. I still remember some of the lectures as if I had heard them yesterday, even though this was forty years ago. Stohr during his lectures made master drawings, and a whole new world seemed to emerge on the blackboard. When he talked about his investigation of the migration of the white blood cells, I felt I was witnessing the birth of thoughts. Or the lectures presented by the famous zoologist Boveri![4] When he spoke about evolution, we could hear the pulsation of intense life. How distant the monotony of the factory city now seemed. Before my eyes, the world of science was being built.

After lectures, I used to stroll through the neighboring forests, usually alone, seeking to find myself. In this search for myself, I must admit, Nietzsche was of great help to me. Later, the Germans converted him into Hitlerism's court philosopher. But he—the poor, disease-stricken man—spoke of yearning for the pure air of summits.

Gradually, the idea of devoting myself to scientific work matured in me. It was not a special interest in something definite but a dislike for the common forms of life and a fascination with the calm meditation prevailing in a scientific laboratory. Had I lived in the Middle Ages, I might have fled life by joining a monastery. In the laboratory, I felt the monastic calm and, at the same time, the creative breath of science. There was one more strange coincidence. I studied anatomy from Rauber's textbook.[5] Most of the pictures contained in it had been drawn by a famous Polish anatomist who was educated in Paris: Ludwik Hirszfeld. This coincidence brought scientific work very close to me. I was nineteen years old then. I decided to transfer to the university in Berlin. This was one of the best traditions of university life in Germany: students could transfer from one university to another. They learned to know life and various schools.

I will not describe Berlin. What could a student who was dreaming of science and was a foreigner without contacts know about the life of a city which could mirror a whole era?[6] However, I must say a few words about the lectures. For one semester, I dropped medicine and preoccupied myself with philosophy. I learned the history of philosophy from Simmel. He presented his lectures like a god of pure intellect. Whenever he said "Die Idee hat sich zugespitzt" [the idea has come to a point], he stretched his black, lean figure and lifted his arm up, and at such moments he looked like a pointed knife directed toward the sky. His words took shape, his ideas would dance and swirl. I mention this because later I endeavored to give my lectures the same plasticity of expression I found in Simmel's.[7] In the winter semester of 1904 I returned to medicine. What lectures those were! When Orth[8] discussed pathology, his lectures often ran fifteen, even thirty minutes over, and yet no one—absolutely no one—would leave. Or the lectures and operations performed by Bumm! He seemed a sorcerer.[9]

In 1905, a revolution broke out in my country, and I felt a strong drive to go back and participate in active combat.[10] I even bought arms. However, I continually postponed my departure until a professor finished a chapter or I completed an experiment. In this way, the revolution was brought to an end without my participation.

Yet the lectures did not give me full satisfaction: I longed for a laboratory and a project of my own. I registered for an additional course in bacteriology given by Professor Ficker.[11] He was deaf, and it was difficult to converse with him; but he taught excellently. He was always so happy when a student succeeded in making a laboratory preparation or completing an experiment. Currently, he is in Brazil. Just prior to the war, when he was preparing to retire, he sent me his photograph with a dedication that touched me deeply: "Herrn Kollegen Hirszfeld, den ich mit Stolz zu meinen Schulern rechne" [To my colleague Hirszfeld whom I am proud to count among my pupils]. I reciprocated with my picture on which I wrote that I was proud to count him among my teachers.

I remember my first experiment on an animal, when I was watching the development of anthrax infection. Such captivation of a young investigator in the face of the mystery of an experiment is worth the life of an animal. I also began to study monographs and special papers. The first was the monograph by Hermann Oppenheim, *Toxins and Antitoxins*, which included the side-chain theory (Ehrlich and Morgenroth).[12] Many objections were raised against this theory. Years later, when in my own textbook I tried to show the beauty of an emerging science to young people, I formulated my thoughts—reminiscent of my first emotion—in the following manner:

> Those who lived at the time this theory was established, know how exciting the thought was that the non-toxibility of the cell is also a guarantee of overcoming that intoxication and that a disease and its defeat represent individual links of one event, successive and closely correlated acts of one drama. It is the task of every theory to incite creative unrest and to captivate. None of the existing theories has accomplished it more forcefully than the side-chain theory.[13]

I remember the night, crucial for me, when I pored over [Oppenheim's] article until morning and, after getting up, I suddenly felt that I wanted to become a serologist. I did not know whether this was a profession or a madness, I only saw the astounding purposefulness of the reactions constituting the body's defense. This surely was not madness. Bacteriology and serology were becoming the basis for combating infectious diseases. What appeared to be an abstract idea—a hope—was gradually becoming a profession. When Poland came into being and created its own laboratories, I was to become one of its few bacteriologists. Thus, beginning as a meditative student who thought that he would have to pay for his longing for scientific work with a homeless wandering in foreign lands, I eventually became head of a research center, in my own free country.

But let us return to my student years. At first I worked for the bacteriologist Wolff-Eisner. He was studying the theory of hypersensitivity, which subsequently brought fame to his name.[14] I remember the first session in the Physiological Society where he discussed his projects. There were a few elderly gentlemen, and the overall interest was small. It was much later that I learned how difficult it is to get a new idea across. The great Koch could not talk about his discovery of the tuberculosis bacillus because Virchow did not believe in this discovery. When Schaudinn was demonstrating his first microscopic slides of the syphilis germs, the chairman terminated the session with the nasty remark that he was postponing it until the next demonstration of another syphilis germ.[15] Wolff-Eisner was a restless soul and quarreled with everybody. Although he lived in peace with me, I could not tolerate the tense atmosphere and left his laboratory. For a short period I gadded about through various lectures and finally, having pawned my watch to pay the laboratory fee, I registered with Professor Ficker. I was twenty-one years old at that time. I was married, and my wife, by giving lessons in Berlin, enabled me to pay for experimental animals.

I completed my studies at the age of twenty-two and requested a topic for my doctoral dissertation. Professor Ficker showed me a long list of topics from which I could choose. There were more than twenty of them: bacteria in telephones, bacteria in the washrooms in the various districts of Berlin, and so on. What did I care for bacteria in telephones? I wanted to penetrate the miracle of the body's defense. With a boyish frankness, I rejected them all, explaining that none caught my fancy. Soon I was summoned to the institute's director, Professor Rubner.[16] He was the most outstanding hygienist of that time, the discoverer of the law of energy equivalency of foodstuffs. I thought that Rubner wanted to dismiss me from the institute for my audacious criticism of the institute's research topics. "This is the end of my scientific work in this institute," I thought. "I took off with an ax against the sun." Alarmed, I entered the professor's study. To my surprise, he kindly motioned me to sit down and said: "I wanted to meet you personally, because you have criticized our topics with such frankness. I must explain this matter to you. Many military physicians are sent here for training. We must offer them topics resembling parade marches. These, of course, are not proper scientific topics, and I am happy to see a young student with more profound interests." He suggested a fantastic but beautiful project: "Effects of Sexual Maturation on the Body's Resistance." I was unable to cope with this project. It was experimentally unapproachable at that time. However, it brought me in touch with Ulrich Friedemann[17] under whom I completed my doctoral work on blood agglutination.

I was twenty-two and he was twenty-nine years old. He was not yet a *docent* and already he was surrounded by the halo of a great talent. Whatever he said was different, original. When lost in thought, he would scribble on a piece of blotting paper; the result was lovely little drawings. He played the violin beautifully. I loved him the way only a young boy can love a person who leads him on a spiritual trip. Many, many years later, I met him in London at the League of Nations Health Organization symposium on antidiphtheria vaccination. He was a delegate from Germany and I from Poland. We were overflowing with joy. We spent whole days together like a couple in love. I was deeply touched to hear from his wife that throughout the long years of our separation, he had been thinking of me, his junior, with tenderness and esteem. At that time, he was no longer a theoretician but probably Germany's best expert in contagious diseases. "You know," he said, "I began to perceive pure science as a fruitless pleasure." Later, I often saw that pure science cannot always satisfy the most ardent minds, so great is the need to be in touch with pulsating life.

During my student years, Friedemann was only a bacteriologist. He was often visited by Morgenroth, the closest associate of Ehrlich, who was the most prominent serologist of those times.[18] For hours they talked about resistance. I would sit in the laboratory and listen to them. This was the first unfolding of the wings for flight: to formulate problems. I learned to do so there. And Morgenroth's sense of humor! I was told that once a half-baked scientist asked Morgenroth how many of his research projects he regarded as successful.

"I guess about 5 percent," Morgenroth calmly answered.

"Well, Mr. Professor, why are you so modest?"

"I am not modest," Morgenroth replied. "Do you actually believe that as much as 5 percent of your research work has been successful?"

That is the kind of people they were. Morgenroth died in the prime of his life. After a few years, Friedemann married his widow and was like an elder brother to his children. His portrait hung in the living room, and all talked about him as if about someone close, still alive, who had just left for a moment. They were beautiful personalities. Kinship of thought was more essential to them than the reality of life. Friedemann now resides in New

York; he had to leave Germany. With his departure, that country lost its best clinician and theoretician of contagious diseases.

I completed my doctoral dissertation in 1907 at the age of twenty-three. Professor Rubner graded it *eximia cum laude*. Soon I ran across many references to my experiments. This lifted my morale. Many, many years later I gave a speech to girls who were receiving their bachelor's degree. I told them to choose a profession that they ardently desired, without paying attention to what this profession might bring about. Even if we must perish, let this happen in the service of our individuality. Everyone has the duty to preserve the spark that he or she received from destiny.

I was lucky. The Cancer Research Institute in Heidelberg was looking for an assistant in its department of parasitology, knowledgeable in resistance problems. My professors endorsed my candidacy. In this way, my recklessness led me onto the scientific path.

Chapter Two

Assistantship in Heidelberg

Heidelberg, Alt-Heidelberg, is a city imbued with legends. He who does not know you, unique city, knows neither the charm of meditation nor the poetry of an entire epoch.

The blue ribbon of the Neckar River winds among hills. The Neckar Valley is beautiful. One longs to repose and dream of love and youth that pass yet return and are eternal. The Philosophenweg [Philosopher's Way] stretches out on one side, while on the other lie the ruins of the castle and Kohlhof. On the Philosopher's Way one might meet a professor walking in contemplation and occasionally glancing at the quiet ripples in the Neckar, which disappears in the distance, leaving behind its peaceful bliss. Thoughts rise above the reality and roam in infinity. What this single moment of inspiration brings may immediately take shape, power, and flight, but also meaningfulness when spoken to the students, who in Heidelberg are different from in other places, because they too walk as if enchanted and intoxicated, not with the lustful youth of undergraduates but with a sweet thoughtfulness of the Middle Ages. Everything is different there. And the people are better. Later, when I was reading about the air raids, I thought, "May they destroy everything, the whole damned land, for the wrong done to us. Just not this one and only town." I believe that those were also the thoughts of the English and the Americans.[1]

I arrived in Heidelberg in the autumn of 1907, after I had passed all my examinations, and reported to Professor von W., my boss.[2] He greeted me coldly and, after a few minutes, he announced: "The working hours are eight to twelve, and then two to six. Good-bye." I thought he might also be hiring a janitor on the same day and had confused me with him. The next morning I reported, and Professor W. informed me: "You will dissect mice and record how often you find certain types of parasites. Furthermore, here you have a dish with amoebas obtained from straw. You must describe them in detail every day. I will show you how this is done." And he began to dictate to me completely insignificant, random observations. It went on like this day by day, week by week, without a single guiding idea. And this was the Cancer Research Institute![3] Were they discussing new growth problems here? My boss would sit in his study till eleven o'clock, and then he would photograph everything he could put his hands on. I did not notice whether he had even the slightest concept of the formation of new growths. Was all of science one big misunderstanding?

I had entered the temple of science to find it empty with priests who were not even skilled workers. I must run away, run away from this hell of platitudes and insincerity, I thought. Yet, I said to myself, this was my first position. If I didn't hold out, I would get nothing else, and my dreams of science would remain the dreams of a green stripling. I gritted my teeth and began to study amoebas. I did not like descriptive sciences, and the staining technique was not attractive to me; yet I had no choice but to get busy on the amoebas.

I was, however, amply rewarded for the pain of the first disappointment, because I made observations that cast a new light on the biology of amoebas. I cut out small strips of the medium on which the amoebas were growing, covered each strip with a glass, and inserted a capillary tube into the medium. A small, microscopic lake would form at that point. Through the microscope I saw how the amoebas changed their shape until at one moment they formed cirri and like sirens, swam into the lake. It turned out that at least one-third of the existing amoeba strains were flagellates. I often demonstrated this remarkable metamorphosis to a surprised audience. I also succeeded in developing staining methods that fixed the various transitional stages. Without exaggerating, my microscopic slides were the most beautiful of all available at that time, and I described completely new forms of division (triple division) and new forms of the nuclei. I gave some thought to the cause of the motion. I advanced a theory explaining the motion of the amoebas by the effects of the changing electrical charges on surface pressure. The famous biologist Loeb[4] was passing through Heidelberg at that time and, hearing about my theory, mentioned it to the Nobel Laureate chemist Haber[5] from Karlsruhe. It turned out that the latter was also advancing a similar theory which, to be sure, he had developed in more detail.

Meanwhile, my boss was photographing everything. I must admit that he produced beautiful pictures. I presented him with my paper on the biology of amoebas, new staining methods, and the transformation of amoebas into flagellates. After awhile, he returned the paper to me saying that it was not up to par either scientifically or linguistically. I understood. "*Herr Professor*, I do not feel capable of writing better. I should like to ask you to publish the paper jointly." He agreed, of course. It took him three months to rewrite the article. Then, one morning—I was already Dungern's[6] assistant at that time—he sent me the manuscript with the request that I return it in the afternoon. I returned the paper immediately with the annotation that I could not accept it because it was not up to par.

And his lectures? I remember a certain student who attended his lectures, the subsequently famous anatomist von Moellendorf. The usher, the draftswomen, and I made up the entire audience. Subsequently, my boss became a professor in one of the North German cities. A dozen or so years later, at an anti-alcoholic congress in Warsaw, I met the head of the health services in Hitler's Germany, Professor Reiter,[7] who was once my boss's assistant. He told me that the students were so bored during that man's lectures that for diversion they hired a hand organ grinder who played under the auditorium's windows during the classes. I am not revengeful, but this gave me satisfaction. Throughout my entire youth, he was the only man who hurt me with his coolness and offended me in the way he worked.

At that time, the Department of Parasitology had on its staff Professor von Kudicke,[8] subsequently director of the State Institute of Hygiene in Warsaw during the German occupation. In Heidelberg he impressed me as a pleasant, intelligent person.

The Cancer Research Institute was located in the newly erected hospital building and also in a number of houses redone so as to serve as laboratories. The institute was made up of two departments: parasitology and serology. This reflected the hope of the director, His Excellency Czerny, that new growths had their parasite and that serology would find methods for early diagnosis. The already famous Emil von Dungern headed the Department of Serology. Warburg,[9] subsequently a Nobel laureate, was his assistant prior to my arrival. The American Coca[10] and I were assistants at the same time. Our friendship has lasted over the years. Coca eventually became one of America's most outstanding immunologists. When misfortune befell me, he was among the first to extend a hand.

They were working on snake venoms under Dungern's direction. I looked to their department as somewhat of a paradise. There, people were thinking, there were no disgusting office hours, and the boss and his assistant—like two brothers—walked together the paths of thought. Once I dropped in there on some business and found only Professor Dungern. We began to talk . . . about women. Thereafter, we talked frequently and much; however, I must admit with contrition, mostly about women. We were interested in this subject for different reasons. He, because at forty-three years of age, was still looking but was unable to find a suitable wife. Myself, because I fell in love at the age of fifteen,[11] married at twenty-one, and still could not understand why my wife had chosen me. I remember that once we were so involved in a conversation that we forgot our dinner. This was probably the beginning of our friendship, which outlasted all wars and will never die, at least not on my part.[12] When Coca was summoned to America, Dungern—presumably on Warburg's advice—offered me the assistantship. This was how I joined him, the laboratory of my dreams, and was able to collaborate with one of the subtlest minds in serology: wise, profound, and independent.

He created only when he felt like it. He would come late to the laboratory, around eleven o'clock. At one p.m., he would usually say: "We must show ourselves to the people." We would go for a walk around the university: the young women students acted like magnets. We would return along the bank of the Neckar and talk not only about women and resistance but about everything. Dungern knew everything: art, philosophy, antiques, which he collected, as well as old Greek coins and pictures he liked to buy. He could illuminate everything in his own, special way. I could not possibly measure up to him. He represented life on the highlands of culture and prosperity for many generations. Dungern came from an old line of barons. His grandmother was a Russian aristocrat, his mother the daughter of a Baltic baron. His grandfather was creator of the Baden Constitution, his uncle the German ambassador to Turkey. He himself was destined to become *Kammerherr* of the Count of Luxembourg. "You know, my friend," he told me, "I had never in my life felt so mortally bored. I fled immediately." He had no notion of haughtiness. When the Duke of Baden, whom he knew personally, visited the institute, Dungern talked to him as he would have to an usher. Or rather, he usually conversed with an usher as if he were the Duke of Baden.

Looking at him I thought of how dissimilar our beginnings had been. I was just learning about things that Dungern encountered in his childhood, and I heard for the first time the names of famous persons whom he knew personally. He lived on the Neckar River. I visited him every evening. The housekeeper would bring beer or wine, and we would chat forever. This was my true training. There my thoughts acquired shape, élan, and courage. Dungern is the man to whom I am most indebted. That our chats were not fruitless will become evident in this very chapter, because they gave rise to a new science—that of blood groups. He created like a poet. With relish, he told me an anecdote about a scientist for whom the Prussian government built an institute. While handing the institute over to him, Excellency Althoff had said: "*Herr Professor,* I hope you will have beautiful results."

"Your Excellency, I cannot promise you that."

Perplexed, His Excellency remarked: "Well, I understand, the results of the research do not depend on us; but I hope you will do your best."

The Professor interrupted him: "Your Excellency, I cannot promise that either." Quite a period that was. Yet it gave science and Germany more fame than all other work done on command.

We worked irregularly. Sometimes till two or three a.m., and sometimes, during office hours, we took our bicycles and rode down the Neckar Valley. If he came to the laboratory and there were no interesting results, he would lie down on the sofa and yawn out loud. How he could yawn! This was the subject of amusement of all volunteers. Finally, he would say: "Mein Bedarf ist gedeckt" (My need is satisfied), and he would let his thoughts roam far in search of new stimuli for his insatiable spirit.

I also formed a close friendship with the head nurse in the cancer hospital, Pia Bauer. She was a beautiful human being. She devoted her life to the service of those who were doomed to death. It is no trifle to be a nurse in a cancer hospital. You must take the dying patient by the hand softly and gently, and lead him into eternity so that his last sensation is that of goodness.

She knew just how to do that. She once invited my wife and me to her sister's house in the Schwarzwald (Black Forest). Her sister was the widow of a professor in Karlsruhe, a chemist and a poet. A monument to him had been erected in the park in Karlsruhe. There I came to know the most beautiful womanly soul I ever met during my pilgrimage through the world. She walked as if on heavenly paths and not on earth. When she was sad, she played beautifully. She went deep into everything beautiful and spiritual that the world had to offer. She was interested in philosophy, art, and literature, and she loved nature. I wrote to my teacher—our friendship entitled me to do so: "In the mountains I have found a woman whom one cannot possibly not love." He replied: "Persuade her to come to Heidelberg: You may have more luck than I."

She came, and they became acquainted in our modest apartment. They fell in love with each other at first sight. Her sister and I were like mother and father to them. I, a twenty-nine-year-old assistant, became the foster parent of my beloved teacher. Shortly thereafter, I was witness at their wedding. We went on many trips together, after I left Heidelberg; I called on them and they on us. They also visited us in Warsaw. We love them, and they love us. Our friendship has endured all wars. Now that I have psychologically broken with German culture, I consider them and only them as spiritual giants. And when I think of them, I must admit, I am ashamed of my hate.

Soon after I left Heidelberg, Dungern moved to Hamburg. However, he was unable to take the clamorous atmosphere of Germany's northern cities. He abandoned science and lived a solitary life on the Bodensee (Lake Constance), preoccupied with poetry and philosophy. Occasionally he would send me a manuscript of his poems. On his sixtieth birthday, I wrote a tribute to him and sent it throughout the world:

> I was especially fortunate to be able to work with Dungern on problems which now constitute the central interest of biology and medicine—the field of blood groups. If I later succeeded in continuing this study, the origin of this pursuit dates back to the years of our quiet work in Heidelberg.

Dungern ought to be regarded as the creator of constitutional serology. I will not discuss the details of this discovery; I would only like gratefully to reminisce on the period of our joint work. Dungern's charm through which he conquered human hearts was the fact that he gave without realizing it. He was full of *élan* and ideas that spouted from an invisible depth without effort or intent. Dungern was reproached for not being able to force himself to systematic work. Those who worked with Dungern knew that he himself was not aware of his singularly great and fervent creative force, providing however, he had been seduced by a problem. Dungern was a spiritual poet and aristocrat who had to fall in love with a problem in order to be able to work on it. No one could perceive as he did the bliss of contemplating abstract possibilities and of attacking new problems. Idleness and the exhausted run of scientists working without inspiration were strangers to him. He was a flame burning from within. This tiresome search for the so-called topics and questions that had not been fully resolved, or discoveries made by others but still requiring supplements or corrections—all that was alien to him. When serology lost the light of its youth, his firm but restless spirit would not stoop to what at that time could be achieved with serological methods. His problems no longer fit into the narrow framework of the science of resistance. It was the logical consequence of his mentality that he then turned to the all-embracing problems of philosophy and art.

In Dungern's laboratory I learned the source of true scientific inspiration. There, where within two to three years we established the foundation of a new scientific field, I came to understand that science is not made in a somber mood nor to order. I realized that an idea arises from the ebullience of a spirit wishing to dance in infinite space, and that a scientific idea represents joy of life, wonder at beauty, protest against death, the aspiration to preserve, a question posed to nature, the desire to experience, and curiosity about depth. There is neither the grandeur of states nor racial hate, there are no commanders, no orders. Scientific mercenaries need orders, but free, creative spirits do not.

Not everyone there had the power of his spirit, but almost everybody had his "little jinn." I remember Dr. Detienne—he was working day and night on blood platelets. He obtained very good results, but he also held some uncanny and unfulfilled hopes. His young wife was thoroughly unhappy. She would come to the institute but was usually unable to get him to leave. Finally, Detienne wrote an article, sent it to the publisher, and returned to his wife. A few days later we met them: she was radiant while he was grim. The idyll lasted a week. Then he recalled his article and returned to the laboratory. "Science is jealous and without temperament," Warburg would say. "He who wants to have children with her must have passion for two."

My wife, however, would always forgive me my folly: that I neglected her, that I returned home at two or three a.m., and that I woke her up at night to tell her: "You know, Pia's serum agglutinated dog erythrocytes," or similar news that my feverish head had to communicate to someone at once. She took everything with the patience of a loving woman, understanding that this was shaping my mind and building the foundation for my life's work.

We did not immediately find the road to blood groups. Our prime objective was to work on cancer, since this was the Cancer Research Institute. I soon realized how difficult it was to keep a researcher going in one direction, because thoughts usually run in the direction where there is a lure of novelty, the possibility of an adventure, and the reality of one's own observation, and not always the initial intention. Planned work yields results when someone wants to determine or describe a definite reality by means of established methods; it is not enough when one looks for new methods and new paths. These require inspiration and good luck.

However, let us return to the research we had undertaken. We began to work on the resistance of normal tissue, with the intention subsequently to use cancer tissue (only twenty years later did I finally implement the second step). We injected nuclear homogenates into the ears of rabbits. To our surprise, the ears of pregnant female rabbits swelled, while the ears of control animals remained unchanged. It was not difficult to find an explanation: we believed we were witnessing the phenomenon for which we were searching—resistance to the tissue of one's own species. Evidently, impregnation produced a hypersensitivity similar to that produced by the penetration of a germ. Looking into the far future, we saw resistance reactions against cancer cells. Our life acquired hopeful colors familiar to the researcher who sets himself far-reaching goals. We began to draw further conclusions from this phenomenon: the cell that penetrates, fertilizes, and induces growth is a spermatozoon. Let us try to prevent pregnancy by building up resistance or immunity with nuclei. We immunized a number of female rabbits, placed a male rabbit in their cage, and began to daydream. One day, we would be able to control the time of conception, which would bring about a revolution in sexual relations, and later we would relate the findings to resistance to cancer, and so on. We were unable to wait patiently till the end of the experiment; so we took our knapsacks and left for the Tyrol mountains. This was my most beautiful vacation. We climbed the steep mountain slopes, I drank the pure air of the summits with deep breaths and was lost in reverie. We were happy.

After returning to Heidelberg, I ran, with my overcoat still on, to the barn housing the rabbits. All the female rabbits, without a single exception, were pregnant.

No matter what the cause, the dream of my first, wide-scale experiment collapsed. But did these frustrated hopes decrease my happiness? They gave me a few weeks of inebriation. From this I drew the conclusion for the rest of my life that we ought to dream and that no one as yet has paid too much for hope. Hope, they say, is the mother of fools? On the contrary, hope is the mother of those who are not afraid to project their thoughts into the far-off future. When a friend of mine, Professor Gąsiorowski,[13] once told me, "You know, Lutek,[14] I teach my students only the essentials in order not to overburden them," I replied, "I, on the other hand, select what is less necessary. The basics can be found in textbooks. The lecture should make the student dream so that he does not see knowledge as a gray wall to be escalated at great cost and under the threat of examinations."

When we want to point out the beauty of a landscape, we do not force others to remember the altitude of peaks. Thus, we should teach the upcoming generations the charm of a youthful gaze and the delight of being lost in what there is to behold. We must not so bore them with the "essentials" that they lose the sparkle in their eyes and the ability to fly, wonder, and admire. I remember a student whom I once asked whether he had ever been in love. His answer was, "*Herr Professor*, we have so many exams right now."

This was the conclusion I drew from my first disappointment in science. The next experiment also brought me a disappointment but with an understanding that science is capricious and likes to play fast and loose with the researcher. This is what happened. We wanted to check whether similar hypersensitivity reactions develop in pregnant women. In our evening chats, we already saw diagnostic reactions and a biological diagnosis of pregnancy. But who could be used for the injection? Women available in the maternity clinic toward the end of pregnancy might have negative reactions, as happens in the last stages of tuberculosis when the tuberculine reaction is often ineffective. At that time, an assistant in the clinic of ear disease, Dr. Beck, currently professor in Heidelberg, was working with us. In the clinic, he had a patient who was three months pregnant with exudations coming from her ears. "Fine, colleague, but we need a control subject." "You can inject this stuff into me; I will have no reaction. I assure you I am not pregnant," he jested. I discussed the planned experiment with Dungern. We decided to inject rabbit nuclei into the rabbit, and human spermatozoa into the pregnant woman and into Beck. But we did not tell Beck; somehow, we were embarrassed.

The next day, we gave the injections. Then we went for a trip into the Neckar Valley and, on the third day, with the assurance of conquerors, we turned up at the oncological clinic. We were greeted by the happy patient with words of thanks: "What a marvelous medicine. For the first time in three years no fluid is running out of my ears." I exchanged an uneasy glance with Dungern. "And your arm at the place of the injection?" Not a trace of a reaction. Well, we thought, evidently human spermatozoa were less irritating. "Where is Dr. Beck?" we asked the nurse. "Dr. Beck is ill in bed, and he wants you to visit him." We knocked at the door of his room: he was lying in bed and yelling, literally yelling at us with fury: "What did you inject into me?" And he showed us his arm: all red and swollen. The outcome was precisely the opposite of what we had expected.

Never in my life have I laughed so much.

Now let us turn to our most important research project: blood groups.

One day at noon we went out "to show ourselves to the people." My professor began talking: "You know, colleague, I just recalled that Ehrlich and Morgenroth injected goat blood into goats and obtained a serum which did not dissolve blood cells in all but only in some goats. It is worthwhile looking into this; it must be an individual characteristic." If the coercion of an order or the anxiety of endless troubles were hovering above us, wc would have forgotten this flash of thoughts. Indeed, of what importance could the agglomeration of goat blood cells be when there are tears and curses in the world, a war and the Jewish problem, and hundreds of other things that are suspended between the researcher's eye and the object of his investigation. But at that time, there was sun, forests, love, and those delightful young women students; so, we could amuse ourselves with goat blood.

We did not have enough goats, but the Department of Parasitology was raising dogs for its own experiments, which it did not carry out. We began injecting dog blood into dogs, and soon we were carried away by the reality of a great discovery. We continued working through whole days and nights. We established that there were definite serological races among dogs, that the traits were inherited, and that they could not appear in offspring if parents lacked them, even though they could disappear. We established basic laws explaining how and when resistance reactions against tissues of one's own species or organism could develop. We published the findings as our first communication, and for me this was the first test of Great Science.

This discovery established the laws governing the existence of serological races. How could we switch over to human beings? We began to immunize monkeys, but this was a tedious path. Once again, during our chats over a glass of wine, we recalled the almost unnoticed research by Landsteiner,[15] and after we studied his papers we found that Landsteiner had discovered a similar differentiation in man to what we found in dogs. However, due to a strange caprice of nature, even without immunization human serum contained antibodies against the blood corpuscles of some individuals. We investigated this matter and found a confirmation of our presumptions. For our own convenience, we formulated a terminology that ensued from our experiments on dogs. We called the blood group whose erythrocytes were not agglutinated by any human serum group O; the blood group that was frequent in Heidelberg group A; and the rare group B. Group O was supposed to represent the absence of characteristics A and B. On the basis of our dog experimentation, we foresaw that we would discover similar heredity laws for human beings. This time, university professors and their families served as our experimental animals. I must admit that they did not duck the role.

For many years, people talked about an odd professor and his assistant who discretely inquired about the marital happiness of professorial families lest a cuckoo's egg might overthrow the scientific law they had established. And the law stated the following. There are A, B, and O traits. They are inherited according to Mendel's laws. Traits A and B are dominant, trait O is recessive. In order to understand the revolutionary significance of this discovery, one must remember that Mendel's laws had only recently been rediscovered, that the possibility of applying them to hybrid human beings had been demonstrated only once, and above all that this was the first case in which one could *easily demonstrate* the great biological laws of heredity in man. In the meantime, thousands of families were examined, and scientific commissions throughout the world accepted our terminology. My mind harbors memories of walks along the Neckar River, nights spent engrossed in the laboratory, and endless discussions over a glass of wine.

I cannot go into the details of this discovery. When we found that trait O yielded to traits A and B, we considered whether it would not be possible to explain this on the basis of a contrariness of factors A and B. At that time, however, genetics allowed only two opposite factors. Therefore, we had to advance a more complex theory—about two pairs of opposite genes. It was only several years later that the so-called multiple opposite genes were discovered, and the German mathematician Bernstein[16] was able to reactivate the original theory in this respect. In our second communication, we pointed out the possibility of using blood groups to determine paternity.

These methods are currently used throughout the world. Unfortunately, they do not promote child protection as yet but only help to exclude the man unjustly accused of fatherhood. Yet I believe that the time will soon come when it will be possible to establish, in a given case, not only who cannot possibly be the father but also who actually is the father.

In one respect, though, this research has been unpleasant to me. The Germans use it for racial purposes. If a child from a mixed marriage proves that the husband of its mother is not its father, it receives all citizen rights, probably as a reward for not hesitating to slander his mother to personal advantage. But what can be done? This is one of the few cases where scientific discoveries are used for unworthy purposes.

I will not write at length about our further research projects. Suffice it to say that they have become the basis for subsequent studies we and others conducted. We established the existence of subgroups, the existence of groups among animals, group A among apes, and finally we were able to demonstrate that there were many group traits and that the individuality of blood was no myth but a great law of nature.

This was the first step toward applying serology to genetics. The next step—application to anthropology—was not taken by Dungern and myself jointly because life separated us.

Sixteen years later I published in German a book entitled *Constitutional Serology*.[17] In it I tried to expound all the consequences ensuing from the ideas born in Heidelberg and developed subsequently.

I dedicated this book to Dungern with the annotation: "With love and devotion." When in 1938 I published the book in French, I recalled that period:

I was fortunate to be among those who laid the foundations for the science of blood groups. When I was writing the individual chapters, the various stages of my own investigative thoughts passed before my eyes. It is my desire that my book reflect the climate of emerging ideas. While writing this book, I should like to be not only an experimenter preoccupied with detail but also a builder who, with his inner eye, encompasses the edifice of the future.

Indeed, sacred are the moments of youth when they are given in the service of a great cause, without burdening the motives with a false phraseology.

In Heidelberg I was given the opportunity to fly. His Excellency Czerny promised me a directorship should my teacher be summoned to another place. Nevertheless, I decided to leave when Professor Silberschmidt in Zurich offered me an assistantship at the university's Institute of Hygiene. I did not want to limit myself to just one scientific branch. Even though I was most fond of the science of resistance, I wanted to become familiar with the entire field of hygiene and bacteriology. I was lured by a new life, the promise of a *Habilitation* examination, and contact with young people.[18] I must admit that, in spite of all the love I felt for my teacher, I wanted to try my own strength. This implied a certain ingratitude from which I subsequently had occasion to suffer on the part of my own pupils. But I understood also that wishing to remain an assistant is not always a good thing. A child leaves his parents to embark on his own journey. The same emptiness remains in the heart of a professor as it prevails in the heart of a parent; and a longing for the young person he had once guided and who had looked to him with the grateful regard of a pupil.

I spent the last summer with Dungern and his wife in the Tyrol on Lake Garda. The sadness of separation was hovering over us. We decided to keep in touch even though each of us was to go his separate ways—he, as usual, with his eyes fixed upon the charm of pure science; and I whom life was soon to draw into a vortex of events and work done on behalf of other causes. When I think about this man who has been spiritually the closest to me, I wonder why our lives and goals have become so different. Possibly because I am the son of a nation that has suffered while he belongs to the summits from which human suffering is invisible.

In the fall of 1911, my wife and I went to Zurich where I became an assistant at the university's Institute of Hygiene and my wife became assistant in the Children's Clinic. Her boss, Professor Feer, for whom she had worked in Heidelberg, received a temporary appointment to Zurich and offered her the assistantship.

Chapter Three

Sojourn in Zurich

Switzerland is beautiful, and Zurich is beautiful. Far off in the distance one sees the Berner Oberland (Bernese Alps), the Jungfrau, Monch, and Eiger in a white gown devoid of earthly filth, and in the evening—the sun setting over the Limat Valley. How often we interrupted our work and walked to the window to immerse our souls in this infinite beauty. At night, when lights go on in the villages surrounding the lake, it all looks like an eternal fairyland. The local people know and appreciate the beauty, and they pray to their mountains. We often went hiking in the mountains. They were most beautiful at sunrise. At first light fogs ascend, then eerie clouds rise up from the abyss as if to release the earth from the embrace of the night, and finally the one and only radiant source of life and all happiness erupts. The crowd looks on transfixed. The worker, the professor, the student—all pray to this supreme beauty that has always been and will always be, and next to which all hates are so petty.

Or the sea of fog. On my way to the institute, I often went up the Zurichberg: above me was the radiant sky without the slightest tarnish and below me the earth was covered with rolling waves of clouds. I often thought that if the human race possessed by hate could be brought to this source of eternal beauty, it would lift its arms in ecstasy and abjure everything that is petty and sticky. How happy I was to be able to experience such impressions.

In Zurich we were most cordially met by my colleagues. There were two of them. Von Gonzenbach,[1] a Swiss, who contracted Heine-Medin disease (poliomyelitis) while he was still a student and, paralyzed, was unable to follow his true vocation—medicine. He possessed the key to human hearts and would have been a perfect physician, one of those who cure the body and soothe the soul. He was forced to become a theoretician, however, because this required less physical effort. Currently, he is professor of hygiene at the Zurich Polytechnic. The name of the second assistant was Klinger.[2] He was Viennese by birth, naturalized in Switzerland, and married to a Swiss woman of great culture of heart and intellect.

There was not a trace of envy nor a shadow of rivalry among the assistants. We immediately liked one another and soon switched over to a first name basis. A most delightful climate of friendship and cooperation arose and, even though we all left the institute long ago, the tradition of those years lives on.

Gonzenbach was filled with a cheerfulness that in view of his serious handicap—paralysis of the lower extremities—appeared as heroism. His physical suffering made his thoughts lofty and even radiant. How often people find refuge from pain in the world of beauty. Later, he became involved in the city's sanitary policy and gave the most exquisite lectures on the radio.

Klinger was an ascetic. He had an excellent, unusual mind. He had a heart condition but, due to a frantic feeling of duty, he did not take care of himself at all. He knew neither a day nor an hour of rest, he knew neither Sundays nor holidays; he worked continuously. He was aware of the threat of a possible early death, and he wanted to give the most of his intellect and spirit in service to the world. I worked with him, and I liked him best of all. After I left, he continued to work and publish in a frenzy. He even published research that I had criticized because it seemed not quite justified experimentally. Today, I believe that he went through visionary ecstasies, because many of his theses were confirmed by subsequent investigations. But he was unable to subdue the ardor of his spirit, and he burned out. When I was in Warsaw, he wrote to me that science did not satisfy his desire to absorb and understand the world. Experimental work forces one to think about one problem, but he wanted to learn and feel the mystery of the world's creation and the source of human thoughts. Soon after our separation, he left the institute and settled in the country in a small cultural and administrative center. His true objective was to concentrate on the study of philosophy, art, geology, and history. Again I recalled Warburg's wise words that science is jealous because it grabs man and draws all the strength out of him, leaving him no free time for other spiritual needs. Therefore, scientists are often one-sided and are unable to talk about anything but their own paths of thought. Klinger wanted to embrace the whole world with his spirit. A few years later, after his children had grown up, even that loneliness no longer satisfied him. He left his family, got on his bicycle, and departed into the world. He wrote to me in jest that he wanted to prove that laziness itself can be creative. Actually, he left like those Buddhist saints who suddenly leave their palaces full of spiritual splendor to search, with an insatiable drive, for the mysterious riddle of existence and for a contact with eternity.

Yes, this was a beautiful soul with regal talents. I do not know whether he is still alive. It strikes me as something very strange that these two persons closest to me, Dungern and Klinger, turned away from life and ended their association with science. I remained faithful to life and science, probably because in addition to my enchantment with science and the beauty of the world, other strings were also playing inside me.

Finally, I will mention someone whom young assistants usually do not appreciate: our boss, Professor Silberschmidt.[3] The attitude of children toward their parents is cruel, because young people identify the natural drive to gain freedom with an alleged necessity to reject all authority. We too often regarded our boss as the personification of coercion. Only later, when I myself had to shape human characters and teach duty and endurance and not only lofty flights, did I understand how good, reasonable, and deeply human this man was. I did not go on journeys of dreams with him as I did with Dungern, but I see now that I am much indebted to him. I owe him a hard training in fulfilling duties and the understanding that a student is like a plant that must be cared for every day—and that thrusts of thought and dreams are not enough, but that familiarity with details and a cult of duty must be developed in young people. If it were not for him, I would know neither the real duties of a pedagogue nor the art of cultivating human souls. If Friedemann and Dungern encouraged my mind to soar, Silberschmidt laid my hands to hard work. When I arrived, he was a widower, and later he married a woman of immense goodness. They both believed that bosses should be like parents to young people, and Silberschmidt often emphasized that we were one family in the institute. For him, this was no empty phrase but a feeling of deep community with those who serve an idea. When many years later misfortune befell me, he was the first to extend a helping hand to me.

Yes, my youth was filled with happiness, because I met good and wise people. No one asked anybody of whom he was born or where he came from. For them, a young person who wanted to work was like a rare flower that ought to be cultivated so that it might bloom for the glory and benefit of mankind. Only later did a time arrive when people who should have been spokesmen for science preferred that great discoveries remain unknown rather than be made by persons belonging to a different "race."[4]

This was the atmosphere in which my research proceeded in Zurich. By and large I worked and published jointly with Klinger. It is not the objective of this book to present a summary of our research projects. They are described in scientific articles in a dry style without the stigma or the charm of personal emotions. However, I should like to mention them just briefly.

We studied resistance [i.e., people's biological immunity to certain diseases] and goiters. The first problem was much closer to my heart. Contrary to the prevailing hypotheses, we became convinced that the body which dissolved blood cells and bacteria—the so-called complement—was a function of the serum, which depended on the changes in the serum's physical state. While these changes were taking place, the serum would become toxic and give a positive Wasserman reaction. In this way we were able to prove that a number of important functions of the serum consisted of changes taking place in the serum's physical state. Then we investigated the correlation between resistance processes and blood coagulability, and we succeeded in developing a coagulation reaction replacing the Wassermann reaction. This was the first such reaction, and now, twenty-five years later, it is beginning to gain citizenship rights. Investigation of goiters was made possible by a grant from the Mianowski Fund, because official Swiss funds financed only research done in Bern's institutes.[5] At that time, the so-called water theory of goiters dominated the scene. Our investigation revealed that this theory was based on completely false data. The author of that thesis accused us of incorrect observations and persisted in his interpretation. Therefore, we brought Zurich families from various remote villages and presented them before a commission of professors. In various places in Switzerland, we fed water to experimental rats from various districts. We spent entire weeks taking epidemiological pictures; a great original, Dr. Dieterle, participated in this work.[6] We succeeded in demonstrating the dependence of the goiter on the geographic area, but not necessarily on the water. Subsequently, in Poland, I became chairman of the Goiter Commission, which introduced iodination of salt in Wieliczka.[7]

I will not write about other research projects we pursued. My sojourn in Zurich was fruitful. If our research there did not acquire the fame of our study of blood groups, it was because genetics was a star rising on the scientific firmament. Yet, our studies prevailed, and others have pursued them. A scientist must reckon with the possibility that the edifice he is building will be inhabited by others. He must be ready to build from the excess of his spirit, without counting on being remembered. Most of his endeavors live as nameless achievements of the era.

And so it was with my life in Zurich. I spent three years there in actual fact although officially I was there longer since I spent the war on a military leave. I was a teaching assistant throughout that period; I set up laboratory work for students and aided in supervising doctoral dissertations. In this way, I was able to become acquainted with the whole field of hygiene, but I also noticed its artificiality as a subject for lectures. Hygiene, like medicine, represents a number of scientific branches that differ from one another in methods and approaches. In Zurich, I became very fond of pedagogical work and, aside from research, I spent my best moments with the university students. I felt like a gardener who walks in the garden of human souls: here he props up a flower, there he plucks out some weeds, and when he notices an especially beautiful variety, he surrounds it with special care and tenderness. When I occasionally contemplated how I would like to survive in the memory of young people, it was never as a professor or—God forbid—as a director, but only as a gardener of human souls. And I believed the task of one's life is to deserve it.

When after two years in Zurich I took the qualification examination, I fulfilled my third aspiration. As a young boy I decided to go through life with a woman I loved, and I fulfilled that aspiration when I married Hanna, in spite of the advice of others to marry only after I had a position. I still bless my recklessness. My second aspiration was to do scientific work. Again, contrary to advice given by rational people that I should first secure my existence, I did not ask what science would offer me and whether I would be able to create something worthwhile. Instead I followed the irresistible voice of an internal compulsion, and soon life offered me the bliss of sowing and harvesting. Now I was able to fulfill my third aspiration: to transform ideas into live words and to lead young souls into the land of ideas. We agreed that I would take the qualification examination first, and Gonzenbach second.[8] Klinger did not want to acquire the qualification degree even though he had the most to offer of the three of us. I chose my investigation of resistance and blood coagulation as the material for my qualification. The department issued the following appraisal: "With his research, Ludwik Hirszfeld has proved that he is able to point out new paths in science."

Qualification in Zurich consisted of two parts: a twenty-minute long presentation for the department and a public lecture given for the professors, students, and general public. The first presentation left an unpleasant impression. While I was talking, the usher handed out money to the professors, which seemed more important than my speech because they kept on counting it. I was even convinced that this lack of interest was indicative of the unworthiness of my presentation, and Professor Silberschmidt had to comfort me. In contrast, the public inauguration lecture was a great experience because it represented the bestowal of officer's spurs upon a young docent.

This is a great festivity in Zurich. Citizens from the whole canton come to see how a new man rises on the academic firmament, because in that country people respect those to whom they entrust their children. The rector and the dean lead the young docent into the auditorium, and behind him follow a long file of professors. The docent must speak in such a way that the student trained in listening is not bored and the layman too can understand. The speaker must prove that he can kindle a thought and express it in words that catch the imagination of the listeners. Not all young persons can take this tension, and therefore many read their presentation. They prefer to fill gaps in the development of the ideas rather than to capture the ideas with a vivid flow of words.

Furthermore, my wife forbade me to read. Once, just once in our whole life she threatened me: "If you read from the manuscript, I will divorce you." What could I do?[9]

I can see, as if it were only yesterday, the beautiful auditorium of the new university: my professors sat in the first row, the audience filled the auditorium, and in the upper row sat my wife with her frightened eyes wide open. The title of my presentation was "The Problem of Heredity in the Light of the Science of Resistance." The subject was our investigation of blood groups and what, according to my anticipation, would ensue from it: the constitutional approach to resistance. I began, and I heard my own words: "From the numerous problems attacked by the study of resistance, I should like to bring up one: Can we, by means of serological methods, distinguish individual characteristics of blood, and how are these characteristics inherited?" When I spoke these words, I recalled the long conversations and discussions I had had with Dungern, our successful experiments, and I felt a strange richness of thought and the happiness of a bird that flies free. As if through a fog, I saw my wife's large eyes looking at me, no longer with fear but with concentration and trust. When I completed the lecture, I knew that I had gained victory over myself and that from now on I would have this unique gift that was neither an expression of knowledge nor the so-called gift of oratory but a grace given from above enabling one to transmit the ardor of one's spirit to others.

When I was later asked the secret of a good lecture, I answered: "I know of only one secret: He who wants to kindle others must himself be aflame."

This is the story of my qualification. It took place long ago, in 1914, a year that terminated an era. I was twenty-nine years old.

I lectured on communicable diseases to students from all departments and gave a course in serology for medical students. The lectures gave me a great deal of pleasure. I usually had a full audience, yet I was terribly frightened before each lecture. It takes a long time for a young docent to learn to control his fear lest he lack material. At the end of the semester and contrary to the custom, I received from my students a big basket with flowers and various sweets hidden among them. This was my last meeting with Swiss youth. I have pleasant recollections of them: they know how to work and they know how to be grateful.

That year offered me one more great pleasure. It was the sixtieth birthday of the two creators of the science of resistance: Ehrlich and Behring.[10] The leading periodical, *Zeitschrift für Immunitätsforschung*, was publishing a special edition, and the editors asked me to submit a paper because they wanted to publish articles by the most outstanding serologists in this memorial edition. I told my wife: "I see how this recently helpless boy without a present or a future was lured by scientific work. Now I will join the world's scientific elite in a joint publication. The main credit goes to you because you did everything to keep this little spark in me alive. How can I thank you for the fact that you, so rich yourself, have devoted the main effort of your life to me?"

In 1913 and 1914 I attended German scientific conventions. At the meeting of microbiologists in Berlin, I presented a paper on the formation of toxic bodies in the serum and on various aspects of blood coagulation. Klinger and I opposed the then modern theory that anaphylatoxins depend on the activity of the complement. Doerr[11] and Sachs,[12] well-known researchers, also opposed this theory. Doerr, however, was drafted in Austria and was unable to attend, while Sachs had to leave because of an accident in his family. Thus, I carried the burden of a discussion with Friedberger,[13] a seasoned researcher and fencer—alone. At the end of the discussion, Landsteiner,

Kraus,[14] and others got up and supported my thesis. On that occasion, I established a close acquaintance with Landsteiner, subsequent Nobel Laureate. He had the most universal of serologists' minds, and his later studies of immunochemistry belong to the most brilliant investigations of the current era. Until one o'clock in the morning we walked the streets of Berlin talking and discussing. Such contact with great men is a source of immense intellectual pleasure.[15]

In 1914 there was a convention of internists in Wiesbaden. I gave a talk on our new reaction for syphilis established on the basis of blood coagulation. The meeting made an unpleasant impression on me. Young assistants canvassed professors for favors, and one had the impression of being on a marketplace and not at a scientific convention. I occasionally read that success is the measure of value. He who has attended scientific congresses must reject this definition of value. A man who knew how to sell a bit of knowledge was successful at a market that was called Europe's scientific life. There, one could see caution, lack of character, and the one and only goal of gaining a professorship.

Finally, I encountered the third proof that I had entered the regular scientific path and that the world was beginning to open its gates for me. Our studies of blood coagulation and resistance were also noticed outside of Switzerland. In 1914, I was visited by Dr. Brandt (subsequently professor) of Latvia and by several Japanese researchers.

Chapter Four

The Great War

The word *war* has a different meaning now than it did when I was young, in 1914. There had of course been various conflicts: the Russo-Japanese War, the Balkan wars, and the Boxer Rebellion. But the world was large, and all this was taking place somewhere far away, in distant lands. Europeans left for these wars as if they were going on exotic safaris. There are always people who like to hunt tigers. But in Europe, tigers were kept in cages. When Wilhelm II brandished his sword in a threatening manner, he looked ridiculous and almost had the appeal of some kind of harmless clown. He was playing at soldiers. No one could imagine that respectable Europe and serious representatives of religion, science, and the well-established order would suddenly jump at each other's throats like young hooligans, leap over trenches, shoot at each other, and carry out thousands of unintelligible cruelties committed when blood floods the brain. Monsters of hate, cruelty, and stupidity began to creep up from the depths and embrace human hearts with their slippery tentacles. Man, who prided himself with having attained the peak of civilization, was converted into a creature worse than a beast, a creature that reversed the history of the world and moved it back several thousands of years. The human mind surpassed itself only in creating new tools for crime. Modern man—the so-called implementer of the ethics of coexistence—proved to be lower than troglodytes. After the war, I met a black man who was returning to Africa. He said he was going home because he could no longer take European savagery. In Africa they killed to eat; in Europe they killed for pleasure.

When German scientists wrote their famous *Es is nicht wahr*, this was sincere to a certain degree.[1] The mind of a professor detached from reality could not understand how well-brought-up people could be so evil. It seemed impossible that a man fond of music could kill children. Nevertheless, it was possible. He who murdered was acting under the influence of a deeply hidden but unfortunately pulsating power. The human soul has various strata. Love for one's fellow man and respect for life and honor were present in a thin layer on the surface. But underneath, hot lava and fire were roiling and surging, searching for an exit through this woefully thin layer of culture to spread out, cover the world, and reduce to ruins this paradise conquered and built with hard work over the millennia.

In view of the abyss that opened up in Europe, national interests seemed petty. One knew only that modern man was losing his paradise as a punishment for not fulfilling the first prerequisite of living in a community: good will toward his fellow man. People spoke in all sincerity of sacred national egoism. Slogans of love changed into their own caricatures. National gods were created—the German, Russian, and other gods—and these gods were immediately mobilized for the needs of the war. In factories, the slogan *Gott mit uns* was stamped onto the buckles of soldiers' belts. They did not understand that this was blasphemy. And there was no one to protest against it.

So what was I to do?

Great dramas, even the poor ones, draw into their whirlpool all who thirst for depth and are unable to remain standing on the edge of life.

I went to the movie theaters and saw how men were being mobilized in all countries. I was often irresistibly attracted to the young enthusiasts shown on the screen who, with their homeland in their hearts, went to their deaths. In the clinic, my wife was training nurses for the legions. I did not want to join the Russian army. To whom could I offer my help?[2]

Thus, the first seven months of the war passed by. I worked, lectured, and published, but when I heard the distant, dull cannon roar that reached Zurich from the Vosges, I felt an overwhelming desire to leave, to throw in my lot, to help.

Many people came to Zurich at that time. The most interesting news was brought by Radek whom I had known as a student in Berlin. He was an Austrian citizen and had been exiled from Germany for antiwar propaganda.[3] He showed me a pamphlet in which he had anticipated that war would break out before 1917 because a mobilization reform was to take place in Russia prior to that date. He wrote that the Germans would never allow the reform to be carried out. He also showed me a paper by Liebknecht (senior), who had written that in the next war, the Germans would march on France through Belgium but would be driven back at the Marne River.[4] This was written toward the end of the nineteenth century. At that time, Switzerland was also inhabited by the creators of the Russian revolution, which was destined to spread a conflagration throughout the world.

Thus I lived torn by conflicting emotions until the terrible news about the typhus epidemic in Serbia reached us. We heard that thousands were dying, that the epidemic had destroyed villages and towns, that there was a shortage of physicians and nurses, and that missions from all over the world were hurrying to the rescue but were powerless in the wake of such disaster.[5] Several physicians, including my brother-in-law Dr. Klocman,[6] left Switzerland, but within days we received the news of their illness or death. Suddenly I felt this call of blood, this call to combat, which I had never been able to resist. Was it a desire to fill the apparent calm of my internal conflict with a manly struggle, or was it a longing for a battle dictated by my conscience? No matter, I thought, by profession I am a hygienist, and the Serbian people are entitled to assistance. I registered, and in February 1915 I received a reply from the Serbian Government in Niš to come. I took a leave of absence from my department. I remember the farewell words uttered by the famous ophthalmologist Haab: "If you want to commit suicide, why travel so far? The struggle is hopeless."[7] But this was exactly what suited me. I left my quiet laboratory, my beloved lectures, my apartment with a view overlooking the distant mountains, the lake, and the students. I said good-bye to my wife, who had to stay in the clinic where she was replacing mobilized Swiss physicians, and I left with a delightful anticipation of future experiences.[8]

My wife accompanied me to Florence. From the train window, I saw the last twitch of her lips. I clenched my teeth and embarked on a new life. I boarded a ship in Brindisi. I remember everything as if it had happened only yesterday: the tension of expectation, a strangely sweet awareness of the danger of submarines, the sea, the soldiers, and my fellow countrymen returning home. On the ship, I became friendly with a Russian named Moiseyenko who was going to Serbia as a male nurse. He was a socialist revolutionary, a nice man with an exquisite mind. When we parted in Niš, I said: "If epidemics don't frighten you, ask to be posted near me. I will try to put you to good use." It so happened that these words subsequently saved my wife's and my own life.

Athens and the Acropolis, which we visited on our way, made a great impression on me. The view from the Acropolis of the wide sea and sky is unique. There we were in the cradle of human ideas and art. But when we left Greece and arrived in Salonica by ship, we felt that we had been pelted from a radiant past into an eastern bazaar. In Serbia, railroad traffic for the civilian population was halted because of the epidemic; one could travel only with a military pass. At railroad stations, soldiers were sprinkled with a stinking liquid; this comedy was supposed to represent disinfection and delousing of the travelers. Finally we arrived in Niš, the city of the East, where what remained of the perishing Serbian state was fleeing in hope of rescue. All apartments and hotels were overcrowded. Extraordinary scenes were described to me: a lodger engaged in a love orgy with a nurse, while a dying man in the same room was begging for a glass of water and several other men were sleeping and snoring. It was hell, a veritable hell. I reported at the Ministry of Military Affairs to Colonel Karanović and asked him to send me to the area where the epidemic was at its worst. Colonel Karanović looked at me with surprise and compassion: "If you wish, I will send you to Valjevo. That is where the epidemic started."[9]

I left completely on my own with only the desire to fight and sacrifice myself, with no mission, assistance, or equipment. I had been promised a laboratory in Valjevo. I reported to the Infectious Diseases Hospital, located on a hill outside the city. I was assigned a room, a small, narrow hallway, and a small box which was grandiloquently called the bacteriological laboratory. It comprised a small incubator for a few test tubes, some fifty test tubes, and a small amount of reagents. Those were the weapons with which I was to attack one of the greatest typhus epidemics.

I set out immediately to investigate how the epidemic had begun. The Austrian army had advanced into Serbia's interior. The First Serbian Army retreated. No one presumed that the First Army was headed by a commander of God's grace—one of the most brilliant commanders during that war, a man who had a special feeling for the soldier: he knew when the soldier's love for his homeland mutated into combat action. This was Vojvoda Mišić with whom I subsequently formed a friendship and who himself told the story of this counter-offensive.[10] Vojvoda Mišić occasionally went through visionary states in his dreams. When he was retreating his heart filled with despair in face of his perishing homeland, he had a dream in which he was commanded to set off against the enemy. When he awoke, he telephoned the position and asked the liaison officers: "Are you ready?" From their voices he perceived that they were. He ordered a counteroffensive and, like David, threw himself at Goliath. The huge Austrian army retreated in alarm, leaving 60,000 prisoners of war in Valjevo. However, Serbia had neither enough food for these prisoners, nor housing, nor means of disinfection. Soon they were all infected with lice, and the sporadic cases of typhus among the prisoners changed into a conflagration that literally burned the whole country. Suffice it to say that out of a population of four million persons, 1 million contracted the disease, and that out of 360 physicians almost all contracted typhus and 126 died. Life came to a standstill. My colleagues who had gone through this Gehenna told me that hospitalized patients received no visits and were not nursed because everybody was sick. Patients who were not completely incapacitated would go downtown to get food. Some patients lay several to a bed; most however were prostrate on the floor. Occasionally, someone in a delirium would get up and run into the city in his nightgown, spreading panic and the epidemic. Several delirious patients ran out of our hospital and drowned in a river nearby. There was not enough time to bury all the dead, and their dead bodies were piled up in the vicinity of the hospital. There were cases when, by mistake, unconscious patients were placed on the piles of cadavers. This was the reality of war and the Serbian epidemic in 1915. How distant were the university palaces in Zurich and the clinics in which no efforts were spared to save every child, even a prematurely born infant. Whole villages and towns were depopulated; inspections were like walking through graveyards. And there I was with my little laboratory and fifty tests tubes supposed to cope with this disaster!

A French medical mission headed by a deserving researcher and an expert on typhus, Conseil, was working in Valjevo.[11] The mission had virtually no contact with the Serbian population, and therefore could perform only hospital functions. I, on the other hand, probably because I was alone, was forced to draw on the energy of the local people. Maybe this is why my work became the center of a new activity.

I asked that the physicians be called to a meeting, and I began with a lecture on typhus. This was 1915. In 1909, Nicolle had published his monumental epidemiological study on typhus, establishing that the louse was vector of the disease.[12] Combating pediculosis [lice infestation] was nothing new to Serbian physicians. Without money or assistance but with some superhuman effort, they began to organize or rather improvise devices for dry disinfection that accomplished more than all the foreign missions taken together. My lecture, which included their epidemiological observations and acquainted them with the results of scientific research, came as a welcome surprise. People who had been in despair, rendered helpless by the epidemic, and discouraged by the lack of resources suddenly felt as if reborn through hope (I have often observed that the secret of energy consists of a breath of hope). We decided upon a task which, I realize today, was not appropriately tackled but it gave strength to everyone involved: the idea was to catch the epidemic hydra by its head, that is, to disinfect the entire town with its 4,000 inhabitants. I requested the corresponding quantity of sulfur and, together with local physicians, disinfected literally all apartments and all items. Whether this disinfection contributed something to the improvement is difficult to say, but it probably did because the epidemic wave began to subside and the situation began to improve. They began to regard me as the victorious commander of the battle against the epidemic. The main credit goes to the Serbian physicians, however; my role consisted mostly of stimulating and coordinating their efforts.

It was on their request that I began a series of lectures on contagious diseases. In this way, a sort of academic center was formed in a remote area of Serbia, with lectures and discussions, and that noble emulation that is at the inception of any intellectual undertaking. In the meantime, I got busy organizing the laboratory. I requisitioned from the army tubes that were used for drugs and used them instead as test tubes. With the help of a local tinker I constructed sterilization devices out of empty naphtha tin cans. I made stands out of cardboard, a large thermostat out of a tin sheet, and so on. I adjusted the bacteriological techniques to the laboratory's poverty, and I saw that such techniques can be much simpler than those used in big laboratories. Laboratory technology often fails to develop because of the far-reaching labor differentiation and division. In order to be able to concentrate on research, scientists tend to delegate the implementation of laboratory work to their assistants who carry out the orders automatically instead of trying to simplify the technique. Soon I was cultivating unknown strains that proved to be paratyphoid A; they were among the first strains of paratyphoid A cultured in Europe.

My modest laboratory became the center of medical life. This was due mainly to a Serbian physician, Major Aca Savić. Savić subsequently played a great role in Serbia's sanitary history.[13] Since he was taking his first steps in my laboratory, I should like to say a few words about him. He was a giant with an athletic build and a severe expression on his face. He had the strength and obstinace of a stripling, but his attitude toward me and the laboratory was downright tender. He studied with zeal but also with the conceit of a pampered child. One day he decided that he had the whole of bacteriology in his pocket, and on the basis of this he demanded that he be given appropriate social and professional functions. He gained his end because later, in free Yugoslavia, he became minister of health and, as I gather, was a good and energetic minister. In 1928, jointly with the Commission of the League of Nations, I attended the opening of the School of Hygiene in Zagreb. Savić was then the Yugoslavian minister of health and began his speech at the banquet not by expressing his thanks to the foreign missions, but by recalling the time when a young docent from Zurich and his wife came to Serbia to rescue those who were perishing. Gratitude is a characteristic of persons made of noble stuff.

But let us return to Valjevo. Within a few months, my wife left Switzerland to join me. She could not take the peacefulness of a neutral country nor the unfriendly attitude of some of our colleagues toward the allies. My wife has a restless spirit. She was unable to calmly tend to one patient, knowing that thousands were perishing in the war without medical care. Against my will she came to Valjevo to fight the epidemic. My heart trembled at the thought that she, the best of all, might perish. Yet I knew that she had the hardiness of a soldier who does not succumb to the threat of an epidemic. I still bless her selfless physician's instinct. The war lasted three more years, and our work not only brought about scientific results but also conquered human hearts. When, during the German occupation [of Poland], Serbia learned of our misfortune, the king proclaimed me an honorary citizen of Yugoslavia, and as such I was to be reclaimed from the German government. This made me realize that there was only one form of immortality worthy of respect: human benevolence.

Major Savić met my wife at the railroad station. He vaguely remembered that it was customary for the king to take a paper from his chancellor and read it to the people. He considered himself the king . . . of his hospital. The role of the chancellor was played by the so-called *blagajnik* (cashier) who with humility and consciousness of his historical role was standing behind Savić with a big bouquet of red peonies in his hand. Major Savić delivered his rather protracted welcome speech and with a royal gesture took the bouquet from the *blagajnik*'s hands and handed it over to my wife. Only after all these ceremonies were over was I able to greet her.

A labor began that my wife and I recall always with great emotion. My wife's brother-in-law, Dr. Klocman, an excellent physician and chemist, was working with us. He had already had typhus and therefore took over the more dangerous functions to spare my wife as much as possible. A physician's functions in a war do not consist only of healing. Customs become loose under wartime circumstances. The patients were pick-pocketed every day. We had to watch the cook, the attendants, and even the patients themselves to prevent them from stealing from one another. We had to get up at night and make surprise visits to the hospital wards to check whether unconscious or dying patients were being robbed. We had no professional nurses, and their functions were performed mostly by prisoners of war who were also fighting for their lives. We had to pay attention to the prisoners in the camp who developed serious scurvy. I should like to emphasize that the Serbs displayed no trace of hatred toward the prisoners but, on the contrary, took a deeply human approach to them. Yet life was stronger. We bought large amounts of fruits for the prisoners. Each morning, my wife personally attended the sick and distributed eggs and milk to them. The elegance of a Swiss clinic was lacking, but a strong overtone of brotherhood prevailed. Not for a moment did we regret the Swiss palaces we had voluntarily left.

People around us seemed close, simple, hospitable, manly, and cordial. We became especially close to Vojvoda Mišić. At that time he was the commander of the First Army, and later, in Salonica, he became the commander in chief of the Serbian Army and supposedly was the creator of the strategic plan of breaking through the south and defeating the Austro-Bulgarian Army. I have already mentioned his visionary dreams. Once he told me that he had the following dream: he dreamt of the Triglav [a mountain in Slovenia]; the Serbian Crown Prince Alexander ascended the Triglav and, with the glance of a monarch, looked down on a state of Serbs, Croats, Slovenians. This happened at a time when no one in the army even dreamed of the future establishment of Yugoslavia and when the only aspiration of the Serbs was to include Bosnia, Backa, and Banat and to create a Great Serbia. Another time, deeply moved, he told me that he had dreamt of an independent Poland. When defeat came, we retreated, following an order against his will. Later he left for Naples because he did not want to partake in a defeat of which he was not guilty. Subsequently, when the Serbian Army reunited in southern Greece, he took command of the First Army and later of the whole army to lead the soldiers back home in a victorious march. When the soldiers learned that he had become commander in chief, they made the sign of the cross and said: "We will soon go home." His name was the symbol of victory for the soldiers. When I was returning to Poland and saying good-bye to him in Belgrade, I made a reference to his visionary states and asked him about Yugoslavia's future. I will cite his words verbatim, and may history judge whether they were prophetic. "A nation is not united by a common language but by a common struggle and shared suffering. Right now, there is an abyss of mutual dislike between the Serbs and the Croats, because we fought in opposite camps. Only a common struggle can unite us. Now that Austria and Germany have been eliminated, it is Italy that threatens us, because Italy—oblivious that the Dalmatian Coast is ours—wants it for herself. A common battle waged against the Italians will unite Croats, Serbs, and Slovenians. There have been serious frictions between the Croats and Serbs; yet I believe that this war will give rise to a full union of southern Slavs."

I should like to say a few words about the director of our hospital, Dr. Bašević, a native of the Black Mountains. He was the most beautiful man I have ever seen: he was slender and buoyant like a poplar tree and had black, ardent eyes, a subtle nose, beautiful white teeth, and a charming smile. He spoke only Russian and Serbian but could easily master English and French women. While we were there, a Scottish medical mission consisting of women only arrived in Valjevo with their own excellently equipped hospital, and the Serbian authorities assigned it to Bašević, probably because his magical effect on women was well known.[14] These women advanced thousands of demands and, to be sure, justified them. For example, they demanded that the Serbian soldiers keep the washrooms clean and that patients receive the rations to which they were entitled. Our dear Bašević was able to explain everything to these Scottish ladies, probably because they could not understand him and yet it was impossible not to like him. But I had the most fun with the French nurses. Why they came I cannot fathom and it was puzzling to me that the French health authorities had let them go to Serbia. I suspected that they wanted to get rid of them in this innocent way. During the epidemic they contracted typhus as all the others did and, because of the shortage of nurses, they were cared for by Serbian and Austrian attendants. This gave rise to the well-known love idylls, the only difference being that the nurses were male while the sick warriors were female. But the ending was the same. The Serbs required no special services from the French nurses; nevertheless, these ladies made unceasing demands. Poor Bašević offered them various explanations in Serbian and Russian; they did not understand a word but were evidently pacified. Bašević and I returned together through Serbia and Albania, and I had ample opportunity to see what a good and noble man he was. In Salonica he was compelled to subdue Greek women. He did so successfully with the same mysterious charm. Finally, I should like to mention Colonel Staić, a surgeon, who subsequently became chief of the Health Service. He observed my work in Valjevo and my ability to improvise even under very difficult circumstances. Later, in Salonica, he allowed me to organize a course in bacteriology, which was to lay the foundations for public health in Serbia.

Our life went on in this way, filled with work and impressions. We became acquainted with diseases that are unknown in central Europe. There, for the first time, I saw the germs of relapsing fever tropical ague,[15] as well as completely new clinical infestations. From time to time we buried a colleague who fell in the battle against the epidemic, and we wondered whose turn would come next. In order to organize a bacteriological service for the whole army, I toured field hospitals. This gave me the opportunity to become acquainted with the charm of camp life: the long evenings around

...ne campfire when stories were told about previous battles, wounds that did
...not hurt in the excitement of the battle, and comrades-in-arms who had
perished but who were talked about as if still among the living. Later I listened to the melancholy Serbian songs. I was deeply impressed by a night I
spent on the Drina River. The enemy was on the other side of the river, but
the Viennese papers reached the Serbian camp with a regularity of which
even the best post office could be envious. That night I learned that the
university had been reopened in Warsaw. This had a great impact on me; it
was like a vision of my future work in Poland. In Serbia I learned many
things, and I made many bacteriological observations, but mainly I learned
to live with people and I realized that previously, in the seclusion of scientific
laboratories, I had not been familiar with the main motors of human action.
I used to live as if in a glass cage, and if it had not been for the war, I would
have gone through life without properly knowing mankind, knowing only a
group of selected persons detached from life. The long evenings at the
campfire and that dizzy Serbian *kolo* (circular dance) gave me an insight into
the human soul, where man is not a single entity—he is filled with earthly
songs, memories, and hope and not just preoccupied with specific topics or
detail.

I came to know the Serbian peasant with his ardent love for his homeland
and his understanding of the causes and objectives of the war. The Serbian
peasant was building his homeland with full consciousness and deliberation
and not as a result of the obligatory mobilization. Serbia had been under the
Turkish yoke for five hundred years. A large area of arable land belonged to
the Turks, and the Serbs united their desire to possess land with yearning for
political freedom. That made me realize one unfortunate and very important cause of our Polish misfortune: serfdom was abolished by the tsars.[16]
The Polish peasant thus received land from the hands of the partitioner.
Such tragedies as turning insurgents into the hands of the partitioners were
unthinkable in Serbia. The Serbs were happy because they were able to fight
for their land under the all-encompassing banner of national freedom. The
fight against the Turks made the Serbian peasant's soul strong and valiant.
Now he felt the duty of brotherhood, but an understandable interest was
also at play here. The great Hungarian landlords, fearing the competition
of Serbian agriculture, frequently put an embargo on the Serbs' main
export items: hogs and sheep. On the other side, Austria tried to prevent
Serbia from reaching the sea, while the Serbs felt that they had to conquer
a greater breadth for themselves. That generation of Serbs was fully aware
of its historical mission. The Serbian peasant knew how to be appreciative,
probably because he had endured a hard life. He admired physicians if they
took care of him, and his attitude toward women physicians was pure tenderness and affection. I never noticed a trace of indelicacy. Striking in the
Serbian peasant was his complete absence of religiousness. It never hap-

pened that a dying man or those around him would ask for a priest. The selflessness of the French clergymen who worked as hospital attendants had no counterpart in the Serbian clergy.

The spring and summer of 1915 went by. The epidemic was under control. However, new blows and a Golgotha, probably the only one known in history at that time, awaited the Serbian people. Mackensen,[17] due to his tremendous technological superiority, broke through the front and pushed the Serbs deep into the country. Again we heard the dull resounding of distant shots, and our sleep was interrupted by the thud of cannons driving by. Wounded persons were brought in growing numbers. A tragedy was approaching. One autumn night, we received the order to evacuate the hospital and leave the seriously wounded behind. The Serbs loyally suggested to the foreign physicians and missions that they stay behind. I said to my wife: "If we stay, we will be able to return to Switzerland and resume a normal life. However, we have assumed the duty to help others. Now they are perishing. Shall we abandon them at this crucial moment, like rats deserting a sinking ship?" I did not have to persuade either my wife or her brother-in-law. So we told the Serbs: "We are soldiers in your service. In good fortune and in bad. If we perish, it will be for a good cause."

Thus our homeless wandering began.

It was a migration of peoples. How often I thought during my various trips that the war also had some beauty to it: a sweet anxiety about tomorrow and a play of community instincts that cannot be perceived in a quiet, individual life. An epidemic does not become a legend; legends are made of heroic deeds and of the trumpeting combat call of the victors. Now we were becoming acquainted with the other side of the story: the suffering of a nation that had lost its homeland and taken to the roads, not knowing whether it would ever see its native land again. We retreated as a hospital formation but were joined by the wounded who did not want to yield to the enemy, as well as by Slovenian prisoners of war who preferred Serbian captivity to Austrian service, and civilian families who preferred the uncertainty of wandering to the certainty of captivity. Finally, evacuated young boys went with us. Pasić, the brilliant Serbian politician,[18] foresaw that the war would last a long time and that these children would grow up during it; he did not want to turn these future soldiers over to the enemy. He knew that they would have to replace their fathers and brothers who had perished or would do so. However, the evacuation of children was not properly prepared either organizationally or politically. The Italians contributed to the deaths of many thousands of these boys.

This migration of peoples was unique in its kind. There were no trucks; besides, the roads were unsuitable. All possessions were placed on peasant carts drawn by oxen. At the cart's bottom were suitcases; on top of them were mattresses, straw mattresses, or loose straw; and all this was covered

Map 1. Hirszfelds's evacuation route from Serbia, 1915 (based also on accounts in Čedomir Antić, *Ralph Paget: A Diplomat in Serbia* [Belgrade: Serbian Academy of Sciences and Arts, Institute for Balkan Studies, Special Editions 94, 2006]). Drawn by O. T. Ford.

with a tarpaulin in the form of a tent. Such a cart was a mobile tent moving at a speed of twenty kilometers a day. In it we spent the nights, and during daylight hours we walked: at first through Serbian towns to the Ibar Valley, then along the beautiful Ibar Valley to Mitrovica and farther on to Priština, and finally across the Kosovo Plain to Prizren, where we reached the border of Albania. This migration to Albania lasted approximately six weeks.[19]

Serbia was beautiful. The mountains were covered mostly with oak forests glistening in the autumn sun with all hues of gold and bronze. We saw torrents and rapids and lush orchards with fruit trees drooping under ripe apples and plums. Curious was the charm of that country in autumn, and curious was the charm of these people who with humility endured great suffering, thinking of their homeland.

Again, I felt the delight of living in a community: evenings spent around campfires, soldiers' tales of past battles, quiet meditation in anticipation of death, and anxiety for the fate of those close to us. I appreciated the deliciousness of military dishes: the aroma of a ram roasted on a spit, and the taste of slivovitz [a type of plum brandy]. How far away the civilized world—how distant Switzerland was! Perhaps I was doomed to perish far from my laboratory, my students, and my colleagues, but I would die with the awareness that I was drinking life with full drafts. Our march through the Kosovo Plain was unforgettable. Five hundred years ago, the Serbs had lost their independence on that very field.[20] The memory of the tragedy lived on in folk songs and stung the heart of every Serb. With a superhuman effort they had gained their freedom, and only a few years earlier they had begun to live as a free nation with the vision of also uniting other southern Slavs. From afar, we heard the resounding of cannons and rumors that the Bulgarians had struck laterally. Serbia was perishing. All around us were people crying: officers, soldiers, and the crowds of the fleeing population. This nation loved its homeland. We said our last good-bye to the Arnaut [Albanian] city of Prizren. We received the order that all military formations were to pass through Albania to the sea so as not to surrender to the enemy, and that all possessions were to be burned.

Everything that Serbia had gained by its own efforts or had received as a gift from its friends was burning along a stretch of several dozen kilometers. Only people and packed horses and oxen continued the march. A nation was ending a period of its life when it had been permitted to have some possessions and the homeless were leaving for unknown destinations. I bought two horses to save our most important belongings, and we gave the rest to Arnauts [Albanian soldiers] in exchange for some food, because Serbian currency was no longer being accepted. Then I learned what a faithful friend Savić was. "Docent Hirszfeld," he said, "you didn't know that in wartime one must have gold. Here, take these three hundred gold francs—

you'll give them back if you come out of this alive. If not, I feel it my duty as a Serb to help you." He was a hard man, to be sure, marked by a great hunger for life. But he also had the dignity of a free nation that does not like only to take. Such were the Serbs.

Albania. The road suddenly ended. A small plank led over a brook, and we entered a path that climbed up steeply. We found ourselves in a completely different world: cliffs, goat trails along precipices, mountains covered with snow, rapid torrents, and unforgettable river passages. The Turks did not know how to build arches, and therefore bridges led up and down steeply, often at an angle of 45 degrees. These bridges had no rails and were covered with ice. People and horses slipped down into rivers and drowned. Two bridges are etched in my memory forever: Dušanov and Vezirov Most. Before we could cross, we had to wait our turn for several hours in temperatures below minus 10 degrees centigrade [14 °F].

I remember trails over precipices at the bottom of which we could see dying horses and hear their painful neighing. Often we forded brooks submerged in ice cold water up to the waist, or we crossed swamps on the shoulders of the Arnauts, who alone knew safe paths. We usually slept under the naked sky: we spread a tarpaulin on the snow, placed a blanket on it, and covered ourselves with an overcoat. So we lay in a temperature 20 degrees below zero [–4 °F], looking at the starry sky because sleep often did not want to soothe our tired bodies. Or the nights in Albanian *chans*, little windowless and chimneyless huts in which smoke from our fire burned our eyes and lungs. Once I spent such a night with King Peter.[21] He was fleeing along with his people. He asked us not to light a fire because he was suffocating. Hard was the heart of the soldier. I remember old Pasić, who was sick and was carried by soldiers on a litter. I remember Crown Prince Alexander,[22] with his face distorted from pain. Thus the crowds of soldiers and civilian population kept on, dying on the road from cold and starvation. The rain, however, was even worse than the snow. I remember when after a day's march, drenched through and through, we wanted to warm ourselves, but the wet wood would not catch fire, while the rain was pelting through our wet clothes with drops as sharp as needles. Many times we were just a step away from death, and it often seemed to us that neither our bodies nor our spirits could withstand the hardship any longer.

One incident has vividly remained in my memory because it showed me that in every situation one must be kindhearted. The physical effort and the terrible sights of human suffering did not leave my wife unscathed. She moved along frozen, hungry, and depressed. The first night in Albania, she lay down on the tarpaulin at the fire and, exhausted, immediately fell asleep. She held her feet close to the fire and did not notice that the shoe soles had been scorched. The next morning, her shoes hardly held together. She was so weak that she could not keep up with our unit. We lagged behind the hospital like two marauders. At one moment, we reached a small meadow with a campfire still burning; evidently our soldiers had rested here. I took advantage of the fire, warmed up some condensed milk I carried with me for my wife, and gave it to her to warm her up. When we were ready to move on, I noticed a heavy soldier's bag. It must have belonged to someone from our formation, because the narrow goat trail made it impossible for anyone to pass by. I thought that this was probably the last precious belonging of a soldier and, in spite of the frost that was stiffening my fingers, I picked up the heavy bag and swung it over my shoulder. A moment later, my wife sat down on the snow, began to cry, and said that all further efforts were useless because her shoe soles had fallen off completely and her feet would soon freeze which, in that situation, meant death. As if guided by premonition, I looked into the bag and found there a pair of soldier's boots. It was not difficult to stuff them with paper so that my wife could resume her march in them. It turned out that the bag belonged to the Russian Moiseyenko, who willingly let my wife keep his boots. She crossed Albania in them. When life hangs on a string, things occasionally happen that are perceived as destiny. Every soldier knows this feeling. Once in Skutari, our colleagues invited us to have dinner with them in the camp. Stirred by a strange foreboding, I refused to go. When they were entertaining the camp, it was hit by a bomb from an Austrian plane.

In Albania, we reached the end of our strength. Again, I was overcome with regret for the lost comfort, quiet laboratory, and lectures. I felt a longing for Switzerland like a condemned man recalling the pleasures he no longer knew and regretting every moment he had wasted instead of using the time to indulge in his greatest passions. The march through Albania lasted twelve days, and each day we were convinced that it was impossible for anything worse to happen and that we could no longer withstand anything. Finally, we began to descend from an altitude of about 2,000 meters [6,500 feet] down to a flower-covered meadow and, instead of the frosty air of the mountain peaks, we felt the warm breeze of the south and spring. Instead of the eternal snow, we saw emerald-green grass, and far away on the horizon beyond the fog we saw the Adriatic Sea. Never before had I had such a strong feeling of being reborn. Totally exhausted, we fell down on the grass and fell deeply asleep: physicians, soldiers, and guards. When after a long period we awakened, we noticed that we had been robbed of everything: our horses and all our things except for a small bag that we were using as a pillow were gone. Yet this did not diminish our delight with our survival. We resumed our march toward Skutari, and soon we reached this unique place on earth. The streets were surrounded by high walls concealing the secrets of harems. The site on the lake was marvelous; the place was filled with gardens, flowers, palms, and cacti; and there was a delicious warmth even though it was December. It occurred to me that happiness consists of conrasts. Life is like lemonade: if it is to taste good, it must be sweet and sour, otherwise it is insipid. We had spent twelve days in the land of death, and in Skutari we were drinking life with full drafts. What bliss it is to sleep not only on a bundle of straw but also under a roof and in warmth. Dinner at a restaurant was sheer enchantment.

In Skutari, however, the Serbian Government did not want to assume the responsibility for the lives of foreign sanitary missions and physicians. It bade us a sincere farewell and entrusted us to the English ambassador Lord Paget.[23] We went with all the missions to San Giovanni di Medua, a small Albanian port, to await rescue. At that time, an insurrection broke out in Albania against the Serbs, and we began to hear rumors of massacres that might also threaten us. The port of San Giovanni di Medua was bombarded on the day of our arrival. From afar, we heard the thunder of bombs, and when we entered the port several hours later, the masts of sunken ships were still protruding from the water.

Lord Paget received the news that an American ship was due to arrive but that it would take women only (this was December 1915, and America was still neutral). We all held a meeting, and the men decided to walk to Durazzo or Balon, while the women were to wait for the American ship. Only my brother-in-law and I opposed this idea, because we did not want to abandon the women at a time of an exploding insurrection and a possible massacre. So the two of us stayed. But good luck rewarded us, because after the American ship had left, the Italian ship Brindisi arrived—by that time the Italians were our allies—bringing food for the Serbs, and it took us all to Italy.

On a beautiful, moonlit night, the ship was sailing calmly, carrying us back to a calm life. After a twenty-hour-long trip, we disembarked in Brindisi. Yet, the first impression this beloved Europe made on us was disastrous: within an hour, the rest of our belongings were stolen. At the railroad station, they fed us with suspicious haste and immediately dispatched us to Bari to the Russian consul. The consul did not inquire about what we had gone through and what the fate of his Slavic ally was. He was interested only in whether we had our documents. Our Russian friends had to offer multiple explanations. I told the consul what I thought of him, and we went downtown to shop for necessities. We had to dress ourselves from head to toe. In Europe, rumors about Balkan cheating were frequent, but our personal experience pointed to something else. The shoes we bought had—as we later found out—soles made of cardboard, and hotels simply fleeced their guests. Yet at night, the city of Bari, filled with the blue light of street lamps, seemed to us—who had come from the land of ice and death—like a fairy tale.

Our English and Russian friends who had left us in San Giovanni di Medua underwent torture for a whole month until they finally reached

Durazzo. The most tragic fate befell the evacuated Serbian youth. The Italian authorities did not let them into Valona, and those poor, starved children were forced to return to Durazzo. Several thousand of them died on the return march. The Serbs can still not forget it. Vojvod Mišić, who had sent thousands to a sure death, told me about it with tears in his eyes. If nations knew how cruel their governments can be on behalf of the so-called reason of state, they would not grant their governments so much freedom. However, what is the significance of several thousand children in the wake of the mass murder committed these days?

In Rome we visited the Sistine Chapel. Immortal beauty again enveloped us with its grandeur. The contrast was striking: in our minds we still carried the image of the Albanian mountains and the dead bodies, while before us stood the perfect beauty of the Renaissance. We returned to Zurich highly inebriated. What a divine pleasure it was to sit down at my desk in my study and talk about the past and dream about beauty, science, and art. This state of divine euphoria lasted one week. After one week, I felt an unbridled longing for distant places and further struggle.

Chapter Five

Armée d'Orient

In Zurich, I roamed aimlessly in the streets. I saw mountains and abysses, and my thoughts were filled with memories of evenings spent at campfires, folk songs, and the fight against the epidemic. I began to study the use of salt for states of shock. But I felt the immateriality of this problem compared with what was happening in the world. In Serbia, I tread on ground where the seeds had been sown for an exuberant and rich future. In Zurich, I walked on asphalt where seemingly nothing new could grow. The Red Cross invited me to speak about my experiences in Serbia.[1] My presentation was the first news to reach Zurich about the fate of an army and a people who perished in the Albanian mountains. The great concert hall where I spoke was filled to capacity. I told the audience about the people who marched across Albania, about their inhuman efforts and suffering, and about an epidemic that destroyed more than cannonballs could obliterate. I painted mental pictures of burned villages, of mothers standing over the graves of their children, and of the heartbreak of parting forever. Tears glistened in the eyes of my audience. Just about that time, I read a novel by a Swiss author. To describe the Swiss army he chose maneuvers. The mawkish story of a little soldier with his little sword seemed to symbolize that period of life in Switzerland.

Much is still being written about the psychology of neutral nations. A petit-bourgeois attitude toward life prevails among them, because only great suffering gives rise to great tension and creativity. This is a difficult problem. One thing is for sure, however: certain types of people must undergo great tension, charge themselves, and then unload their burden. This is associated with a feeling of immense bliss proceeding on a different plane from the common happiness of prevailing in immobility. Even pain causes a certain bliss. Ostwald once expressed the idea that every experience gives rise to a feeling of bliss that is magnified when the experience is pleasant and diminished when the experience is unpleasant, which is why strong experiences attract people into their whirlpool.[2] Travelers know the unrest drawing them to distant countries. Occasionally, this unrest overwhelms entire communities who then abandon their native land and take off for the unknown. In fact, it is a kind of illness, the disease of the far yonder. And I had caught it.

My wife felt the same way. She went to Naples to join the Italian army. She had to pass the state board examination, and she did. I wrote to the Colonial

Office in London offering my services. I was especially interested in exotic countries to combine my thirst for adventures with a study of tropical bacteriology. Suddenly, my dreams came true: the Serbian Government in Corfu telegraphed me to buy a laboratory in Switzerland and come immediately with my wife.[3] I did not hesitate for a moment. I met my wife in Rome and after a few days, we were both aboard a ship. Again, we were surrounded by the sea and the sweet uncertainty of tomorrow. Time and again we noticed a submarine periscope in the distance, and death glanced into our eyes; we pulled our life preservers over our heads. Sharks followed our ship in anticipation of food. Still, all this seemed better than stagnation in Switzerland.

In Corfu, we learned about the fate of the army. Its survivors marched through Albanian ports—San Giovanni di Medusa, Durazzo, and Valona— and were finally evacuated by the French to the Corfu Island and to Bizerta. The Serbian Government was in Corfu. Great sufferings were awaiting the Serbs. A cholera epidemic broke out. Sick people were placed on the small island of Vido soon to turn into a hell well-known to the Serbs: each day hundreds of dead bodies that could not be buried were thrown into the sea. The soldiers were so starved that often after eating their first rich meal they died in violent pain. At the same time, new life was waking up. It was spring, Corfu was covered with orange and lemon blossoms. Olive groves resembled Gardens of Eden. The snow-covered Epir Mountains were visible in the distance, while the blue sea was right next to us. At night, there were billions of stars in the sky, while other stars shone on earth: thousands and thousands of glowworms flitted around like so many dancing stars. There was Queen Elisabeth's castle, the Achilleon, located on a hill in an enchanted park and haunted by the shadow of its empress who was able to satisfy all desires except longing. Here on Corfu, the Serbian Army was replenishing itself. I will not deal with political and military history; rather, I should like to reproduce the mood that prevailed. In spite of all, it was mainly a mood of hope. It was on Corfu that the beautiful Serbian song arose:

Far, far beyond the sea
Is my village and my beloved one.

Such a frame of mind and song of a homeless soldier went straight to a Pole's heart.

Corfu had an active French laboratory directed by two able bacteriologists, Burnet and Lisbonne. Later, Burnet was named successor to the famous Nicolle in Tunis, while Lisbonne became professor of bacteriology in Montpellier.[4] My laboratory was to be dispatched with the first military transport to Salonica and was supposed to become the central laboratory of the Serbian Army.[5] I was assigned to the largest hospital of contagious diseases located in the vicinity of Salonica in the village of Sedes, under French administration. My wife was to direct a department in the same hospital. We boarded the English military transport ship *Elloby* and sailed across the Isthmus of Corinth to Salonica. Again, we were overcome by the anticipation of the far yonder with the adventures it had to offer. The ships' captain was an old sea wolf. He was moved by the sight of a woman joining the war. He invited us to his table for the duration of the trip; he offered us various delicacies and sang English songs. This whole idyll took place amidst mines and submarines. One can never perceive life as strongly as in moments of danger.

As we were approaching Salonica, we caught sight of minarets and white-walled houses and, outside the city, a small, beautiful park, ruins of a castle or a tower, magnificent cypresses and, on the other side of this miracle, the snow-white Olympus.[6] All this looked so beautiful with the Macedonian valley in the background that my wife called out: "Oh, how I wish I could live there!" Fate favored us. The tents of the hospital to which we were assigned were located right above the park, while the chief physician, Dr. Damond,[7] lived in the beautiful park itself and offered housing to us. Once this building had been a Turkish pasha's castle, the tower housed the harem, and nearby was a stone-walled water reservoir that was supplied with springwater and in which *houris*[8] had bathed not so long ago. Dr. Damond did not eat at the hospital canteen, but rather in separate dining quarters that he shared with a few other military personnel and to which he invited us. As for my laboratory, I received a barrack up the hill, with no floor to be sure, but spacious and bright. In this barrack, I was destined to make several discoveries that were subsequently rated among the most important scientific discoveries made during that war.

Thus, we started a new life. The hospital—with beds for up to 1,000 infectious disease patients—consisted of large tents each able to accommodate up to forty patients. From the hospital we could see Olympus, the distant sea, and our park. Beyond, stretched a mountain chain with the highest peak called the Hortiak. Located between these mountains and the sea was a valley, swampy in the fall and winter and parched in the summer by awful heat. Occasionally, a hot wind blew from the Vardar, and then clouds of burning sand would whirl in the air. Such a wind was able to overturn tents and knock people down.

The personnel comprised French military physicians, three Russian physicians in French service, one Greek physician, my wife, and myself. The nursing service was French, while auxiliary functions were performed by Serbs. The French male nurses were mainly clergymen and teachers; their intellectual and moral standards were high, and their selflessness and disinterestedness were tremendous. Later, French nurses arrived.

We would get up at five in the morning, because in the summer, in view of the heat, work in the hospital began at six. The work was so absorbing, however, despite the heat we stayed in the hospital and in the laboratory the whole day. I had a few Serbian soldiers and one student to assist me. I trained my soldiers perfectly; they were willing, skillful, and intelligent and they appreciated the work. One should not use the intelligentsia for simple tasks, because then they work carelessly. The first prerequisite of conscientiousness is respect for one's work. From the physicians, I received instructions as to what examination should be performed on whom and, depending on the case, I either took the blood samples myself or, in simpler cases, sent my assistants. I consider this approach better than receiving samples collected by someone else, because then the bacteriologist is deprived of the stimulating effect of the patient who is to be examined. Thanks to this organization, I established contacts with the physicians, described blood sedimentation, cultivated unknown paratyphoid germs, and so on. The hospital bacteriologist should function as a consultant to the clinicians and not just mechanically implement their assignments.

When Major Damond returned to France due to illness, we liquidated the barrack and arranged a room for ourselves in the stables. An old Serbian soldier was assigned to us as our cook, my orderly was our servant, and in this way we organized a life that was not only industrious but also rather graceful. The beautiful park and the sort of family life we established drew to us many military physicians who did not have a home of their own. Another attractive aspect was the fact that at our table everyone could converse in his native language, because the *Armée d'Orient* was a conglomerate of various nationalities, and one could often hear at our table simultaneous conversations in French, English, Polish, Serbian, Russian, and Italian. Various types of people appeared as if in a kaleidoscope; we were able to get to know national psychology, customs, and mutual relations. This was not only interesting, valuable, and pleasant but in the end it was thus that we were able to study the blood of various races and nations, which under different circumstances would have required many years of work. I should like to say a few words about several interesting persons. We were frequently visited by the sanitary chief of the French Army, General Riotte. He was a military physician, a surgeon, who suffered undesirable consequences inasmuch as he did not foresee the main calamity of the eastern army—the ague.[9] He was a humanitarian, a gentle man, and an epicurean who had no understanding of discipline. He liked good wine and he used to say: "When I see a poor man add water to wine, I pity him; when I see a rich man do that, I despise him." Like all Frenchmen, he had a weakness for women and often visited our nurses who were piously devoted to their chief. He evidently liked a feminine atmosphere. Rumors about his weaknesses reached Paris, and a special parliamentary commission arrived to check the matter on location. He told us the story with great amusement. He was told: "Mon général, on dit que vous donnez les médailles des épidemies aux infirmières avec lesquelles vous couchez" (General Riotte, I gather you give medals to the nurses with whom you sleep). To which he frankly replied (he was about sixty years old): "Mais vous exagérez mes forces, je donne les médailles des épidémies a tout le monde" (But you exaggerate my strength, I give medals to every one). This argument fully satisfied the commission, and the matter ended at that. He did not like and had little respect for his bacteriological consultants.

The hospital director, Major Damond, was a charming man, an excellent physician, and a sincere and independent person. He felt crowded in France and so spent his life in the colonies. At that time, Duchess Narishkina was an influential person in Salonica. At first, Dr. Damond was the chief physician in her hospital; however, he could not tolerate the specific feminine, meddle-in-all management. He had what is probably the most essential feature of the French spirit: independence of thought. He was a sincere democrat but knew how to impose his will on others. We established a close friendship with him as well as with other French physicians. We often think of them as good human beings and good physicians. The thing that was a novelty for us was the great influence of women, or rather of femininity. The nurses who arrived from France were well educated, belonged to the best social circles, and were beyond comparison with the pseudo-nurses I had met in Valjevo. Military automobiles soon began frequenting our hospital; women acted like magnets. I will not mention the gossip; however, one anecdote was too good to be forgotten. Our nurses had by and large sharp tongues. Once, we received a list of women who had received a decoration. My wife wanted to congratulate one of the nurses; however, it turned out that she was bypassed for the medal. Yet, she put a bold face on the matter and said: "Cela ne fait rien, au moins ma mère saura que je suis restée vierge" (No matter, at least my mother will know I have kept my virginity). Another woman was ugly but very popular. Once she explained to my wife at length that it was not difficult to be popular when one was pretty. To be liked when one was ugly, that was a proof of value. Contrary to English women, the French women were not eager to join the war, at least not the eastern front.

I will not make special mention of my other colleagues, all of whom I recall with pleasure. However, the Greek physician in French service was unusually amusing. For a while, he was in charge of our dining room. Once, a sheep was stolen from our kitchen. Our doctor took two soldiers, went up the hills, and took a sheep from the first flock he saw. He could not understand why he was forced to return the sheep. He lived with a woman whom he called his sister-in-law. Once, I visited his room and saw only one bed. "Where does your sister-in-law sleep?" I asked. He proudly pointed to the bed. "Well, and you?" I went on indiscreetly. "I," he replied, "sleep under the bed." We maintained the most cordial relations with the Serbs, particularly with the chief of the health service, Colonel Roman Sondermeyer.[10] He was a native Pole from Cracow but left for Serbia as a young physician as a result of an unhappy love and had advanced to the highest positions. He was esteemed and loved. He married a Serbian woman and had three charming children: two sons on the front and a daughter. One of his sons was a pilot. They visited us frequently. Once we spent Christmas Eve together. I found books by Mickiewicz and Słowacki.[11] We sat around the Christmas tree and sang Polish carols and read Polish poems. He cried in reminiscence of his country and countrymen. Repeatedly he took my wife as an interpreter to the front where she met Prince George, who was once the Crown Prince but, displaying little self-control, was subsequently eliminated from the succession to the throne. He was a brave soldier although somewhat irresponsible. He hated diplomacy. Each time he visited us with Voivod Mišić, he complained about the government representatives as if he were the greatest revolutionary. Once, Colonel Staić, later health chief of the Serbian Army, came to me greatly worried. "Mr. Hirszfeld, tell me quickly what antianaphylaxis means. The Prince wants to know, and I do not want it to show that I've never heard the word before." Those were the troubles they had with him. To reward my wife for performing the function of interpreter, he personally bought for her the medical Larousse [dictionary]. He was evidently impressed by my wife's courage. At one time he asked unceremoniously: "Which one of you wears the pants?" As a bacteriologist, I visited his brother Alexander, the later king, several times, and he visited us in the hospital named after him where I later had my laboratory. He was a very captivating and intelligent man.

After a while, a Serbian bacteriologist, Dr. Buli, was assigned to my laboratory. He was excellent at telling jokes, but his cheerful and garrulous spirit did not tolerate bacteriological plodding. He was a good companion, and therefore the Serbs liked him, but they preferred not to saddle him with any responsibilities. We also met several Serbian painters whom the government surrounded with great care. They were individuals with great talent and, by and large, children of country folks. We formed a close friendship with Milovanović[12] who painted a beautiful portrait of my wife (unfortunately, it burned with our other possessions). His father died when he was six years old. When he was nine, his mother remarried. This was such a blow to him that he ran away from home, attended school living off his wits, and eventually studied in Munich. He told us a touching story of how, as an adult and famous painter, he returned to his native village and took care of his mother and sisters. I bought beautiful paintings from him, which were very dear to us as a memory of that war. Later, the Germans took the paintings away from me, because I was not entitled to possess anything, and least of all works of art.

Serbian soldiers were accommodated in Serbian, French, and English hospitals. All these hospitals were located close to one another, thus enabling us to establish contact with physicians of various nationalities. I will begin with the English women.[13] Some hospitals were managed exclusively by women. I remember the directress Dr. McIlroy, who later became a professor in London, Dr. McNeill, Dr. Taylor, Emslie, Dr. Hutchinson, and many others. They were the incarnation of soldierly and womanly virtues at once. Wherever English women appeared, the atmosphere of a home immediately arose. With that, they had the courage of heroes. Once, when a bomb hit an English hospital, neither the women physicians nor the nurses interrupted their work. Some English women worked as chauffeurs driving wounded from the battlefield; they would come up to the line of fire. After the war, many took Serbian orphans with them and raised them. Even now, after so many years, we still maintain a deep friendship with many of them. Nor do English men have the stiffness and reserve generally ascribed to them. They were very sincere and hospitable to us and the Serbs. And they can play like children. English schools purposefully do not overtax the intellect with excessive instruction, but they educate the will and self-control. In sports, Englishmen are the only ones who acquire this unique fair play and gentlemanly attitude so characteristic of them. The only thing that shocked us foreigners was their attitude toward Hindus. An English physician would never sit at a table with a Hindu physician, and yet we met there several highly educated and cultivated Hindu physicians.

The *Armée d'Orient* did not comprise white races only. We met Hindus, Malaysians, Negroes, and Arabs. We began to learn their way of life and mentality. However, I would be carried much too far if I wished to go into detail, especially since studying human psychology out of the natural habitat offers no insight into the true mental motivations. A visit to Salonica's Cafe Floca at four o'clock presented a unique kind of a kaleidoscope of races and nations. How distant was the monotony of Swiss psychological types, and how alien was the figure of the so-called émigré who saved his peace and property in a neutral country but failed to realize that he lost one of the rare moments when one can participate in creating a new world.

Let us now turn to matters that were the subject of my scientific interest and professional activity. At first, Macedonia had almost no typhus, and cholera died out in Corfu. Typhoid fever appeared, however, and by the end of the summer, in the fall of 1916, ague broke out with a tremendous force. It did not destroy the Balkan front only because on the other side, the Austrian and Bulgarian troops were suffering just as much. There was almost no typhoid fever in Switzerland, while here I became familiar with various types of this disease. Soon, I succeeded in cultivating paratyphus A germs, which were almost unknown in Europe prior to the war and which probably came in with the Hindu troops, swamped Europe for a certain period, and then disappeared for unknown reasons. I also discovered a new germ that I called paratyphoid C.[14] I communicated this to the headquarters allies in a confidential letter, but I did not publish it immediately because the discovery of a new germ seemed quite immaterial in view of the historical storm. It was only after two years, when I demonstrated these bacteria to my students in a discussion of their significance for epidemiology in the Balkans and saw their interest, that I realized the significance of this discovery and sent a note to the *Lancet*. The English gave the strain my name; it turns out that it plays an important role in the epidemiology of the East.

The ague was the most interesting of the diseases. Even experienced French colonial physicians did not know about its great diversity of symptoms. A physician usually believes that the signs of ague consist of regularly recurring fever and chills. However, there we saw patients who fell down as if struck by lightning and died displaying atypical signs. We saw people who died in a coma or with signs of dysentery, typhoid fever, or encephalitis. Without a bacteriologist it would have been impossible to guess that these were all various forms of ague or tropical malaria. The ague (malaria) slides I made at that time were marked by such a variety of form and such beauty that I could look at them for hours. Many years later in Poland I used them to teach my students.

I too experienced a severe form of ague. This is one of the most painful recollections I have of that war. I remember the first signs of unrest, a feeling of imminent danger, and a general irritation. I mustered enough willpower to examine my blood, and I found tropical malaria germs in it. I did not want, and I could not afford, to be sick, and so I decided to cure myself radically by means of a large dose of quinine. Soluble preparations of quinine and urethane were in use at that time.[15] Jointly with Dr. Damond we decided to inject a large dose, 1.5 grams, intravenously. My wife carried out the injection. The needle was still in my vein, when I felt a strange taste of camphor and fell down unconscious. When I woke up I felt a terrible rumble in my head, and I had a queer vision of five numbers of which the even represented the rhythm of death and the odd the rhythm of life. The uneven numbers won, and I began to return to earth. I was surrounded by a deep darkness. From a distance, I heard my wife weeping spasmodically. I realized that I was perishing, that I was dying in insanity. I uttered a few words: "Darkness prevails, I am losing my mind, save me," and again I fell into a deep faint. This fight between life and death lasted seven hours, and I still have the impression that it was given to me to visit the other world and feel the strange rhythm of death. This impression was probably caused by the fact that due to quinine poisoning I lost all feeling, hearing, and sight. This induced the sensation that I was somewhere far in a different world and in a different existence. I was saved by an immediate hemorrhage produced by Dr. Damond with his pocketknife and a voluminous, intravenous infusion of a salt solution. This was one of the rare cases where quinine-induced blindness subsided.

When I came to my senses and back to earth, I had the sensation that I had seen a terrible reality stronger than life on earth. After that, I was sick for six months; fever lasting for several days recurred once every two weeks. Often, feeling the approaching paroxysm, I had to continue my laboratory work. For a long time I had the feeling of indifference toward worldly affairs, like someone who had glanced into the eyes of an awful mystery. I was close to a breakdown even though I was aware of my pathological state. Horseback riding and possibly the stimulating French wines pulled me out of the depression. The strength and lust of the horse on which I was galloping over the Macedonian wastelands filled me with the joy of life. The horse as it were suggested its strength to me.

This was the second illness I went through. The first was paratyphoid A to which I succumbed shortly after my arrival. I remember a bad headache and a strange malaise. A Serbian soldier took a blood sample. I tested the sample myself, and when I established that I had paratyphoid A in my blood I felt—I am embarrassed to admit it—a certain satisfaction that I was ill with a disease so rare in Europe. I was ready to die and asked Dr. Sondermeyer to take care of my wife; however, I somehow scrambled out of it. Despite the disease, I did not interrupt my work except for the periods when I had a very high fever. Work was plentiful. The Vardar Valley was heavily infested with ague, and the several hundred thousand soldiers increased the number of susceptible individuals. Local people go through this disease in childhood, and therefore adults have a hidden ague without major symptoms. On the other hand, soldiers who came from countries with no ague suffered very badly. Headquarters received a report about mysterious cases of sudden deaths on the front, which occasionally occurred after a battle but usually without cause. I took my utensils and left for the front to discover right in the trenches the cause of the mysterious deaths. I will never forget the examinations performed to the accompaniment of cannon shots. I succeeded in discovering the cause: it was tropical malaria. For the second time, we witnessed the tremendous effects of epidemics on the progress of military operations. The first time it was in Serbia when typhus almost put a halt to military operations; the second time it was in Macedonia when the ague overpowered both armies. I will not write much about combating the ague, but I should like to point out that today treatment is based on completely different principles. At that time attempts were made to cure patients with large doses of quinine in order to implement the postulate "therapia sterilisans magna."[16] This was also the reason why such a large dose of quinine was injected into me. Currently, thanks to Jameson's work and the League of Nations Health Commission, relatively small doses are used to induce a chronic infection and thus produce resistance. New, effective drugs have also been found: Atebrin and plasmaquinine.

Dysentery was a plague to the army. Two forms of the disease prevailed: the bacillary and the amoebic. Since the treatment of each form is completely different, the bacteriological laboratory, by making identifications was rendering great services. The tests are subtle and difficult. I worked especially hard on bacillary dysentery and discovered germs that could not be identified with most of the serums that we received from English, French, and Italian laboratories and which agglutinated with the serums of the patients. Unfortunately, I did not have the time to publish these findings. I often watched the autopsies of those who had died of dysentery: the large intestines, swollen and gangrenous, formed one large wound. Dysentery was the only disease that frightened me. Epidemiological studies were of little use since the germs were transmitted by billions of flies. Dysentery did not spare the royal house: the hoary King Peter succumbed to it. In his stool, I found germs of the Flexner type in amazing quantities.

I will mention one more disease that was widespread in southern Europe, the so-called pappataci fever.[17] It is not lethal, but it causes unbearable headache, pain in all muscles, fever for several days and, as an aftereffect, a prolonged depression. Both I and my wife went through this disease and this was the illness that finally undermined my strength. Thus, after two years of work, we decided to take several weeks' vacation in a convalescent camp in the mountains of Vodena. There, I began to regain my health. We met many nice people, but we were most impressed with an elderly English woman, Miss Tubbet. She worked in the camp as a laundress. In spite of her education and social manners, everybody believed that she came from a poor and simple background. However, when it became necessary to set up a dental office, Miss Tubbet donated a large amount of money for this purpose. Soon we found out that she was a wealthy woman who wanted to help the Serbs and who believed that washing laundry was just as necessary as other services. Socialites often have an undefined picture of the role of angels, and some of them, knowing nothing, impose their selflessness on others. The joke was circulating there that when such an angel of a lady asked a soldier whether she might wash his face, he answered: "Maybe tomorrow. Today, my face has been washed twenty-two times."

I have outlined the epidemic situation and the work in my laboratory. However, in my floorless barrack I accomplished one more important thing for the army: I developed a vaccine against typhoid and one against cholera. The army was first vaccinated in the spring of 1916 with a French vaccine. Evidently, very toxic strains had been used, because after the injections, whole regiments lay in windrows with fever and edema. I personally saw several deaths among elderly persons. The soldiers rebelled, and the commanding officers informed the head staff that they would allow no vaccinations for military reasons. Moreover, the vaccine did not contain paratyphoid C, which I had only recently discovered and remained so far unknown to the French. So, I went to the front to see Voivod Mišić. "Mr. Mišić," I said, "you have come to know me as a man who can improvise. I am ready to manufacᴄure a vaccine which will induce no reaction and will produce resistance against the local strains."

"If you make such a vaccine, I will be the first one to be vaccinated and I will make the soldiers let themselves be vaccinated too."

I embarked on a project in which I had to improvise everything. I ordered large incubators to be made. Since they had no thermostats, soldiers had to keep pulling the lamps to and away from the incubators, night and day, so as to keep the temperature constant. Instead of special bottles, I used empty Vichy water bottles; I gave up using agar and introduced dozens of other improvements. I only used the local strains, the toxicity of which I checked on myself and my soldiers. Most important, I inoculated my assistants, simple soldiers, with an understanding of the importance of our task. The work was in full swing day and night. I was assisted by one student and a number of simple soldiers, and I won. The vaccine was first tested on our soldiers and then was sent to the front. Voivod Mišić was vaccinated first, and after him, the staff members and higher officers. With the exception of the first vaccination made with the French vaccine, all vaccinations till the end of the war were made with the vaccine produced in my laboratory. There were almost no reactions, and typhoid fever disappeared. For this, I received a very high decoration—the Order of Saint Sava Third Class. For many years to come, this decoration was precious to me as a memory of work and duty accomplished.

So, time went on in the hospital laboratory in Sedes. The Serbs wanted to establish a medical center. They founded a large hospital in Salonica named after Crown Prince Alexander. The Serbian Government transferred my laboratory there and assigned me a separate building with five rooms. My wife and I lived in one room, and the other four I reserved for laboratory tests, scientific projects, and bacteriological courses. With the sanitary chief of the army, Colonel Staić, we decided to train Serbian bacteriological cadres and establish a laboratory in each army, because our laboratories that were associated with hospitals were unable to service the front. We had to select suitable men. The first to register was Dr. Ivanić, a young physician who had studied in Vienna and who wanted to acquire a further specialization. I found the second man in the following way. Colonel Staić told me that while sitting in a trench during bombardment he had met a young physician who was studying tropical medicine. "Colonel," I said, "this is our man; send him to me." Thus, Dr. Kosta Todorović arrived.[18] Finally, Dr. Ranković and several other physicians and students were assigned to me. Among the foreign physicians, one Russian, Dr. Kossovitch,[19] also joined us. Thus, a bacteriological course began; it was unique with regard to the zeal of the participants and the results. In the morning, I gave a lecture in Serbian (I had mastered the language to a sufficient degree). Then we visited the patients, discussed the cases, pointed out what examinations should be performed and how the results should be interpreted, and took blooc samples. Each participant performed the bacteriological examinations on the patients he had previously studied from a clinical viewpoint. Every cultivated germ produced tremendous joy. When I showed them new things I discovered, such as paratyphoid C, the enthusiasm was so great that I was finally induced to write a report and submit it to an English journal.[20]

There I learned a single important lesson: that cheerful study increases intellectual capacity, and compulsion and roll calls are unnecessary when young people are shown new worlds and horizons. Nowhere else was my teaching so fruitful. After the war, Dr. Ivanić became director of the beautiful Belgrade Hygiene Institute and then chief of Yugoslavia's Health Service. Dr. Todorović became professor of contagious diseases in Belgrade, while Dr. Ranković became director of the diagnostic division. Considering that one of my students in Belgrade was Dr. Šimić, who subsequently became professor of bacteriology, and that my pupil in Valjevo, Major Savić, became minister of health, I can proudly say that my endeavors yielded a rich harvest. At one session of the Commission of the League of Nations, a great organizer of Serbian hygiene, Dr. Štampar, said that I had laid the foundations for Yugoslav bacteriology and public health.[21] I was deeply moved by this compliment. In addition to this first course, we organized graduate training for physicians. I would often return from the city, tired by the sight of eastern bustling to find my dear boys bent over their microscopes intoxicated with enthusiasm. Such teaching is possible only in an ardent atmosphere of struggle and aspirations to rebuild one's country.

I also introduced blood transfusion into the army. After the lectures, we would often collect blood from one of us and infuse it intravenously into patients by means of simple, self-made instruments. The blood donors, my pupils and soldiers, would not stop working even after they donated their blood. I still have disdain for anyone who is stingy with his blood and is afraid to give it. The Serbian Army was one of the first to use blood transfusions on a large scale. I remember one case in my wife's department. There was a dead-pale soldier with chronic malaria and just 15 percent of hemoglobin. He was shaking from cold and suffocating because of lack of oxygen. His physicians said that he would die. I succeeded in saving him. We infused my blood into him for several days. For a week, he was hanging between life and death, but slowly he started to regain color and strength. When I was leaving Salonica, he was working in the garden. At that time, he had 60 percent hemoglobin. I thought of this case many years later when I was opening the First International Congress on Blood Transfusion in Rome.

After a few months, the Americans supplied us with field bacteriological laboratories, which we sent to the front. Efficient work was done in them. My pupils became the soul and heart of the sanatoriums in their armies. Doctors Ivanić and Todorović later joined me in Warsaw to receive further training in our institute. Dr. Todorović began his qualification lecture with a reminiscence of his first teacher of bacteriology who had joined a perishing nation. He prefaced his beautiful textbook on contagious diseases with an introduction in which he referred to me with gratitude. Kosta was one of my favorite pupils, not only because of his unusual abilities but also because of high zeal and integrity. After the war, Dr. Štampar developed an outstanding activity as a hygienist in Yugoslavia, established a number of hygiene institutes, and gave them significant executive power. When I visited, with every step I saw traces of work done by my pupils. In 1928, jointly with the League of Nations Health Commission, we attended the opening of the School of Hygiene in Zagreb. I gave a talk on behalf of the Polish, Czech, and Hungarian institutes, and I said the following in Serbian: "Not so long ago, we came here to teach you. Now, we have come to learn from you. And believe me, this is the greatest reward for a teacher." Later, I gave a speech in Belgrade in which I discussed the development of scientific ideas born in the modest military laboratory. Dr. Ivanić preceded my speech with a review of the course that left such deep marks on the history of Serbia's hygiene and bacteriology.

Dr. Kossovitch, my Russian associate, returned to Paris after the war, joined the Pasteur Institute, and gained a very good name in French bacteriology. He continued the research he had begun in Macedonia, France, and Africa, and made important sero-anthropological contributions. Whenever I visited Paris, he met me, and I spent pleasant moments at his house with him and his charming Norwegian wife. We would recall our exotic Macedonian period and say that life was more interesting for those who were not afraid of it.

Now, I ought to discuss our scientific work. I have mentioned my discovery of the spurious typhoid germ that I named paratyphoid C. When I was collecting blood samples from malaria patients, I noticed a faster sedimentation of blood cells, and I developed a precise method for studying this phenomenon. I performed a thorough analysis and distinguished this phenomenon from erythrocyte agglutination at very low temperatures. This I described later in Warsaw. The accelerated blood sedimentation was described in 1918 by the Swedish researcher Farreus, and it is generally believed in the West that he is the discoverer of this phenomenon. The fact is it had been discovered long before the war by a Pole named Biernacki, and my findings were published one year prior to Farreus. I published this article during the war in 1917 in a Swiss journal that was little read, and therefore my article went unnoticed. I observed this phenomenon in malaria, a disease which was rare in Europe. Farreus applied his findings to tuberculosis where they are of great significance. I still consider the method by Farreus insufficiently precise, because it does not take the anemic factor into account, which I emphasized and which greatly affects blood sedimentation.[22]

The most fertile study was that of blood groups in various nations and races.[23] Even when I was with Dungern, we dreamed of investigating the world from the serological aspect. Here we were amidst a unique agglomeration of various races and nations, and a project requiring many years of study and travel could be accomplished within several months. We asked the head staff for permission to perform this study among soldiers, and we received wide support. Thus, we began a research project that took a fantastic course and which, in addition to scientific results, offered us unknown emotions. In the company of an English officer assigned to us, we went to the Struma front where Hindu troops were deployed. On our way back we were escorted by the Hindu cavalry riding white horses while a fight between two planes was taking place in the air. We investigated Negroes, Annamites, and Madagascans, and the contact with each race proceeded in a specific atmosphere. We had to speak in a different way to each nation. It was enough to tell the English that the objectives were scientific. We permitted ourselves to kid our French friends by telling them that we would find out with whom they could sin with impunity. We told the Negroes that the blood tests would show who deserved a leave; immediately, they willingly stretched out their black hands to us. Only the tests made on Russians were unpleasant. This was after the revolution and the Peace Treaty of Brest-Litovsk. The Russian troops in Macedonia no longer wanted to fight, and they were employed in road-building. The nervous condition of the soldiers was so bad that they fainted when blood samples were collected from them. The procedure consisted only of puncturing a finger and collecting a few drops of blood. But, let us turn to the results.

We found that factors O, A, and B were present in all races and nations under investigations. We tested Englishmen, Frenchmen, Italians, Serbs, Bulgarians, Russians, Greeks, Jews, Hindus, Annamites, Madagascans, and Negroes. However, these factors are present in uneven proportions. We found that type A was present as a majority only among north European nations. On the other hand, group B, rare in Europe, was more frequent among Asian and African nations. This discovery suggested the conclusion that a group factor had been formed on two opposite ends of the world: the A factor somewhere in northern Europe, and the B factor in Asia, perhaps on the distant highlands of Tibet or in India. Evidently, driven by eternal unrest, people migrated from the West to the East and from the East to the West, mixing with one another and giving rise to racial hybrids that were smelted in the foundry of joint history into a single nation. A study of the blood of individual nations enables us to decode their distant past. It is difficult to describe with what emotion we tabulated our observations. The study was announced under my and my wife's joint authorship. She gave a talk at the English Medical Society in Salonica, while I talked at the French Medical Society.[24]

The fate of this study was fantastic. At first I sent an article to the *British Medical Journal* but received no answer for several months. Growing impatient, I sent a letter to the editors stating that the significance of that study exceeded a simple medical communication, and that it was our duty to take advantage of the agglomeration of various nationalities and races in the war. In reply, I was sent back my manuscript with a short letter that the study was not interesting to physicians. At that time, I represented a small nation and small nations were regarded as culturally inferior. It never had happened to me before that the editors returned my papers. However, I remembered how Professor Rubner had told me that he discovered the energy equivalency of foodstuffs when he was assistant to Voit. Voit considered it such a nonsense that he threatened to dismiss him if he published it.[25] For a long time his manuscript lay in his desk. The second event was related to me by the secretary of the Biological Society in Zurich, Dr. Schoch. At one conference, a professor made the following announcement about one of his assistants who wanted to make a presentation: "This is nothing special, but perhaps you have a free evening; I should like to accommodate this young man." That presentation proved to be the first communication by Einstein about his theory of relativity.

Immodest as I was, I tried to comfort myself with such analogies. But I felt bitter in my heart. I sent my manuscript to another medical journal, the *Lancet*; however, this time I secured for myself the support of my English friends. For a long time, I did not know whether my paper had been accepted. In November 1919 I came to Warsaw and passing by the publishing house of the *Kurier Poranny* I glanced at a newspaper on display. I felt as if I had been struck with a hammer on my head. I noticed an article stating that the sea of blood shed during the great war was not lost in vain, because two English physicians . . . and so on. There followed a description of our discovery and our names. Indeed, it was difficult to find a study with more far-reaching consequences.[26] It became the basis of a new branch of anthropology, and its significance was later compared with Retzius's discovery of the differences in the structure of skulls.[27] Special expeditions were dispatched to investigate perishing races, and special congresses were devoted to this problem. I experienced my first great satisfaction in 1928 at the International Congress of Anthropologists in Amsterdam, where a special section was formed at which I presented the results of my own as well as international studies. I maintained a lifelong correspondence with investigators in all countries due to this research, and I experienced great emotional satisfaction. Now, whenever I am at congresses and present the results of my studies, I feel that the true culmination point is not the moment when the implemented idea is being transmitted to others but the moment when the idea is conceived. Writing about these international congresses, I am thinking of the endless chats with Dungern, the war, and the outstretched white and colored hands. I am most moved when I recall a scene in Salonica when we heard a terrible explosion one night and believed that a bomb had hit our hospital. We jumped up, and my wife tried to save—in the first impulses which is always the most essential—the notebooks with the results and protocols of our research. This was my wife's combat christening. Our research was dearer to her than her life.

Thus we lived in Salonica in tension and in dreams. I had left a splendid laboratory in Zurich to throw myself into the whirlpool of life. I was assisted by a faithful companion. Yet fate allowed me to establish a new branch of science, find pupils, and form strong ties of lasting friendship, which weave an individual into community life. Furthermore, I learned to speak with the average man and to hear a song that is not heard by all—the song of an earth that wants to bear fruit and of a nation that is waking up and wants to live. Many years later I faced a similar problem when, possessing nothing but the desire to help, I had to build a new life from scratch. To a certain degree, I succeeded in this last attempt. Unfortunately, it was a song of men doomed to be exterminated. It was not granted me to enjoy the fruits.

When the days of the final offensive on the Macedonian front arrived, I took laboratory equipment and went to the front hospitals to make bacteriological examinations of the wounds. I was particularly interested in studying the differences in the progress of infections spreading from them, depending on the bacterial flora. I saw grave struggles and terrible pictures of gas gangrene when limbs were turned black right before my eyes and their shriveling up announced the soldier's death. I cultivated many strains, but unfortunately I did not investigate them thoroughly because I was carried away by other tasks in my country.

I remember the day of victory and truce in November 1918. I will not write about the frenzy that overcame everybody and about the emotional excitement we experienced at the news of an independent Poland. Colonel Sondermeyer came personally to tell us the news, and moved to the bottom of our hearts, we cried, all three of us. How we wanted to go home—immediately. However, it was impossible for me to leave the center. It was only in the spring of 1919 that we left with the laboratory and the hospital by ship to Dubrovnik (Raguza). Toward the end of all these escapades, we were thoroughly tired of the East and the monotonous azure of the sunny sky, and we longed for clouds and coniferous forests, and for something elusive which is neither the air nor the earth nor a plant but is everything, one's whole life under conditions one grew up with where each breeze, each tree, and each bird is a memory. When the ship was approaching Dubrovnik, from afar we could see stone pines and cypresses resembling our pines and poplars. It was a delightful, reinvigorating, fresh greenery. The soldiers began to sing the song born in Korfu, "In the far yonder," and everybody without exception was crying. It was a song of a life that suffers and is reborn. It was the voice of the homeland one greets after long years of separation.

I stayed in Dubrovnik for several days. A beautiful city combining Slavic contemplation with splendor of the Renaissance. The people are beautiful, too. They speak a caressing language made up of Serbo-Italian sounds. We were given three freight cars which were attached to a train headed for Belgrade. The trip lasted several days; we passed through Bosnia and visited Sarajevo. In Belgrade, we were given a building belonging to the military hospital, and there we established our residence. Demobilization was not announced as yet, but the combat tension began to subside. There was no longer a kaleidoscope of races and nationalities. There was only a pleasant and simple nation that desired to rebuild its country and heal its wounds. We were no longer in the center of great events. Again, I felt an irresistible longing but not just for the blue yonder. Something new stirred in me. In my youth, I was not interested in having my own home, and I did not like children. Now, I began to long for the peace of one's own home and family life in my own country where I could feel like a tree spreading its roots wide and deep.

Chapter Six

Home Again

During my four years of service and in spite of the serious diseases I had undergone, I had taken only two weeks' sick leave. So, we decided to go for ten days to the lake of Bled in Slovenia and then for an additional two weeks to Vienna. During the war I had become familiar with new fields of medicine, and now I wanted to learn something about peacetime problems in bacteriology (production and titration of serums and similar matters). Finally, I must admit, I felt that I had become somewhat of a "peasant." In hospital laboratories work consists mainly of making observations; in large theoretical institutes it consists of creating new concepts and verifying them. There are certain mental gymnastics that give loftiness to one's thoughts, but one loses them during "observational" work. I prefer to think than to observe, and after several years of "observational" work in complete detachment from great scientific centers, I felt intellectually *déclassé*. I believed that human ideas had advanced far and wide, while I was left with registering existing things. When I arrived in Belgrade, I picked up the belongings I had left in Valjevo, including reprints of my papers. I read them as if someone else had written them—so distant were they and so clever. It was a queer feeling to read my own papers as if they had been written by someone else and someone smarter.

In Vienna, I decided to familiarize myself with the [scientific] literature so as to present myself in Poland in a less "peasantlike" state.

We stayed for a while in Zagreb. It was the first city that did not resemble the Orient. I noticed how much I was longing for Europe, the Gothic style, and its counterpart in nature: pines and poplars. I was yearning for clouds. That was how weary I had grown of the Byzantine style, the eternally blue sky, and the Oriental bustle. I remember a morning in Zagreb: the market place was full of fruits and flowers, women were neatly dressed, and some long-forgotten memories came back to me. We went sightseeing to look at the powerful sculptures by Mestrović, the city, the opera house, the museum, and the university. Eventually we arrived in Bled located on a beautiful lake. There we saw the first evergreen forests; we felt the coolness of the brooks; and we enjoyed the luxury of a hotel. This was no longer the war bustle of Salonica. But what bad luck I had: I went through my last malaria paroxysm and spent ten days lying in bed or in an armchair.

Vienna . . . I will not describe the hunger that then prevailed there nor the psychology of human strata who had once been the rulers and were now realizing they were defeated. I shall speak about the scientists only.

In Vienna, I met Paltauf, Pick, Landsteiner, and Bacher.[1]

Paltauf, the famous director of the Sero-Therapeutical Institute looked at me with displeasure. I had returned from the enemy camp, and he did not want me to work in the institute. I went to see Professor Pick, a famous pharmacologist, and told him so. He was indignant. "How could he! We want to improve our department, we want to attract international spheres again. We are no longer the capital of a big country, but we want to be the spiritual capital of the world. I will see to your case." I felt offended and requested a written invitation. My next conversation with Paltauf was proper but cool. I began to work under Bacher in a laboratory that belonged to Paltauf's institute but was located in the suburbs of Vienna. Bacher was an outstanding expert; he received me kindly and showed me everything. Currently, he is a professor in Turkey, and fortunately he evaded persecution.

Landsteiner made the biggest impression on me. I had met him at the Congress of Microbiologists in Berlin in 1913. At that time, he was a very good-looking man: tall, a high forehead, a manly posture. In Vienna I saw a wreck. He was obviously starved. With pain in his voice, he told me: "When my child asks for a slice of bread and butter, I am unable to give it to him." I promised him that in Switzerland I would try to get an invitation for his child. Landsteiner showed me his first investigation of compounds of proteins with inorganic bodies. Continuing Pick's line, he incorporated various chemical substances into protein molecules and, by means of a special method, he showed that under certain conditions even simple inorganic particles were able to provoke the formation of antibodies. He discovered new worlds for serology.

Expansion in this direction gave rise to a new science and enabled us to understand the quintessence of serology. I was totally captivated by the wide horizons of this research. Later, Landsteiner received the Nobel Prize for blood groups. He deserved not one but two prizes; however, not for blood groups but for chemo-antigens. Blood groups, though discovered by Landsteiner, became the foundation of a new science because they were introduced into genetics and anthropology. I told him about our sero-anthropological observations, and they captivated him just as much. We could not stop talking. I spent several hours at his place, and a few days later he visited me in my hotel. Conversation with him was satisfying some sort of mad thirst for thinking in me; a thirst I had been unable to gratify during the war. This tossing of continuously new ideas made me ecstatic, and I thought how happy were those who had not been in the war but stayed in their laboratories throughout that period. When we were taking leave, Landsteiner said to me: "My dear colleague, how happy I am that I finally had an opportunity to talk to someone about serology. I am completely alone here." I understood then that those who walk on the summits of thoughts are always lonely. I understood it again when I later visited the capital of serology, Frankfurt on the Main. At that time, Landsteiner had obtained a position in the Rockefeller Institute in New York. He received all possible awards in bacteriology in addition to the Nobel Prize, and he is rightly considered the greatest genius in the science of resistance. Strangely enough, he was not completely satisfied. When I congratulated him for the Nobel Prize, he wrote to me that I had something he did not possess: my own institute. Probably it was not only the institute but a whole complex of impressions and life in one's own country.

I began to study the professional literature. I approached the journals with a feeling of personal inferiority. I read that Sachs, Ehrlich's most brilliant pupil, not only confirmed the studies Klinger and I had conducted on complement but also accepted our findings, rejecting the complicated hypotheses of Ehrlich.[2] I felt like someone who thought he was poor and suddenly noticed he was not. Evidently, my work was not unwanted. The greatest tragedy for a scientist is the feeling that his efforts are purposeless. Working in science means sacrificing one's life for it. It was a mortal blow for Pettenkofer that Koch did not even notice his study of cholera.[3] Every investigator produces some research which, though valuable, remains unnoticed. Even the discoveries by the great Mendel had to be rediscovered for a second time. Among my publications there are also articles that I respect but which have gone unnoticed. American science, for example, did not notice my investigation of the complement. Such work and results are like children who die prematurely. It is painful to walk through the cemetery of one's ideas. When reading Sachs' articles, however, I realized that the work accomplished in my youth had not remained echoless.

I bought piles of books. I had hard currency and felt like Croesus in Vienna. Two weeks later, I left.

Finally, we had reached the Polish border. The emotions that rose up within me are indescribable. The first calls: "Herbata, bułeczki" [tea, buns] brought tears to my eyes. I was returning to my native country after fifteen years of absence. Leaving the railroad terminal in Warsaw, I could have embraced every passerby. Women's voices with their characteristic alto were like music to my ears. I went to my mother's apartment but did not find her there; she was taking care of my dying grandmother. My mother did not know exactly when I would come, but the table was set and she had laid out all the foods I had favored as a child. Not a detail was missing. In her memory, I lived as her one and only little boy. When I realized that, I forgot the war, science, and Switzerland, and I felt like a little boy who was happy that someone knew that he liked veal. And I began to cry like a child.

Such was my return to Poland. At first, my heart was moved when I saw Gothic forms and pines, then when I heard the first calls "Herbata, bułeczki," and finally when I found proof on the table that I still lived on in my mother's heart. I was home.

There were not many Poles who had worked in science laboratories abroad, and among these still fewer had returned to their country. Narutowicz was one who did.[4] Several years after his death, I met a friend of his in Switzerland. He told me that Narutowicz was going to Poland in order—as he put it—to do the military service he owed his country. But he also said that after completion of his service he would go back to Switzerland, which he considered his second homeland. The zoologist Janicki and botanist Basalik returned from abroad. Rajchman returned, at least for a certain time, and so did I. However, neither the great Skłodowska [Madame Curie], nor Danysz, Babiński, Pożerski, the brothers Minkowski, nor yet Mutermilch came back.[5] The motivations of human actions are complex. Perhaps, he who has found a second homeland yearns less for his native country. Or perhaps man is like a tree that can be transplanted into a different soil only when it is young.

I left Poland when I was eighteen years old and returned at the age of thirty-five. It was the autumn of 1919. I was lured neither by a scientific career in a foreign country nor by the regenerating life in Yugoslavia that had become dear to me. Each impression induced a hidden resonance in me. I felt as if I were continuously growing deep roots and finding the new nourishment my whole entity was yearning for. I was like one who had lived by his intellect only; while here my heart opened widely, and I was listening intensely to once-heard sounds, dictates, and memories. In Switzerland and Serbia I was strangely alone without a past. By my own efforts, I had to fill the space around me. Whereas here, whenever I looked back, I would recall the thoughts of my youth, the underground life, and then the longing of a homeless wanderer; and in front of me, there was Poland, the beginning of a new life and the generations to come in the future. It was then that I understood the difference: in one's native land one has a past and a future; in a foreign country one only has the present. A young, exuberant man bent on conquests mainly desires to be, while a mature man desires to be a link between that which was and that which will come.

Thus, my wife and I decided to stay in Poland.

Chapter Seven

Life in Warsaw

I began to look around for a laboratory. I saw two possibilities: pursuing an academic career or working in a research institute of the Ministry of Health. It struck me that the organization of university laboratories was strangely loose. Although a specific bureau had been created at the Ministry of Education for just that purpose, it seemed that no one was in charge of attracting back those Poles who had acquired knowledge abroad and could determine the scientific and organizational needs of the country. This was not true, however, for health matters, which had been taken over by a man who looked far into the future and took to heart all issues involving public health. This man was Dr. Ludwik Rajchman.[1] He had spent his youth in London, becoming familiar with Anglo-Saxon hygiene, and he had established valuable contacts both in England and America. When Poland regained its independence, Rajchman immediately gave up London—he had no military obligations as I had—came back to Poland, and took charge of the Central Epidemiological Institute, that is the entire field of hygiene in the Ministry of Health. As the German troops were retreating, he followed them to the front accompanied by a young country doctor, Czesław Wroczyński—subsequently chief of the Health Service—and in spite of the danger, bought several military laboratories from the Germans and brought them back to Warsaw.[2] Next, he purchased for the ministry an asylum for the mentally ill that was still under construction and designed a plan for transforming it into an epidemiological institute. In the meantime, he set up a laboratory in a small house nearby, a building known as *Ciglerówka*. That was where I visited him. We came to an agreement concerning a work place for me: an Institute of Serum Research would be established with the task of controlling both state and private production. This was so because the ministry had decided to nationalize the existing production plant of the Scientific Society; although it also permitted private production.

Thus, my institute was to correspond, albeit on a smaller scale, to the Institute of Experimental Therapy in Frankfurt-on-Main. We delineated the scope of activity and the personnel: I was to have three assistants, which was enough to take care of my needs. Later, the scope of my activities became considerably greater. The then minister of health, Janiszewski, insisted that the position be openly advertised, which was quite improper because at that time I was the only Polish *docent* in my field.[3] Unable to wait for the appointment, I left for Switzerland and Belgrade to liquidate my affairs and say good-bye to my former associates. And I decided to return to Poland determined that somehow everything would turn out all right. I stayed two weeks in Switzerland, took leave of the department, and cleared out my apartment. I sent the Swiss money I saved during the war—13,000 francs—to Poland. My country needed foreign currency. I did it during the inflation, with the result that I was hardly able to buy a modest apartment. I had no money left. The salary of a state employee during the inflation was not enough to see to the most modest needs. We were expecting a baby at that time. It was necessary that my wife find gainful employment immediately. She took the position of a school physician and then of an insurance agency physician.

I returned to Poland in January 1920. I found that my recruitment had been signed by the [new] minister of health, Chodźko, without having recourse to competition.[4] The war had not yet come to a close. The Ministry of Military Affairs wanted to mobilize me and assign me the function of central epidemiologist. However, the Ministry of Health did not agree, because it wanted to build peacetime work laboratories as soon as possible. Construction of the institute's main building was rapidly approaching completion. I began to work in one room, but soon I had to take over other functions. Paderewski[5] was going to Geneva to attend the [first] session of the League of Nations, and he was taking with him Dr. Rajchman who had a perfect knowledge of the sanitary condition in Eastern Europe. Epidemics of typhus and cholera were raging in Russia; the world war and the civil war had provoked there the greatest epidemics the world had known. The epidemics coming from the East were to be stopped in Poland.[6] Thus, I had to replace Rajchman in his function as the director of the Epidemiological Institute. I had also to purchase the necessary equipment and familiarize myself with the serum titration technique in the Institute of Experimental Therapy in Frankfurt where, prior to the war, standards were manufactured and sent to all countries.

It would be difficult to describe postwar scientific life without outlining the background situation. Therefore, I will interrupt my narrative and turn back to other events. In the first half of 1920 I went to Germany several times to make purchases and visit the institute in Frankfurt. My reminiscences reflect the mood prevailing among German scientists. In the summer of 1920, a tremendous dysentery epidemic broke out in Poland, and the Polish Government purchased from the serum manufacturer Behring-Werke a large quantity of anti-dysentery serum and paid for it. The socialist German Government, however, regarded the therapeutic serum as a military resource and halted the whole transport, even though the serum was ordered by the Ministry of Health and not by the Ministry of Military Affairs.

In Berlin, I visited the Hygiene Institute wishing to obtain various bacterial strains that had nothing to do with the war. The director of the institute,

Professor Flügge,[7] did not allow these samples to be given to me, a representative of a hostile country. To visit the institute in Frankfurt, one had to have a special permit issued by the ministry. I was forced to visit this institute, because at this time it was the world's only serum standardization center. Later, the Hygiene Committee of the League of Nations transferred this institute to the capital of a neutral country, Copenhagen, to make it free of political factors.[8]

Kolle,[9] the director of the Frankfurt institute, I knew only by the reputation he had acquired in Bern where he taught for many years. His behavior was so aggressive that once the Swiss students and physicians set up a shivaree[10] for him and, supposedly, his former pupils came from all over Switzerland to participate in the party. Thus, I reckoned with his being tactless and prepared myself for an acute defense. Indeed, he greeted me with the following words: "*Herr Kollege* Hirszfeld, have you heard about the Treaty of Versailles?" The entire personnel of the institute was present. Very calmly I replied: "*Herr Professor*, the Germans and the Poles have been fighting for ages. If we were to weigh exactly all the injuries and injustices you have suffered at our hands and those we have suffered at yours, then you must admit that the greater measure lies on our side." Evidently, this impressed him, because he became polite. He did not, however, find it necessary to invite me to his place as was customary when highly regarded scientists visited from abroad. Only Professor Sachs—one of Ehrlich's former associates and who represented old traditions in the institute—invited me. A close friendship developed between us. He tried to put me at ease: "Kolle," he said, "is actually not a bad man, only he feels bad if he doesn't pick a fight first thing in the morning." Later, I regularly saw Kolle at the sessions of the League of Nations [Biological] Standardization Commission—which I will come back to later when I outline the development of postwar scientific relations. [During my visit,] I was very much struck by a conversation I had once alone with Professor Sachs' wife. She told me how happy her husband was finally to be able to talk about serology. Yet the monumental work accomplished by Ehrlich had come from the Frankfurt Institute, and one would have thought that just a few years after his death his tradition would still be alive. I, for one, had gone to Frankfurt, as to the serological capital of the world. And there I found the same feeling of loneliness of which Landsteiner had complained in Vienna. I remember also my conversation with Rona, a well-known biochemist in Berlin.[11] Some fifty physicians of various nationalities were working under him, and yet he complained to me: "They all want something from me, and no one wants to give anything to me." Later, I too suffered from this state of loneliness when I was surrounded by a whole staff of associates.

On my return from Berlin, I remember the border between Poland and the Province of Posen [Poznan]. The Posen gendarmes tore up Polish documents and confiscated all the food articles the Poles were trying to bring from the Province to starving Warsaw. I prefer not to think about it. Only in the wake of the mutual bitterness of nations and scientists and the raging threat of epidemics coming from the East can one understand the exquisite endeavor of Rajchman in trying to reach an agreement among scientists and encourage a joint effort in combating infectious diseases.

My life was so intricately involved in this international effort that I cannot ignore how it unfolded, even though I did not participate in all the projects.

Warsaw hosted a series of international lectures on infectious diseases. Their organization in Poland was entrusted to Colonel Szymanowski, subsequently named professor.[12] Scientists from various countries that had taken part in the world war were invited to present lectures. As a result, we had the opportunity to become acquainted with a number of scientists. We were impressed most by the well-known Romanian researcher Professor Cantacuzène[13] who was an outstanding scientist and a very pleasant person. He frequently visited us at home and noticed that our little daughter was attracted only to colorful dresses. After returning to his country, he sent her a beautiful doll in the national costume to satisfy, as he put it, her need for beauty and colors. Professor Barykin from Moscow also gave interesting lectures. Once, we organized a general presentation in the presence of Minister Chodźko; Barykin spoke about the research pursued by Russian scientists, while Dr. Sparrow,[14] our associate, spoke about the work carried out by Polish scientists on typhus. I will mention a small detail that was characteristic of that period. One day after that meeting, I received a confidential inquiry from the attorney general about whether it was true that our institute had contacts with the Bolsheviks. Yet this was an official international action organized under the auspices of our government! Dr. Berger, an official from the Gesundheitsamt [National Bureau of Health] also came from Berlin to lecture; he was a quiet and cultivated man. Then there was Professor Abel from Jena to whom we later sent, for several years, our Polish scientific papers complete with summaries in English, French, and German. After a while, Professor Abel suddenly sent us a letter asking us not to send him any papers that were not published "in allgemeine verständliche Kultursprache" [in a generally understandable language of culture].[15]

The epidemic threat coming from the East was such that we required international assistance in organizing aid.[16] Poland was too poor to carry the burden of combating typhus and cholera on her own. The League of Nations dispatched its Epidemics Commission headed by Rajchman. Poland received large funds, which made it possible to build bathhouses and hospitals. A quarantine was organized in Baranowicze, and our institute established there a bacteriological laboratory to detect cholera carriers. In spite of the various cholera foci among Bolshevik prisoners of war, the cholera epidemic did not spread to Poland or the West. I must admit that from a bacteriological viewpoint this was a puzzle, because tens of thousands of samples taken from refugees in Baranowicze did not reveal a single cholera carrier. On the other hand, the quarantine halted the spread of typhus, even though many physicians and nurses died there. Soon, the commission's work took a different course. Dr. Rajchman was summoned to Geneva to assume the post of medical director of the League of Nations. In our institute, he was replaced first by Dr. Sierakowski and then by Dr. Celarek;[17] the scientific directorship was in my hands. In 1926, in connection with the opening of the School of Hygiene, our institute was issued a new statute. My Serum Research Institute was officially annexed to the Epidemiological Institute, which also included the new School of Hygiene. The Food Control Center, and the Pharmaceutical Institute, the Department of Diagnosis, and other related organizations were merged into the Department of Bacteriology and Experimental Medicine, while my institute was annexed as the Department of Serum Control. I was appointed director of the Department of Bacteriology and Experimental Medicine, head of the Department of Serum Control, and deputy to the director, Dr. Rajchman. The scope of my duties increased tremendously, especially when one considers that my department also included subsidiaries.[18]

Before I discuss the scope of my work, I must briefly mention events that were to draw my institute into the orbit of a great international public health movement. In 1922, Warsaw hosted the first international congress, convened by the League of Nations, devoted to epidemic control in Eastern Europe.[19] Dr. Rajchman presented an impressive paper pointing out that the question of Eastern Europe was now one of epidemic containment. The various countries were represented by scientists and health services directors. I invited all the scientists to my home, a gesture I considered of utmost importance because this was the first attempt to bring together, on a social basis, scientists who had once belonged to hostile camps. My guests included Castellani, Cantacuzène, and Otto.[20] The School of Hygiene was opened in 1926,[21] and the ceremony drew numerous scientists from all over the world. Again, I considered it my duty to bring the scientists together on a social basis, even though I did not have an entertainment fund at my disposal. I would not have been able to afford it if my wife had not been working.

Soon, the American Jewish Joint Distribution Committee ("Joint") donated forty thousand dollars to the institute to purchase *Amelin*, a property where Poland's first Health Center and a dormitory for the students of the School of Hygiene were established. The objective was not only to facilitate the life of the students but also to exert more educational influence on them, which was possible only on a campus. At the same time, the Rockefeller Foundation granted the ministry a number of scholarships to train health officials in America.[22] Dr. Rajchman organized an exchange of health officials and visits to various countries. Thanks to the Rockefeller scholarships, our physicians went abroad, while foreign physicians visited our institute and even worked in it. In this way, a definite international exchange arose with our institute and School of Hygiene as its Polish base. Our institute also became the center of international exchange in science. The League of Nations Health Committee established a Commission for Standardization of Sera and Vaccines; I worked with this commission from the very beginning and remained its permanent member. The chairman was Professor Madsen from Copenhagen, a scientist of great measure. Later, the commission took over the titration of hormones and vitamins under the chairmanship of Professor Dale, a Nobel laureate.[23] It also initiated certain team projects in which I participated, recruiting my associates not only from the institute but also from the subsidiaries.

The first project was titration of anti-dysentery sera and standardization of serological diagnostic methods in syphilis. This work was paid modestly but in Swiss francs and was therefore of great aid to us scientists. In addition, the commission held annual conferences abroad, which enabled me to travel and maintain direct contact with scientists abroad. In those years of inflation, travel abroad without special funds was impossible. Dr. Rajchman came back periodically to visit us from Geneva, and we would organize meetings at which he told us about his international activities. These meetings were usually followed by social gatherings to which we invited Warsaw's scientific medical elite. Many foreign scientists also came and presented papers at our institute. Thus, the Institute of Hygiene became the center of scientific and public health life, all the more so since the completion of the School of Hygiene—thanks to the Rockefeller Foundation—transformed it into an unusually elegant building.[24] The Institute and School of Hygiene were pleasant to look at also from the outside: they possessed a certain grandeur yet were of modest design that struck the imagination both of Poles and of foreign guests. The students often told me that the atmosphere of the Institute and the School of Hygiene filled them with a feeling of respect, which made student rivalry unthinkable. All this created a feeling of community so that the institute's staff members considered themselves a sort of family. Meanwhile, colleagues who had been away on foreign fellowships came back to Poland:[25] Ludwik Anigstein, who graduated from the London School of Hygiene and Tropical Medicine, took charge of the Department of Parasitology; Feliks Przesmycki returning from America became my adjunct and then head of the Department of Diagnosis and inspector of the subsidiaries; Stanisław Sierakowski brought back many new methods from America; Jozef Celarek developed the production division with great success; Marcin Kacprzak, brought an ardent heart and was an outstanding writer; Brunon Nowakowski, became professor of hygiene in Wilno and Stanislawa Adamowicz[26] belonged to the international women's movement;

Helena Sparrow was to introduce modern vaccination methods; Edward Grzegorzewski, a subtle and exquisite mind; Jozef Lubczynski; and Aleksander Szniolis, Poland's best sanitary engineer. The head of the Health Service, Czesław Wroczyński, was a former Rockefeller fellow and thus closely associated with our institute, while the minister of health of long-standing, Dr. Chodźko, was named director of the School of Hygiene. Thus, a true elite came together with the capacities needed to direct public health administration, teaching, and science in Poland. All this imposed tremendous organizational and representative duties on me, because we now had a lever enabling us to lift Polish public health and science off the ground. And that is how I achieved what I had been dreaming of: I was able to work for my country possessing the most wonderful laboratory a Polish scientist ever had.

Work in the institute satisfied some of my aspirations but not all. I wanted to be active also in teaching, because I felt greatly attracted to young people and believed that my true vocation was not administration and public health, but teaching and science. In 1924, I was appointed by the Free Polish University to give lectures on immunity. The students, mainly girls, were eager and industrious, and therefore I lectured with pleasure and liked to spend time with the other professors. The Free Polish University, being a open school, offered great possibilities for shaping teaching activities, regardless of one's profession. It was there that I acquired many valuable students, both men and women. Still, I was also longing for an academic life and the possibility to lecture to medical students. But the university statute had no provisions for lecturers other than regular and irregular professors. To gain associate professorship one had to make numerous applications and pass department examinations. I was not particularly anxious to apply for associate professorship, much less to take the examinations, since I believed that my scientific past and my position proved me to be beyond that.

I still believe that this was a great shortcoming in the university statute, because it made it difficult and even impossible to attract scientists as lecturers, bypassing the usual *docent* career that by and large was more suitable for young persons. However, Roman Nitsch—professor of bacteriology—acted with great tact, and sent me an invitation on behalf of several professors to assume *docentship*.[27] Referring to this letter, I wrote to the department stating that I would be happy to participate in teaching. The department in turn exempted me from all examinations with one exception, and incidentally a very pleasant one: a lecture to be given to the department. That is how I obtained in 1926 a *privatdozent* in the Medical School of Warsaw University. At the same time, the School of Pharmacy asked me to take over recommended lectures in bacteriology for pharmacy students to which I happily agreed. The School of Pharmacy did not impose any special requirements on me concerning the teaching program or the number of hours. Thus, I was able to present the whole of bacteriology to pharmacy students. I gave courses in serology for medical students, my lectures were attended by students of the Free University, and finally I gave lectures on immunity to country doctors who attended the School of Hygiene. My teaching obligations were equivalent to a normal teaching position, but on top of this I was also in charge of the institute. The diversity of my work gave me an insight into the problems of teaching public health, which I will discuss later. Yet, all of this was still not enough for me. We needed to establish scientific societies, publish scientific journals, and offer the whole of Polish science an opportunity to express itself in foreign languages. Articles published in Polish usually escaped the world's attention, and publishing papers in foreign journals required knowledge of foreign languages and certain contacts that not everyone had, especially if he had been educated in Poland and never worked abroad. First, I will discuss the former problem, namely how I succeeded in establishing the Polish archival journal.

Initially, Poland had only one medical journal, the weekly *Polska Gazeta Lekarska* [Polish Medical Gazette]. Designed for clinicians, the journal did not publish long papers with protocols, hence, a special archival journal was required. Dr. Rajchman secured a grant for publishing an epidemiological review—*Przegląd Epidemiologiczny* [Epidemiologic Review]—with articles in the areas of bacteriology and epidemiology. Stanislawa Adamowicz was the first editor, followed by H[enryk] Raabe. After two issues, the Minister Grabski[28] withdrew the subsidy—owing to the stabilization of the zloty—such that the journal was in the lurch. I then turned to my university colleagues, and first and foremost to Professor Hornowski.[29] He was a man of unusual intellect and integrity. An excellent speaker and outstanding pathologist, he loved young people, and the fate of science as a whole was equally close to his heart. I suggested to him that we establish a joint journal for experimental medicine. And he set out to implement the idea with great zeal.

We invited a number of professors from various scientific branches to join the editorial board and called the journal *Medycyna Doświadczalna i Społeczna* [Experimental and Social Medicine]. It became the organ of the Institute and School of Hygiene, as well as the theoretical departments of the Medical School in Warsaw, and subsequently of other medical schools in Poland as well. The next question was how to acquire subscribers. We had zero funds for publishing the journal, so we proceeded as follows. At that time, Poland had three pharmaceutical companies manufacturing Salvarsan: Spiess, Kiełbasiński-Pozowski, and Knoff. Their products were regulated by my institute, such that I had relations with them. We asked them to buy jointly three hundred copies a month and distribute them to students through the medical students' clubs. In this way we hoped to teach students how to use original writings, and to secure a certain amount of guaranteed income for the journal. I remember that the decisive meeting took place at the Institute of Pathological Anatomy in Professor Hornowski's office. Otolski, Kiełbasiński, and Knoff were all present. They agreed immediately. Later, I repeatedly observed the selflessness and social attitude of our pharmaceutical and apothecary industry. This was subsequently reflected in the formation of the University Institute of Pharmacy.

The next step was publicity. Professor Szymanowski, who became coeditor of the journal, wrote a winning appeal. At scientific meetings of medical societies, Professor Hornowski, who enjoyed the greatest popularity, regularly presented appeals to support the first Polish theoretical journal. We circulated lists of our subscribers and had great success. We acquired more than five hundred subscribers, which together with the three hundred subscriptions paid for by the pharmaceutical industry amounted to a considerable number. One thousand copies were printed each month. After several years, we were able to go without industry support and stand firmly on our own feet. All assistants at the institute were required to subscribe to the journal; this was considered the proper thing to do. In view of the small number of scientific institutes in Poland, it was the prerequisite for the existence of each scientific journal that each employee subscribe, regardless of whether or not the journal was carried by the institute's library. Professor Szymanowski and I were the editors and publishers, and later we asked for the cooperation of Dr. Marcin Kacprzak who, incidentally, assumed the main burden of publishing. Linguistic and scientific corrections were made by my assistant, Miss Róża Amzel.[30] This was indispensable, because the years of partition had left their mark; most authors had studied abroad, and therefore their Polish contained "Russianisms" and "Germanisms." I should like to emphasize that the publishing, editorial, and correction work was not paid. Moreover, in spite of the work they were putting in, the editors and the members of the editorial committee were paying for their own subscriptions. Nevertheless, the number of subscriptions began to decrease, which unfortunately was true of all scientific journals in Poland. Older persons who considered it their duty to support Polish journals were dying, while the spirit of younger people was not very generous. They found Poland in the stage of reconstruction, and they did not feel responsible for it. From time to time, when we published special issues on scientific meetings, we were helped by Health Services Department. In this way, about thirty volumes were published prior to World War II. *Experimental and Social Medicine* rendered a great service to Polish science. Summaries of original papers were also given in West European languages, and the journal was sent free of charge to all large institutes throughout the world, such that our scientific results were known [outside of Poland]. When I look back at its modest beginnings, I see with great pleasure that some credit is due to my optimism and faith, even though more work was carried out by Kacprzak and

Szymanowski. When the journal was created, few believed that it could have any future.

Let us now turn to scientific societies. In Warsaw, I encountered the *Towarzystwo Lekarskie* [Medical Society]. By and large, it was of a clinical nature, and the purely theoretical presentations enjoyed no popularity. Moreover, the society was imbued with an anti-Semitic spirit. Accustomed to scientific societies in the West, I felt that such politics were inappropriate. I confided my thoughts to Professor Hornowski, who was of the same opinion. We decided to establish a new, purely scientific society devoted not only to theoretical medicine but also to biology. Professor Sosnowski, a biologist and later rector of the Central School of Agriculture, joined our cause. We decided to found a biological society and simultaneously expand its tasks. France had its *Société de biologie* with subsidiaries in many countries associated with French culture. This society issued bulletins dispatched all over the world. If our projected society were to join the French one, a window to the world would be opened for Polish science. Our authors would not have to make individual efforts to publish their papers abroad, because they would have a publication guaranteed by the mere fact of presenting their papers in our society. The French Biological Society made a suitable proposal to us, and now it was only a matter of finalizing the agreement. We succeeded in interesting a number of professors in Warsaw and in other university cities in our plan. In spite of the obvious advantages, however, it turned out to be difficult to organize such a society in Poland. Cracow, for example, took the stand that Polish scientific laboratories were inadequately developed and that compared with international science we would cut a poor picture. Hornowski, Sosnowski, and I did not share this pessimism. In fact, we suspected that perhaps the Akademia Umiejętności [Academy of Sciences] in Cracow was afraid that the Biological Society might divert authors from publishing with them. We, on the other hand, took the stand that publishing with the Academy of Sciences was onerous because access was limited to members only; nonmember authors, who represented a considerable majority, were unable personally to defend their papers at conferences; and finally, because journals issued by academies were in fact an honorary grave for articles published in them. Lest they perish, scientific investigations must be published in journals that are read and that enjoy popularity. The delegate from Poznan created all sorts of difficulties strongly smacking of sabotage. Professor Hornowski once told me that we simply had to put that delegate's objections on the agenda because, according to private information, his objective was to kill the idea of forming the Biological Society. Lwów [section of Polish Academy of Sciences] presented different difficulties. It wanted to establish a society modeled on an academy; only outstanding scientists and professors would become members, while the mortal masses would be entitled to present their papers only through the mediation of members. Our Warsaw colleagues, on the contrary, took the stand of democratizing science and attracting as many participants as possible. We were able to come to an agreement with the Lwów negotiators. We established a compromise statute providing for editorial members and ordinary members, with each being entitled to make presentations; only an editorial member could, although he did not have to, be appointed to evaluate papers. Soon the difference between editorial and ordinary members disappeared, and the Biological Society in Warsaw and Lwów displayed, at least at the beginning, excellent growth. Professor Hornowski was appointed chairman of the society and I, its secretary general. Subsequent chairmen were elected from among professors in theoretical departments, including biologists and physicians in alternation. After a while, researchers from Poznan and Wilno joined us, and only Cracow remained recalcitrant. The Biological Society has a beautiful chapter in the history of Polish science. We often invited foreign scientists to give papers. I remember that in the initial years, the auditorium in the Institute of Physiology, where the sessions usually took place, was filled to capacity. The meetings were attended by several hundred persons. This was that early period of our independent life when the older generation was still strong and happy that our science was developing. Later, unfortunately, the intensity of scientific life in Warsaw began to diminish, which was also reflected in the attendance at the Biological Society. Much of the fault lies with us. When a staff member of an institute is presenting a paper, all professors and other colleagues should be present. If we want to create a scientific atmosphere, we must be able to enjoy and cherish the work done by others. The young researcher must feel that he is gaining something with his effort, and this encourages him to further his work. If he sees empty benches before him, he has the impression that no one cares for his research, while instead he should feel that the earth's very axis goes through his investigation. It is our fault that we do not feel responsible for creating a scientific atmosphere in our country, and that we are ready and willing to be donors but refuse to play the role of recipients. The fact is that we must be both.

Thus we established laboratories, our own journal, a society, and the means to publish both at home and abroad. The formal framework for scientific life was laid. I could conduct experimental studies; and I was in touch with public health, medical, pharmacy, and biology students due to my lectures. I had, it would seem, a load greater than one man could carry. However, this still did not satisfy a strange urge in me to fill the gaps in Polish life. I now turned my attention to two matters. Our institute was a consulting organ of the Health Services Department on matters of epidemic control. Periodically in Poland, we had to deal with large-scale epidemics. In addition to typhus, we had outbreaks of dysentery and typhoid. During one dysentery epidemic, I went to Krzemieniec and to Stryjski County. Not only the primitive living conditions but also our country physicians' ignorance of epidemic circumstances and their inability to cope with these problems made a dismal impression on me. Every year, some eighty-thousand pilgrims visited the monastery near Krzemieniec, and yet there was not a single toilet in the whole village. Eighty-thousand persons would relieve themselves literally behind the fences. The local population was obscure, superstitious, and fought epidemics by chasing out the devil, that is, exorcism. They held somber processions at night, prayed to the moon, and did other things of that kind. There were dysentery-induced death cases in almost every hut in Stryjski County. There were billions of flies. I saw myself that the only way for a mother to help her dying child was to chase flies away. Chamber pots were an unknown commodity. Straw with bloody stools was thrown onto a moist pile swarming with flies. Stools were covered not with milk of lime but with powder lime, which was completely useless. The chief of the Health Services Department and I asked the county physician whether he was attending to these matters. "Of course," he said, "I issued an appeal." We began reading: "It is important to observe personal hygiene . . ." and so on. For heaven's sake, what can the peasant understand when he hears the term "personal hygiene"? And thus I realized the mistake we were making: we were teaching epidemiology the way it looked in the world's large scientific centers inhabited by scientists. But we did not teach the realities of epidemics. To develop common sense among country physicians, to become familiar with local conditions and superstitions, and to develop a skill for improvising was much more important than to teach the theoretical foundation of bacteriology. After returning to Warsaw, I submitted a request that an epidemiological seminar for county physicians be organized in the School of Hygiene. For this job, I recruited experts in fieldwork: Drs. Stypułkowski, Grzegorzewski, and others. County physicians were assigned the task of studying individual cases of infectious diseases and preparing reports. Then we held discussions in which the physicians shared their impressions and experiences from fieldwork, while I was illuminating and expanding the epidemiological problems that were brought up. I consider this type of training as indispensable. I organized a similar training for students. Under the supervision of epidemiologists, they made trips into epidemic areas, returned after a few days, analyzed the collected samples, and made reports on the overall significance of the problem. I remember one student who, having returned from one such trip, told me that within a few days he had learned more hygiene than during the lectures at the university. Here we have touched upon a sore spot in the teaching of hygiene and bacteriology in universities: the departments have no contact with real life. What the hospital is for the clinician, the region is for the hygienist. Only he who has gone through an epidemic can have this internal zeal in describing an epidemic. Is it conceivable to discuss diseases without patients? And this is exactly what is required of a hygienist.

Trips to various institutions are not enough. Some countries, aware of this discrepancy, make students engage in public health work in rural areas, as for example Romania and especially Yugoslavia. It is absolutely necessary to reorganize teaching in Poland. In tsarist Russia, professors went often with students to regions infested with cholera. This is an excellent idea, because it trains the young student to combat contagious diseases, it creates an esprit de corps and the awareness of a higher mission, which every physician should have. I was contemplating developing the teaching of epidemiology along these lines; however, I lacked the appropriate organization that would enable me to implement my aspirations on a larger scale. Yet I still believe that my lectures were successful because while discussing contagious diseases, I would talk about matters in which I had practical experience.

Observations of this kind forced us to reorganize the way we taught bacteriology. Combating bacteria is one of the many tasks of county physicians; usually they must have contact with laboratories, but often they do not know how to use them. We therefore decided to change the structure of our subsidiaries. We created a new profession, that of an epidemiologist who was to spend half of his time in the field and half in the laboratory. We hoped that in this way our laboratories would become involved in regional problems. In addition, I was dreaming of working out the country's epidemiology not only from the statistical viewpoint but also taking into consideration the differing customs in the country's various regions. The Health Services Department very willingly approved our projects, and they were soon implemented.

I should like to emphasize the great contribution made by my associate, Dr. Feliks Przesmycki. He was full of excellent organizational ideas and energetic in implementing them. To a large degree, credit for reorganizing our subsidiaries, as outlined above, is due to him.

But from where could we draw associates? How many bacteriologists and epidemiologists did the country need? I am approaching here a problem which, in my opinion, is very poorly handled in states unaccustomed to planning. They often estimate the amount of required materials, but they do not estimate the amount of required manpower. Our ministry created the Supreme Health Council. I was a member and attended the meetings of this council. At the annual plenary meetings, the Department of Health Service usually supplied me with a report on the government's plans and achievements in my field. This forced me to take a broader approach to the problems encountered in the country's bacteriological service. I saw the mistakes made by medical schools. These schools believed that we had a surplus of physicians, hindered the influx of students to medical schools, and discouraged the applicants from medical studies by pointing out the difficult living conditions of physicians. Dr. Kacprzak conducted a poll which proved that as far as the number of physicians was concerned, Poland was in the second to last place in Europe; on the average, we had one physician for every three thousand people. We were right behind Albania or Bulgaria. At the meetings of the Supreme Health Council, I personally saw how representatives from medical schools suddenly became conscience-smitten and condemned themselves for their ignorance of the facts. Incidentally, I might mention that the professional qualifications of members of the Supreme Health Council, this democratic institution, began to decline, and subsequent representatives from medical corporations and even medical schools were unfamiliar with the main sanitary problems. Even being an outstanding clinician does not mean one knows something about health policy. The remarks made by the representatives from medical corporations and schools showed a lack of initiative and a poor understanding of problems. It is my opinion that professors of hygiene and bacteriology should serve in the Health Services as permanent representatives of medical schools. This would bring them closer to fieldwork, while the sanitary management would be able to take advantage of expert advice from theoreticians. I worked out exact data on Poland's hygienists and bacteriologists, and I arrived at the following conclusions:

> The crisis which university hygiene is undergoing must currently be regarded as catastrophic. The facts speak for themselves: the Department of Hygiene in Poznań has been closed, while the Department of Bacteriology in Lwów was opened just a few days ago. Out of five Polish universities, only Warsaw University has one associate professor of hygiene; there are no associate professors of hygiene in Wilno, or in Lwów, or in Cracow, or in Poznań. The departments of bacteriology in Cracow and Wilno have no associate professors. The slightest hitch in the life of a professor, even a short-term illness, will bring about a standstill in teaching which, incidentally, in several cities, due to a long-standing lack of appropriate departments, does not correspond to Ministry programs. It is difficult to analyze the causes of this catastrophic state of affairs; but one conclusion is obvious: the budgetary policy of universities must account for the country's health requirements. For a country as outdated in sanitary standards as Poland, the existence of hygiene centers is of paramount importance. Closure of departments not only halts activities for a short period but also undermines the general direction taken by research for prolonged periods.[31]

I wrote these words in 1936. World War II converted the crisis into a tragedy. When we now talk about Polish bacteriology, we roam over cemeteries. Professors Bujwid, Padlewski, and Nitsch all died during the war as well as Professor Gąsiorowski who was still actively teaching. Professor Gieszczykiewicz was tortured to death in prison. Professor Eisenberg,[32] Róża Amzel,

Bronisława Fejgin, Landesman, and many others were killed.[33] Will the weakling plant of Polish bacteriology ever be able to grow into full bloom? But let us return to my efforts to organize bacteriological service in Poland prior to the war. Where was I to find suitable personnel? It was clear to me that we had too few physicians anyway and that it was a waste to let them undergo long periods of training in performing medical analysis. We therefore organized special courses, to which we also admitted chemists, pharmacists, and biologists. The objective was to carry on this work during peacetime but also to ensure bacteriological services in case of war. Our courses provoked a veritable thunderstorm. All medical chambers and medical schools protested. However, this did not prevent many biologists and chemists who applied for admission to the course from presenting letters of recommendation written by university professors. The cause of that storm was the professional interest of physicians in large cities where analytical laboratories yielded good incomes. In smaller towns, analysis offered smaller revenues than medical practice, and therefore, the physicians had no intention of performing analyses. I pointed out that we did not have enough physicians to carry out medical analysis and that if we did not admit other professions, then in smaller towns it would not be possible to carry out analyses even on urine. My opponents presented a bill in the House of Deputies proposing that medical analysis be regarded as an integral part of medical practice. I discreetly supplied the representatives of the apothecary profession with the announcement of a law that had recently been passed by the French Senate entitling apothecaries to perform such analysis. My opponents' bill was killed, and other professions acquired the right in Poland to perform medical analysis even though this required a special permit from the Department of Health Service. To check which private workers could be relied on in case of war, we later conducted a poll. It showed that most were laboratories run by nonphysicians. I must admit, though, that by and large the students taking the medical analysis course were not to my liking, even though I had several outstanding pupils among them. I was dreaming of young eagles who could become our friends in scientific or social work, but among the students I often found persons interested only in learning how to perform profitable analyses. I had no desire to fight over this; large- and medium-sized towns acquired bacteriological service due to our course.

My appeals did not go unheard. General Rouppert, the army chief physician, asked us to train bacteriologists for the army.[34] Each year, ten to fifteen young physicians from the Military School arrived at the State Institute of Hygiene. They were by and large very properly selected. In these courses, my ideal of pedagogical work was implemented. I had brilliant boys in the laboratory all day long, and I was able not only to train them in bacteriology but also to influence and educate them more generally. The program did not restrict me in any way; I was assisted by my associates, some of them docents.

I was able to distribute the topics so that each could talk about his own research and experiments. Finally, these courses allowed me to make something of an "experimental department": I was able to test various teaching and organizational methods.

At 11:30 we all—that is, staff members and students—gathered to have a sort of brunch together and to imbue the students with the specific atmosphere of the institute. On Saturdays they assembled in my office, so that they could ask me various questions, and I sought to expand and develop the issues. The library prepared scientific articles from various areas. It was the librarian's duty to record the areas covered by the articles that were read. This was a test for the teachers, showing which ones were able to arouse interest among the students. This course approached my ideal of cultivating human souls. Every time I began a lecture, I felt that I was embarking on an expedition for young souls. And indeed I succeeded in conquering them. We were opening three new subsidiaries simultaneously; we had to have epidemiologists and assistants for all subsidiaries; and finally, we were not immortal—we had to think about our successors. The Department of Health Service was favorably attuned toward us: it let us open new permanent positions and gave us special funds for teaching. When we ran out of money, we received stipends from the Potocki Foundation enabling us to work on cancer.[35] In this way, we created a demand for bacteriologists and the possibility of covering this demand.

The above comments show the duties of a man who wanted not only to do science but also to serve his country faithfully. It was not enough to do research; scientific journals and societies were needed. It was not enough to employ assistants; centers for their further professional advancement were required. Forgetting one of these links would inevitably halt the progress of the whole.

The technical level in our Warsaw institute and in most of our subsidiaries was high; we made sure of this by sending our assistants on various special courses. However, there was still the bacteriological service in the rest of the country, in municipal laboratories, and in state insurance health centers. For several years, I conducted polls to find out what tests were being performed in Poland's public bacteriological laboratories. I noticed, for example, that the laboratories of health insurance centers were not used for epidemiological purposes. In 1934, we organized a graduate course to which all military and civil laboratories sent their workers. In this course, we were able to demonstrate new methods to all bacteriologists and to create the esprit de corps so necessary for cooperation. In 1938–39, we published a small booklet on the uniform bacteriological methods used at the State Institute of Hygiene. We checked our methods against the institutes of Eastern Europe and against the standards of the League of Nations Health Committee. The meeting of Eastern European Laboratories took place in

Warsaw under my chairmanship, and it greatly improved laboratory techniques in the appropriate institutes. After all these projects, I intended to effect a functional coordination of all public laboratories in Poland, which would guarantee a rapid introduction of improvements. But even so, we were able to raise the standards of bacteriological methods in the country. We tested new methods in our laboratory and then announced them, and the authority of our institute was so great that most Polish laboratories would use them. Docent Przesmycki wrote a very good outline of bacteriological techniques, and we always ran out of the brochures we published in no time. I believe that laboratory heads regarded our cooperation not as a control but as a useful aid. I never noticed any other attitude toward the State Institute of Hygiene.

My next task was to introduce new vaccination methods.[36] Jointly with the Children's Clinic, our institute introduced vaccination against diphtheria and scarlet fever. The vaccination campaign in Warsaw was initially conducted by Docent Helena Sparrow. A broad organizational framework was required to ensure assistance from state health insurance centers and municipal governments in covering the expenditures. Dr. Wroczyński found an excellent solution. We established the Society of Preventive Medicine. Dr. Chodźko was the chairman, while the heads of the Warsaw Health Department and I were vice chairmen. In addition, I was the chairman of the Vaccination Section, while Dr. Palester, former chairman of the Health Department and an experienced man extremely devoted to the cause, was director of the society. The vaccination campaign in Warsaw, conducted with the cooperation of Dr. Mikołaj Łącki,[37] proceeded efficiently. For the actual vaccination, we recruited members of the medical students' club who in this way acquired good training in socio-medical work. I reported on the vaccination results to the Supreme Health Council. On the basis of our experience in Warsaw, anti-diphtheria vaccination was made obligatory throughout the country. This is a beautiful chapter in the history of Polish public health. Later, I will discuss the scientific aspect of these problems.

Interesting developments and observations resulted from these activities. In the first place, I should like to mention the court expert testimony I made in connection with blood groups. At the time, the method establishing paternity, developed by Dungern and myself, had been introduced in most European and non-European countries as the mandatory method in this type of investigation.[38]

In Poland, possibly because of the Napoleonic Code that did not allow the investigation of paternity, this method was not as popular as in other countries, but nevertheless hundreds of such cases went through my hands. I often gave talks on this matter in judiciary and medical circles, and I could write an entire novel about the trials and tribulations of the individuals I investigated. I reported on the most interesting cases in my books on blood groups. I will mention here three of these cases, two of which involved exclusion of maternity. The first case involved an exchange of children in Riga; in the other case a divorced woman presented a child from her second marriage as her own. In both cases, I succeeded in proving that these women could not have been the mothers of those children. Those are the only two cases in world literature concerning exclusion of maternity. The third case was probably the only one in which it was possible to establish with full certainty that the given man was the father of the given child, because usually we are only able to exclude the man who is unjustly identified as the father. In this particular case, we found a very rare blood anomaly in the child and in the putative father.

By and large, I encountered great interest and friendliness on the part of judges and attorneys. I wrote the book on blood groups not only for physicians and biologists but also for jurists.[39] There was only one case that offended me, and I should like to recount it because it showed the impropriety of the defense. The matter concerned exclusion of maternity. I felt very sorry for the defendant, but I could not conceal the truth. Her attorney asked me: "Have you read the book by author X?" I had never heard of the book. The attorney turned to the court: "Please record that the expert has not read the book by author X." Then, he asked me again: "Have you ever heard of the law of great numbers?" I could not understand what he was driving at, while he turned to the court again: "Please record that the expert is not familiar with the law of great numbers. . . ." and he continued in this manner. After a while, I guessed what he had in mind and I said to him: "The law of great numbers is used when the problem involves the probability of the appearance of a certain trait. For example, I can never be certain whether a given woman will give birth to a boy or a girl, but I can always be certain that she will not give birth to a bulldog. It would be very advisable for attorneys to understand their own questions." Subsequently, this attorney visited me in my institute and we became friends. He was a clever man, but he believed that the expert should be treated as the witness for the prosecution—that is, that he should be discredited. I went through the same story during the famous court proceedings involving Mrs. Gorgon, which I should like to recount in a few words since at one time this case shook the whole country.

From Lwów, I received an expert's opinion concerning the bloodstains of Mrs. Gorgon who was accused of killing her lover's daughter. The Lwów expert had concluded that the investigated bloodstains were the victim's. From experience I knew that group properties could also be discovered on underwear and on objects touched by the individual. The Cracow court asked me for a new opinion, which revealed that the handkerchief belonging to Mrs. Gorgon was imbued with group components regardless of the bloodstain, and that the accusation should not be based on the results of the group opinion. My reply, which to a certain degree disqualified the verdict of the previous experts, literally shook up the country, because it involved confidence in legal proceedings in general. Unfortunately, the court proceedings degenerated into something resembling a tournament, which was contrary to my will and intention. I became as popular as an operetta diva. This was very unpleasant for me. I stopped answering the telephone because newspaper editors assailed me with questions; my secretary in the institute and my wife at home gave no information to anyone. Once only was I caught. "Professor Hirszfeld, we should like to ask you for a photograph." "Unfortunately, I must refuse; I never give my picture to be shown in newspapers, because I don't care for that kind of popularity." To which they replied: "We quite understand, please accept our apologies." I said to my wife: "For once, a tactful reporter." My optimism was misplaced. The next day, a newspaper showed my photograph, supplied by God only knows whom. There was also the incident at the railroad station when I was going to Cracow. My assistant brought me copies of the expert opinion papers from the institute. A crowd of reporters was waiting for me at the train. "I'm sorry, but I cannot give you any information because the court has priority." The next day, I read in the *Ilustrowany Kurier Codzienny* (Illustrated Daily Courier): "The distracted professor lost all his papers which at the last moment were found by his assistant." When I arrived in Cracow, I found in a newspaper a picture of an elderly gentleman with a beard who, even with the greatest stretch of imagination, did not resemble me at all. Under the photograph was written: "Professor Ludwik Hirszfeld." Probably, this was Magnus Hirschfeld, a well-known sexologist. I later asked about the purpose of such rubbish and was told that this was shaping public opinion. Poor public opinion! What disdain the reporters must have for it. In Cracow, I met a number of nice persons: attorneys, physicians, and journalists. Mrs. Irena Krzywicka[40] impressed me as a subtle and pleasant woman. Her court reports were jewels of psychology. Similarly, the reply given by the psychiatric expert has vividly remained in my memory. "Doctor," Attorney Axer asked, "do not feel offended, but I should like to ask you . . ." The expert in psychiatry interrupted him: "By profession, I am a psychiatrist, and a psychiatrist never feels offended." This was true class. Unfortunately, not all experts displayed this level of sophistication. The behavior of Żmigrod was especially unpleasant. I believe that he was psychologically deranged at that time, and this may be the reason why he disappeared after the proceedings.

After we introduced regulations for the standardization of serums, I served as the expert in a case concerning bad serums manufactured by a private plant. I do not wish to go into it; I prefer that such things be forgotten. The man responsible for the output of the bad serums died long ago, and undoubtedly the consciousness of his fault or negligence was a greater punishment for him than the court verdict.

The introduction of obligatory control of vaccines and serums cleared our market of worthless products and completely rehabilitated Professor Bujwid. After unfortunate court proceedings that took place in Cracow, this man, accused by the Austrian court, was condemned and deprived of his professorship. It was the Polish Government that eventually gave him back his retirement pension. When I returned to Poland, I met with a negative opinion on the serums manufactured in his private plant in Cracow. For many years, I had the opportunity to become convinced that his serums were excellent and could well compete with the best European serums. Eventually, I became closely acquainted with Professor Bujwid. He was a very kind, benevolent man, and during his lifetime he had helped a number of young bacteriologists. Bujwid made great contributions to Polish bacteriology, especially in the area of organization. He established the first plant that manufactured serums and the first pasteurizing institute for combating rabies. He was a pupil of Pasteur and Koch, and I saw in Paris how much he was liked and respected. He represented Poland abroad with dignity. We made him an honorary member of our Society of Microbiologists, and published an issue of *Experimental and Social Medicine* for his eightieth birthday. We were very upset by the disfavor that surrounded him in Cracow. It reminded me of *Les Maîtres* by Duhamel. How sad that scientists have such a capacity to hate.

Finally, I should like to mention my participation in the introduction of blood transfusion in Poland and the legal aspect of this matter. The beginnings were modest. I appealed to my students taking the serology course, encouraging them to organize a donor center on a partly paying and partly philanthropic basis. The idea took hold and soon not only Warsaw but all the "Medics Circles"[41] organized donor centers. Later, the Red Cross took over. We also saw to the legal aspect of this problem. Poland was doubtless the first country on the continent to establish legal regulation of the problem of blood donors.

Thus, my life went on full of tension and an unfulfilled desire to construct. Had I been living in a foreign country, I would not have devoted so much work and sincere effort to organizing science. I consider the initial years of our independence as the happiest period in my life. It seemed that many persons filled with the same happiness were creating the framework for Poland's future. Those were wonderful years. Every now and then, however, ominous rolls of thunder foreshadowed the gathering storm. I remember the coffin covered in mourning bearing the Republic's first president who was killed because he believed that all citizens of the Polish State had equal rights.[42] The hydra of nationalism was raising its head. I remember the degenerative factionalism which at first was an expression of exuberance and creative unrest and then became a manifestation of decline, a source of interest and a cause of violence against individuals, and it began to penetrate everything, including universities and research institutes. I was under the impression that science and the cultivation of human souls could lead a separate life and that a certain "autonomy of good" prevailed in the temples of science, which had to be preserved even in the flurry of political conflict. Thus, I often did not become involved in matters requiring intervention and struggle; I wished to preserve my strength and defend this sanctuary of which I considered myself to be a guardian. Perhaps this was a mistake.

Until 1933, I was deputy to the institute's director, Dr. Rajchman, who was working in Geneva. Minister Slawoj-Skladkowski did not want to extend his [Rajchman's] leave nor did he want to nominate me as the director. Therefore, he appointed Colonel Dr. Szulc, docent of hygiene, to this post. Minister Piestrzyński informed me of this in a polite manner, giving as the motivation for the nomination the government's desire to use the institute more for military purposes, and he stated that instructions had been issued to facilitate every one of my projects and that I was to remain in the position of section director and department head.[43] I had known Docent Gustaw Szulc previously; he was a kind and good-willed person. I had never been particularly eager to direct a large institute; on the contrary, I found administrative work tiresome. Nor did I wish for attributes of power to erect a wall between myself and my pupils. I always remembered the almost brotherly relationship I shared with Dungern. One cannot cultivate souls when breaking the will, and a director at times must do just that. Yet I must admit that when I was showing the institute and transferring authority to my successor, I felt bitterness in my heart. My associates tried to display their good will toward me even though I had no authority. They presented me with a beautiful palm plant and a beautiful album by Wyspiański.[44]

Gustaw Szulc entered a well-knit and well-integrated group, and must always have felt like an alien representative of authority without emotional ties. I was a frequent guest at his house and can vouch that he was a man of goodwill, who desired to do what was best for the institute and to harm no one. We spent much time making plans. During the [German] occupation he did not register as Volksdeutsch [German by nationality] although he could have. However, he could not command authority either among Poles or among Germans. When later, as an exiled man, I visited the institute, he was never afraid to be friendly to me. He told me several times how frightful it was for him to behold the ruins of Polish dreams, the ruins of his own work, and that he would rather have died were it not for having to care for his family. Fate was kinder to him than to me; he died during the war beholding his children.

As of 1933, I thus became director of the Section of Bacteriology and Experimental Medicine and the head of the Department of Serum Investigation, and I often replaced the director. Administrative work did not absorb too much of my time and attention, because I was efficiently assisted by Docent Feliks Przesmyski, who also supervised our subsidiaries. I stayed in the institute from morning until night and did not go home to dinner. I still believe that such complete devotion to one goal is an indispensable prerequisite for earnest, efficient work. Holding several positions is not a proof of diligence; it only dissipates one's attention. Of course, I was able completely to devote myself to my work only because my wife was working. A state salary, even with an additional income for lecturing, is not enough to maintain a certain standard of living. Before I discuss our scientific work, I should like to say a few words about the selection of our associates and about the organization of research and teaching.

Chapter Eight

Life within the Institute

I had spent my youth in institutes that were small but of high quality. The director could train his pupils and assistants while daily attending to them; I had the unique fortune of being daily in touch with my teacher. Here, I was part of a great machine. I had to train many pupils and find them an adequate outlet. There was no time to bring up a favorite student who would cherish memories of his teacher and his ideas. Instead these youngsters required rapid training so that they could perform specific social functions. The difference was analogous to that prevailing between a master artist who teaches a young apprentice what art is and how to plan a picture in an individual manner, and a factory where the objective is to supply the market as quickly as possible. When I recall my life in Warsaw, I see that it was entirely caught up in this conflict. I was dreaming of shaping the spirit, making joint plans, and establishing a fatherly relation toward the pupil through whom I would live on. Life granted me many pupils, but usually without such a rapport in the calm of the laboratory, because most of my "children" were forced to take employment in other centers or, due to financial hardship, were forced to give up scientific work altogether. And when I look back, I do not see many who pursued my ideas in Poland. My activities in the Institute of Serum Investigation carried me back to the atmosphere of my youth.

Dr. Feliks Przesmycki came to us from the Production Department with the wholehearted intention of engaging in thorough work. Later, he left for America for a year as a Rockefeller fellow. He always worked efficiently, passed his qualification examination, and gave good lectures, displaying a great organizational talent. Scientific work, however, did not satisfy his temperament. He later gave vent to his energy as inspector of our subsidiaries, and in this role he rendered our country good service.

Janusz Supniewski had a far more adventurous past.[1] I was told about a student who "inherited" a laboratory from the Bolshevik Army and discovered certain new bacteria. In my search for a congenial mind, I requested that he be brought to me. He arrived agitated, looking somewhat strange, and brought along some of the strains he had cultivated. It was the paratyphoid C strain that I myself had discovered! I had never been particularly proud of this discovery, because it was no great trick for an expert to notice that he had a new germ on hand. However, this discovery was an impressive achievement on the part of a student who had never taken a course in bacteriology and had mastered the method by himself following instructions from a book. I immediately offered him an assistantship, since there were no regulations as yet requiring definite qualifications. Supniewski graduated from the medical school while working as my assistant and passed all examinations with top grades although, I daresay, he attended no lectures. He did some mysterious chemical experiments, and once an explosion tore out a section of the ceiling in the laboratory. I was never sure whether he might not blow the whole institute into fragments. After graduation, he went to America for two years, and now he is professor of pharmacology in Cracow—and, so I have heard, a good and just teacher. I am glad that I had the opportunity to help him move forward, although he would also have made it without me.

I did most of my work in collaboration with Wanda Halber. She, too, had no formal qualifications, and many wondered why I chose her for my special, rather intricate projects. Therefore, I should like to emphasize that I classified her among the pleiad of higher intellects in spite of her one-sided education as a student of biology at the Free Polish University. She died of cancer after fifteen years of working with me. I was at her side almost to her very end, I read to the poor dying woman the poems she liked, I tried to create in her an emotional disposition that says "yes" to suffering, and I considered it my sacerdotal duty to carry her over into eternity. The commemoration I dedicated to her after her death may reflect the value of a person to whom I was associated with ties worthy of a teacher and pupil.

On 28 January 1938, after long and severe suffering, Wanda Halber, my assistant and associate of long-standing, passed away. I remember our first meeting in 1922. She came asking me for an opportunity to do scientific work. It took me awhile to notice what an unbridled yearning for work and what a capacity for tremendous effort this young girl possessed. She had a subtle and critical mind and an unusual zeal for work. Scientific work was everything to her: her greatest passion, delightful intoxication, and complete absorption. She was able to work the whole day long and, after completion of experiments, to contemplate, discuss, and read. She worked without counting on—even without desiring—a reward, and probably the only reward she desired was friendship and kindness. Her tragedy was that she did not believe in those feelings with regard to herself. She was a solitary her whole life long.

Her curriculum vitae? From the time I met her, she participated in most of my experimental work. Under the aegis of the Ministry of Military Affairs and jointly with Docent Mydlarski, she carried out a serological survey of Poland. Those were the first investigations of this type, and they brought worldwide fame to Poland. All her research projects were also mine, and together we experienced great hopes as well as frustrations with regard to many unfulfilled plans.

After Wanda's second operation [to remove the cancer], we wrote our papers on the immuno-biology of pus [the thick opaque usually yellowish-white fluid matter containing white blood cells, tissue debris, and micro-organisms] and new growths. We both knew that this would be our last joint study. With extraordinary moral strength she wrote and compiled protocols on the tests when we discussed the results and planned further investigations. The paper on the immunology of pus and new growths published in the foreign press was thus written by myself and a human being doomed to [imminent] death. She knew it. Yet in those last moments when death had already cast its shadow on her, I came to know the innermost parts of her psyche. During that last period of our work, I saw more clearly than during her youthful, impulsive actions, the moral and intellectual grandeur of this lonely woman. While working on a topic that for me was but a stage in the long road ahead yet for her marked the end, she displayed the peacefulness and stoicism of calm heroism.

When I look back at her concentrated and lonely life with no hope for personal happiness, a life which was a chain of the noblest endeavors, and a life which proceeded on summits where ideas were projected far into the future, I think of the boundless enthusiasm and energy not only found in Wanda but also hidden in the multitude of nameless, shy, and devoted workers. What a mistake we make in so seldom reaching people with ardent hearts whose only desire it is to serve a great cause. Thus, may her only reward after death be the fact that those who looked at the endeavors of her life will always remember her with great kindness and sincere friendship, because she was able to devote herself to science without reservation.[2]

Róża Amzel also was a close associate of mine. She was a highly talented woman: she spoke most European languages and was a brilliant lecturer and an excellent writer. The scope of her scientific interests was wide, and she would become enthusiastic over every new project and idea. Her femininity and good-heartedness even hindered her in her work. She helped all her friends, corrected their papers, and taught students. Being abroad on the eve of the outbreak of the war—I had obtained a Pasteur Institute fellowship for her—she came back to Poland on the last train. During the siege [of Warsaw], she rendered invaluable service to me in organizing the blood transfusion action, displaying great courage. Later, in the district [the Jewish ghetto in Warsaw], she was my right hand as an organizer of our laboratory and as a first-class lecturer. She perished in the Aryan sector after she got out of the ghetto; she was killed together with her old mother.

Henryk Brokman had worked with me in Heidelberg.[3] In Poland he specialized in pediatrics but did not neglect the study of resistance and even performed a number of discoveries in pediatric serology. He is now one of our best clinicians.

Jerzy Morzycki had great intelligence and a certain boyish charm.[4] Later, as head of our subsidiaries, he proved to be an outstanding organizer full of ingenious ideas.

The picture of this early period would not be complete if I did not say a few words about my first secretary, Zofia Mańkowska. There was a certain coarseness and a critical attitude toward me in her, but I did not mind it at all. By half past two, she would already have become restless during dictation. She often interrupted her writing, paying no attention to my chronic inspiration. I had full confidence in her. Whenever she told me: "You're repeating yourself," I would change the paragraph I had just dictated.

I will not write at length about my colleagues and the institute's staff members who did not work on problems that interested me directly. Many of them became outstanding professionals. Sierakowski was interested in methodology, and after his stay abroad he introduced new methods in the institute. Anigstein had the restless spirit of a traveler: in Warsaw he yearned for the tropics, on the equator he longed for Poland. His nostalgia for distant lands alienated him from many and finally, for reasons unknown, he was dismissed. This happened immediately prior to the war. He went to the United States where he acquired a professorship. His dismissal was a scandal and a veritable sin, because he was Poland's only expert in tropical medicine. Helena Sparrow, a persistent researcher and a woman of great beauty and personal charm, efficiently directed the whole antidiphtheria action. Many of our clinicians had a crush on her, a fact which greatly benefited immunization programs. Zofia Milińska was extremely skillful in the Wasserman method: at the Copenhagen meeting she was regarded as a phenomenon of adroitness. Julia Seydel specialized in diphtheria.[5] Tadeusz Sprożyński became chief of the department manufacturing smallpox vaccine.[6] Helena Rabinowicz was very successful in teaching students. Jadwiga Goldberzanka was a universal bacteriologist. These people were the old guard: I will not write about the younger ones because there were too many of them.

In my opinion, with regard to ability, Polish youth do not lag behind that of other countries. If the work done by Polish youth is less productive, this is the result of special circumstances and a peculiar atmosphere that I will describe presently. During the last several years, after the introduction of doctorate degrees, sixteen of my students completed doctoral dissertations, most of which were published abroad; many of the graduates received special honors from the medical school. I distributed doctoral topics—in a different way, however, from how it once had been done in the Hygiene Institute in Berlin. I always tried to become familiar with the candidate's mind and interests, and I would select the topic according to his individuality. The versatility of the institute and the cooperation with our docents and assistants—Anigstein, Przesmycki, Sierakowski, and Sparrow had all passed their qualification examinations—enabled me to take this approach. I must admit that I had a certain weakness: I liked the youngest ones best. My older workers would occasionally reproach me for this, and then I always replied: "Some children are good and some are bad, but grandchildren are always wonderful." Evidently, I already considered myself the grandfather of these young people, and I admit that I liked being in their company; also, a successful lecture given by my associates caused me more pleasure than one of my own. There are two types of directors: some have the complex of a crown prince and are afraid of their successors, while the others are like parents who want their children to outgrow them. I belonged to the second type; I abounded in paternal instincts. Some believe that a leader must consume all minds in order to become an authority. Such authoritative spirits usually have no pupils; the ground is frozen around them. It was in a book by Flexner,[7] I believe, that I read about an interesting conversation between a student and the director of an institute. Asked about a method, he sent the student to his assistant. When the student expressed his surprise, this veritable scientist replied: "My assistant knows even more methods than I do; but those which I do know I invented myself." In the same vein, Professor Moro, an outstanding scientist and teacher, once told me: "If I assign a topic to a pupil of mine and after three months he does not know the subject better than I do, he is a blunderer." Occasionally, the scientific director must give up his idea in order to promote the advancement of his assistant. He ought to assume the most difficult tasks, formulate problems, develop new methods, and plan the whole. An easier conquest should be left to the young person so that he may learn to climb and rejoice in his results. Ideas should be suggested in such a way that the young person can perceive them as his own. The director who reminds the young man, "I'm the one who gave you this idea," forgets that he too was once young, poor, and desired to have something of his own. Such methods can be dangerous when applied to ruthless characters who elbow their way forward. Therefore, it is best not to admit unsuitable persons and to surround promising individuals with care. Moreover, the example set by the director has edifying effects throughout the institute.

The most important task of a director is to create the proper atmosphere. Not everybody can do this. I tried to create and preserve the proper atmosphere for twenty years. My comments are hardly the fantasy of a man cut off from life, but arise from actual experience. They may sound quite paradoxical, yet they will be understood by those who feel as I do. One important objective is to let most associates pursue their own interests and work on their own projects and to allow each to believe that the world's axis runs through his research. A well-functioning scientific institute somewhat resembles a house of maniacs. A soldier's courage stems from his belief that in his knapsack he has the marshal's baton. The same may be said of a young scholar: he has the right to overestimate the significance of his research and his observations. The scope of one man's ideas is narrow, and the supervisor must not only offer ready-made topics but also arrange it so that they spontaneously arise in the minds of young people. How can this be accomplished? One must rejoice when a young man gets an idea or finds something. The main functions of the director are to arouse excitability in young people and to rejoice at the harvest. I am proud that I have never told a young person that his or her premises were unwise. I always say, "try," and if he does not succeed I console him that one can acquire wisdom only through trial and error. One must be optimistic at the beginning and critical toward the end of a project. I will later have the opportunity to cite Nicolle's wonderful words about the need to have a certain chaos that facilitates proud, radiant thoughts. Unfortunately, it is the opposite that often happens: budding ideas are smothered by excessive criticism, while appropriate control tests are not then demanded to complete the project.

An atmosphere can be easily poisoned. It suffices to change somewhat the attitude and the interests of the young investigators. An idea is as bashful as young love. It is easily destroyed when one discourages the young person, oppresses him with excessive criticism, or puts too much pressure on him by requiring premature publication. It is desirable to create a competitive atmosphere, but one must take care that it does not degenerate into opportunism and envy.

Academic atmosphere vanishes when we cease to respect the effort of a creative spirit. An early distinction becomes apparent between those who are strong minded inasmuch as they want to become managers at a rapid rate and those who are contemplative and somewhat awkward. Usually, the latter grow into scientists, and therefore they must be well guarded. Unfortunately, it is easier to satisfy successfully a thirst for life in areas other than science. A true man of science is a poet doomed to eternal dissatisfaction. If the driving, strong-elbowed figure appears on the scene and is praised, the atmosphere of calm and contemplation may easily be dispelled. This was well understood by Alexis Carrel, who in his book *Man, the Unknown* discusses the role of monasteries in the Middle Ages.[8] They isolated monks from the flurry of struggle and enabled them to concentrate their thoughts on abstract yet loftier matters. A scientific laboratory ought to be such a monastery. Life's necessities and theoretical interests should be correlated. Unfortunately, in our country as all over the world, the hunger and brutality of life have penetrated into the tranquility of a laboratory. This is beautifully presented in the story of the American bacteriologist Martin Arrowsmith.[9] Some institutes resemble monasteries without God or factories manufacturing scientific papers in series. I will recount later how the atmosphere of a temple began to disappear in our institute. Who will have chased the Pharisees out of the churches of science?

I believed that the actual task of my life was to create the right atmosphere, because I felt that neither my own writings, nor the laboratory, nor scholarships or awards would create science: this could be accomplished only by this elusive breeze of enthusiasm. There had to be someone who gave and someone who rejoiced. I desired to be both. Unfortunately, another reality was growing; it was more powerful than my aspirations, it was building a wall between the teacher and the pupil, and it was tearing down the edifice of my dreams.

I have discussed my endeavors as a director of a research institute. Now, I should like to discuss my teaching activities, which I enjoyed most of all. The scientist and the artist delivers himself in teaching, and finally this activity also reflects paternal instincts. I lectured to county physicians, medical students, pharmacy students, and students of the Free University. My lectures for county physicians did not give me much pleasure: I saw faces of older persons, tense and fixed, betraying their inability to forget their worries and duties. Evidently, the period of high perceptivity when man likes and is able to absorb and assimilate new facts is short. As a whole, I most enjoyed teaching general bacteriology. In addition, I derived much pleasure from teaching graduate students, officer cadets, whom I could initiate to the methods and to the mystery of investigative reasoning. It is challenging to talk about a subject as a whole without making too many generalizations while not delving into too many details and thus overshadowing the main concepts. How should a lecture be constructed in order to capture students' imagination?

A lecture should be like a concert. The *andante* sets the basic tone. The tonality is different for tuberculosis, and different for childbed fever and syphilis. Each contagious disease has its own dramatic tension and musicality. One quality of voice should be used for diseases that roll over humanity like avalanches, and another for diseases of the individual, even the most dramatic ones, such as rabies. This introduction must act as a reveille, as the beat of a drum. All listeners must tense up in expectation, desiring only an answer to the question: What kind of a disease is it that afflicts and destroys millions?

Then comes the *presto*—the clinical course presented in the most illustrative way possible so that the students see the first elusive signs, development, and merciless consequences of the infection. From the progress of the disease, he should deduce the location of the germs and thence draw conclusions as to the epidemiological laws. The problem must be presented as a battle of two worlds, in which the germs want to prevail, and not perish with their host, thereby saving themselves by infecting other individuals. Next, the teacher should discuss the specific statistics at the world level, country level, and in a given city. If possible, he should mention a personal experience, still colored by memories.

Adagio. This is the etiology of the disease, presented not as an accumulation of facts but as an effort to fight for the truth. If possible, a detail from the life of a researcher should be tied in with an important historical moment or with the methodology. *Microbe Hunters* by de Kruif contains many such solos.

Finale. This is prophylaxis; the great hope and a fanfare announcing victory. All this should be presented as dynamic action and not as a dry accumulation of facts. One must never require the students to climb a smooth wall of compiled information. Everywhere there must be notches on which the thoughts can latch themselves, and there must be a play of questions and answers, of tension and relaxation. While talking, the professor must not look at the blackboard, or the floor, or his notes, but at the listeners. He must watch the expressions on their faces. The pedagogue quickly becomes the focus of the student's thoughts and emotions, and he becomes aware of their reaction. The eyes of the listeners are the test for the professor, showing whether he has absorbed them, because aroused emotions magnify the capacity of the intellect. One can even talk in a very difficult way, provided one has the full attention of the listeners.

One must have mastery not only of the text but also of one's voice. One should speak in a straightforward manner without oratory. "La vraie éloquence se moque de l'éloquence" [True eloquence needs no eloquence], wrote Rodin. The main facts should be compiled in a table. I personally prefer tables to an epidiascope [slide transparencies]. The student looks at them for a longer period. One should not cite too many details, because these are available in textbooks. One should present the overall picture, not omitting the newest discoveries. The student must have the confidence that the professor is leading him to the very summits. One should not be afraid of subjective illuminations, while making a clear distinction between individual approaches and objective facts. The student appreciates when the professor expresses his own opinions.

Whenever possible, the lecture should be tied in with actual life. Even with ethics. My lectures about venereal diseases were a test for me as to whether I had gained control over the mentality of my listeners. I spoke about problems concerning love, sexual instincts, and morality. A lecture can and should make the listener better: people who listen become better, because what they hear attenuates their hunger for life. One must establish contact with the students even beyond the lectures. The professor must converse with them in the laboratory, organize field trips, and even attend student balls. I attended mainly the balls of the pharmacy students because all students registered in that school attended my lectures; as far as medical students were concerned, only those who were interested in the problem of resistance would come to my lectures.

As soon as I entered the ballroom, I would be surrounded by young people: "Our dear professor." They knew that I liked to dance, and they took care that I should never be alone. I did not like to go to balls in foreign countries. I must admit that I do not particularly like excessively elegantly dressed women. At student balls, I had the impression that I was walking in a flowery meadow. One girl was like a violet, another like a pansy, still another like an opening rosebud. A violet of a girl comes up to me and tells me about her hopes, her little troubles, and about the lectures. She tells me which lecture she liked best and for what reason. She also tells me that I have nice assistants and that the atmosphere in the institute is so pleasant. Everyone behaves in a simple manner, and the assistants do not put on the air of ministers. One girl tells me that even her fiancé, who is not a pharmacy student, attends my lectures. "Because of me or because of her?" I speculate. And I see how my words are maturing into a harvest and how this child is looking around with a wondering gaze that the world can be so beautiful. I am no longer a professor who only supplies the information required in a pharmacy. I am leading these children to the summits and showing them distant horizons. You see, the world is not enclosed in a pharmacy. Heidelberg[er] was a chemist, but he discovered the biochemical basis of the virulence of germs.[10] A great scientific experience may also be waiting for you. "You want to work for your doctor's degree under me? I will be happy to give you a topic. But now, let us talk about your fiancé."

My youth comes back to my mind. I met my wife during dancing lessons. "We were just children. On a calm, moonlit night, two children had a date."

> On a blossoming apple tree
> A nightingale is singing
> So lovely and so silver,
> Of love and of springtime
> Of youth and of honey.
> Do you hear it, Hania?

A colleague asked me how to dance the tango, not being acquainted with modern dances. "Oh, it's very simple," I said. "Just walk around and think of your first love."

I reminisce about those moments in my present solitude, separated from life, knowing that they will never come back, never. Do my spiritual children remember me the way I remember them? I feel something choking me, and I feel the taste of salt in my mouth. "Unshed tears are burning my eyelids."

Chapter Nine

Scientific Activities

Our institute was under the mandate not of the Ministry of Education but of the Health Services Department, a fact in itself indicative: we were to fulfill a function of utility. The question was how much room there was for science while rendering our activities ever more useful. Yugoslavian institutes, for example, enjoy independent executive boards, as well as their own outpatient services and sanitary engineers. In Poland, an organization of this type was unnecessary, because *voivodship*[1] offices were in charge of implementing health programs. Still, we had to find some kind of cooperation, because as long as we were separated from the Health Services' offices, no one could take proper advantage of us. Expansion of the regional laboratories had to be established—according to our plans—in every *voivodship*, which required close contact with the Health Services Department. This is how we went about it: managers of regional laboratories were also official counselors to the chiefs of health service departments, while the epidemiologist of a regional laboratory was also an official in the *voivodship* health service office. Thanks to this scheme we not only participated in combating epidemics but also took the best road toward determining Poland's epidemiology. It was more difficult to expand the scope of laboratory work, because there arose a certain conflict between pure and applied sciences. I should like to emphasize that the authorities did not particularly interfere with our work; they supported us and did not try to change our course of work. I presumed, however, that they had less understanding of pure science and wanted practical results as quickly as possible. While fully appreciating the significance of applied science, I defended pure science as being indispensable. "Do you think, gentlemen, that if Galvani had received the assignment to make the streetcar, he would have discovered electricity? No one could have foreseen that the contraction of a frog muscle would represent the world's greatest technological revolution. Similarly, combating syphilis is unthinkable without the Wassermann reaction. Yet the origin of that reaction was an abstract discussion about whether there were one or several complements."[2] Some health official once said to me that the researcher should always keep the "welfare of the people" in mind. I told him that his perspective was slightly skewed. Love's objective is to preserve the species. Yet it is unthinkable to propose to a woman, referring to the fact that the nation needs children; rather, you tell her how beautiful her eyes are. The same is true of science. It is not important what the ultimate goal of the scientific effort is and what the spiritual impulses stimulating us to thinking and creating are. These impulses, by the way, are: inquisitiveness, curiosity, admiration, and delight. Rather, a good researcher is in love with his project, and it was good that Galvani was interested in the frog muscle and not in the streetcar. Therefore, the first casts of thought must be free of utility considerations. Special institutes such as cancer or tuberculosis research institutes, which lay down a specific direction for research, are often characterized by the fact that the work for which they were created does not take place within their walls. On the basis of his findings concerning the effects of some stains on cancer, Wassermann received the Cancer Research Institute [sic] and he literally did not know what to do with it. He complained that the constant visits by Emperor Wilhelm the Second, who demanded new discoveries each week, only fettered his spirit. Ehrlich introduced experimental cancer research methods at the Institute of Experimental Therapy, while in Heidelberg, at the Cancer Research Institute, the most productive work did not concern cancer at all. The discovery of the carcinogenic effect of some hydrocarbons makes it possible to carry on planned work requiring a special laboratory. When this subject is exhausted, the Cancer Research Institute will again resemble a dead-end alley. Good ideas usually arise on the fringes. The methods of studying filterable germs were developed in the Colloid Research Institute. Once, I was to receive an invitation to Japan to study leprosy. I pointed out that I knew nothing about leprosy, and I received a very significant answer: "We have many investigators who are specialists in leprosy. Now we are looking for a researcher who is not familiar with leprosy but has proved that he has good ideas." Such specialized institutes are usually formed because it is easier to obtain money for them from rich people; but experts regard these institutes as dead-ends. Science should obviously be applied to social needs; however, it is life that renders the researcher's mind perceptive as the practical consequences of Pasteur's or the Curies' research have shown. Completely pure science with no application was impossible in an institute of our type, because we had to respond to needs in public health by establishing scientific foundations for preserving health. My task was to conciliate the two. Since I had the instinct of both a researcher and a social worker, I was able to take up both tasks with a pure conscience.

I will begin by discussing our practical tasks. Our prime task was to perform bacteriological tests for the health service as well as for some clinics and hospitals. The growth of this department was excellent: 85 percent of all bacteriological tests required to combat epidemics in Poland were performed in the State Institute of Hygiene and its regional laboratories. The number of tests approached half a million. Our next great task was to introduce methodological improvements not only in our laboratories but throughout the country. Two examples may illustrate this work. The League

Nations Health Committee initiated international projects to improve and standardize the serological diagnosis of syphilis.[3] Several of the greatest European institutes, including ours, received unknown sera; the results were subsequently compared. There were two meetings in Copenhagen during which staff members of various European and American institutes performed tests; the results were later compared to find out which institute had the best method. I will add that the results presented by our colleagues were very good. We tested most of the methods and modifications in our laboratory and subsequently announced the results. We prepared the reagents for the best methods and distributed them among all Polish laboratories. In this way, not only our regional laboratories but all Polish laboratories were able to take advantage of the technical improvements introduced by our laboratory. This was no trifle. A 10 percent improvement in the results meant that tens of thousands of individuals were properly diagnosed and treated. Great credit is due to the League of Nations Health Committee for making possible an international agreement on these matters.

Currently, by means of certain skin reactions, susceptibility to certain diseases can be ascertained. We had to check the correctness of premises, immunize susceptible children with various vaccines, verify the existence of resistance, observe the course of diseases in vaccinated and nonvaccinated individuals, and so on. The existence of various types of diphtheria bacteria with various degrees of pathogen status as well as of various types of pneumonia and meningitis germs had been recently discovered. All these investigations required teamwork. In addition to the staff members of my department, personnel in the production department, the Children's Clinic, and the Karol and Maria Children's Hospital also participated in these projects. As a result, it was possible to introduce obligatory anti-diphtheria vaccination and, as a crowning of our efforts, I published a volume on our work in the area.[4] It would take too much time to cite here the names of all those who participated in these projects and whose endeavors culminated in the introduction of new vaccination methods and in the use of new serums in Poland.

This was the approach we took in our work on typhoid fever. The disease is a public health problem mainly in Western countries. Again on the initiative of the League of Nations Health Committee and with the support of the Rockefeller Foundation, various studies were carried out by appropriate institutes, and the best methods were selected at a conference held in Warsaw under my chairmanship. This considerably improved the efficiency of bacteriological investigations not only in Poland but also in Czechoslovakia, Hungary, Romania, and Yugoslavia. I should like to point out that an outstanding method of testing water for the presence of paratyphoid germs was developed in my institute by my colleague Miss Szper. In fact we achieved many valuable results of this type: Amzel, Anigstein, Fejgin, Milińska,

Przesmycki, Sierakowski, Sparrow, and others all participated. The agu
played a great role in the initial years of our institute. We carried out combat
action which was necessary even in Spala[5] and the president's palace. I will
not discuss our participation in the fight against various other epidemics nor
the growth of the production department headed by J. Celarek and Saski.
All these projects and achievements prove that the institute not only did not
avoid projects of a collective and practical nature but, on the contrary,
expanded them.

But beyond such practical preoccupations remained the domain of
purely theoretical investigations, and these were correlated with the research
projects I had pursued in my youth. After we published our findings about
the uneven distribution of blood groups in the world, two other articles
appeared that expanded our observations. Verzár discovered a similarity
between the blood of Gypsies and Hindus, while Coca established that most
Indians had the zero group.[6] This suggested the possibility that man's initial
property was the O group, which was subsequently transformed into the A
and B groups at the two opposite ends of the world. The German mathematician Bernstein advanced the hypothesis that the O, A, and B genes were
alternative alleles.[7] This had certain interesting implications concerning the
blood groups of expected offspring. This hypothesis needed to be verified
by studying families. Thus, anthropological and genetic investigations began
all over the world. They were also reinstated in Poland. The Ministry of
Military Affairs performed an anthropological survey of Poland; the serological studies were done in consultation with me. Published by Halber and
Mydłarski, the findings attracted everybody's attention, and Poland again
became the center of international investigation into blood groups.[8] I was
able to supplement this research in one aspect. Professor Prawocheński
asked me for information regarding animal blood groups. Colleagues' studies had established the existence of the A group property in sheep, hogs,
and oxen, which was similar to the human A group. The main difference
consisted of the fact that human sera always contained bodies agglutinating
other group factors, while animal sera did not always have them. We checked
the heredity of this absence of antibodies, and found that this was a recessive
inherited property. In view of this, we began a new investigation series on
the heredity of other immunity properties. These studies, carried on for
several years, became the basis of the constitutional approach to the property of resistance. This also required a different approach to infectious diseases. According to the traditional views, the germ was always harmful while
differences in individual susceptibility were of secondary importance.
According to our findings, the relation was precisely the reverse: latent
infections, dependent on constitutional resistance, play the main role.

Gottstein, the famous German statistician, once told me that our findings had elucidated for him a number of problems in the epidemiology of

diphtheria. At that time, blood group investigations became a very fashionable medical research topic and were even exploited for racist propaganda purposes.[9] The well-known publishing house Julius Springer turned to me wishing to publish my monograph. The book came out in 1928 under the title *Serologia Konstytucjonalna* [Constitutional Serology], and in it I not only proceeded on the basis of international investigations of blood groups but also developed various theoretical consequences and problems ensuing from this new science.[10] Resistance properties were regarded as constitutional traits that could be magnified artificially but which also followed a definite developmental pattern as anatomical features did. Thus, parallel to the morphogenesis, I pointed out the serogenesis as a certain internal compulsion in the growth of resistance properties. Latent infections could magnify these properties but were not a sine qua non prerequisite. These ideas spurred a heated discussion, and again the whole world began to study the susceptibility and resistance of the various races and to ponder over the question of whether resistance was associated with latent infections or whether it was inherent. In polar and tropical regions, diphtheria was a rare thing; we wanted to know why. Greenland was a country most suitable for such research, because it has a reserve of people isolated from external factors with ships calling on its harbors only twice a year and then under strict medical supervision. I was able to check my concepts in a fantastic way. The Dane Bay-Schmith came to visit me and subsequent to our discussion carried out a research project on the resistance of Eskimos and the presence of germs in Greenland.[11] Thus, in my book, I was able not only to discuss the splendid growth of the study of blood groups among various nations and races and compile the results of a huge international research of the heredity of blood groups but also to pose new questions open to discussion.

I prefaced my book with the following introduction:

> In the current era, the study of blood groups represents the most universal and perhaps the most profound application of immunology, which considerably exceeds the scope of individual pathology. I have given this book a general title in order to emphasize that this direction is the beginning of a new concept of resistance properties. At first glance, the concept of constitutional serology seems to be contradictory, because we regard the biological system as an instrument able to effect endless serological changes. The structural abundance of antigens which are constantly being produced by the organism should, it would seem, be manifested in diverse serological reactions. Constitution, on the other hand, implies something fairly stable and immutable. Correlating serological phenomena with constitutional factors means accepting preformed, hereditary properties of resistance. Therefore, the body's immunological response not only reflects

the variety of stimuli acting on the body but also proceeds along predetermined paths. Thus, the concept of constitutional serology represents serological differences within a species and their effects on the body's serological reactivity. Normal antibodies are regarded as predetermined biochemical agents whose formation and development are governed by laws similar to those governing anatomical organs. Therefore, the appearance of antibodies depends on the inborn ability of the cell; their quality and quantity depend to an equal degree on the immunized animal and on the type of antigen used.

Many feel the need to adjust our immunity concepts to the constitutional approach. The outbreak of an epidemic is not determined only by the entry of a germ acting as an external enemy. Epidemics should often be compared to a revolution, because the microorganisms which are dormant in the body gain an opportunity to develop. The constitutional basis of some forms of susceptibility is beyond doubt. I might mention the susceptibility or resistance of some animal races, the possibility of selectively breeding resistant individuals, and the significant discoveries of experimental epidemiology. A number of problems raised here are still in their beginnings, while some studies have been initiated in my laboratory. May the reader forgive the author that, while discussing these matters, he indulged in a certain inevitable physiological subjectivism.

For me, writing this book was a delight. In experimental work, results must ensue from experiments. A researcher promulgating too many hypotheses dilutes his findings. A book, on the other hand, allows one to embrace wide horizons. In the last chapter entitled "Foundations of Constitutional Serology," I attempted to separate the hereditary factor from the extrinsic factor. I compared the organism to a drawn bow, the degree of tension of which and readiness to propel a defensive arrow are inherited. The normal antibodies might represent the normal biochemical instruments whose development ought to be investigated. The motto I gave to this chapter was: Formulating a problem is the first step toward its solution. My book was very well received by critics as an announcement of a new direction in science. The critiques were detailed, and the interest aroused by this new direction was reflected in some reviews.

For example, the *Warszawskie Czasopismo Lekarskie* [Warsaw Medical Journal], no. 12, 1928, wrote: "We are dealing here with a solid scientific achievement of prime quality; not an abstract theory but a creation pulsating with life in full bloom, irresistibly stimulating one to thinking, and with each step revealing new, wider horizons. Nor should one forget that this work was initiated and pursued in postwar Poland under immensely difficult circumstances. This is a beautiful example of Poland's regenerating scientific ideas."

Ten years later, the Germans eliminated me from my institute, because I allegedly belonged to the "parasitic race." But when the book was published, they wrote the following reviews:

The author to whom first and foremost we owe the utilization of the discovery of blood groups has compiled and critically explored current publications, establishing the foundation for further research. His concept of constitutional serology is completely new, and we are sure that it will prove as fruitful as his study of heredity and anthropology. (*Zentralblatt für Bakteriologie* 92, no. 1–2)

Whoever picks up Hirszfeld's book is overwhelmed by new and thoroughly fundamental problems the significance of which for biology can simply not be predicted. (*Zentralblatt für Innere Medizine* 49, no. 13)

It is thanks to the group investigations that a new impulse has arisen in research which will fertilize all clinical departments. (*Zentralblatt für Neurologie* 50, no. 9)

This book will be the basis for all future group studies. (*Zentralblatt für Gesamte Hygiene* 17, no. 11–12)

The problems I advanced in my book were a treasure trove from which we drew ideas for many years. And the studies were all the more pleasant for me since I was able to correlate them with my wife's clinical work. Eventually, the sum of the accumulated findings and experience matured to such a degree that it became necessary to compile all results in a special monograph. My wife was the one to do so. The monograph, entitled *Rola Konstytucji w Chorobach Zakaźnych Wieku Dziecięcego* [The Role of Constitution in Pediatric Infectious Diseases], was published in Polish and, two years later, in French. Professor Debré[12] from Paris supplied it with a beautiful introduction in which he emphasized the achievements of Polish science.[13]

I will not discuss in detail the consequences of my findings from the viewpoint of legal medicine. I will only mention that after Mrs. Gorgon's court proceedings, very many cases were referred to me for evaluation. My court cases were the most numerous in the world. However, in view of the presence of group properties on items not stained with blood, I came to the conclusion that the scope of group investigation in that area was limited. Moreover, I became convinced that in view of foreign experts' ignorance of our discovery, namely that items in close contact with a human being were saturated with group properties, they committed a number of mistakes in their evaluations. One Belgian scientist announced that he was unable to confirm our results. I invited him to come personally or to send his assistant to Warsaw and see the correctness of our observations. Assistants in legal medical institutes in various Polish cities had the opportunity to learn the technique of these tests in my laboratory.

But probably the most interesting course was taken by the investigation of so-called subgroups. When I was still working with Dungern, we had demonstrated the existence of two types of A groups: a stronger factor and a weaker factor. This observation, though confirmed by many, evaded elucidation for many years. The studies I performed with Kostuch and later with Amzel finally clarified this matter.[14] The weak A factor was a transition form between O and a strong A factor. We regarded this as a manifestation of an incomplete mutation. We were able to demonstrate the existence of a whole gamut of similar transient forms with regard to the A as well as the B property. With these facts available, we were able to explain the absence of antibodies for the O group. I cannot go into the details of this problem, but I would like to point out that this gave rise to a number of questions requiring distant travels and international cooperation. I talked about our observations in Paris in 1937 and in Rome in 1939. A translation of my book on blood groups, published in Polish in 1934, appeared in French in 1938. I ended the book with the following remarks:

> Encouraged by the significance of the science of blood groups, I should like to raise the following question: did serological mutations occur only in primordial times or are they also possible today? By investigating the degree of agglutination of blood cells with our special methods, will we not be able to get a feeling for the creative rhythm of our era? I should therefore like to conclude this book with a question prolonging the problem my wife and I raised some 20 years ago. At that time, we wondered to what groups the blood of various nations belonged. Now, I ask to which pleiades it belongs.[15] While such a question may seem too bold, it has been provoked by the thrust forward characterizing the science of blood groups. To all the problems which have been solved by this science with an outstanding precision, I should like to add yet another: *What kind of serological mutability exists in our era?*

The constitutional approach to resistance was of great consequence for the interpretation of vaccination. We had to assume a priori that susceptibility to immunization would be different according to age group, class, and race, and that we would obtain varying immunization results depending on the intensity of the epidemic and the spread of the germs. This viewpoint ensued from a sharp differentiation between extrinsic and constitutional factors. Vaccination against diphtheria carried out on a large scale in Warsaw made investigation of this type possible. The Warsaw medical inspector, Dr. M. Łącki, the statistician Dr. Grzegowski, and Dr.

Jakubkiewicz who in addition to vaccinations also performed laboratory tests, as well as many others set themselves to the task. We disclosed many statistical and immunological errors. Even after they had been eliminated, we found that the vaccination decreased diphtheria morbidity more than fourfold. Following Warsaw's example, other Polish cities having implemented anti-diphtheria vaccination carried out similar statistical evaluations. I reported on this project to the Medical Academy in Paris. Gundel, who was in charge of diphtheria vaccination in Germany, declared that our report was the most critical of all.[16]

Now I will discuss another research series. The serological distinctness of new growth tissue has been suspected for a long time, but no one had succeeded in proving it. The reason was the serological complexity of the tissue, which also contained a number of other properties. In order to establish whether or not cancerous degeneration represented a change in the serological structure of the new growths, these various factors needed to be separated out. To perform the study, I received a grant from the Academy of Sciences and was able to collaborate with Halber as a serologist, Laskowski as a pathologist, and Floksztrumf and Kołodziejski as clinicians. We established that new growths indeed contained other bodies than those found in normal tissue. Demonstration of this difference entitled us to a certain amount of hope that it would be possible to make a therapeutic serum to treat new growths. Unfortunately, this problem proved to be more complex. I talked about this for the first time at a meeting of microbiologists in Cracow. The news of a certain, justified hope of obtaining a therapeutic serum against cancer leaked into the daily domestic and foreign press, and for a period I was snowed under with letters asking me for the serum. I was aware of the danger that would arise if descriptions of our research became available to [manufacturing] plants financially interested in taking advantage of this discovery. I refused to distribute samples of the serum until it could be tested by my associates. The outcome was unfortunately negative. Investigation of the serological diagnosis of new growths yielded more interesting results: serums from people who had new growths often reacted with alcoholic extracts of the new growths. However, a similar reaction was also obtained by using the serum of pregnant women or of patients with syphilis. Control experiments in other institutes were less fruitful; this was in my opinion because they differentiated in a less subtle manner the reaction associated with the instability of serum.

On the basis of this research, I was invited to present the lead paper at the International Cancer Congress in Brussels in 1936.[17] This study was expanded to various aspects of tuberculosis, experimental new growths, and tissue necrosis. We were able to establish that tuberculosis tissue contained other serological bodies than those found in normal tissue. We discovered a serological specificity of normal lungs and a change in this specificity induced by the tuberculosis process. Curiously, during the course of tuberculosis, these bodies appeared in the liver, which seemed to be unchanged; evidently, they circulated with blood and were picked up by liver cells. Most curiously, we found similar bodies in pus, in white blood cells, and in new growths. My pupil Dmochowski performed a series of investigations of the specificity of experimental tumors. Miss Kraushar discovered these bodies in the kidney, which died after ligation of blood vessels. Thus, these bodies indicate that a cell is dying; white blood cells normally contain them probably because these cells live for a very short period. We believed that the appearance of these bodies foretold the cell's death. Our findings have opened new paths to investigating tuberculosis and tumors, and they should also be considered in the study of viruses. I also expressed the presupposition that the morphological changes of blood could be induced by the formation of immune bodies directed against various blood cells of one's own body. These investigations, confirmed in Holland, have not yet found a wide echo. A certain period must usually elapse before the findings of one laboratory can be picked up. For example, our work on blood groups began in 1910 but became the stimulus for the work of other researchers fifteen years later. I had hoped that our study of cancer would mark the beginnings of a new direction, but I had to interrupt my work. Miss Habler's death, Dmochowski's departure to London, and Dr. Floksztrumf's to America deprived me of my main associates.

Those were our main experimental projects. I will not write about our numerous other projects, because I would only tire the non-medical reader. I will just touch upon a few publications not without sociopolitical significance. As everyone knows, the Germans introduced obligatory sterilization of individuals afflicted with certain mental diseases. For religious reasons, the church opposed this law. The Eugenics Society raised the issue at the Health Council, and we had to consider whether obligatory sterilization should be introduced in Poland. Such a law could unleash the opposition of the church. I pondered this problem with my wife, and we came to the conclusion that this measure would be of no social value, because such traits were inherited as a recessive property, and over one-half of the population were carriers of diseased genes. The church was right in this conflict, even though it did not operate on the basis of a genetic analysis as the advocates of sterilization in Germany did. My wife published the results of our analysis in the medical press, and she also published a pamphlet entitled *Z Zagadnień Genetyki i Eugeniki* [Problems in Genetics and Eugenics]. Supported by these publications, I took an appropriate stand before the Supreme Health Council. I consider the resolutions of this council, which required not only eugenic but also social indications for sterilization, as the most fortunate that have ever existed.[18]

All of the projects I have mentioned deal with specific problems. But as a result of an inner maturing process, other impulses were awaking in me.

I believe there is a relation between scientific and artistic creativity that manifests during the synthesis of various developments. This I felt keenly when I was reading Zweig's book on Freud. I was struck by his vivid manner of discussing seemingly dry, scientific matters.[19] Around that time, the Julius Springer Publishing House asked me to write a textbook on bacteriology in German. I should add that the publishers entrusted this task to me because in their opinion the available German textbooks were not quite up-to-date. I did not have enough time to complete this textbook, and it was only after I was dismissed from the institute that I found time to write the manuscript in Polish. I decided to write an introduction of a rather literary nature; I wanted it to express a biologist's credo of sorts. I had pangs of conscience that in presenting infections and resistance, we subconsciously exercised a certain anthropomorphism. We always talk about a struggle against infectious diseases and about bacteriocidal bodies and destroyers, making the a priori assumption that this struggle is the principal aspect of life. Yet it may very well be that this is not so and that the concept of struggle as a basis of life is only a reflection of our own quarrelsomeness. Problems of resistance must be considered not only as a conflict but as a symbiosis and not only as opposing interests but as the desire to establish coexistence which, on a certain level of evolution, may assume the form of ethical impulses. Should ethics be introduced into biological phenomena as an a priori premise? It seemed that only the victory over the religious approach had established the basis for pure science. However, if we have introduced the instinct to struggle as a basis for biological thinking, why should we not introduce the instinct of solidarity which—in its sublimation—represents ethics?

Thus, I tried to reanalyze resistance, not on the basis of a conflict between macro- and microorganisms, but on the basis of a drive toward coexistence. I wrote an article which, judging by the number of reprints requested, aroused more interest than many of my experimental investigations and even provoked protests against my introducing a religious factor into pure sciences. I published the study in German in 1930 [sic] under the title "Prolegomena zur Immunitätslehre."[20] I attempted to systematize the various aspects of resistance, not on the basis of a struggle but as a trend toward coexistence of macro- and microorganisms. Instead of talking about bacteriocidal antibodies, I pointed out that the antibodies induced mutations and that bacteria, instead of perishing, assumed other forms of existence. Taking the trend toward harmony and coexistence as my premise, I discussed latent infections as the result of this trend. I fancied a law, which I would call the law of least suffering: inter-species conflicts proceed in such a way as to enable the largest number of species to survive with the smallest amount of pain. In the introduction to the *Prolegomena to the Study of Resistance*, I wrote:

When we look at the scientific data on resistance, we notice that they have not been put together into a system which, much like the periodic table in chemistry, would reveal the gaps in our knowledge. The immunological data currently available make a synthesis possible; yet they lie fallow, because they are not correlated with other observations and are not illuminated with data from general biology. It would seem that the science of resistance blushes in its youth to be beset with hypotheses, and with unjustified depreciation rejects deduced conclusions. Yet it is the task of an idea to arouse creative unrest. The conflict between a hypothesis, and an observation gives birth to truth.

Analyzing resistance by deliberately ignoring axioms and expressing the desire to form a synthesis seems justified for theoretical cognitive reasons. Our systems and generalizations are not always the result of observations. Our general outlook on the world often cuts in between the observation and the interpretation, so that when we examine a contagious disease and explain the various aspects of resistance, we are subconsciously dominated by the concept of a struggle as the only possible form of encounter between macro- and micro-organisms. Regarding constant struggle as an inevitable biological state and acknowledging a chronic tension and mobilization of resistance forces cannot satisfy him who perceives the formation and the subsequent equalization of defensive forces as a waste of energy.

Let us analyze the basic theses in bacteriology. The world of microorganisms, which play such a tremendous role in the world's metabolism and create a bridge between living and dead matter, gives rise to pathogenic bacteria which, when they are maximally parasitic, grow only in living tissue and destroy it simultaneously. Their victory is a veritable Pyrrhic victory, because the bacteria perish along with the victim. Even if they escape total extinction as a species because they infect other individuals, this only postpones the moment of death. Thus, the acquisition of pathogenic properties by bacteria represents, if not the death sentence, at least a diminution of their chances for survival. Pathogenicity is like excessive bellicosity during peacetime which eventually must lead the fighter to the gallows. If increased virulence cannot represent increased chances for the bacteria's survival, the usefulness of various resistance reactions is equally problematic. There are forms of resistance that depend on the presence of microorganisms in the body. Killing them is completely senseless: it would be like destroying live sources of defensive energy at the height of combat.

Let us have no illusions that the study of resistance will be able to explain the body's protective reactions according to our common sense. In order to preserve the species, it would be senseless to create pathogenic properties that deprive the bacteria of the realm of dead matter and then give them, instead, a live body in which they encounter defense. It would be senseless to kill microorganisms where their existence sets defensive forces free. Before we assume that nature might have created useless arrangements and unnecessary conflicts, however, we must examine whether our interpretation of natural phenomena and our image of the world are false. Indeed, our generalizations and interpretations of resistance reactions ensue from our primitive belief in the sacred egoism of the species. The assumption that one of the two opponents must be conquered should yield to the understanding that it would be infinitely more reasonable for both species to remain alive. Symbiosis is not only ethically higher than parasitism but also biologically more reasonable. Getting accustomed and adjusted to each other is as reasonable as destroying the so-called hostile species. The concept of bacteriocidal bodies is a projection of the concept of a struggle among species onto resistance reactions. Pathogenicity should be comprehended only as a stage in an evolution toward symbiosis. Epidemiological and immunological data, individual and communal resistance, and the attenuation of a germ through prolonged contact with a sensitive species indicate that pathogenicity in only an unfinished process of adjusting to each other by means of symbiosis. The world displays a trend toward an equilibrium between macro- and microorganisms. It is trying to attenuate and eliminate combat readiness. Therefore, our systems and theories of immunity should not be built only on the assumption that there is a struggle and mutual destruction, but we should try to enlighten them and, if necessary, find defensive forms and mechanisms that make symbiosis between the microorganism and the host possible.

At a certain age, this desire to establish a synthesis arose in me with vigorous force. I deferred many experimental projects, trusting that they could be done by someone else, whereas I attempted to understand the essence of the conflict between the visible and invisible worlds. I was convinced that the researcher's psyche was subconsciously reflected in his biological concepts. My pupil Chwat[21] had just completed a translation of a beautiful book by Nicolle entitled *Narodziny, Życie, i Śmierć Chorób Zakaźnych* [Birth, Life, and Death of Contagious Diseases]. I was asked to write a preface and present the author to Polish readers. Since Nicolle had created the concept of a latent infection, I tried to bring the results of his experimental work into agreement with his spiritual figure. I will cite the main paragraphs from my preface and excerpts from Nicolle's books, which beautifully and profoundly describe the main driving forces of his experimental work. Several weeks later, my preface was published as an obituary and this is the form in which the excerpts are given below.

Nicolle's Obituary[22]
Charles Nicolle died on 29 February 1936. Friends and admirers anticipated his death with painful sorrow while he showed the stoicism of a philosopher. In sketching the portrait of this unusual researcher, a man of

exceptional personal charm, I should like to review in brief what I wrote in the introduction to the Polish translation of his book entitled *Birth, Life, and Death of Contagious Diseases*. Not that I would be unable to find fuller tones in the chart of his life, but because he fully approved of my preface and sent me a reply which touched me deeply. He wrote that this was the last flash of happiness which befell him. I am deeply convinced that by presenting his biography in this form, I am fulfilling my duty of piety toward the deceased. I will only slightly change the sequence, primarily emphasizing what is interesting to physicians. I could never resist the belief that a certain manner of perceiving and feeling is expressed not only in social concepts and one's attitude toward ethics but also in biological concepts. Nicolle created the concept of asymptomatic pathology. The premises and observations that led him to this concept which, in my opinion, represents the greatest revolution in bacteriology since the period of Robert Koch, I have characterized as follows. Nicolle looked upon the conflicts prevailing among people with disgust. In his book, he devoted a special chapter to this problem. "Have we not suffered enough undeserved defeats from irresponsible and evil fate? Why should we create new, inhuman, cruel suffering?" If the budget of the last war had been allocated to scientists, it would have been possible to drive away and even eliminate forever the specter of many of our cruel diseases. Unfortunately, the prejudices, the fury, and the perversity of people would like to mobilize them sometimes and convert them, if possible, into instruments of conquest and destruction. . . . The only worthy conquerors are the teacher and the physician. Only their work justifies the attempts of strong nations to gain rule over weak nations. "Was it by accident that this mind, so sensitive to all tragedies, reacted differently to the conflict between species and took a different approach to the suffering caused by infectious diseases? Nicolle made a seemingly minor observation: he noticed that the typhoid organism could circulate in an animal without inducing pathological signs. With the intuition of a great biologist, he immediately perceived the different mechanism of coexistence of micro- with macroorganisms and outlined asymptomatic infection and asymptomatic pathology. Should the various species exist only for the purpose of combating and destroying each other? This idea, which probably ensued from his general outlook on the world, has created a new vision of the invisible world. Disease, pain, and suffering are only a transitory effect of the conflict between the visible and the invisible worlds. Cohabitation, and not just conflict, is a great law of nature. In this, the author has developed marvelous, profound ideas of how infectious diseases might have been formed, what the biological motives were for the pathogenic germ to wish and to be able to penetrate the system, what consequences of this struggle created pathological signs, and how in the perspective of millennia is symptomatic pathology changing into asymptomatic pathology. When comparing textbooks of modern bacteriology with this marvelous attempt to synthesize problems governing the existence of species, it seems that only a researcher with the soul of a poet could summarize the science of infectious diseases.

These citations reflect the results of Nicolle's work in research, but they do not reflect his attitude toward the world. They say what Nicolle did but not what he felt. I know no researcher who so strongly desired to present the depth not only of his thought but also of his emotions. In his writings, Nicolle let loose his instincts of poet and philosopher. On the subject of a clinician's mentality, he wrote: "His brain should be like a mirror: freed of prejudices, it should have the sensitivity and distinctiveness of faithfully reflected ideas, educated and enlightened ideas." He continued: "There is a discrepancy between the instinct for teaching and the instinct for doing research. The need to possess broad knowledge creates a mental state which separates the pedagogue from the researcher. The researcher should not overburden his mind. It is good, and occasionally even excellent, when there are gaps in his knowledge. Without such gaps, the researcher's intelligence would not dare to combine bold, radiant ideas which are rejected by orderly and excessively disciplined minds." His following marvelous words characterize his view of the motives which draw people to research: "Scientists do not devote their lives to research in order to boast of their success, but only to satisfy their infinite thirst for knowledge and to feel the pleasure of studying. Is not instability toward one's own achievements the trait of genius? Take some away from a scientist and he will find others. For heaven's sake! What is it all about? Fantasy. And there will be plenty of fantasy for all. Each of us can take as much of it as the heart desires."

Nicolle received the Polish translation of his book *Birth, Life, and Death of Contagious Diseases* prior to his death; he was bedridden and aware that he would not see the year through. He thanked me for my introduction in his letter of 26 December 1935, a letter of farewell which I would like to share with the Polish reader:

> My Dear Friend. I am infinitely grateful to you for the preface you wrote to the Polish edition of my book, *Le destin des maladies infectieuses.* Your praise is excessive, but I accept it coming from a man like you who saw in my text that it was as much a work of art as of science. As such, we are well disposed to understand each other and nothing separates the Frenchman from the Pole. You are aware of the seriousness of my condition. Your preface will have been one of the last gleams of happiness for me. I as well as those who surround me thank you—with all my heart,
> Charles Nicolle

His letter betrays unusual modesty. The secret genius of the man was that he was able to look at pathological phenomena through the eyes of a biologist and an artist. When he was beholding a forest, he saw not just the trees but an abundance of animals and plants living together and struggling with each other. His letter shows also that he valued himself highly as an artist and knew well that the basis for all creative work is imagination and a passion for searching: there is no boundary between true science and true art.

This is evidenced by the deceased man's life and creativity.

Let us now turn from the researcher's soul and from his impulses and concepts to the content and meaning of his work. We see that an infection is not only a conflict but also a trend to establish a coexistence of two worlds. Disease is a fairly rare breakdown of this coexistence. The immunity acquired after a disease represents not only a victory over the germ that has perished but often a coexistence on a different plane and with a different degree of mutual adjustment. When in 1938 the Committee of the Congress of Biologists and Physicians in Lwov asked my wife and myself to present the leading paper in our area, my wife spoke about problems of constitutionalism in contagious diseases, and I spoke about the role of infections in nature. I also discussed the nature of suffering, the probable evolution of infections, and the role of infections in nature.

In the first place, let us ask whether all infections are not evidence or the beginnings of conflicts manifested in the form of infectious diseases. Can it really be that on this globe, the battlefield of species for millions of years, the initial conflicts between the visible and invisible worlds have not mutated into an advanced interdependence of interests and advantages, that the concept of accidental hostile and definitely harmful infections has not lost its essential meaning? What are the simplest examples of infections? A number of protozoans contain intracellular green or yellow inclusions: zoochlorelles or zooxantelles. This is a symbiosis between the host which needs oxygen and algae which supply this oxygen. A process which is the basis of life in plants and animals is here introduced into the cell. Here the infection represents coexistence brought to perfection; in fact it resembles intracellular gardens. . . . The fluorescence of the sea is caused by microorganisms living on the surface of the body of sea animals. The microorganisms illuminate the depth of the sea for their hosts and function as lanterns of sorts luring animals of the opposite sex. . . . Termites breed various microorganisms as food. The latter produce cellulose-splitting enzymes which tremendously facilitate the termites' living conditions by, as it were, enabling them to gain new ground. . . . Thanks to coexistence with microorganism-splitting wood, certain strains of ants had to grow organs enabling them to crush wood. Here the infection is a driving force which conditions the creation of new forms and thus is the basis of evolution. In insects, we observe special organs composed of microorganisms and a tissue produced by these microorganism. . . . Bacterial symbionts often supply the necessary splitting enzymes. . . . In all these cases, one has the impression that the macroorganism is the partner which pushes the microorganisms into the sphere of interests of its own species and which seems to possess a deliberate will to create forms of coexistence.

While discussing viruses, I wrote:

Are we witnessing the formation of germs from the alien cell of the host and are we implementing the old dream of creating life out of matter? Or have we disturbed the balance between the cell of the macro-organism and the intracellular, almost ubiquitous germ, and have we thus loosened the symbiosis and established for one of the symbionts a fairly independent life within our alien system?

It seems improbable that the macro-organism or its individual cells would always be oriented toward defending their biochemical sacrosanctity, the way it is presented by modern immunology. Biological analyses suggest rather a vision of coexistence and a balance between macroscopic and microscopic creatures; a disease seems to ensue from a disturbance of this balance. If this is really so, we must form a different image of the world to displace the vision of macro- and micro-species fighting with each other for death and life. The need for a different picture of infections also comes from the need to widen our views with regard to the struggle for existence.

Biocenosis as an elementary unit of existence. Modern biology has understood that we cannot regard the so-called struggle for existence only as eternal conflicts and battles. Life endurance is associated with the mutually controlled vital élan of the various species and individuals. Our picture of the world should encompass the conflicts as well as the harmony of coexistence; with such a view, disease and death of the individual become the successive foci for regenerating life. It is not the individual who constitutes the prime element in nature, but a group of species living in a definite spatial unit. Botanists have named this phenomenon *biocenosis*. This concept transcends the species. Botanists cite an example which explains the essence of the concept. The forest forms such a biocenosis with its own specific rhythm of life and its own will, as it were, to prevail. A forest is not just a group of trees; it is a tremendous agglomerate of the most diverse creatures. A forest contains an abundance of plants which make up its underbrush and soil, and it is inhabited by innumerable animal species, beginning with wildlife and ending with caterpillars on tree branches, beetles nestling in tree bark, and mites which populate the soil by the millions. All these coexisting and combating species comprise the concept of a forest. In a healthy forest, the individual systems are mutually adjusted, and their élan, will to live, growth opportunity, and growth rate are mutually regulated in such a way that all members of this group called a *forest* find the possibility for their species to survive. A biocenosis is situated in harmony, an equilibrium. Yet this harmony, which guarantees the survival of species, can be preserved only through constant destruction, struggle, and parasitism. The struggle for existence, heat, and frost guarantee a mutual control to the members of the group. Starlings control beetles by snatching pupae out of the bark. The caterpillars in tree crowns are controlled by flies and wasps. What we have called mutual control is often, from the viewpoint of the individual, a boundless pain, eternal fear of the enemy's assault, a wound, or premature death. Yet, at the same time, this control is the prerequisite for the survival of the whole. The slightest shift toward the victory of one species may disrupt this harmony. Elimination or diminution of the number of birds, which frustrates the control of beetles, may induce such a senseless proliferation of the latter that they will eat up all the leaves and the forest will perish, burying itself under the species of insects which has run wild in the spurious victory. Thus, the victory by one partner in the biocenosis does not mean that the latter has been transformed in such a way that the harmony can be re-established on a new basis. Nature is not interested in eliminating the suffering, disease, and death of the individual, but only in maintaining inter-species equilibrium, because this guarantees continuity and the greatest polymorphism in life.

Therefore, we must ask whether an infection does not have a deeper meaning and whether pathology is not a manifestation of a disturbed or not yet attained equilibrium with microorganisms which preserve biocenosis. He who assesses natural phenomena in this way must see certain advantages in infections, even those involving pathogenic germs.

Thus we see that infection existed at the beginning of development. The aspiration to represent the evolution of coexistence between the visible and invisible worlds is understandable. We must investigate which biological or social factors cause a disturbance in the equilibrium between man and bacteria. The problem presented in this way reaches the field of biology and sociology and surpasses the interests associated with the analysis of individual epidemics. Yet I regard such an analysis as absolutely necessary, because I see how incorrect presuppositions direct our perspicuity onto false paths, how they distort our image of the universe and often force us to make unjustified and erroneous moves while combating epidemics. Epidemiology has degenerated into a police investigation with which we try to establish a certain banditry on the part of bacteria. Epidemics are an integral part of a great natural event. We will be unable to shift the equilibrium toward man's advantage if we do not come to know the laws governing the coexistence of the visible world with the invisible world. . . .

The unilateral, pathological approach to infection has done harm to the study of resistance. For this study constantly operates on the basis of a fiction: the fiction of a total fight against the germs which have penetrated the body and the concept of preserving the biochemical sacrosanct aspect of tissues and cells. But it has failed adequately to illuminate developments dominating life as a whole, such as physiological infections and the immunology of the carrier state. . . . An epidemic is part of a great clash, one of those eternal clashes between the visible and invisible worlds. Future epidemiology will discuss these problems the way paleontology and comparative anatomy discuss the natural history of animals and plants and the way history discusses the disappearance of old and the appearance of new nations and political states.

The concept of the coexistence of the interwoven visible and invisible worlds of creatures destined to walk through life together reflects the contemporary degree of evolution more profoundly than the concept of non-infected species able to lead an independent life. Man is trying to tip the balance in his struggle with pathogenic germs to his advantage. He often succeeds. However, will it be possible to preserve life on the basis of an artificially created equilibrium? Whether we will be able to eliminate all infectious diseases and implement a unilateral control of pathogenic germs in the biocenosis represented by modern society, only the distant future will show.

I also published this essay in a Viennese medical journal[23] and I discussed these problems in a talk I gave to medical students in Paris. Similar ideas were expressed by two or three other authors who wanted to break away from the concept of a struggle and bring to the forefront of prevailing problems the issue of coexistence and the trend toward a symbiosis between the visible and invisible worlds. I am certain that this approach will be the basis of future epidemiology. I expressed this conviction of mine in the preface to my book on resistance.

I will not discuss in detail our other research projects. Each year we published from thirty to forty-five scientific papers. Every second year I sent a volume with the papers published by my department to all larger scientific institutes throughout the world.

Homo sum; humani nihil a me alienum esse puto.[24] I wanted to express my views on matters concerning the organization of science and sanitary needs in the country. For a commission of the Ministry of Education, I wrote a memorandum about reforming the teaching of hygiene, about the bacteriological service in the country, about the needs of Polish science, and so on.[25] I believe that these articles did not remain without useful practical effects. An exact report on these matters would carry me much too far.

One of our institute's duties was to inform the public about scientific progress. I encouraged my associates to do this, and I personally gave general lectures several times a year. I presented these lectures not only in Warsaw but in almost all of our university cities. Our institute actively participated in the meetings of Polish microbiologists; I was proud to see that our little group gave good talks and everywhere met with interest and response. Almost every year, I traveled abroad to attend the meetings of the [Biological] Standardization Commission of the League of Nations, of which I was a member. In addition, I presented leading papers at meetings abroad which I always attended as a delegate from Poland. In 1939, the International Congress of Microbiologists was to take place in New York. A great honor befell me: I was invited to be vice chairman of the bacteriological section and vice chairman of the congress. I was to go there as Poland's official delegate. Incidentally, several years ago I received an invitation to present a paper before the famous Harvey Society in New York and to give a series of lectures in the American Science Extension. However, I was so absorbed by my work in Poland that I postponed the trip. I might also mention that I was cofounder, chairman, and secretary general of the Biological Society and cofounder and chairman of the Society of Polish Microbiologists. I am writing about this to point to the prevailing conditions and to show that I did everything to fulfill my resolution to serve my country. In 1924, I was offered the position of director of the beautiful Epidemiological Institute in Belgrade. In 1937, my former teacher, Professor Silberschmidt in Zurich, nominated me as a candidate for his successor and asked me to submit my papers and reprints in the Zurich Medical Department. I turned down both offers without a moment's hesitation. Was the decision wise? Man follows the voice of an internal compulsion even if he is to pay for it with his life.

Chapter Ten

Scientific Meetings

This book is not designed for experts; it is meant to reflect an epoch in one of its aspects and the role played by scientists in the wake of conflicts agitating humanity. That is why I will not describe the scientific content of the meetings I attended but rather the general impressions they conveyed. We established a Polish Society of Microbiologists and Epidemiologists. Meetings were held once every two years, and every second meeting coincided with the meeting of Polish biologists and physicians. The scientific level of our meetings was high, and an atmosphere of mutual friendliness prevailed. There was none of the canvassing for favors of professors and distributors of chairs and assistantships, there was none of the unpleasant feeling of competition and envy and, of course, there was none of the cramming of politics into science, which was so typical of Fascist regimes. Attempts of this type were harmless and, with a little sense of humor, they were easily brushed off.

I remember that we once decided to have a simultaneous meeting of microbiologists and pediatricians in Łódź. Leading papers were to be presented in the overlapping areas of these two disciplines. Professor Michałowicz was chairman of the Pediatric Society, and I was chairman of the Society of Microbiologists. At the banquet, he asked me: "Don't you think we ought to drink a toast to our Commander-in-Chief Rydz-Śmigly?"[1] I replied: "I could correlate 'rydz' as a species with bacteriology, but I don't know what to do with Rydz-Śmigly. [In Polish, *rydz* means a kind of mushroom, and the expression 'healthy as a *rydz*' means someone who is in the best of health. *Śmigło* means propeller. Hirszfeld was playing on the different meanings of these words.] Perhaps this toast should be suggested to Szulc?[2] Since he is a colonel, it is easier for him to make a reference to our commander in chief, and furthermore, he is used to drinking toasts to a command." Szulc agreed, and we drank to the health of Rydz. I felt that under the influence of these statehood-promoting toasts, the atmosphere became somewhat chilly. I felt those gathered round the table needed a little warming up so, having asked for the permission to speak, I delivered the following oration:

> In his greeting address, the Voivod has asked why we love Lodzianers [inhabitants of Łódź] and to what the general sentiment for this factory city should be ascribed. The Voivod suggested that we love Lodzianers

because of the work they do, because of their diligence, and because of the poor sewage system in this city. I can assure the Voivod that neither the poor sewage system nor the proverbial diligence of the Lodzianers has anything to do with our sentiment. We love you because Lodzianers have respect for those who are in love. As a man of science, I must support my thesis with convincing evidence. Well, twenty-five years ago, as a high school student, I was in love with a girl who lived on Piotrowska Street.[3] Right opposite her house was Mr. Słonimski's little bookstore. I hung around the shop day in and day out. In order to be able to do so, I regularly sold my high school books which, incidentally, probably saved my intelligence from being completely hebetated in the tsarist *gimnazium*.[4] When twenty-five years later I again visited Łódź, I found the same little bookstore in the same old place and, deeply moved, I walked in and asked for the owner. An elderly woman was there who told me that her husband had died and asked me whether I had known him. "Indeed, madam, twenty-five years ago I constantly hung around this store, because I was in love with a girl who lived right opposite." To this she exclaimed: "Oh, so it's you?" The memory of a student in love had lived on in her for twenty-five years. To me, this is characteristic of the Lodzianers' mentality, and that is why I can say: we love Lodzianers because they have respect for those who are in love.

My speech melted the ice. Everybody asked my wife who the girl was. But she, alas, claimed not to know her. She did not want to admit that it was she.

Let us turn to the meetings held abroad. In the first place, I should like to describe meetings organized by the Hygiene Committee of the League of Nations. As I have mentioned before, Dr. Rajchman was the medical director of the League; Professor Madsen from Copenhagen, a well-known scientist, was chairman of our commission.[5] The commission's tasks were not purely scientific but methodological with certain ideological-political implications. Our aim was to ensure cooperation in the area of public health among scientists of discordant countries. The committee has a record to be proud of in the history of science and culture; however, it is not the purpose of this book to go into detail in that area. Our commission worked on standardization of serums, serodiagnostic methods, and so on. Even though these meetings were quite useful, in view of the limited topics and the small number of experts invited, they did not offer as much as special scientific meetings did. Professor Madsen combined great science with great diplomacy; he performed his presiding and representative duties with charm. He had a happy and radiant nature that, I believe, could not deal with people in misfortune. When misfortune befell me, a helping hand was extended to me primarily by the friends I had acquired in my uth and by those with whom I indulged in meditations in privacy.[6] One ling struck me painfully: in spite of the tone of reconciliation and cooperation, the representatives of the various countries considered themselves to be the diplomatic representatives of national science and often attempted to make a showdown of national interests on the scientific forum. The introduction of political factors into matters concerning standardization of serums was often amusing. For example, the problem of assaying dysentery serums. This assaying was done on animals. Doerr[7] and I preferred to use rabbits, while Kolle preferred mice. Actually, the difference was not essential, and since the objective was to establish a standard method, we agreed to use mice. As I was told later, a radiant Kolle returned to Frankfurt and triumphantly announced to his institute: "Victory along the whole line: mice will be used."

The first meeting of our commission took place in London in 1921,[8] the second in Paris in 1922. Germany was represented by Kolle, Neufeld,[9] Wassermann, Sachs, and others. In spite of the formal concord, a strict differentiation of the scientists representing the various belligerent countries was still made. For example, there were two banquets: one for all attending the meeting and the other for allies only. The German delegation placed a wreath at the Tomb of the Unknown Soldier but even this political move did not pass by without a ridiculous incident. The daily *Matin* sent its correspondent to interview Wassermann.[10] "Professor Wassermann, we are asking you for an interview as the most outstanding member of the German commission." Wassermann did not fail to boast about this in front of his countrymen. Kolle was unable to tolerate this slighting, got up, and announced with a flare: "If someone must be the most outstanding man, then it is I." Through indiscretion, this little story reached us and caused understandable amusement. By and large though, the tendency to forget past wrongs prevailed. The German commission was represented best by Sachs due to his modest and most dignified behavior. I formed a deep friendship with him. By way of indirect routes, I received news about the fury and bitterness of German scientists. One prominent Swiss scientist, a Viennese by origin, was successively a candidate for a professorship in four German universities. In Vienna, governed by the Socialists at that time, he was rejected because he was a Semitophile; in Munich, because he was Protestant; and in Bonn, because he had cooperated with the League of Nations (in assaying serums!). When the fourth university made him an offer, this proud and adamant scientist replied that he would not accept because he did not consider it an honor to be a professor in Germany. All this was happening under the Socialist government, but academic autonomy protected the conservatism and the hostile attitude of German scientists toward international democratic undertakings. When contact with the League of Nations entered the scope of German political interests and Germany became a member of the

League of Nations, we were invited to attend the meeting in Frankfurt-a. Main. We were given a sumptuous reception; there were more than tw hundred people at the banquet, and the speech given by Germany's repre- sentative was full of reverence for the work done by the commission. Later, Germany organized scientific meetings under the motto of a community of Nordic nations. Still later, the so-called interests of the white race were for- gotten, and the German-Japanese spiritual community was emphasized. When it became desirable to discredit Poles, the German Society of Legal Medicine summoned, in 1939, a meeting under the official slogan "The Pole as a Criminal," at which supposedly authentic photographs of Polish cruelty were shown. How such motion pictures are made is generally known. At the same time, the gentlemen professors did not hesitate to supply the most beautiful scientific motto: a community of spiritual efforts for the welfare of mankind—each time, as the needs of the moment required.

I will deviate somewhat from the congresses of the League and will recount my meeting with Sachs in Heidelberg in 1938. As mentioned before, I did not accept the offer from Zurich University, but I wrote a voluminous treatise on the reform in the teaching of hygiene. On my way back from Geneva from a commission meeting, I stopped in Zurich to visit my former chief, Professor Silberschmidt. The dean of the Medical School called on me to discuss the filling of the open position in the Department of Hygiene, and I recommended Sachs. Since I was going to continue my trip through Heidelberg, the dean asked me to visit Sachs and invite him to give a lecture in Zurich. Upon arriving in Heidelberg, I went to the Cancer Research Institute which, at that time, was headed by Sachs. He also was professor of serology and, in view of his great scientific achievements, he had not yet been dismissed even though he was Jewish.

When I arrived in the institute, however, I found everybody in tears: on that very day, Sachs had been dismissed from all posts he had held.[11] He was informed about this in a brutal manner, and the dean did not even consider it his duty to express his regrets. Sachs lived in the same apartment building as the professor of pathological anatomy. They had known and visited each other for years. When Hitler came to power, the pathologist donned a brown shirt and told Sachs: "My dear Colleague, I think the same as you do of all this rubbish, but you must understand that on the street, we must act as if we don't know each other." The assistant who replaced Sachs was unable to cope with his duties. Sachs, a man without malevolence, wanted to help him. However, the assistant talked to him only by telephone: "Professor Sachs, I am sure you understand that in a period like this, I cannot meet with you personally." Jews were not permitted to write joint papers, and Aryans were not permitted to cite Jewish authors, unless it was inevitable and then only with an annotation that the author was Jewish. This state of affairs was probably represented best by the Sachs's German housemaid. She robbed

Sachses and subsequently several other Jewish families. Sachs was present the court proceedings as a witness. The judge asked her why she indulged n stealing. Her reply was: "But they were Jews. I thought that robbing Jews was allowed." At least she was sincere.

The German scientists, on the other hand, rationalized the same impulses on the basis of world outlooks. The things that happened were so grotesque that it is beyond me as to how the German scientists did not burn from shame. An organizer of international courses in Switzerland told me the following story. Julius Bauer, a well-known Viennese clinician, once wrote an article to the effect that science was objective and international.[12] Wagner, the official leader of German physicians, called this generally accepted thesis a Jewish impertinence and forbade the German physicians to attend the course in Switzerland since Bauer was one of the lecturers.

In the early 1930s, before Hitler gained power, I was visited in Warsaw by the well-known German clinician von Bergmann, who was summoned by the Turkish ambassador.[13] He complained that professors had no control over young people, who were completely Hitlerized. I do not know whether he ever expressed his opinion in public. In his book, he wrote rather with esteem about the government's interest in problems concerning [serological] constitution, even though he must have known what motive guided Hitler's government when it emphasized problems of race and constitution.[14] The only objective was to collect evidence for the superiority of the German race. The tragic consequences of this thesis should have been foreseen by the German elite. It was during the Hitlerite period that Professor Kafka from Hamburg visited me in Warsaw. This researcher, a Jew by origin, was married to a German woman. Nevertheless, he was dismissed from all his posts and put into a concentration camp. He said to me, heartbroken: "The worst was when I finally understood that I had spent my whole life as a stranger among aliens."

I must mention the Society of Blood Group Investigation, established in Vienna prior to the Hitlerite era. An anthropologist from Leipzig was the chairman, while Steffan was the secretary general.[15] They worked out a rather unrealistic project for the study of blood groups, founded a journal, and elected an improbably large number of honorary members. Pointedly, they did not elect the most prominent Jewish scientists: Bernstein, Sachs, and Schiff. Initially, Landsteiner and I were also bypassed. The *Klinische Wochenschrift*, Germany's most serious medical weekly, protested against the omission of Landsteiner and me. Probably under the pressure of public opinion, they offered us honorary membership. I rejected the offer, however, explaining that I had no confidence in a society that ignored scientists because of creed or race. Later, this society published a tendentious textbook on blood groups and, in its journal, published articles which displayed a culpable lack of objectivity. I was surprised that Danish researchers published their papers in such a journal. It would seem that scientists shou. boycott such journals.[16]

I will not discuss such matters as the burning of books, the firing of Jews from editorial committees, and so on. These matters pale in comparison with subsequent cruelties; besides, they are generally known. I might mention, though, the stand taken by scientists of goodwill. In 1933, the famous surgeon Sauerbruch[17] issued an open letter to all scientists throughout the world, in which he stated that the German Government had peaceful intentions, and therefore he demanded that all scientists support the aspirations of the Hitlerite Government. I knew Sauerbruch from Zurich; he was a great, frank, and courageous man. His letter must have been dictated by his good faith. I personally received Sauerbruch's appeal, and I wanted to issue a public reply on behalf of Polish scientists, characterizing the racist policy as incompatible with the conscience of a physician and of a scientist. My letter was to be signed by a number of Polish scientific societies. Just at that time, the German-Polish cultural agreement was concluded, and it became impossible to publish our reply in the press.[18] Therefore, I sent it in my own name to Professor Sauerbruch.

Such was the atmosphere prevailing in Germany. Scientists who were against the policy of violence kept quiet, or emigrated, or were put into concentration camps. However, they were few, while most displayed weakness of character. All this buffoonery was soon to be transmuted into an extraordinary crime: the murder of millions. The government motivated the killing with the proposition that Jews were the carriers of pathogenic germs. The representatives of the medical world listened to the government with submission and reassured it of their loyalty. At one time, science had succeeded in gaining independence from the church, and the period when champions of the theory of the earth's rotation around the sun were burned at the stake belonged to the days of yore. We are still proud when we think of those visionaries of truth who went to the stake to support a science that was independent and unyielding to the outlook of men who ruled the world at that time. Modern science will not be able to say that it assumed a worthy stand during this despicable period.

Chapter Eleven

International Congress of Anthropologists in Amsterdam; Opening of Schools of Hygiene in Budapest and Zagreb

An international congress of anthropologists was convened in Amsterdam in 1928, and the organizing committee asked me to give the keynote lecture on blood groups.[1] The other keynote was to be presented by Verzár, a well-known Hungarian physiologist.[2] I willingly accepted the offer: it was the first time that the issue of blood groups was to enter the forum of an international congress of anthropologists. Blood group studies were being conducted throughout the world; in Holland, the investigations were headed by Professor van Herwerden,[3] with whom I corresponded. The fact that blood groups had been chosen as a leading subject at an international congress and the keynote paper entrusted to me represented a victory of our research pursuits. Not that long ago, after all, the *British Medical Journal* had refused to publish our article. Immediately prior to the congress—to my great dismay—I received the program and saw that my paper had been rescheduled as a regular communication rather than a keynote speech. I am not vain, but I was hurt and inquired of the committee whether this was not a mistake. I received an answer that irritated me thoroughly, since it was an evidence of politicking and fighting for national prestige in the area of science. The point was that the French had demanded to be given the leading paper. Since I had been already invited, the organizing committee had decided to stick their heads in the sand like an ostrich and have no keynote paper at all. I informed the committee that I regarded this as a withdrawal of the invitation and that I would not attend the congress. I received a telegram summoning me to come and stating that the programs would be printed anew to announce that mine would be the leading paper. In view of this arrangement, I could not possibly refuse, and I must admit that our Dutch colleagues spared no efforts to show me their friendly feelings. A special Blood Groups Section was established, and I was its chairman throughout the congress.[4] At the banquet, only the chairman of the congress and I sat at t presiding table next to the Prince Consort. I should like to emphasize th; the atmosphere prevailing among scientists belonging to small nations, such as Holland, Denmark, Sweden, Norway, or Switzerland, was completely different from the tense atmosphere dominating in Germany.

Evidently, theirs was a purer atmosphere. They did not twaddle about national missions, and they did not dissimulate conquest-bent desires with the cliché that they wanted to make mankind happy. Their simple cordiality, profound humanitarianism, and culture were impressive. During the congress, I was able to tell how much my subject matter interested every one. Some of the talks dealt also with the crossing of races. Most of the anthropologists were opposed to the idea of crossing the white race with colored ones; crossbreeds were supposed to be of lesser value. Yet, the Dutch do not oppose mixed marriages in their colonies; in the Dutch East Indies, crossbreeds fulfill important social functions. I must admit that I remain somewhat perplexed with regard to this question. If the tradition of a specific group or class declares itself against mixed marriage, then that tradition risks breaking either independent spirits or else people who feel no limits at all, [such as] alcoholics. Since those of the latter category are more numerous than those of the first, it is hardly surprising that a superficial survey of crossbreeds will appear to indicate that their children are less worthy. One also hears that a crossbreed is from the onset beset with a feeling of inferiority that hampers him throughout his life. This again reveals a poor knowledge of psychology since on the contrary overcoming a feeling of inferiority is often conditioned precisely by a much greater effort. And thus I believe that it is possible that a certain patriotism of the white race subconsciously influenced scientists' objectivity on the subject.[5]

During the Amsterdam conference, I grew friendly with several scientists and especially with Verzár. He later was named professor of physiology in Basel and carried out a series of beautiful experiments on the adrenal gland and vitamins. I visited him on numerous occasions while traveling through Basel, and he also came to Geneva. He enjoyed excellent working conditions and yet he was homesick. He was one of those who stretched out a helping hand to me when I was in trouble.[6]

After the Amsterdam congress, I traveled along the Rhine through Nüremberg to Budapest for the opening of the School of Hygiene. A great gala was held, the school having been founded by the Rockefeller Foundation. The League of Nations Health Committee sponsored such schools and organized a large expedition in which participated—aside from the United States—representatives of all nations belonging to the League.[7] We were received with great pomp and introduced to Prince Horthy[8] and his wife. Next we went with a group of invited scientists to the opening of the School of Hygiene in Zagreb; then later, invited by the Yugoslav Government, ᴇ traveled for a couple of weeks admiring the wonderful effort the nation ᴀad put into rebuilding the country. There are two explanations for such an energetic reconstruction: the suffering endured during the epidemics of the war and the decision that they should never occur again, as well as the exceptional organizational talents of Dr. Štampar.[9]

We saw beautiful institutes, health centers, medical and dental offices, antimalaria institutes, sanitation work, traveling museums of hygiene, and, most important, a staff of devoted persons and a remarkable sense of togetherness between the intellectual elite and the simple people. They knew how to play these heartstrings. We were taken to a village where the standards of hygiene had just been raised. Not only had medical affairs been attended to but also improvements had been made in cultivation of crops and tobacco, cattle breeding, and so on. The Serbs understood that hygiene could not be improved only by means of medical and sanitary measures but that it required raising the standard of living in general. We were invited to a folk feast. Scientists from all over the world were at ease communicating with the peasants in a perfect manner, and finally all were united in the *kolo* [circle] folk dance. When I saw elderly scientists filled with emotion and dancing with peasants, I believed I was witnessing a new era of class fraternization. What a wonderful task might await scientists if instead of playing at being great politicians, they used the halo which justly or unjustly surrounds them to raise people's morale and their standard of living. For I was undoubtedly the only one in that group who had known the Serbs before and was able to appreciate their efforts. That effort was unique in the world. In other countries, such as England or France, the level of hygiene rose almost of itself, simply because the standard of living was higher. Here, hygiene was the motto. Personally, I experienced an immense satisfaction, because I met my former pupils who had become the most active promoters of this new trend and I noticed that a legend had arisen around life in Serbia during the war. At that moment, I felt more strongly than at any other time how pleasant it was to gather the harvest after good sowing. As mentioned before, my former pupil Savić was minister of health,[10] and he tried to show me his appreciation in every aspect. He very much wanted to give me one more decoration. I explained that friendship was dearer to me than decorations, but finally it was impossible to refuse. So, I asked for a decoration that could be gained for military service only. The next day, he solemnly decorated me with the Order of the White Eagle.

Serbia has a beautiful past, and the future of Yugoslavia will—I hope— also be beautiful. The soil on which one treads is robust; the sap of strong plants flows also in the people. The main characteristics of the country stem from its youth. The parents of great leaders and poets often could not read. By contrast, the subtlety of the Polish soul, a certain feeling for the aristocratic, and the age-old identification of the Polish nation with noblemen

have created a certain social grace and charm. But this also has its dra
backs. In the Serbian intellectual, one can still feel roots grounded in th
soil; the positive effects of this dominate over the negative.

There is one more thing that is typical of young nations: respect for science. For them science is the symbol of higher life, a stamp of nobility. That
is why I am convinced that a veritable temple of science will arise on the soil
of these young, strong, and simple nations. That this requires time is inevitable. He who has lived with young nations does not believe in the racial
supremacy of nations, temporarily claiming to be on a higher level of culture. Fluctuations are eternal and the peak of a wave may precede its
decline.

Chapter Twelve

The 1935 Blood Transfusion Congress in Rome

This was the penultimate act of a historical drama, the year of the war with Ethiopia,[1] and that year the Italians organized the First International Blood Transfusion Congress.[2] It was to be the culmination point of the great efforts they had made in that area. They had established a [blood donor] society on a secular and charitable basis. Seven thousand registered donors were giving blood repeatedly. The donors received no payment; at best, the society issued certain certificates to them. Poland sent an official delegation to the congress composed of three people: Colonel T. Sokołowski as the delegate from the army, Dr. J. Kołodziejski as the delegate from the City of Warsaw,[3] and myself as the delegate of the government and the Medical Department. The congress presidium asked me to open the meetings in the name of foreign scientists.

We experienced unforgettable moments. The congress was opened in the Capitol in magnificent halls where the echoes of the speeches delivered by senators still seemed to resound. The statues had preserved the atmosphere of ancient Rome, and the windows gave out onto the Roman Forum and the Eternal City. The presidium was composed of Professor Alexander, a surgeon and representative of the government; Professor Morelli, representative of the Medical Chamber and the secretary of the Fascist Party;[4] Professor Lattes, chairman and author of an outstanding book about blood groups that appeared in Italian, German, French, and English;[5] and finally myself. All Italians were in high boots and black shirts, which did not make a very festive impression on those who were seeing this outfit for the first time. They rather resembled Russian peasants. The speeches reflected pride in the work accomplished. Admiration for the Duce was still discreet. It was still devotion and respect and not a divine cult. After the government delegate and the chairman completed their speeches, I took the floor and delivered the following address (which was included in the album printed for the society's tenth anniversary):

Notre science exprime non seulement le progrès intellectuel, mais aussi les valeurs morales d'une nation. Donner son sang au prochain, c'est faire acte de compassion, c'est imaginer et souffrir la souffrance [d]'autrui.

Pour cette raison l'organisation des donneurs du sang, l'intérêt qu'une nation porte à ce problème, permet de juger non seulement de la culture de l'esprit, mais de ses forces morales. C'est dans notre domaine que fut créée cette organisation unique au monde des donneurs anonymes offrant leur sang au souffrant inconnu.[6]

Whether it was caused by the tone of respect for our hosts or by the appeal to an impartial selflessness in transfusing blood without causing pain to others—in any case, my speech was received with great enthusiasm. I later met a number of pleasant Italians, scientists and politicians, and everywhere I found a warm feeling of friendliness and esteem. At the final banquet, I sat next to the wife of the society's secretary general. She talked about how she was impressed by the congress and how deeply her heart was touched by the foreign countries' esteem, as expressed by me. I had seldom met a woman of such beauty and spiritual culture. Her face was a cameo; her velvet eyes were dark and radiant.

After the congress was opened, the hosts handed us wreaths. Placing wreaths on the Grave of the Unknown Soldier was in the program. We were willing to do this. It was to be a symbol of international concord. We were to pay our respects to a soldier who died for his homeland. There were no more enemies, only those who had once been blinded. We went to the Venetian Palace, with the foreign delegation heading the procession. The Germans were still referred to as enemies, and the Soviet delegation as a spokesman of an enemy ideology. The Yugoslavs were probably thinking about the recent conflicts, and we were thinking about Colonel Nullo. With a feeling of performing an act of respect and a symbolic ritual of unifying nations under the aegis of Science, we were led to the Grave of the Unknown Fascist. Before we realized it, the act of international concord changed into an act of reverence for the regime. There was a moment of silence. The hands of the Italians rose up in the Fascist greeting. Before we were able to control our reflex, which was a reflex of courtesy in response to a gesture which apparently was the custom in the country playing host to us, there was a brief snap. All foreign delegations were photographed in the fixed pose of the Fascist greeting at the Grave of the Unknown Fascist. Before we became aware of it, the congress was used to propagate the praise of Fascism by science. We were unable and, unfortunately, unwilling to defend ourselves. The idea had not yet matured that this was no international courtesy but a profound symbol and that we were witnessing the subordination of science to politics.

The congress was organized by a committee established by the Fascist Union of Physicians. Was this a confession of faith? Were there nonpartisan associations? We were told that all associations were encompassed by Fascism and that therefore this represented no *confession de foi*.

We visited hospitals and scientific institutes. First we saw the Forlanini Hospital, which developed the pneumothorax procedure and made a permanent mark for itself in the treatment of tuberculosis.[7] Even those who came from the richest countries had never seen anything so beautiful as this facility. There were huge halls all in marble and gold, and there were terraces with screens driven by power in the direction of the wind. The wards, beds, and gardens were equally impressive. All walls were covered with pictures resembling Christian legends, but in a Fascist paraphrase: the Duce surrounded by children, the Duce distributing bread, and the Duce pointing to distant horizons. On the wall in the lobby was the foundation table with the foundation date: the 12th year (if I remember correctly). The Italians explained to me that this was the date of the establishment of the hospital, the years being counted from the Fascists' march on Rome. Almost the whole world, in order to emphasize that a new era has begun, counts its history from the birth of Christ, an attempt to introduce love as the driving force in life. In the consciousness of the Italians, this moment was to be replaced by the march on Rome, which was supposed to represent the beginning of a new era. Even to irreligious persons, this was blasphemy.

I asked the Italians where they had obtained this tremendous amount of money [to build such lavish establishments]. They admitted that this had cost them millions and had depleted all funds of the newly introduced tuberculosis insurance. A friendly Italian professor went on to explain:

So far, we do not feel the consequences, because we do not yet have to pay indemnities. However, in a few years, we will feel the aftereffects when we see that the walls, though very beautiful, have been built with money intended for the treatment of tuberculosis. The Duce ordered us to build much and fast. The objective was to leave landmarks in our history. Regimes wanting to amount to something must express themselves through architecture. This is a longing to preserve great historical moments. The Egyptians built pyramids. Actually, we should be grateful to the Duce that he wants to immortalize his era by building edifices dedicated also to medicine and science.

I asked whether this would not ruin the country's economy.

This is the only way in which great things and edifices can be created. It was public injustice which gave rise to the magnificent architecture which is still proclaiming the grandeur of the times gone by.

Nevertheless, a scientist could only feel embarrassed. In the Dental Institute, we saw large offices with electrically driven chairs and superb lighting installations; however, the offices were almost empty. Involuntarily, we were thinking of how much public energy was being wasted to subordinate science to the leaders' efforts to guarantee their own political longevity. Cannot science provide its own impetus to create buildings whose beauty will consist of intellectual power? Must the temples of science drip with gold? In a moment of frankness, an Italian scientist told me: "You see, when I need large sums for special instruments, I am sure to get them. But believe me, I often have no money to buy a couple of rats."

The congress attendees were taken to a movie theater to see a propaganda film. What a marvelous and subtle motion picture it was. A laborer is working in a mine. His little daughter is carrying food to him. She is hurrying. She does not see the train. She falls. Blood is gushing. The scenery is changed. We see a colorful meadow that is growing pale and disappearing from the eyes of the dying child. The scenery is changed again. A telephone is ringing. A blood donor is notified. A car is driving at full speed. A transfusion is performed in the hospital. Again we see the meadow growing bright and colorful in the sick child's consciousness. The finale presents a flower-covered meadow with children playing on it—the eternal symbol of beauty and life.

Everybody was deeply impressed. With a film like that, one could gain thousands of donors. And this is how one should understand the annotation later entered in the memorial book by the French representative, Professor Tzanck:[8] "The day will come when the only form of shedding blood will be donating it to a sick person." Maybe it was a vision of a future not so far off.

We went to another theater to see a motion picture from Ethiopia. We saw the primitivism of countries located deep in Africa: overpowering work, captivity, poor huts, and strange customs. The Ethiopians were actually presented without nastiness, but as an immensely simple and underdeveloped people. Then, from a distant dawn, Roman columns began to emerge. Ranks of Fascist youth appeared, led by a huge shadow of the Duce. The Italians were the spokesmen of culture, which they were carrying into Africa. The crowds began to cheer.

This was the way in which they created a peculiar, heroic legend and a parody of a longing for the blue yonder.

We went to our embassy. All professors had been instructed to check their representative speeches with the ambassador. Thus, I also brought my prospective speech, but no one in the embassy knew what to do with it. All Italian newspapers carried announcements that the congress would be opened by the delegate from Poland. Only Poland did not know anything about this. The embassy had not informed our Ministry of Foreign Affairs. So much the better. Why should an affair of national prestige be made out of the efforts or personal happiness of a researcher? At the same time, there as a meeting of boy scouts or gymnasts in Rome. Tea was served jointly for all at the embassy. Ambassador Wysocki was very tired, and he did not conceal this from us. There was a great difference between the tense will of the young Fascists and him. The figure of a tired aristocrat must have appeared alien to those young people.[9]

The congress proceeded smoothly. We were impressed by the voluminous material presented by Russian researchers who had expanded blood transfusion in a manner unknown in the West and who in many respects had greater experience than we had.[10]

This feeling of belonging to an era must either be inborn for the Italians or they are taught it. Cicerone[11] showed us the Palace of the Invalids of the Great War. He pointed out the differences in architecture and the composition of details in the nearby Palace of Justice. He continuously emphasized: this is our era and our style, and that one over there is outdated. The English and French visitors looked at him in surprise: Can one not have more veneration for a recent tradition? Must artistic creativity ensue from imperialist motives?

Professor Lévy-Solal[12] from Paris and his wife invited me and Professor Lattes to spend the evening together. It was a charming evening. We knew each other from our articles, and this was the first occasion to establish a friendship. Later, I met Lattes's wife and daughter. We agreed that he would come to Warsaw. I was really touched when I saw my photograph on Lattes's desk in his laboratory.

The congress ended on a Saturday. The banquet was at six o'clock in an old Roman tavern, and at seven we were to be received by the Duce. We entered the Venetian Palace without being particularly inspected. It did not seem difficult for an opponent of the regime to steal in.

We walked into a large, empty hall and stood at the walls, forming a circle. The presidium asked France's Professor Tzanck and me to come up to the center. Only the two of us were to be personally introduced to the Duce. After several minutes of waiting, the door to the drawing room opened, and the Duce, dressed in a white suit and taking heavy steps, walked up to the center of the hall. We were introduced to him, shaking hands briefly. The Duce asked in a low voice: "Chi parla?" Professor Lattes, chairman of the congress, came up and delivered, in Italian, a speech that was sheer adoration of the leader: that the organization of blood donors was only one aspect of the great desire to shed blood for the homeland under the direction of the leader, that he would like to thank the queen, the organization's patroness, and the Duce for their care, and so on.

While Professor Lattes was talking, the Duce stood motionless with his heavy gaze fixed on the talker. He gave the impression of a large, wily cat purring with pleasure. Evidently, this was as indispensable as was the burning of incense in honor of the ancient Caesars.

Finally, the Duce spoke up. He spoke in French. The essence of
speech was; "Gentlemen, when you return home, tell your countrymen th
the Italians are suffocating." When he was uttering the words "are suffocat-
ing," one could feel a great power that was encased in his chest and trying
to free itself. This was Saturday. The next Tuesday, war was declared on
Ethiopia.

On Monday, I was in Geneva, attending a meeting of the Hygiene
Committee of the League of Nations. When others heard that I had been in
Rome where I had seen and heard Mussolini, they asked me whether he was
still able to talk. They had heard that he had a progressing paralysis; the
Wassermann reaction was supposedly made on him in Lausanne.[13]

Several years later, Professor Lattes was forced to emigrate from Italy. It
turned out that he was a Jew.[14]

Chapter Thirteen

The 1937 International Congress in Paris

Two years later, in connection with the international exposition, a conference was convened in Paris.[1] The well-known surgeon Grasset was elected chairman, while the hematologist Tzanck was elected secretary general.[2] The program specified four sections and plenary sessions. Theoretical problems were to be discussed in sections, and clinical problems in plenary sessions. The distribution was somewhat artificial, probably because both the chairman and the secretary general were clinicians. The first section was headed by Professor Coca from America, creator of the first blood donor organization; the second section by Lattes from Italy; and I was chairman of the blood groups section, at which I presented a long inauguration paper.[3] At first the sections met, and then the congress was officially opened and plenary sessions were held. The Polish delegation was composed of Colonel Sokołowski, representing the army; myself, representing the Medical Department; and Docent Gnoiński.

We were traveling through Germany. In Berlin, a tall gentleman boarded our car and spoke in a sharp voice to a student with Semitic looks. He immediately asked me about the Jewish problem in Poland. In brief, he struck the harsh, merciless tenor of the Hitlerites. We passed the Belgian frontier. My German fellow traveler began to change and soften. We started speaking about German culture. I expressed my opinions point-blank and cited a poem by Liebknecht, who had been murdered by the counterrevolution. Liebknecht wrote this verse as a sixteen-year-old boy. In 1914, he was the only deputy protesting against the war. The verse went as follows:

Ich kann nicht wägen, kann nur wagen,
Nicht ernten, sähen nur und fliehen,
Ich kann den Mittag nicht vertragen,
Nur Sonnenaufgang—Abendglühen . . .
So sei mein Tag.

Once, when I was in Zakopane,[4] I translated this beautiful verse as follows:

Nie umiem się wahać, raczej w ogniu się palić,
Ani płonem sie cieszyć, raczej posiać i zbieć.
Żar południa nie dla mnie, raczej zorzy świtanie . . .

[I cannot hesitate, I'd rather burn in fire,
I don't enjoy the harvest, I'd rather sow and flee.
The torridness of noon is not for me,
The dawn is what I do admire . . .][5]

When I was reciting this verse, a spring seemed to come loose in him. He told me his story. Prior to Hitler, he had been a socialist and was married to a German aristocrat. He had two daughters. One had married a Jew, an engineer with an excellent position. When Hitler had come to power, his daughter and her husband had been forced to emigrate, and he had had to join the party, lest he be turned out of doors. His second daughter had become ill with schizophrenia and had been sterilized. Thus, two infamies had befallen the family. The mother had developed a suicidal mania and had to be nursed constantly. "You see," he said, "I am going to Paris to see my daughter—secretly, because as a state employee I am not permitted to talk with Jews. My only peaceful moments are in France at my daughter's. Germany is a veritable hell. It is a cursed land."

There was one more German in the compartment. He was holding an illustrated newspaper in which I noticed a picture of an athletic young girl and a monstrously crooked old Jewess. The headline read: "This is what the German race and Jewish race look like." A physician could easily make the diagnosis: the Jewess suffered from a softening of the bones—osteomalacia. "Sir," I said, "even a medical student could diagnose this. Cannot the Germans contrive smarter arguments?" My fellow traveler was a manufacturer; he did not like the anti-Semitic decrees, but he praised Hitler because he made order, because there were fewer strikes, because unemployment had decreased, because workers had become more efficient, and so on. I could not take it any more and transferred to another compartment. The Russian delegation was there, including Professor Bogomoletz, chairman of the Medical Academy in Kiev. He said that once he had been a rich man, but that he did not crave wealth because he had a great institute, a large apartment, and many assistants.[6]

The congress was opened in Paris; our French colleagues were nice and friendly, as always. The famous hematologist Weil greeted me with the words: "Mon vénéré maître" [My revered master].[7] I felt like a candidate for a deceased man. I had always thought that this pompous term was used only with regard to very elderly persons. The section conferences were by and large interesting and rewarding; my numerous tables prepared by Polish draftsmen were liked by all. Finally, the plenary session began. In his long ᴑening address, the chairman, Professor Grasset, praised the merits of only ˙rench researchers; not only of the deceased but also of those present at the congress. He did not say a word about foreign guests even though they included creators of blood donor organizations from America, England, Holland, Italy, and Spain. After the opening, the chairman came up to me and apologized for not having mentioned my name. He had had it written on a slip of paper but had forgotten it. I told him that it did not matter, but I was wondering why a scientist should emphasize only the merits of his own nation; simple politeness called for praising the guests a little bit too. When will scientists consider themselves priests in a single, international temple?

The congress proceeded smoothly. The Russian delegate presented a report on transfusion of blood collected from cadavers. At the banquet, we heard a touching talk given by a Spaniard, a famous hematologist who had been forced to emigrate from his country. In beautiful French, he spoke about the nostalgia of exiles. This was the first indication of a tragedy approaching scientists.[8] Later, I met many of them, and eventually I personally went through the tragedy of losing my laboratory. I was relatively lucky, though: I was mistreated not by my countrymen but by the enemy.

At the first plenum, the Italians spoke in Italian, and therefore they were understood by almost no one. For this reason, most Italian scientists tried to present their papers in the sections where they could speak in French. Why should people be forced, for reasons of prestige, to speak a language unintelligible to the majority, and why should the meaning of international congresses be warped? Again, we faced the collision of nationalism with science.

I will not describe the exhibition. However, in view of the critique of our pavilion, I might mention that the French liked it very much. The only thing that struck me unpleasantly was the fact that the exhibition of the Alcohol Monopoly was right up front. This should have been played down, because vodka was not the exponent of our culture.

I visited the big Masson Publishing Company and suggested they publish my book entitled *Blood Groups*, which had been published in Poland several years before, and my wife's work entitled "The Role of Constitution in Children's Infectious Disease," which had been published in *Medycyna Doświadczalna* and in the monographs of Professor Michałowicz.[9] The publisher asked me to give him a day to think it over. He then informed me that he was willing to publish these books; however, I would only be paid a royalty: 1,500 French francs. This was about 200 złotys; less than the payment due the typist for retyping the book. Indeed, one could not make a fortune on scientific books in France. I was deeply impressed by the conversation I had with the director and owner of the company. He reminded me of Anatole France: he had a similar intellect—flexible and universal. He spoke about the crisis of French books, ensuing from the shrinking South

American market. He made subtle comments on the role of the school. F
was a true artist. "Sir," he said, "this popularization of knowledge has it
negative aspect; it satisfies the creative unrest but is rather harmful to great
minds. Have you ever seen a school produce a genius of a painter? Great
talents are always inborn. The truth is that we are clipping the wings of our
young people by constantly teaching and drilling them." I mentioned the
primitive condition of French scientific institutes and said that French science was definitely lagging behind.[10] As an explanation he pointed out the
mentality of Frenchmen, whose gift of improvisation awakens in times of
need. Again, he said: "Civilization overburdens the creator with knowledge
and technology. We want to force the researcher to improvise. Perhaps this
will not deepen knowledge as in Germany, but it will force us to have ideas.
This is more valuable."

Nevertheless, there was a crisis in French scientific literature. I remember
that in 1928 in Germany, I published a book on constitutional serology. The
reviews were thorough and gave me great satisfaction. For years, I had felt as
if I had conducted a conversation with my German readers, and I was able
to evaluate the effects of the ideas they suggested. I gathered all these
reviews in an album, and when my heart feels heavy I open the album,
glance through the cutouts, and think: *non omnis moriar*.[11] This preservation
of an idea in print is marvelous. My French book was published in 1938
while I was in Paris. In 1939, I received the first reviews. I was struck by the
similar expressions used by almost all critics. "Why do these words sound
familiar to me? It's my own introduction." Apart from this, there were several generalizations. I felt deeply hurt and complained to Professor
Chevallier,[12] but he laughed at me. "You talk like a young poet. At your age!
Do you really think that the critic would read more in your book than the
introduction and the conclusion? Anyhow, your book will live if it reaches
the reader." I mentioned my royalty and asked whether Masson might have
apprehended a financial loss on my book. Professor Chevallier laughed
again: "He must expect very much from your book if he paid you anything
at all. Generally, we have to pay a certain amount for the publication of our
scientific books."

Later, it became obvious to me that Professor Chevallier was right. The
French public feels a sense of contact with the author. I received many letters from my French readers, with requests and often with very subtle comments. Evidently, the review of scientific books is not taken very seriously in
France. The elusive inclination of readers is much more important.

In the context of this experience, I recall another conversation I had in
Paris. I was not sure whether Masson would publish my wife's book. At that
time, she did not have the title of a professor, and the subject she discussed
was less popular than blood groups. I decided that it was necessary to gain
some patronage. I went to see Professor Fabre,[13] who had come to Warsaw

nce and presented a beautiful lecture. But what was I to tell him? That the ɔook was interesting, that it contained the results of our joint research, and that it compiled the results in application to medicine, while my book discussed mainly biological problems? He would reply [to himself]: Oh, well, he is praising his wife; he is probably dominated by her; it is said that priests and scientists make the best husbands. I had to approach him differently.

"Dear Colleague," I said, "I would like to tell you something: I love my wife." He looked at me as if to check whether I had not lost my mind. Being a courteous man, he replied: "Well, I understand; I too love my wife. But why are you telling me this?" "You see, I met my future wife when we were still children. We have gone through life together. Whenever I get a good idea, I have the impression that I have read it in her eyes. She is to me what Daimonion was to Socrates. Now if I may say so, I am conquering new continents. This is so because while with a scientific communication one can gain the reader's attention, with a book one can gain his heart. Am I to proceed alone, without my companion? Believe me, you will not regret it, because the book is worthy of recommendation."

He shook my hand and said, with great feeling: "I understand you. I too love my wife."

He was a true Frenchman. Foreigners think that to Frenchmen love is an adventure, a pleasure, a relaxation. If you really get to know them, you see the heartfelt profoundness of their family life.

Masson published both books even before receiving the recommendation: Professor Fabre's letter and the publisher's affirmative reply crossed each other. I asked Professor Debré to write the introduction to my wife's book.[14] He wrote it with that delightful French exaggeration but, at the same time, he described with great elegance the new directions medicine was taking and emphasized with feeling the ties between French and Polish science.

Chapter Fourteen

The 1937 International Cancer Congress in Brussels

In 1937, an international congress devoted to cancer research took place in Brussels.[1] Sometime during the 1930s, Dr. Laskowski, my assistant Wanda Halber, and I had established that new growths had a specific serology.[2] We worked hard on developing a serological diagnosis of cancer. In spite of promising observations, the method finally proved useless in the wake of the information we had on that disease at that time. Nonetheless, these studies again aroused various scientists' interest in the biological specificity of new growth tissue, which was palpable by means of serological methods. I was invited to present the main paper. One of the accompanying papers was presented by Professor Rondoni from Milan.[3] The second one was to be presented by a Spanish scientist; however, because of the Civil War, he did not show up.

This time I did not go to the congress as a delegate of our government. Unfortunately, the Polish Government could not be officially represented, because our ambassador in Brussels had taken offense with the congress and had withdrawn his name from the list of the honorary committee. Thus, the honorary committee of the congress included representatives from all countries except Poland. The story behind it, as I was told in the Ministry of Foreign Affairs, was as follows. In Madrid, where the last cancer congress had taken place, it had been decided that at the next congress the languages employed would be English, French, German, Spanish, Italian, and Russian as the Slavic language, since it was spoken by the largest Slavic nation. It was also decided, as I was told by the chairman of the Cancer Society, Dr. Weiner, that the Polish language would be used at the following congress. Our ambassador did not know this and took offense in the name of Poland. But it was impossible for our Ministry of Foreign Affairs to disavow our ambassador *post factum*. As it turned out, our truly charming Ambassador Sobański gave our little delegation a most friendly reception. The delegation included Professor Pelczar; Dr. Lukaszczyk, director of the Warsaw Cancer Institute; Dr. Floksztrumf from the Cancer Committee; and myself. Apart from the official delegation, there were a few other scientists, including Dr. Flaks, who had had interesting research results. For the sake of comparison, I might mention that the German delegation included

.bout thirty-five people. This limited interest on the part of our researchers will eventually result in a qualitative deterioration of our representations abroad. For reasons of prestige, we often apply for membership in presidiums. This is a totally platonic [not merited or real] honor. The only effective propaganda is honest work.

The congress was opened in the presence of the king. It would have been difficult to imagine a more beautiful face. His was the face of young god, sculptured in stone. The greeting address was delivered by the minister of health, the famous Socialist Vandervelde.[4] There followed other speeches and meetings of individual delegations. All this proceeded in six languages. Are scientists not aware of how ridiculous they are making themselves with this incongruous patriotism? The monotony ensuing from six repetitions of each statement was deadly.

As an aside, I might add that if scientists who do not attend international congresses knew how unnecessary such a launching of their languages is, they would not put so much effort into making conquests of this type on the international arena. I remember an incident in Warsaw. When we were discussing plans for an international organization of blood transfusion congresses, one member of the group demanded citizenship rights for the Polish language. "My national pride," he said, "demands that I express myself in my native tongue. Whoever is interested in my ideas should learn the Polish language." "Professor X," I said, "no one will learn Polish because of us; on the contrary, the harvest of Polish ideas will perish. In this overproduction of scientific communications, most scientific results are lost anyhow. There are a few journals which monopolize the results in a given scientific field; he who does not gain access to those journals has little hope that he will ever be noticed." It is too bad that scientists do not use Latin. If it is impossible to have just one language, let us take advantage of those few which have become international. Let us not fight over trifles.[5]

In the evening, a party was given for the main speakers in the Vanderveldes's private apartment. The hostess was a physician, and a nasty rumor had it that she was actually the minister of health. That would not have been bad, because she made the impression of a most unusual woman. We talked about the effects of intellectual work . . . on the beauty of women. She argued with intelligence that spiritual work preserves youth better than cosmetics.

Both in Paris and Brussels, I was struck by the large number of Russians, mostly emigrants cast abroad by a historical storm. They included professors but were mainly assistants, laboratory technicians, and individuals on grants. I was told they were excellent workers. One Russian showed me a cellular operation he had performed under a microscope of his own design. I was really impressed. The Russian colony lived its own life which, in the manner of students' colonies, was only loosely associated with the local society; but the Russians were completely devoted to scientific work. This reminded n. of my assistantship in Heidelberg and Zurich. Even at that time, it was diffi cult to find autochthonous assistants for theoretical institutes, which were therefore staffed exclusively by foreigners. Poland, after its partition, sup- plied foreign countries with outstanding scientists, and France most hospi- tably took in the emigrants, including for instance, Danysz, Hirszfeld (the anatomist), Babiński, and Pożerski, not to mention the great Skłodowska.[6] Currently, Minkowski, Mutermilch, and among the younger men Chwatt and many others are still working.[7] The Pasteur Institute employed many foreigners: the great Metchnikoff, Besredka, Weinberg, and Levaditi.[8] The French did not suffer any losses due to their hospitality. Those scientists considered themselves Frenchmen and were the pride of French science. I asked why the Russians made up such a large percentage of the assistants, and I learned that the reasons were the same as before the war. Science was not profitable. With the exception of very contemplative minds, tempera- mental persons could not find their fill of life in science. Studies were pro- tracted; not until the age of twenty-eight to thirty could one take one's first independent steps. Yet life was rushing forward: one could become a pilot, a traveler, or an officer. The world was full of adventures for a man with zeal. There was also this uncertainty about tomorrow: if an assistant was not pro- moted to a professorship, he might suddenly find himself out in the cold. I heard about a Hungarian who, having lost his position after the Pasteur Institute was reorganized, committed suicide. Finally, there is this constant searching for new paths and ideas. In the end, it can be said that science is pursued by those unable to satisfy their hunger for life otherwise. This does not mean they are not contemplative intellectuals. I believe that the secret of the calling to science can be expressed as the algebraic sum of passion for thinking and the hunger for life. If the hunger for life is too great, the pas- sion for thinking does not suffice to satisfy it. Therefore, science and art attract mainly people who are somewhat derailed and have fewer opportuni- ties in life. Science and art, if they are true, calm, contemplative, and yearn- ing, do not like statehood-promoting clichés. This reminds me of Martin Arrowsmith, the hero of American scientific ideas, who fled the large insti- tutes to the lonely stillness of a laboratory.[9] If autochthones devote them- selves to theory, they soon desire to become managers; their willfulness and hunger for life require a dynamic expansion and do not tolerate a restless hunt for scientific ideas. That after a certain period these meditative assist- ants know more than the managers and directors of institutes is understand- able. Madsen, a scientist of great measure and a brilliant organizer of the Serological Institute in Copenhagen, said to me: "Large scientific institutes are often the death of a creative mind. A docent become professor hunts for grants and seeks to expand his institute, but unfortunately his last publica- tion is often reduced to a description of the institute."

There comes a time when these émigrés become candidates for directors, and then a racket about foreign intrusion is raised.

To this is added a nostalgia that undermines one's psyche and leaves one to another possibility of satisfaction and fulfillment such as science. In his beautiful book about teaching, Flexner says that theoreticians should not be given very high salaries.[10] The effort required to spend this money absorbs them to the detriment of science. Axel Munthe wrote that in Capri, people prick the eyes of canaries to force them to sing more beautifully.[11] Those who cannot fulfill themselves and are yearning and restless usually make the best scientists. Strong and noisy individuals become managers and directors; they gather without sowing. True scientists often sow without gathering.

I believe that this is the secret of the Jews' enthusiasm for science, not the special ability inherent to race, nor the desire to gain positions in which the autochthones are not interested. Those Russian émigrés were by no means Jews; they were simply uprooted individuals overwhelmed by longing. In Poland, the role of the lowest intellectual stratum in my field of work was filled in my day by nonphysician Jewish women. Those women who were physicians were able to satisfy their hunger for life in a more buoyant manner. I never hesitated to work with them; they worked excellently and self-lessly. Wanda Halber, whose character I have already outlined, represented this type of devotion to science. In Poland, as in other countries, an individual with full rights, a man who was a physician and a Christian, would seldom abandon himself to theoretical work. I was forced to square the circle, to recruit scientific workers with a monthly salary of 250 złotys from among young physicians who were facing splendid opportunities in state agencies, the Health Insurance Organization, hospitals, and private medical offices. I had to take those whom I could find so that the laboratories for which I was responsible could function. Besides, the shortage of theoreticians was universal. When the chairmanship in the Department of Hygiene became vacant in Switzerland, it took them several years to find a suitable man.

But let us return to the Cancer Congress. I was struck by one more tragedy, that of émigré scientists and the danger threatening their objectivism. Blumenthal presented a paper;[12] he was a famous researcher and formerly director of the Cancer Institute in Berlin, who had had to emigrate because of his own origin. In Switzerland, he launched a new anticancer remedy, which was neither properly investigated nor particularly effective. A young roentgenologist from Zurich, Schnitz, treated him roughly, like a half-baked, unconscientious young boy. From the viewpoint of pure science, he was probably right. Blumenthal should have been less optimistic, especially in view of the commercial companies that had often made big money on the optimism of scientists. However, it also occurred to me that this seventy-year-old scientist had had to make a new start in his exile. He had been unable calmly to complete his research in his laboratory. He had had to create something great immediately to embarrass the Germans for his exile and to create for his new hosts another motive for their hospitality than just pity for an old man. In this situation, how easy it is to overestimate one's results. Inevitably, the immortal words come to my mind: "Let he who is without sin cast the first stone." Therefore, when I was speaking with Blumenthal, I wished him good luck and success. I could not help it. Evidently, I am no Cato.

I presented my paper in German.[13] Usually I prepared my speeches for scientific meetings in French and, since I did not have much practice in that language, I always read from the manuscript. And yet, I had spoken fluent German when I was lecturing in Zurich. That had been twenty years ago. But the memory of my youth made me languorous, and I decided to speak in German. It proved that I still knew how to. The chairman thanked me "für den formvollenden Vortrag" [for the form-perfect presentation]. I had the impression that I had conquered the hearts of my listeners. How happy are those who can speak in their native tongue which, at the same time is understood by all scientists. We representatives of small nations are handicapped *a priori* as far as the expansion of our spirit is concerned. We cannot captivate the audience with our internal fire, because its intensity abates while reading from a manuscript.

I talked in part about studies that had not been published. We had probably succeeded in elucidating the nature of the serological specificity of cancerous tissue. We had found similar bodies in pus, tubercular tissue, and necrotic tissue. These bodies were probably not a manifestation of growth but of regression; in a sense, they announced death. I pointed out the analogy to tubercular tissue and the possibility of explaining the decomposition of tubercular tissue on the basis of the effects of the antibodies. Biehl, a Dutch researcher, became very interested in my findings and subsequently sent me one of his articles confirming our results. Nevertheless, during the presentation, when I felt most elated, I was also thinking of the hard work done in the laboratory, the efforts exerted, and the unfulfilled hopes, because no scientific work ever fulfills all the hopes of a scientist. The essence of scientific work is an imagination that embraces more problems than those suggested by observation. It is a lottery: whoever has more tickets has a greater chance that one of them will win. I arrived at the conclusion that the truly happy moments in a researcher's life are those spent in his quiet laboratory. All the meetings, presentations, and successes are an unnecessary superstructure. Those for whom this superstructure is an attraction and the main motor of the work eventually sell something very precious for a bowl of rice.

The Belgians displayed extraordinary hospitality. Gala official receptions were given by the chairman of the Council of Ministers and by professors. The invitation cards called for appearing dressed with medals and orders of merit. I had mine with me, but I could not force myself to pin them on. I recalled the words of a brilliant and witty scientist, Professor Doerr from Basel, who once was the main hygienist in the Austrian Army: "You know, the only night when I could put all my orders on my chest and still not feel ridiculous is at Shrovetide." Indeed, why should one stress nonscientific achievements at scientific meetings? One should rather eliminate everything extraneous to let the intellectual power shine all the brighter.

I will not write about the results of the congress. It is not the objective of this book to present the history of science but rather, to describe how a man who was dreaming of a sacerdotal role in science saw the era in which he lived and worked.

Let us turn now to congresses that took place in 1939, that crucial year which brought an end to the old life, when all the false assumptions taken on by scientists would be shown up by the outburst of international combat and the monstrosity of the occupation.

Chapter Fifteen

The 1939 General Pathology Congress in Rome

I received an invitation to the General Pathology Congress in Rome. I agreed with the committee that I would talk about problems of heredity associated with my recent studies of blood groups.[1] We had succeeded in discovering transition forms of blood groups, which was probably of great significance for genetics and anthropology. The topic was relevant in view of the growing science of genetics. The congress was to convene in mid-May. Immediately thereafter, I was to go to Paris to talk in the Medical Academy about antidiphtheria vaccination in Poland and to give a talk to the students in the Parisian Bacteriological Institute.

It was in 1938 that I committed myself to present the paper in Rome. In the meantime, Italy began to support German policy, publishing various articles against Poland and issuing various decrees against Jews.[2] In light of that, should I go? Would it not be better to protest against the suppression of scientists because of their race? I discussed the matter with French scientists who—in spite of the political campaign against France—decided to go. They wished to attempt making men of science independent of politics. The fact was that in spite of the Italian-French conflict, numerous Italian delegations had attended the fiftieth anniversary of the Pasteur Institute. It was said that there was no anti-Semitism in Italy, but rather compassion for the unjustly treated Jewish scientists. An agreement must be found that rises above politics. I went to see our minister of foreign affairs, who stated: "We do not want the Polish delegation to be large, because we want to make it clear that we have been offended by their press. Yet we must be present, because we do not want to sever our cultural ties with the Italians. It is convenient to us that you want to go, especially since you do not go on your own initiative but as a main speaker who has been invited by the Italian presidium of the congress." The Scientific Commission of the Academy of Sciences asked me to represent it; therefore, I decided to go. The presidium informed me that since some countries had difficulties with foreign exchange, the main speakers would be considered guests of the Italian Government and have their expenditures reimbursed. This did not suit me. Under the circumstances, I did not want to be a guest. I replied that the Polish

Government, which had officially delegated me, had given its scientists adequate funds for the trip.

The congress did not even resemble a friendly handshake over the heads of politicians. The Italian congress committee had no feeling for the intentions of the scientists coming from democratic countries. The whole congress was set up as a great tool of Fascist propaganda.

The congress was opened in the presence of the King and the Queen. She was a personification of majesty, while the King was rather small and had an ailing face. They sat in large armchairs, and behind them in the front rows were diplomats and generals. By chance, I was sitting right behind the Queen and was able to observe her behavior. The German delegation was not at all favored. Several scientists came from France, England, and Sweden. The United States was absent as a protest against racist policy. The inauguration address was delivered by the chairman, Professor Rondoni, director of the Cancer Institute in Milan, who had presented a paper on the serology of cancer with me at the Cancer Congress in Brussels (he had also given lectures in Warsaw). He was a scientist and politician put together, being a senator, an "Excellency," and so on. His gala inauguration address was dedicated to the achievements of general pathology and the contribution made by Italians. The attention of the audience not used to listening to a foreign language was soon dissipated. We heard only the word Duce being periodically repeated. We did not know what this had in common with general pathology, but every time the speaker uttered the "sacred" name, he would raise his arm in the Fascist greeting.[3] This exhibition of adoration for the Duce in front of scientists who had overcome their reluctance for the regime in order to reach out a friendly hand to Italian scientists made a tragically grotesque impression. And those repeated Duce . . . Duce . . . the congress was making no headway. In the morning, the main papers were presented in international languages; in the afternoon, communications were made in Italian only. No scientist had come from abroad to present papers; only those who were giving main lectures had come. I spoke in French. The presentation by Ramon from Paris enjoyed great success.[4] A young Russian scientist, the experimental geneticist Timofeeff-Ressovsky, who was working in Berlin, made an excellent impression.[5] One could see an inborn talent in him, and he had the reputation of being a very interesting man. As usual, the northern researchers' stance was striking. I invited Sweden's representative to Warsaw. The conferences were held in university buildings and in the newly built marvelous institutes. The Italians had every right to be proud of this university city. Nevertheless . . .

We visited the Cancer Institute. The first floor formed a hospital which, obviously, was full. It is not difficult to fill a hospital ward; sick people are always plentiful. But the second and third floors housing laboratories were empty. We passed from one laboratory to another and looked at splendid treatment rooms and instruments, but could not find a soul. I inquired about this of an acquaintance, a young Italian docent, and learned that two Jewish scientists expelled from Germany had been working there until recently, when they had had to leave; thus, he stayed alone in several of a dozen laboratories. The English looked on, bemused. "We lack the means for such laboratories at home," they said, "but work is in full swing. And we attract people from all over the world. What do we care whether or not he is a Jew if he wants to work?" We went to see the Malaria Institute.[6] The government's great project to dry the Roman Campagna was well known. However, an institute should be the domain of scientists and of pure thought. The director gave a talk on the institute's work. He did not say: we have discovered this and that, or we are interested in this and that. Instead, he spoke about the Fascist Government whose interest in science has become manifest in this and that. You would have thought that mosquitoes were being destroyed on the Duce's orders and for love of him. The Poles, Englishmen, and Frenchmen exchanged glances. Did not our hosts see that Fascism, even though it had built beautiful edifices, had degraded the scientist's mind?

Lunches were served in the student cafeteria. That suited us well. We saw nice, cheerful, intelligent young faces. We were warmed up by their gay chitter-chatter. But there were signs in the cafeteria reading: "May the book and the rifle be your pledge" and "[obbedire], credere, combat[tere]."[7]

I left with the friendly Italian docent whom I had known from France, where he had worked in the Pasteur Institute with my former pupil and assistant Dr. Kossovitch. Like most Italian scientists, he was nice, subtle, and intelligent. When we went sightseeing through the port of Ostia outside of Rome, and along the charming, old Roman one-storied houses surrounded with joint gardens, where from the distance we saw the sea, he told me: "Just look, we were building this way 2,000 years ago; and now the Germans have come and moved us back centuries. And that is what is supposed to be having a talent for building an empire." "Well," I said, "we foreigners are not surprised by your intelligence, creative initiative, and impetus. We all know how much culture owes to you. We also remember the speech your Duce gave when he was fighting for Austria: that you had a great culture when the Germans were still climbing trees. What disturbs us is your constant reference to your national heroism and your constant incongruent saber-rattling. Can't you talk of malaria without mentioning the "sacred" name of the Duce and of the Eternal City? You should at least attenuate your pathos. What if your Duce dies—may the Lord give him a long life—what then? Will you cease to work on malaria? Life is eternal while a dictator is transient." "You see," he replied, "we are not bothered by dictatorship. The Caesars were dictators, and we fared pretty well through the ages. There is a kind of continuity even in dictatorship. Only one must become attuned to a certain note. Besides, our Caesars gave the people a greater freedom than the senate. Brutus was a spokesman for the privileges of the aristocracy and not of the people. Don't you see an analogy to your democracies? Money, not people, is gaining power. As to the Duce constantly rattling the saber and talking about Rome's grandeur at every occasion? The point is that we have been liked everywhere, but no one has been taking us seriously. There you have the Italian; he is nice but something of a clown, an artist, a tenor, and a man in love. We want finally to be taken seriously. Unfortunately, this is possible only when others fear us." "I understand," I replied, "but do you believe that this is the path to power? A child from a truly fine family does not talk about his ancestors. All this German talk about their culture and superiority is only the result of their inferiority complex. Have you ever heard the English say that they are the master nation? Yet they are just that in every meaning of the word. And your dictator? It is typical of your regime that no successor will arise, because no proud spirit can grow mature in this atmosphere of obedience. *Credere* and *obbedire*—this is no climate for creative, independent minds. Such dictatorships have the characteristic feature that they leave no posterity and terminate in scandal. Just remember Napoleon the Third. A nation must dream of great goals, and it must also feel united. But these are two different things. One can dream of science without making references to national grandeur. You know, science is jealous. You cannot pray to two gods at the same time and in the same temple. You will see that this will end badly. The English may not have your charm, they may even not have your quick intelligence, but they have something you lack—please don't take offense—moderation and common sense. A wise Frenchman once told me: 'Out of 99 smart Frenchmen, the one stupid man will be elected to the parliament; out of 99 stupid Englishmen, the one smart man will be elected.' They do not need to adore their leaders; they are ruled by traditions and not by clichés. When it is necessary, they find the right leader. This nation does not need artificial stimuli; it believed in the power of the ideas of the future and is not afraid of traditions. When the English rule, they do not claim they have the right to do so."

I began to inquire about internal academic relations. It was simple shop talk. One young assistant spoke about choosing one's profession. There were two men in one scientific area: his chief and another scientist. The latter was a senator and an influential man, and therefore the chances of this assistant were small. That was how scientists were being selected in Italy. Those who reproached the Polish Government for having made science addicted to politics evidently did not see the rest of the world. To make the career of an assistant or docent dependent on whether or not his chief belonged to the party! Supposedly, an Italian could not even become a porter if he was not a Fascist. I was shown the science and propaganda journal in *La Razza*. It was a queer mixture of *Das Schwarze Korps* and

Zeitschrift für Gesellschaftsbiologie,[8] containing articles about Mendel's laws and blood groups and, at the same time, caricatures of Jews and crooked noses. Finally, there were photographs of blond Italian men and women, with names and addresses stated, presumably as a proof that the Italians were Nordics. To think that there were people who wrote for these journals! I heard that Pende had joined this trend.[9] A journal of this type represented not just a breakdown of human integrity but an ultimate, hopeless stupidity.

I opened a daily [newspaper]. There was the Duce as a miner, as a farmer, as a peasant—the Duce, the Duce, and the Duce everywhere. Invectives against Poland were numerous. The only serious paper was the Vatican's *Osservatore Romano*. We visited hospitals. The walls in all wards and offices were covered with photographs and portraits of the Duce with a boxer's face and a protruding, brutal jaw. For heaven's sake! A dying man wants to look at a mild face that speaks of eternity. It corresponds to certain internal needs that hospitals display pictures of Christ and their patron saints. Had the Italians become blind from a constant visual strain? At the same time, what nice people they were who I felt had a general friendliness toward Poland. I felt that the cult for the bellicose Votan weighed heavily on them and that they would have preferred to perform the Dionysian rituals of beauty and gaiety.

I visited our ambassador, General Długoszowski-Wieniawa,[10] whom we all liked and knew from Warsaw. I had last seen him in Warsaw in May 1938 at the high school graduation of his daughter and mine. During a chat with him after the graduation exercises, I reminded him that he too was a physician who had abandoned his profession. "My dear Professor Hirszfeld," he said, "I am probably the only physician in Poland who has done harm to no one." I visited him in Rome in the embassy during visiting hours. The embassy made a pleasant impression. There were many busy, good-looking, very nice girls. I smiled: we all knew the general's weaknesses. He received me with a diplomatic earnestness so little corresponding to the face and countenance of a modern Kmicic.[11] He spoke of politics. "You know something, I believe the French do not desire peace with the Italians. The Italians' infiltration of and their natural growth in Tunisia is so great that in fifty years they will gain complete control over Tunisia if they are not halted by a war." He spoke in a friendly manner about the Italians. We agreed to meet at the congress ball, which was to be attended by the diplomatic corps. In the evening at the ball I introduced French, Italian, and Swedish scientists to him. He did not want to meet any Germans. What a charmer! Before we had time to look around, he was surrounded by a group of elderly gentlemen. Inevitably, I compared him to the tired aristocrat Wysocki. True enough, the general was "dry" [sober]. I do not know how successful his diplomatic talents would have been if he had recalled his "wet" period in

Warsaw. At that time, I would not have given much for his diplomatic discretion.

The Italians' hospitality was extraordinary. Every day, at the invitation of a large pharmaceutical company, we went to visit beautiful institutes, buildings, and the port of Ostia. There were several unforgettable parties: in the new excavations, in the Papal academy, and in the gardens of the Vatican. What a marvelous atmosphere of calm and bliss prevailed there! Those people knew the psychology of work bordering on ecstasy. And everywhere there was this nice, cordial tone. I met a woman professor from Sicily. I have forgotten her name, but I remember her sculptured face and brilliant intellect.

At one time, I sent the French edition of my book on blood groups to the Vatican Library, in response to the Pope's appeal to combat racism in science. Appropriate chapters are included in my book. I received a beautiful note of thanks in Latin. I discussed this with our hosts, who emphasized the Pope's objective attitude toward the work done in his academy.

I recalled the University of Bologna with Europe's oldest lecture room. Above the professor's stand, there was another one for the priest. If the professor said something unsuitable, he was lashed. Fascism took over this lashing in a more radical way. To give someone a beating—what a trifle! To destroy a scientist, to starve him if he were recalcitrant, or to prevent him from speaking—is more effective. That the ideas which cannot be born hurt most—that matters little. Science can only be a part of a great national effort, a servant to the state. But to function as an independent power—what for? With this painful memory of a valuable nation that had betrayed the idea of an independent science, I left for Paris.

I had a small intermezzo during the trip. The Orbis Bureau gave me the wrong ticket; Rome–Paris through Marseilles. In Genoa I was asked to leave the train. I had no money, because it was forbidden to carry money out of Italy. Fortunately, an employee of the Papal Academy who was going to Paris was in the same compartment and lent me some money. I was waiting for the train bound for Marseilles and learned that the Duce was to pass by. The Italian youth gathered to greet the leader. Those black shirts created a festive impression, in spite of everything. On the streets, flags were displayed from and between windows. Nonetheless, the citizens did not suspend their custom of hanging out washed laundry. As a result, the pell-mell of flags, signs, portraits of the Duce, and ladies' *dessous* created a queer scene. "Vincere, obbed[i]re, combattere." I boarded my train, and soon we reached the boundary: Ventimiglia. *Egalité, fraternité, liberté*. A different world. I breathed deeply as if I had left a charming prison.

Chapter Sixteen

Medical Academy in Paris; French Youth

Paris. How well I remember those moments, but how distant they now seem, almost unreal. A radiant, unforgettable city. A city full of contrasts: meditation and gaiety, the most subtle family happiness and debauchery, hard work and artistic laziness. I was met by Professor Debré who had recently been lecturing in Poland and now was introducing me to Paris. We agreed as follows: on Tuesday, a lecture from three to four p.m. at the Medical Academy; on Wednesday, a lecture at five p.m. for medical students in his institute, followed by a small party with champagne and sandwiches to give me an opportunity, as he said, to establish personal contact with French youth. Afterward, I wanted to spend a few days with my daughter.

The Medical Academy. The world capital of medicine. The echo of the dramatic speeches delivered there by Pasteur, Dr. du Bois-Reymond, and Charcot had not faded away.[1] I entered the building filled with emotion no less intense than when I had been speaking at the Capitol [in Rome]. I was going to speak at the very same spot where the greatest intellectual powers had made their presentations. Involuntarily, I was flattering myself, because I was going to talk about antidiphtheria vaccination in Poland. The method of vaccination was French, but we were one of the first European countries to introduce obligatory vaccination. In some cities we had vaccinated up to 99 percent of all children. We had succeeded in halting a growing epidemic wave. Our method of making calculations represented great progress, and German scientists had called the Polish report the best and most critical of all those existing. This was all the more important since Ramon's investigation, excellent in the bacteriological aspect, required better statistical elaboration. What a pleasure and honor it was for a researcher, who had applied, expanded, and supplemented in his own country the results of a foreign scientific idea, to present these results before the Areopagus of famous scientific authorities.

Such ambitious, elating thoughts were going through my mind while I was entering the academy building with Professor Debré. Busts of famous scientists lined the walls. Professor Debré introduced me to several scientists: all had great, well-known names. I was somewhat disturbed, however, by the fact that so many persons were strolling in the hallways; the session was already in progress when we had entered the building so I turned to my *cicerone* with my enquiry. Professor Debré explained to me: "You see, the Tuesday meetings in the academy are like a Mass: to attend is the proper thing to do. However, the topics are not always interesting." I understood this completely. I had often been bored at scientific meetings. Nevertheless, I wanted to enter the auditorium. The room proved to be rather small and had double-seat benches each supplied with a small headphone. The audience was unceremoniously conducting private conversations. I asked what these headphones were for. "You see," I was told, "when the members of the academy are not interested in the lecture, they like to talk to each other. It would be embarrassing for the chairman to call these 'immortals' to order. So we found a way out: those who want to listen have their headphones, and those who prefer to chat are not disturbed." I admired this liberalism. With shame I recalled how at a meeting in Warsaw I momentarily interrupted my lecture when I heard a loud conversation in the audience.

From the podium someone was talking about submarine hygiene. He was showing a motion picture of how a boat should be scrubbed: one picture, a second, a third. . . . What did this have to do with science? The presentation lasted a fairly long time. The chairman—but perhaps I was mistaken—appeared to be taking a nap. Next, someone read out a long list of pharmaceutical compounds that were supposed to be approved by the academy. No one was listening, but the compounds gained the honorary seal—"approved." Then, the elections of "immortals" meant that the audience began to wake up. Finally, every one left the room except for the chairman, Professor Debré, Ramon, and Weinberg. And it was to this indeed exquisite yet somewhat small group that I presented my report. I was not very happy, being used to larger audiences at home. Especially since only an hour earlier I had had such exalting thoughts about national efforts and the significance of science. Professor Debré tried to comfort me: "Our scientists devote exactly one hour from 3 to 4 in the afternoon to the academy and not a minute more. Please don't feel vexed that as a guest you were not invited to speak first. The Parisian Academy is a tradition unto itself. Here where world science was created and the greatest intellects made their presentations, no one pays attention to hospitality. The main thing is that your paper will appear in the academy's bulletin."

The paper did indeed appear during the war with the honorary annotation: "Presented at the Medical Academy in Paris."

It was then that I understood the good and bad sides of historical maturity. Paris will not be moved by a foreign guest. They have seen many from among the greatest. As such, these often tired men will not sacrifice one unnecessary minute to make an extra effort. No longer must they seek to win friends from abroad. By force of its tradition, Paris is the scientific capital of the world.

And I thought how beautiful was the peace of the autumn. But it could become dangerous if it spread to other aspects of life.

The next day, I gave a lecture for students at the Institute of Bacteriology. I talked about new trends in epidemiology, the role of latent infections, and the need to change our concept of germs that struggle and our struggle against germs. I explained that the researcher's mind was unable to free itself from social bonds and that the struggle of man against man was incorrectly projected onto man's struggle against the world of invisible foes. I spoke about the natural phenomenon of symbiosis and its significance in microbiology. I was imbued with a yearning for harmony and a disbelief that fighting promoted progress. Evidently, my lecture appealed to the French mentality. The students were listening attentively. A subtle play was visible in their faces, which seemed to reflect my ideas like a mirror. The lecture was followed by a reception. I met many ladies and gentlemen from the society, including the young Nicolle whose father [Charles] I deeply admired. The students were asking me various questions. My daughter, a Sorbonne student, had attended the lecture.[2] I was happy. I heard that my book was widely read and that three-quarters of the editions were sold. I visited Professors Panisset, Besredka, Lévy-Solal, and Kossovitch.[3] In the evening we gathered at Professor Debré's. We agreed that he would endorse my candidacy for a lecture at the Medical School in 1940. I wanted to come back to Paris to be with my child and to start working on the biochemistry of blood group factors with the prominent French biochemist Boivin.[4] The next day we had lunch at Professor Ramon's in Garches. After lunch, we went for a walk in the park and sat on a bench where Pasteur once liked to sit. Ramon promised me a scholarship for my assistant Miss Róża Amzel so that she could begin her preliminary studies. I visited Professor Chevallier who was married to a Polish woman. Everywhere, I met the social elite. What pleasant people they were; what wide interests and what a humanitarian approach they displayed toward social problems. They were sincere, good-hearted human beings. They took care of my daughter as if she were their own child. She regularly visited them, and they understood the concern of a father who was leaving his only daughter on her own. I met Valery-Radot, Pasteur's grandson. When he heard that my daughter was studying in Paris he invited her, too. He said to me, "Vous nous faites honneur en laissant mademoiselle votre fille à Paris" [You honor us by leaving your daughter Miss Hirszfeld in Paris]. Pasteur's grandson! How courteous these people could be. My daughter lived in the Cité Universitaire in the British-French Pavilion. The Cité was formed by buildings built by various countries for their students. In the park there was the central building with recreation rooms, a bar, a canteen, and a restaurant. The building was funded by Rockefeller. I read the inscription: "I am donating this building to the world's young people so that they can get to know and love each other." I went for lunch with my daughter to the canteen. I saw faces from all nations, white and colored races—at the same table. There was none of the *vincere* and *combattere;* one could feel the spirit of great communion. Perhaps these young persons will be able to implement the dreams of our generation; perhaps this exuberant youth will unite peoples who could not be united by the greedy voracity of adults and the pseudo-international flurry of scientists. I went about with my daughter and those charming young students, and it seemed to me that I was seeing the beginning of a new life.

Then came the storm that was to take my child away from me. Now my thoughts roam over a cemetery. I entrusted my happiness to the world and I was deceived. But I remember those moments as a promise of happiness for others. May it be so. To my French colleagues, I would like to say the following: "You gave my child moments of happiness and a vision of noble life. When bad luck befell us, she lived with memories of you. A day before her death, I was reading to her Eve Curie's book about her mother. She listened to the story of her love for one of your scientists, the story of a student who was to become the symbol of the cooperation of our countries. She was listening holding her breath. She came to love your country as her second homeland. I remember you as thorough, true human beings. Because you are friendly, wise, and gentle. And, as my child did to her last moment, I believe that in spite of your fall, you will rise again."

Chapter Seventeen

A Home in the Sun

We had spent our youth not only in the most beautiful but also in the most pleasant spots of the world, surrounded by the charm of nature and culture: in Heidelberg and Zurich. During the war, through the window of my laboratory, I beheld the sea and the snow-covered peak of Mount Olympus. It was different in Warsaw. The windows of our apartment gave out on the gray wall of a workshop, while offending sounds of phonographs and quarrels intruded into our rooms. It was not easy to live under these conditions, especially when we were thinking of beauty lost.

My daughter was born in September 1920. It is too painful for me to recall those moments. Eight years later, my wife's former schoolmate came to our home with her son who was the same age as our daughter and had lost his father. We decided to raise the children together so that they would have siblings.[1]

We were choking in the city. In the evenings I would often go for a walk to the Botanical or the Łazienki Gardens to peer at the sky and drink in the scent of trees and the earth. He who knows the charm of the night also knows how poor man is when he walks only on the pavement, with the street lamps dimming the luster of stars. One hot July night in 1927 my wife and I took the tram to the Poniatowski Bridge. We got off at the end of the bridge on the right bank of the Vistula River and descended into some kind of an enchanted valley filled with the scent of blooming lindens and meadows. The moon's silvery luster was reflected in the water. The inebriating scents induced wistful emotions in us. "It is night. The murmuring brooks are singing louder. My soul is a murmuring brook. It is night. The hearts of lovers are singing louder. My soul is the song of a lover." We were overwhelmed by the desire to commune with eternity in the ecstasy of this starry calm. We noticed a cluster of small houses to the side; they seemed to us like paradise. I inquired as to their administrator and pulled him out of bed at eleven o'clock at night. Several days later, we owned a lot among the blossoming orchards of Saska Kępa.

It became my passion to have my own little house with a garden. I had never been able to resist my desires. But this was the first time that my passion was directed toward possessing something. New, previously unknown impulses awoke in me. I was walking around as if hypnotized. I was engrossed in building and in selecting architectural shapes that would correspond to our spiritual needs. Those who buy old or ready-made houses know nothing of the true bliss of building. I fell into debt, for I was building in the most expensive period and made dozens of foolish mistakes. The construction lasted three years, but finally, in 1930, I was able to move into my own house. My passion was touching to others: some friends brought me their savings. My wife gave up her well-established medical office in the city. Our actions were not rational in the usual sense. Because of the mistakes made by the contractor the first year was veritable torture. But at last I could live the way I wanted to, and I bless my recklessness and the goodness of my wife who agreed with every one of my follies. Only life in my own home with my own garden revealed to me the warmth and depth of family life. The windows of our bedroom faced east, because I wanted to begin the day with the sight of awakening life. In the morning I would pick roses covered with dewdrops and give them to my wife. This was my thanksgiving song. I would spend the whole day at the institute, and in the evening, having come through the vine-covered gate, I would catch sight of the heads of our children playing on the terrace. Then, we would have supper together. All the adults in our family were working; each brought home the memory of the day's events. I took care of what was most elusive and most important, namely, that a spiritual atmosphere inhabit the walls of our home. Family is the most beautiful form of social life, but family life requires constant tending. One must never regard family life as only a rest after a busy day's work, where tense nerves relax and the soul can walk around in a bathrobe. Family life is a renewal for the old and a source of future memories and prelude to future life for the young. Adults submerge in the eternal spring of instincts, while children shape their souls. It should never happen that a person associates the memory of his youth with that of boredom or violence.

I had a beautiful family life. My wife knew how to create a hearth of modest elegance and community. Whoever entered our home was quickly imbued by the feeling of conscious happiness of people who had not lost their capacity to dream at a mature age. In the evening, prior to going to bed, we would go out into the garden, look at the starry sky, and say: this is a blissful heaven. My home was also often the source of inspiration. A warmth would go through my whole being, and my thoughts would fly off and take shape and power. Someone was standing behind me, setting my soul into vibration and whispering unknown ideas and associations to me. In Saska Kępa, I wrote many papers and prepared many lectures. The best lectures I gave were when I would not pull out my notes and books, but rather, when I could sink into a state of impersonal existence. The intellect is but a small part of one's personality. It must be founded on an unconscious melody of the soul that withers if it is not regenerated during states of utmost concentration.

We did not aspire to an active social life, but I often invited scientists who were visiting Warsaw and our institute. I did so not only to maintain contacts but also to fulfill my duty of hospitality and to let our children feel the spirit of science and the distant world at an early age. The French scientists were especially numerous in our home, due to the activities of the Biological and Pediatric Society. The French Institute in Warsaw and its charming directors, the Mazeaus, often functioned as the connecting link.

This was the atmosphere that shaped our children's minds and souls. In 1938 they graduated from high school. I was requested to give, on behalf of all parents, a talk to the girls at graduation. I took the occasion to speak about what was the innermost motor in my life, my credo, and my philosophy of happiness.

When we look into the hearts of parents, we find there but one desire: happiness for their children. However, let not the desire for happiness be the motor of your own actions. Happiness is not a goal in itself. The feeling of happiness grows like a flower when life is reborn. When we see marvelous changes taking place in nature, we are overwhelmed by a feeling of happiness. He who wants to be happy must not seek happiness: he must seek life. Yet, the life of an individual is an illusion. Only working toward great social goals gives us powerful feelings of life and creates an intoxicating feeling of happiness. We must participate in the efforts of mankind and the nation, project our dreams into the future, and think of the distant one who will come and for whom your love, like the love for your neighbor, will give further strength to the thrust your activities.

Which profession to choose? Your main objective should be not that which offers you advantages, but that which satisfies your innermost desires. For our motivations are represented by two instincts: the instinct of taking and the powerful instinct of giving which in the best of us can never be satisfied. We parents probably commit the greatest mistake when, thinking of your happiness, we think of taking and using. We experience the most intense feeling of happiness when we give something of ourselves. That is why a soldier dying at his post can be happy. He who has the opportunity to alleviate the suffering of others can be happy. He who clears new paths can be happy. Even he who loves unhappily can still be happy. Only he who has no aspirations and no desires is unhappy; unhappy like a leaf torn from its branch and chased by the autumn wind. The prerequisites for happiness are to aspire, to yearn, to think and to work.

To love more than you are loved—let this be the need of your heart and the splendor and pride of your femininity. Let the last words which you will hear in this school be the synthesis of our love for you and of your longing for life:

"May every day of your life be marked by the eternal desire to give."

When I was talking, I saw the glowing eyes of my daughter. She was happy to be in such harmony with her parents. She soon left for France where she spent a happy year while studying in Paris. What came later was an abyss of despair. The gods are jealous. Too much happiness is not allowed.

Chapter Eighteen

The Autumn Draws On

My family life was cloudless, radiant, and profound. The situation at the institute was different. I had my ups and my downs, depending on the surfacing of ideas, on the one hand, and the breakdown of Polish life, on the other. Dungern's and my research had marked the beginnings of a new branch in science and was a center of interest for many years. It was rare to open a scientific journal and not find an article on blood groups. But the same fate befell blood groups as befalls every idea. A scientific idea is like a child prodigy. We see in it at first not just what it can accomplish there and then, but the music of the future: we place in a child prodigy the hope for unknown sensations. An adult artist, no matter how great, has a certain finite repertoire. There are two types of scientists: those who search for new ideas and new projects and those who seek shortcomings and gaps in others' research and observations and try to patch them up. Only the former will understand me. Ostwald calls such researchers romantics and juxtaposes them to the classicists.[1] Such romantics yearn for ideas the way one yearns for something that is young and full of prospects for the future. However, one cannot live by prospects only. Even the greatest researcher has but a few principal ideas, and his work consists of expanding them. Few investigators are fortunate enough to see their research become the beginning of a new scientific trend. And even those fortunate few who have succeeded in catching and illuminating a great secret must be prepared for the possibility that their theories will not be correct in all aspects. Their very own children—their pupils—rebel against them. I heard from Dungern that Metchnikoff, in a moment of frankness, told him that Pasteur's death was a happy occurrence for the Pasteur Institute. The great visionary had begun to oppress everybody with his attachment to his old ideas. Such disappointments befall the greatest men, and then they complain about the ingratitude of their pupils. The great hygienist Pettenkofer committed suicide when he noticed that he was standing outside the main current of new scientific ideas.[2] Ostwald left his institute when he saw that it was able to function without him. Koch retired at the age of sixty, keeping for himself just one laboratory to have a quiet room where he could reminisce about his active years and meditate about his favored topics, while being free of the burden of managing the institute.

The internal conflict that arises when ideas outgrow their creator is described on a different plane by Werfel.[3] Verdi was close to abandoning music when he noticed that Wagner's music appealed more than the melodious Italian music to many people. I too went through similar disappointments. Serological methods set the beginning of a new period of blood group research. Later mathematicians began to analyze our observations with methods less familiar to us. In this situation, one feels like a father who sees his children living under circumstances other than those prevailing in his youth. This feeling is made still stronger by another matter. Scientific methods have their youth and their old age. What was revolutionary at the beginning may subsequently become outdated. This is what happened to serology. Two new branches appeared: immunochemistry and virology. The new methods made it possible to identify filterable germs; the invisible became measurable. The methods I had used or developed were no longer the high fashion in science, yet my research problems were still beset with certain secrets I wished to elucidate. But I was also attracted by new methods. The desire arose in me to take a leave and go somewhere to study these latest methods and enjoy the bliss of learning and assimilating something new. I associated this plan with my daughter's, wishing to go spend some time in Paris.

The desire to renew my research projects was associated with still another factor: the feeling of loneliness in the institute. I had put all my energy into creating an atmosphere of thinking and planning. The call to life was stronger, however; most staff members were preoccupied with current business and had to earn an additional living outside working hours. Around three in the afternoon, I would note with sadness that the institute had emptied while I stayed alone or with just one assistant. I saw the bad side of our institutes: too much administration and a lack of individuals for whom raising and analyzing issues would represent the main purpose of their life. At best, our scientific workers ascertained problems but seldom formulated new ones. In English institutes all staff members meet to have tea together and exchange their ideas. In Germany, after scientific meetings, they go together to drink a mug of beer. During our lunches I often tried to channel the conversation toward scientific problems; yet, interest in activities outside the institute prevailed. The older generation held managerial positions and was predisposed to think about organizational moves. All institutes know this conflict. A director with administrative interests does not enhance a creative atmosphere and frequently even damages scientific work by requiring rapid action. The scientist is often unable to cope with administrative tasks.

In Poland, these tasks were difficult because of the constant struggle for a budget. We received grants from the Rockefeller Foundation, the League of Nations, the Potocki Foundation, and the academy. But securing grants required great effort and often restricted our research projects. I was turning a great wheel and it seemed that I was at the peak of activity when I was overcome by feelings of discouragement and loneliness. I mentioned before that most scientists feel that way; however, in Poland, this feeling was especially warranted. The pulse of scientific life was weakening in Poland, particularly when the most recent years were compared with the initial years of independence. This trend was increased by the difficulty of attracting young people to theoretical work. Even when I was young, most assistants in scientific institutes were foreigners. These foreigners did not come to visit Poland though, and so their role was played by our minorities, especially Jews. At the same time, the anti-Semitic movement increased and considerably hindered the influx from our minorities. I was pleased to hear that our institute was regarded as one of Europe's best functioning institutes. Yet, I felt that our rate of progress would not be maintained without new organizational moves and fresh inspiration on our part. Again, I was facing a conflict so well known to many researchers: to push for the whole group or to reserve for myself the privacy of a smaller laboratory and plunge into the realm of a single idea.

The poor salary offered young assistants was one reason why Poland's intellectual life declined. Another reason was the channeling of the national energy toward political problems. This was carried on in Poland in an especially naïve manner. Dr. Szulc, the director of the institute, received the order to establish an OZON Circle in the institute.[4] I did not attend the organizational meeting, but I freely expressed my opinion in face of such grotesqueness. The next day, Szulc came to my laboratory and advised me to be more careful with my statements. This meant that among the institute's staff members there were people who watched and reported on me. This intensified my feeling of loneliness and distaste.

When people talk about Poland's politics during that period, they are thinking mainly of Jews. Poland was literally sick with this problem. When I think back about the German atrocities, Polish anti-Semitism seems mild to me. Nationalism was young and, seeking to discharge itself emotionally, fell upon the Jews as the nearest suitable object. At that time I did not understand the anti-Semitic problem in full, because I came from a completely assimilated family and had spent much time abroad. Nor was I familiar with Jewish nationalism. I believe that if the energy of young people could find an outlet in distant travels, sports, or a true struggle for the native country, it would not be vented in this pitiable manner. Unfortunately, young people found a more drastic form for their nationalism. Anti-Semitism vented in economic boycotts is unpleasant, but the fight for markets has generally assumed more cruel forms. However, this bellicose nationalism became monstrous when it encroached upon sanctuaries.

I once read the following story. A Negro became a Christian but was not admitted into the church because there was supposedly no room for him there. He cried bitterly, and at night he had a vision that he was visited by

Christ. "Why do you cry, my son?" "I cry because they say that there is no room for me in your church." Christ replied: "Console yourself, my son, in this church there is no room for me either."

Schools should enjoy the sanctity of temples. People who want to study are like those who want to pray. To beat persons listening to a lecture is as monstrous as to beat those who are praying. For several persons to attack one student is a shame that weighs heavily on our universities. I believe that the fault also lies with professors who wanted to teach but did not have the ambition to educate. The leader of pharmacy students once asked me whether there would be separate benches for Jews. I replied that there would be no such benches, that I would not observe this instruction issued by the rector, and that I would feel hurt if that matter were raised again. "I understand, Professor Hirszfeld, and I will see to it immediately." There was no mention of this. Nor were there any improper reactions. The students knew how deeply I had always been concerned for them, and they did not want to hurt me. Many of the professors I knew exerted a similar influence on students, and in Warsaw attempts were made to attenuate the stormy waves of young people's nationalism. In contrast, the stand of some medical schools, which introduced an almost *numerus nullus* (zero quota) for Jewish students, brought disgrace on Poland and its universities.[5] It was contrary to the country's interests because of cooperation with Western democracies. It was inhuman, and it was un-academic because it degraded universities to the level of a factory for diplomas reserved for Poles only. Finally, the introduction of the *numerus nullus* was unlawful: the constitution guaranteed equal rights for all.

The various departments issued such decrees, taking advantage of the academic autonomy. This concept was twisted in a ridiculous way. The point of academic autonomy is to protect universities from political control, and not to convert them into a place where politics are being made. If politics are introduced into universities, they will ultimately become the politics of those who are in power. I never liked the term "autonomy"; it concealed deceit, backwardness, or helplessness. Autonomy should protect a researcher's scientific individuality and not freakishness. Students beat to death their fellow students, and democratic professors were forced to flee through windows lest they be beaten by whippersnappers, while rectors went on and on in their speeches about our beloved youth and yet did not allow the police in to keep order. There is no autonomy for militants. My reproaches are directed not just against Polish universities. In Germany, academic life had decayed still further, and German professors displayed neither pride nor moral strength. When scientists become true soldiers of truth, they will deserve autonomy, even though by that time they will probably exert influence on students and be able to keep them under control.

The opening of the last academic year before the war has filled me with painful memories. In 1938, Polish universities bestowed honorary doctoral degrees on Minister Beck[6] and Marshal Rydz-Śmigly. In this way, they wanted to manifest their participation in the country's political life. I believe that today most of our scientists regret it and have decided to reserve this title for scientific achievements only. I remember well the so-called historical moment of awarding this degree. The rector invited foreign diplomats and in their presence delivered a speech which he concluded with an appeal to Polish youth to work with more diligence because 10 percent of students entering universities were Jews but 25 percent of those graduating were also Jews. In the presence of representatives of foreign countries, the rector declared that Poles were doing a poorer job than Jews. Later, a young boy spoke on behalf of the students. I do not know whether his speech was censored, but he said that the young people were the nation's leaders, that they demanded complete elimination of Jews, and so on. All the while, the "honorary doctor" and the "nation's leader" was listening to the propositions of how—not he but this infantile speaker—was to guide the nation.

I was utterly shocked by this scene, and as I left in the company of a diplomat I asked him: "What do you think about all this?" "It's all very awkward," he replied. "Soon, reports will be dispatched abroad that Poles are less able and diligent than Jews and that, in order to eliminate the competition, Poland intends to introduce Hitlerite legislation in universities. Politically, this is very inconvenient for us. Our people must finally get to understand that democratic countries are interested in defending, not the Eastern Jews, but the principles of equality which constitute the basis of their political life." We pursued our conversation. Eventually, I asked him what he thought of Beck's policy. "I don't understand it," he replied, "but I would feel more secure for your country if I felt sure that Beck himself understood."

When I compare the situation that developed in Poland with what was going on in Germany and Italy, including mass murders, the former seems less awful to me. Nonetheless, one's own wounds hurt most. The worshipping of Hitlerism's external forms by our right wing parties and by OZON and the introduction of Aryan clauses in associations of engineers, lawyers, and physicians—all this was not only ridiculous but also tragic. In Germany, this trend at least had a theoretical basis even though it was a false one. In our country, this was a thoughtless imitation and, since our democratic allies were important to us, it only proved a lack of political sense. Furthermore, it proved that our intelligentsia hated Jews more than it loved the country. Or that it understood nothing, absolutely nothing, of what was going on in the world.

These circumstances were bound to reflect on our institute. Once, a young physician arrived from Cracow with a letter of recommendation from Professor Gieszczykiewicz. He wanted to get special training in immunology. I accepted him because he was a nice, intelligent boy. Several days later, the institute's director, Szulc, notified me that since the newcomer was Jewish, I had to dismiss him, and that this was supposedly demanded by the government. I was unable to protect the boy because the regulations required that every applicant, even a volunteer, be approved by the ministries. I had to dismiss him under the pretext of a military secret. The only alternative left to me was to submit my resignation.

That day I went home with a heavy heart. My life's work was in bankruptcy. My dream of cultivating human souls now seemed grotesque. I tried to set my ideology and attitudes against the prevailing current, but I knew that this was hopeless. For the first time, I felt that my efforts were useless. If I submitted my resignation, my colleagues abroad would soon find out and think that Poland was following in Hitler's footsteps. I had no right to provoke such appearances. Rejection of one Jewish volunteer should not be identified with Nazi legislation; yet, my departure, even though voluntary, would be understood in this way. In addition, I was in charge of so many persons, and the fate of so many students was tied to me. I decided to grit my teeth this time and, at the same time, start looking around for a new position abroad, somewhere in Australia or Africa where there were none of these cursed racial or national conflicts and where I could keep company with the intellectual and moral elite I had frequented in my youth. My wife and I wanted to conclude our life the way we started it—as free, unhampered researchers living in an artistic and scientific bohemia, unencumbered by institutional and administrative duties; as two free birds keeping company only with other free birds.

This was how I intended to start a new life and new scientific research. In the autumn of my life, I wanted to make a new ascent. This time, though, only in the service of pure science.

Fate had prepared something quite different for me. Instead of the bliss of pure science, a thorny path of suffering was awaiting me. I never had great ambitions, only the feeling of a mission to be accomplished. In the autumn of my life; it became my ambition to implement this mission in full. I thought that my tasks concerned only science and students. As it turned out, it also was my mission to comfort those who were perishing.

Chapter Nineteen

Before the Storm

I had a foretaste of the war in October 1938.[1] The Standardization Commission of the League of Nations was holding a meeting in Geneva. I left Warsaw on a Saturday evening, went through Berlin, and Monday morning I was in Basel to see Verzàr. Switzerland was already up in arms: ditches were being dug along the roads leading to the Reich. The situation was so tense that after two hours I decided to take a train to Zurich. Professor Silberschmidt met me at the station and we went to his villa. Old memories of Zurich, the mountains and sunsets surged up in me, but this time the atmosphere was quite different. We turned on the radio: Hitler was speaking against the Czechs. The hate emanating from his voice was monstrous. The next morning, I left for Geneva and went to the League of Nations. The panic prevailing there made it seem as if the League were one of the combating parties. The meeting did not take place—except for a delegate from the United States and myself, no one showed up.

In Geneva, I consulted with my friends as to what I should do with my daughter if war were to break out. Her former teacher, Mrs. Langrod,[2] invited her to Geneva. Friends in England and France also invited her to stay with them. They advised me, however, to take her home to Poland. The conflagration would spread all over Europe; there would be no safe place anywhere, and if we had to perish, then it might as well be together. That afternoon I went to Paris to fetch my daughter. The next day we went to the Polish consulate, where everybody was convinced that war had already broken out and telling us that we would be unable to leave France. I pointed out my official functions and asked them to make airplane reservations for us. But then I learned that a train heading for the border was leaving in the afternoon. I was not sure whether we would reach Warsaw, but I took the risk. In Akwisgran,[3] we learned about the meeting of the heads of state in Munich. There was a remarkable difference in the atmosphere: there was a mobilization in France, and determined crowds were walking calmly but without enthusiasm. The Germans evidently knew that there would be no war this time, and everything was peaceful on the railroads. After we arrived in Warsaw, the situation cleared up so much that in a week we decided to send our daughter back to Paris. She spent a happy year there among good, gentle, and wise people. During our misfortune, she always recalled her stay in France. In 1939, the storm was gathering. In March of that year I was delegated by my government to go to Paris to attend the fiftieth anniversary of the Pasteur Institute, and in May I was in Rome and Paris attending meetings, which I have described in earlier chapters. I knew and felt that war was inevitable and close at hand. In September, the International Congress of Microbiologists, the [World's Fair] exhibition, and the Cancer Congress were to take place in New York. I was to go as a delegate of our government and, as mentioned before, I was elected vice chairman of the congress. If I went, I would be in America in the event of a war and could easily get my family out. What was I to do? When I think back on what was going to happen and the terrible loss I was to suffer, I see that I made a mistake not to leave. I might have saved my child. But, at that time, governing me was the same inner voice that had previously forced me to leave a neutral country and willingly plunge into the whirlpool of war. Human suffering affects me like the wake-up call of combat; this was all the more so when my country was suffering. Was I to leave my country when it was threatened by war? I knew how limited our anti-epidemic resources were, I also knew that individual moves make up the effort of a nation. If everyone were to leave, emptiness would remain.

Many did leave the country, abandoning their posts. May history resolve which was the better thing to do.

In July, my daughter returned from Paris for vacation and we spent the first half of August in Wisła. Those were my last moments of happiness. On the German radio I heard that the occupation of Slovakia was inevitable because of the Central Industrial District. Poland was to be overtaken from the south. In the mountains, I saw our Boy Scouts digging ditches. These children were supposed to be our defense against the world's largest army. Nevertheless, they were "strong, coherent, ready."[4] We returned to Warsaw on August 25. I sent a wireless to the congress chairman to say that I would not come to New York. The reply came after the war had broken out: "Your friends and admirers will miss you." Perhaps the words were written by a nice secretary, or maybe it was just a form of politeness. Yet, when I soon found myself outside the circle of life, condemned and rejected, these words kept up my morale. I still had friends and admirers in far-off America, and I should like to tell them: "These words comforted me for many years." And I should like to tell everybody: "Let us not be stingy with warmth; it may do much good to those who are freezing in loneliness."

With these words I conclude the period of my youth and adulthood, which was a period of building, and I will turn to the world tragedy that destroyed my happiness.

Chapter Twenty

The Siege of Warsaw

When I strive to recall the course of the war, what comes back first was the way people reacted: their selflessness, their keen sense of responsibility and awareness of being at the heart of what their country would become. Next, I remember the defeat, the hopelessness, the loss of self-respect, and the ghastly questioning: What was our country worth? What was its will to live? And then I think of the night that followed and the abyss of pain and destitution.

Poland had known for ages that its western neighbor was threatening its existence. It knew that the "corridor" was not the main objective.[1] Hitler had explained in his book [*Mein Kampf*] how to go about crushing a nation. It should not be deprived of its independence all at once. At first, only a little should be demanded and with some sort of justification. Next, when the country gets used to yielding bits of its own entity, it thereby loses its pride and defensibility. By that time, it is in the right psychological frame of mind to be deprived of its independence. Czechoslovakia was a glaring example of how to rob a nation of its independence, under the pretext of liberating one's fellow countrymen.

Mein Kampf contains another revealing statement: an open admission that agreements should be observed only as long as they are advantageous. Otherwise one may break the agreement, and this breach of contract is not immoral. Italy's betrayal offered the Germans a good opportunity to test this maxim.[2] There was a nonaggression pact between Poland and Germany that obligated both countries to abstain from aggression against each other for ten years. It is said that strict adherence to agreements is a Germanic or Nordic trait. Perhaps so. In any case, it was not a national-socialistic trait, as per the Führer's own words.

When Poland was attacked, the nation was admittedly politically divided, but in matters of foreign policy it was united. Everyone wanted to defend the country. Rydz-Śmigly's slogan was "Strong, Coherent, Ready!" The people believed in the power of the Polish Army. But the "strong" and the "ready" portions of the slogan turned out to be an empty phrase. We were neither strong nor ready. The only truth in it was that we were indeed united in our love for freedom and our determination to defend it.

We knew that the fight was uneven, but we hoped that we could withstand the assault for a few months and that our allies would then surely come to our aid.[3]

Only for a few days did a certain degree of optimism prevail in Poland. At first news came of the planned maneuvers by the army and the great battles, and there was hope that the allies would rapidly offer their aid. When we realized that our army had been knocked off the battlefield as swiftly as if it had never been there, a tragic sobering-up took place. A few days later, we heard on the radio that the government was leaving the capital and going into exile.[4] After playing the national anthem *Jeszcze Polska Nie Zginęla* [Poland Has Not Yet Perished], the news about the government's departure was announced, and then the tragic hymn about rebirth was again played. The lyrics of this song were the only thing our hearts could cling to for years. Were we Poles condemned by history only to dream about independence but never to experience it? To lose it after only twenty years and be unable to protect this beloved dream! How will we face future generations?

At the beginning, the government ordered everything to be evacuated: institutes, offices, and the press. However, it was obvious that this order was unrealistic. When, in addition to this, Umiastowski[5] issued the order on the radio that all men were to leave Warsaw, a veritable migration of peoples began. Where to? What for? Yet, thousands and thousands of men went with their wives, children, and a few possessions in their hands or on their shoulders, heading south or east like a live, overflowing wave. All east-bound roads were jammed. Soon, all wells were exhausted and food eaten. Endless chains of automobiles and horse-drawn carriages carrying people and possessions were moving along the highways. Since that time, almost all European countries have gone through this Golgotha and learned the pain of defeat and flight into the unknown yonder. At that time, however, we thought that only we had failed the exam in history.

And the Germans? Seemingly convinced of their chivalry, they once reproached the English for dropping bombs on civilian populations. Yet, at the peak of their power, they did not hesitate to throw carpets of bombs on unarmed people and pelt fleeing pedestrians with machine-gun bullets. Along the roads there were burning villages, towns, and cities and piles of dead civilians. All this was done according to the theory that the more cruel the war, the more humanitarian, because it will be thus all the shorter. As I write these memoirs, I see the Germans fleeing from their cities which are now being bombarded. I have no compassion for them. We will not invalidate the concept of the "humanitarian" blitzkrieg until those who first killed unarmed people go through the same suffering. Evidently, in order to have compassion one has to go through the same agony. Wars will not cease to plague mankind until all learn the consequences and the pain. *Chacun à son tour.*[6]

When the tragic truth about the defeat penetrated our consciousness, some kind of an inner spring broke in many of us. Those who because of their profession were all set for a battle and victory were the first to break down. Officers were seen requisitioning private automobiles and fleeing with their families. Abroad, to start a new combat? Who could tell at that time. The army was crushed, communication disrupted, and the air force incapacitated in the first days of the war. The soldier wanted to fight but he no longer received any orders. Whole regiments were led by young officers. Something that had kept up the nation's spirit now broke down: self-respect. "A country of seasons"—resounded the infernal laughter from the West.[7] Rather die than survive one's homeland. But, die with honor and dignity: in battle.

This was how we felt until the extraordinary tension found its political and military expression: Warsaw would be defended even if it meant the city falling into ruin and burying us all. We could no longer endure the sight of a beaten, retreating army, of soldiers roaming in search of their units and the enemy killing women and children with impunity. Our swollen despair, love for Poland, and our hate had to be converted into action.[8]

There followed unforgettable moments of immense distress and the most noble effort. We were determined to sacrifice ourselves and die leaving a legend to sustain the nation in the twilight of a captivity which, God only knew how long it would last. We knew we were exposing to destruction a cultural heritage built over centuries. However, the rapidity of the army's defeat was perceived by all as not only an infamy but also a moral burden for current and future generations; may this beloved city be destroyed, we thought, rather than swallow the insulting appellation of "country of seasons" that collapsed without fighting. When the population learned that Warsaw was to be defended, they felt relieved. At three o'clock in the morning, the radio aired the call to build barricades. I went into town and saw in the women and men a devotion, determination, and hate as I had never beheld before. They were overturning street cars, knocking blocks off the pavement, and erecting barricades. No orders were necessary. Every street crossing and suitable large house was being converted into a fortress. Workers formed battalions, physicians reported to hospitals, and Boy Scouts performed the function of messengers. They all knew what a siege meant. Oh, how easy it is to sacrifice one's life with the vision of victory. How easy it is to die when one has to sacrifice only oneself and a comrade who is ready to die. But, how infinitely difficult it is to make this decision when one has to sacrifice those one loves and seeks to protect with all the power of the inherited instinct: wife and children. To sacrifice them without a hope for victory.

This was the mental and emotional state of Warsaw's citizens in this one and only historical moment. Most were overcome by a surge of enthusiasm, loyalty, selflessness, and readiness to die as one would never suspect in peacetime. While I had desired to be in the whirlpool of great events all my life, here I was experiencing the greatest and most powerful of them— complete engrossment in the common cause and complete fusion with the whole.

This was the historical background of those unforgettable September days.

Let us turn now to my personal memories.

The first of September. The first bombs in Warsaw and the first wounded. An immense excitement prevailed. A great wave of patriotism swept the country. Those who knew how to look saw the treasure of inextinguishable love for one's homeland. The Polish people had matured and were worthy of independence. But unfortunately, the phrase "strong, coherent, ready" was a curse. At first, we had completely believed all communications. Within a few days it became apparent that we had misunderstood our readiness and strength. The government was going into exile. Offices received orders to evacuate. What was I to do? I made up my mind when Warsaw was surrounded and decided to defend itself. I would stay. At the institute, staff members kept going in and out of the bomb shelters. I was suffocating in the atmosphere of passive expectation. The radio announced the order to build barricades and for the men to leave Warsaw. The order did not make much sense.

The wounded were now countless. A woman physician from the Ujazdowski Hospital came to the institute and told us that the military hospital had been evacuated, leaving soldiers without medical assistance, and no supplies. I asked her whether blood transfusion had been organized. It had not. I immediately collected blood from myself and my assistants and reported to the hospital commander. He was extremely grateful and the blood was all used up within a few hours. My proposal to organize a blood transfusion station was accepted with enthusiasm. I must mention that I had been thinking for many years about the matter of organizing blood transfusion in case of a war. With this thought in mind, I had persuaded students belonging to the Medic Circle to organize blood donor centers. For many years the institute had performed an examination for them free of charge. In the years immediately preceding the war, I supported similar plans undertaken by the Women's Cadet Corps. When the crucial hour struck, however, there were no Women's Cadet Corps, nor medical school students' club, nor the Red Cross Blood Donor Center. Telephone calls remained unanswered. There and then I understood that there was a job for me.

The hospital administration handed the bacteriological laboratory over to me. On the radio and through the press, I appealed to Warsaw citizens, mainly to women, to donate their blood for the wounded. My faithful associates were Róża Amzel[9] and Zofia Skórska. Dr. Olgierd Sokołowski saw to the clinical examinations. (Róża Amzel and Olgierd Sokołowski were later killed by the Germans; Zofia Skórska was sent to the Auschwitz concentration camp in 1939.) Zofia Mańkowska and Stanisława Adamowicz[10] were in charge of the office. Irena Nowak, a laboratory technician in the institute, also came to assist us. And a young Boy Scout, Piotr Osiński, offered his services as messenger.

Those were unforgettable moments. Again, there arose in me the strength to stimulate those around me and guide them. In response to my appeal, hundreds of women arrived. In a few days I had enough blood donors to supply not only our hospital but all hospitals in the city. We set up blood transfusion desks in the various hospitals and supplied surgeons with the addresses of blood donors living nearby.

Streetcars were not running, and people had to run from one point to another avoiding the bombs and missiles by taking shelter in building entrances. And under such circumstances, women from all social strata came to the blood donor center and other hospitals. Once I received a telephone call from the Hospital of the Elizabethan Sisters with the request to send blood donors; the person on the phone warned me that the hospital was in the center of the battle and that coming to that area meant risking one's life. I asked the ladies who were present whether someone was willing to take such risk. All stood up with an attitude full of determination; I selected two. Another time, my telephone rang at ten at night: the Holy Ghost Hospital had admitted a large number of wounded persons and needed more blood donors. I sent out a radio message asking blood donors living in the vicinity to report to the hospital. It was night, houses were on fire, missiles were exploding. Yet, a crowd of blood donors arrived in the hospital. Women from whom I could not take blood because they were ill wept and asked whether they could offer something else, for example, money for the wounded. The need for sacrifice but also the respect it evoked were so great that those women who had a blood donor certificate did not have to stand in line for food but were let forward. How happy those women were when they could offer their blood to a wounded man. Sometimes, they would go back to see and take care of him. He was "their patient." The women of Warsaw have left a beautiful page in the history of that war. I can testify that they donated their blood with infinite devotion. I directed my appeals mainly to women, because blood donation had been previously initiated by the Women's Cadet Corps. However, men also reported. I remember three of them. The first was a priest who wanted to make his contribution by donating blood. The second was a murderer. While in military service, he once killed a man in a moment of passion, served his prison sentence, and was discharged from the army. Now he wanted to do something definitely positive for his fellowmen. He was an extremely courageous individual: he carried wounded under bullet fire, several times completed dangerous missions, and often voluntarily accompanied me in my passages across the Poniatowski Bridge, which was extremely dangerous. I have a pleasant memory of him; he was a selfless man. The third donor was a Jew, a veteran of the Great War. He was a poor worker who joined the war in 1920 as a volunteer, lost a leg, and now would come every day on crutches from the Jewish quarter to the hospital to offer his blood.

His name was Jakub Zylberberg. He spoke beautiful Polish and loved Poland. Later, he got in touch with me in the district. I did everything I could to save him from death by starvation. Eventually, he probably was murdered. He deserved a better fate.

I will not write any more about blood transfusions during the siege. This memory is very dear to me. I realize that compared with the millions of dead, the handful of people saved does not amount to much; however, one cannot reason that way. Every life saved and every tear spared has its value. But there was still another factor: honor. How could we have borne to remember the siege of Warsaw with the feeling that we did not do everything to save the lives of our soldiers and citizens? A man who dies because he received no aid remains on the conscience of those who survive. Honorary badges mean little to me; however, the words of thanks I received from the head of the army health services are dear to me.

My work during the siege was not limited to directing the blood donor center. The hospital administration also put me in charge of the military bacteriological laboratory. Given the lack of gas, supplies, and personnel, not much could be done. Eventually, the laboratory assistants reported back to work. I often visited the hospital wards. They looked terrible and reminded me of the worst moments of evacuation in Serbia during World War I. The patients lay in rows on the floor; there were no drugs, dressing materials, nor any linens. Everything had to be improvised. I am still terrified when I remember the order to evacuate the hospital one day before the defense of Warsaw was to begin. But I am always deeply moved when I recall the selflessness of the nurses and physicians. During the worst bombarding of the hospital, Professor Melanowski, an ophthalmologist, and his daughters were carrying wounded men out of the burning barracks.[11] Professor Grzywo-Dąbrowski was performing autopsies, while Dr. Kacprzak was helping to dress wounds. It took about two hours to get from the laboratory to the hospital barracks. While walking through the corridor, one could easily lose one's life, because bullets were flying and bombs exploding all around. The physicians displayed much courage, never stopped working, and never allowed panic to develop.

I did my share to preserve the proper atmosphere in the laboratory. The blood donors arrived under a shower of bullets and entered the laboratory as if it were a temple. I did not want them to find an atmosphere of panic or to see staff members hiding to take cover. Only in moments of the greatest danger did we all, staff members and blood donors, go down to the shelter. Most of the time we did not interrupt our work. I tried to preserve a cheerful, even humorous atmosphere and must admit that it was not difficult to keep calm with such selfless personnel and blood donors. Whosoever had the good fortune to participate in any way in Warsaw's defense will always retain beautiful memories. We felt that the country was perishing but a legend was being born. Warsaw's defense not only saved Poland's honor but also created a legend which, I hope, will encourage future generations. What I saw in my sector was only a fraction of the nationwide patriotism and enthusiasm. Indeed it is not my place to describe the courage of our children who were throwing burning gasoline bottles under German tanks, the bravery of our soldiers, or the selflessness of the civilian population.

During the siege I lived at my sister's in the vicinity of the hospital. The way to the hospital was very dangerous, but the most treacherous was crossing the bridge on my way home, where I went every second or third day. "Are those I love most still alive?" I asked myself continuously. Surely, I would feel their death or even their wounds from far away. When I saw the glow of fire in Praga, I would approach my home with painful anguish.

My wife was put in charge of the sanitary and rescue center in the clinic; in fact, she found herself directing the entire clinic because almost all male physicians had been mobilized. She soon had to transfer this function to her women colleagues, because the State Health Center in Saska Kępa lost its physicians and kept on asking her for help. In addition, the nursery near our house was left without medical care and she also had to help there. Most important, the number of wounded soldiers and civilians was growing, and it was impossible to transfer them to the city center. At the request of the Saska Kępa Citizens' Committee, she enlisted the help of Dr. Leskiewicz, a surgeon; Dr. Jabloński, a veterinarian; and Gromczewski, a medical student and recruited the Red Cross ambulatory and personnel. Within one day she had converted a nearby school of economics into a hospital. Since there were neither beds nor linens to begin with, she appealed to the inhabitants of Saska Kępa and, owing to their great generosity, the most necessary items were soon obtained. Saska Kępa was crisscrossed with ditches. One of the front lines ran some 1,000 meters from our house. Our daughter and Jacek's mother[12] would steal across to carry food to the soldiers in the ditches. There, the danger of bombing was less great, but artillery missiles were raging. Crossing the street or even getting to the hospital, which was in the next building, was very risky. Nursing was carried out by two Red Cross "sisters" while several volunteer ladies performed auxiliary functions. All were working with great selflessness. It often happened that even before the end of a battle, the nurses and stretcher-bearers would pick up wounded men from the field. One stretcher-bearer was killed while so doing. Missiles fell into patients' wards. The operating room was transferred into the basement, a windowless room. When there was no more electricity, surgeons operated by the light of oil lamps. The number of wounded was growing, and there was no means of transportation. It was only after my vigorous intervention that we succeeded in organizing transportation. Our house was hit by more than a dozen shrapnel missiles, its front wall was partly demolished, and the roof had forty holes. Our friends who had lost the roofs over their heads came to our house. Eventually, our house, including the basement, kitchen, and two servant's rooms, accommodated fourteen persons.

This was the way my family lived. This was the way most women and men physicians as well as students worked. Happy were those who found an outlet for their generous readiness to help. Many of those who spent all their time hiding in shelters avoiding the air strikes ended up dying. Such an attitude of taking permanent refuge provoked permanent resentment in many. One day I was informed that Saska Kępa and the right bank of the Vistula River would have to surrender. Only then did my wife agree to move to Warsaw. I got hold of a small sanitary truck with which I evacuated my family and several other persons. A few men and women remained in our house. "If we survive the war, I would not want to reproach myself and you that we abandoned helpless people," I said to the driver. "Are you willing to make one more trip?" "If you go with me, I'll go back," he answered. "Of course I'll go with you." So, we returned under a shower of bullets and bombs being dropped on the bridge as we drove over it. But we managed to bring all our friends over to the other side. I wanted to give the chauffeur an extra tip, but he refused. So I only shook his hand. A somewhat different atmosphere prevailed in the Warsaw shelters: egoism, fear of death, and elbow-pushing. Yet, he who did not see the immense selflessness of Warsaw will have missed much in life. Whenever I think of those moments, a poem by Tetmajer springs to my mind:

Nothing will ever rob my heart of the memory
Of those moments when I saw your tears.[13]

And then came the memorable day when four hundred planes relentlessly bombarded the city. It was dark from smoke, the buildings began to sway and collapse. People ran crazily from one house to another in search of shelter. Dead and wounded men lay strewn on the streets beside horses. This is how it will probably look the day the world ends. But, despite the terror and the death that could befall any one of us, people lived in a state of passion. We were all aware that we were fighting for something infinitely more important than individual life, that we were fighting for the nation's honor. No, we would not allow others to call us a "seasonal country." We had sinned by being haughty, frivolous, and anarchic. Yet, he who knew how to look saw the devoted eyes of those soldiers and officers who stayed at their posts.

The radio continued broadcasting only in the early days. The dignified and ardent speeches given by City Mayor Starzyński corresponded to the needs of the moment. I do not know whether a monument will ever be built to President Starzyński; however, in the hearts of Warsaw's defendants he already has his monument.[14] The trite songs that the radio frequently played were unpleasant. We needed heroic symphonies and not plain songs about little soldiers with their shining swords. By accident, I once tuned in a German station. It was playing the heroic "Fire Song" from Wagner's *Valkyrie.* A song of heroism would have been more suitable for us, too.

Warsaw's heroism was in vain, and the city had to surrender. We hoped to the very last moments that we would not live to see it. But we were not spared that moment. We saw the German divisions march in; we heard the thumping of boots and the songs of victory. We saw our soldiers lay down their weapons. They were distressed and broken, their faces as gray as their uniforms. It was obvious that something essential in them had been shattered. Two images of that period have remained in every Pole's mind's eye. The distressful sight of fleeing high officers and officials, the exodus of masses indiscriminately following Umiastowski's order, and such a chaos that on the second day of the siege it was necessary to appeal over the radio for people to turn over wound-dressing supplies. The second image is that of the heroic defenders of Warsaw. I wonder which image will predominate and shape the minds of the younger generations.

After Warsaw's surrender, I went downtown. The sight was ghastly: there were fragments of destroyed barricades on almost every street corner, some houses were still burning, and the windows in partly destroyed houses resembled eyeholes in skulls. Buildings hit by demolishing bombs were the worst to behold—the interiors were disclosed, such that the intimacy of family life became the neighbor of death. I came to the city to find out what happened to my relatives and friends. It was terrible to approach a home uncertain of whether I would find my friend or only his dead body.

I handed the military laboratory back over to the military bacteriologist, and I returned to the Institute of Hygiene. It was spiritless. Besides the fact that the building was seriously damaged, I noticed deep psychological injuries among the personnel. If I felt lonely before the war, this feeling was magnified now. I was happy during the siege, because I saw heroism and was modestly able to alleviate the suffering of others. The siege pulled a hidden trait out of me: some kind of a pathological pathos and aversion toward cold and indifferent persons. This aversion became so great that an abyss opened between me and those who did not suffer or, while suffering, thought only of themselves. I believe that many people who have gone through hell feel that way. They will always be lonely. Others will forget them and indulge in their little pleasures and conflicts. Those who have peered into the abyss of suffering are subsequently unable to enjoy themselves. It was my fate to visit hell and, with the rest of my waning strength, comfort those who were doomed to death.

Photos courtesy of Joanna Kiełbasińska Belin, except as noted.

Figure 1. Jenny Ginsburg, Hirszfeld's mother, circa 1900

Figure 2. Hirszfeld at age 4; at age 18; in Zurich in 1915; and in Serbia in 1917

Figure 3. Bolesław Hirszfeld (1849–99); Emil von Dungern; and Ludwik Hirszfeld in bacteriology laboratory in 1917 in Salonica

Figure 4. Hanna Hirszfeld in Salonica, 1917

Figure 5. Hanna Hirszfeld, various dates; Marysia (bottom, right) in 1938

Figure 6. Ludwik Hirszfeld, oil painting by Roman Kramsztyk, circa 1925. Courtesy of the National Museum (Muzeum narodowe), Warsaw.

Figure 7. Ludwik Hirszfeld with students before World War II. *Source:* Archives of the National Institute of Hygiene (Warsaw).

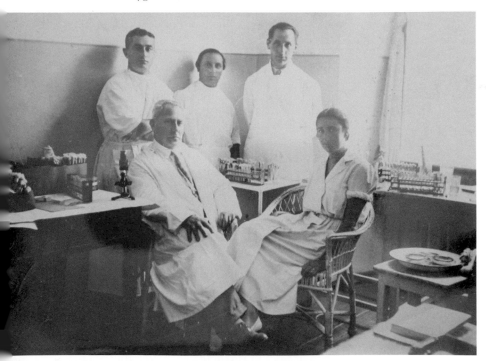

Figure 8. Lab researchers, including Róża Amzel, 1931

Figure 9. Hanna Hirszfeld, pastel portrait by Roman Kramsztyk drawn in Warsaw Ghetto shortly before artist's death. *Source:* National Museum (Warsaw).

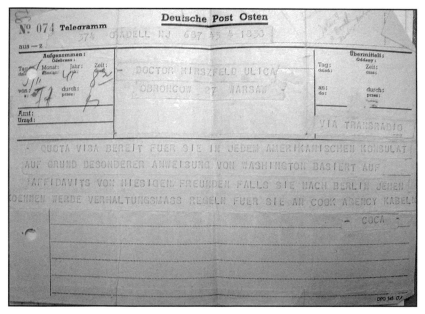

Figure 10. Telegram from A. F. Coca to Hirszfeld in 1941 offering visa for escape to the United States. *Source:* Archives of Polish Academy of Sciences (Polska Akademia Nauk), Warsaw.

Figure 11. Ludwik and Hanna Hirszfeld on boat to the United States, 1946

Figure 12. Hirszfeld with Anigstein (?) during his trip to the United States, 1946

Figure 13. Hirszfeld in lab with microscope, 1947

Figure 14. Hirszfeld's portrait, 1948

Figure 15. Hirszfeld with students, from left to right: Stanisław Dubiski, Maria Grabowska, Halina Liberska, Krystyna (?), Tadeusz Luszczyński, 1950; Hirszfeld reading, 1952

Figure 16. Funeral procession of Hirszfeld, March 10, 1954

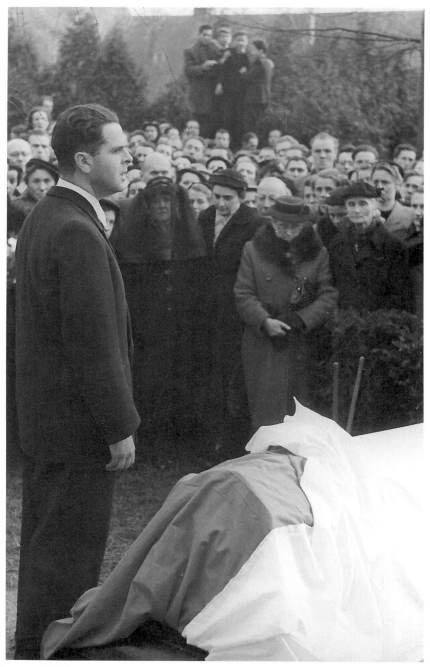

Figure 17. Felix Milgrom speaking at Hirszfeld's funeral; Hanna, with veil, in background and Joanna Kiełbasińzska Belin to her left, 1954

Chapter Twenty-One

Ousted

In the institute I found external and internal chaos. The institute was without management and life. I began to carry out rubble and glass. This moved the attendants, and they began to help me. The professional personnel were morally shaken and broken. Their recent experience had engendered a temporary atrophy of the will. Staff members who had left returned a few days later, including the director, Szulc. The institute at least regained its official manager. I knew that the victors would order me to leave, and I wanted to tidy up everything as anyone leaving for a long trip would do. I cleaned up my office, carried away laboratory equipment that was my personal property, and hung up photographs of scientists who were my friends. I wanted these pictures to stay there after I was gone and to remind my associates of the work I had done. This way they might remember me throughout the war. One day I saw through the window the director accompanying some Germans to their automobile and I thought: I'd rather be thrown out than have to behold such a sight. Another day Professor Schmidt, director of the Behringwerke[1] and a good scientist, came to visit; I had only recently met with him in Paris. A human instinct inspired him to ask after the health of my daughter whom he had met at a banquet in the Pasteur Institute. Finally, one day at the beginning of November, Szulc came back from the department and, greatly embarrassed, informed me that the Germans had ordered all non-Aryans be dismissed. He comforted me that this would be of no consequence and that I should regard it as a vacation. I accepted the news calmly, called all staff members together, and spoke to them. I told them that the enemy wanted to destroy everything that represented the Polish spirit and to convert us into a nation of servants, and that though I was the first one to be hit, this blow was directed against every creative thing and every creative individual. I said there would be more victims, but that we should endure the enemy with pride and hope and only those who would agree to play this role would become servants. I promised that I would visit the institute once a week, and I advised them to get busy reading and writing, to resist melancholy, and to believe in a better future. I no longer felt like a stranger at that moment but like a father speaking to his children.

This was the first time I went home as a scientist without a post. I knew this was the most difficult game to be played. An institute is to a scientist what a canvas is to a painter and what the chessboard is to a chess player.

All they have to do is to spread paint or move figures. Now, my imagination had to create everything that was not available: the vision of a picture, the colors, and the canvas. I was like a chess player who had to move his figures without a chessboard. At the same time, I recalled the insulting words spoken by Łódź's Gauleiter [district commander] Greiser:[2] "Die Polen sollen nicht verdienen, sondern dienen." [The Poles must not earn but serve.] No, they would never see the day of my breaking down. I was strangely proud to remember that I had done more for German science than they. And now they were throwing me out like a dog and did not even consider that I might die from starvation. An obscure soldier was ordered to dismiss me. His name was von Espe. I met a friend and told him in a strange, visionary state: "So, he dismissed me; that's all right. But the fact that he did not even ask whom he was dismissing, that I was a dead object for him and not a human being who had worked all his life for science and in part for their own science—that I will neither forget nor forgive. I feel that he will be punished—that he will soon die." Indeed, three weeks later he died of a shot in the stomach.

I felt that I had to put a firm grip on my emotions. It was cold at home because windows were broken and the central heating system was out of order. My heart was filled with the vision of a perishing nation. Scientific problems had sunk deeply into some secluded recess of my mind. Now I had to bring them back to the surface. I must try to recall the role of infections, the significance of blood groups, and dozens of other problems. I allowed myself one day of rest—Sunday—and on Monday I set to work. I was accustomed to directing a staff of associates. Now, I was a commander without his army. I had only Róża Amzel who had been dismissed for the same reasons. She was my faithful and devoted companion and she knew my thoughts. She alone was to replace the atmosphere of the institute, of my associates and pupils. The two of us had to create the vision of a pulsating scientific life. Where should we begin? We decided to work on data concerning the heredity of the transitory forms of blood groups. After that, I would fulfill my old dream of writing a textbook on immunity. I had repeatedly begun to do so, but had always been interrupted by my passion for experimentation. Chapters in the making had given rise to scientific work. Now, circumstances were forcing me to write. We installed ourselves in the basement next to the kitchen, in the only place where the cold was less penetrating. We reviewed the protocols and gradually wide horizons began to unroll. We defined the various forms of transitional groups and formulated laws of heredity in a manner considerably expanding the significance of observations Dungern and I had once made. And so it was that sitting in a cold room next to a kitchen, rejected and aching, I was building the future edifice of the science of blood groups and making plans for new trips to distant lands where the first blood group mutations had probably taken place. We wrote a short article for the Swiss press and succeeded in sending it out due to the courtesy of a diplomat I knew. I sent two large monographs [*sic*] to Paris through Switzerland.[3] The articles I and my assistant Morawiecki sent to the Swiss press were published during the war;[4] I later read long reviews about these two papers in the German press. I believe those were the only two papers written by Polish scientists that appeared on the international forum during the long night of occupation. Maybe they have proved to the world that we did not break down morally. I began to prepare the second edition of my book on blood groups. Every Wednesday I went to the institute and took my former secretaries something to type. I no longer had any authority over them; were they to type my manuscripts they would do so because of the unbreakable ties of memory that bound us. And I remember with pleasure that by and large, they willingly did this work for me. Whenever I entered the institute my wound would reopen. Every careless word or gesture hurt and, at the same time, every manifestation of friendliness touched me deeply. Once, I met Dr. C. in the courtyard; he hardly greeted me. I had previously liked him, and now I threw him out of my heart. One assistant was absent several times—which was hardly subtle. Yet overall I was surrounded by warmth and loyalty. Intellectually I might have been lonely, but ties of joint experience and interests still existed between me and this group of people. After a while, an order was issued prohibiting non-Aryans from walking on the Aleje Ujazdowskie.[5] Thus, while walking to the institute I could have been arrested; however, I considered it my duty to carry on something that I regarded as my mandate. Once, Dr. Kacprzak visited me and mentioned that by coming to the institute, I was endangering myself and others. "Only myself, Mr. Colleague," I replied, and I kept coming until a German was named director of the institute. This was Professor Nauck from Hamburg. He did not consider it his duty to visit me or to ask through someone else whether he could help me in any way. He acted as a porter who was ordered to throw out a stranger. He should not be surprised if, after the war, the world will treat him not as a scientist but as a porter.[6]

After we completed the manuscript, I started working on the textbook. Again, Róża Amzel was my loyal companion. For two days I dictated a chapter, then we corrected it, and for the following two days I dictated from the manuscript while she was typing. On Saturdays we met with Szymanowski and read what had been written.

This was a curious dictation. I believe that everybody who writes a textbook sees his students with their eyes fixed upon him. I called upon the power of my imagination to evoke the image of my pupils: those of the past, those of the future, and those who would never see me and yet would live in the atmosphere of my words. A textbook is not an exposition of someone else's ideas; a textbook reflects the author's mentality and his attitude toward the younger generation through which he wishes to live on.

I recalled how I had defined the secret of a good speaker: he who wants to kindle others must be himself alight. But how was one to keep the flame burning without seeing those who desired to absorb the author's or the speaker's warmth? How difficult it was to kindle fire in oneself when one did not know whether those to whom the words were directed would survive. Would Poland and Polish science survive? Would not the nation's spirit break down in the overwhelming tortures; would the people preserve the resilience required in order to absorb new ideas? It is difficult to radiate when the rays have no one to behold them. Writing a textbook is not difficult for a man who has been thinking and working all his life. But this was a difficult task for a man who was separated from life and, in a cold kitchen, had to evoke the image of a university auditorium, attentive students, and creative colleagues. I do not know whether my textbook reflects this atmosphere: the state of mind of a man who is praying in a demolished temple.

From time to time I walked through the city and thought of the last moments of the siege. I was particularly haunted by the vision of mercilessly bombarded hospitals. The Holy Ghost Hospital had been burned to the ground, while the Christ Child Hospital was partly in ruin.[7] I remembered how we had hung up a red cross over the clinic of the Ujazdowski Hospital; it only attracted airplanes, and soon the clinic was destroyed. I saw low-flying planes dropping bombs on our hospital and on the wounded patients being carried out. How often it happened that crowds waiting in front of food shops and people fleeing on highways were machine-gunned. Now, when I read descriptions of how German cities are being bombarded and see the Germans' dismay that our allies are destroying their hospitals, I say over and over again: "You asked for it." On the streets of Warsaw, I saw people waiting in lines for a bowl of soup. This was a service organized by the city, but the people were told that this was the occupier's gift. These manifestations of charity were photographed, and the films were sent throughout the world as proof of the occupier's good-heartedness.

To describe the people's amazement at the sight of what was going on, I must use a metaphor. If a mad dog runs out on the street and bites pedestrians, it hurts but causes no amazement or revolt. However, when a human behaves like a mad dog and bites without gaining any advantage but just to satisfy his internal compulsion to bite, this evokes indignation and disdain. If at the same time this biting individual pretends to be a good-doer and maintains that he introduces a new order for the sake of higher principles, he induces the utmost disgust in others.

On October 17, 1939, the Dresden Philharmonic Orchestra arrived in Warsaw to give a concert. The newspapers were writing about a lofty spirit which, even in the fire of combat, did not neglect spiritual needs. On the same day, the first roundup took place on Warsaw's streets, and the astonished population saw that it was possible to catch people like wild animals and kill the defenseless. The parents of the captured children came to me asking me to write applications for their release, especially since some were only children and others were sick. A few days later they told me that all efforts had been in vain, that they had been notified of the death of their children. After five days! People were often informed that the body had been cremated. The price for the ashes was five marks. All libraries and reading rooms were closed. After a while they were reopened, but only for Aryans and after the most beautiful works of Polish literature had been removed. All universities were closed. The library of Poznań University was partly deported and partly burned. Polish bookstores had books requisitioned and destroyed. After a while, Polish bookstores were prohibited from carrying cosmopolitan literature. The occupier understood that this type of literature was shaping the nation's mentality. You might think that having understood this, he would raise another torch and attempt to spread his own vision through new books. Not so—only the master nation could have a cosmopolitan worldview. The same was true of science and education. German scientists visited Polish scientific institutes and removed everything they saw. In the Central Agricultural School, they even took the coat hangers. A certain German professor who had once visited the Institute of Experimental Physics returned: he remembered all the nice appliances he had seen and took them away. The Institute of Mineralogy, the Institute of Anatomy, and many others were stripped in the same way. The professors of Cracow University were invited to a lecture on the relation of National Socialism to science. The lecture was set up in such a way that all professors who arrived were sent to a concentration camp in Oranienburg. There, they were beaten and maltreated. Eighteen of them died.[8] All scientific societies were dissolved and their assets confiscated. All scientific journals were discontinued. All newspapers were abolished, with the exception of one that represented the enemy's interests. A commentary appeared in the German press: "What an uncivilized nation these Poles are; the capital city has only one newspaper, and poorly edited at that." The occupier was musically inclined and once every few weeks organized concerts in the Brühl Palace. But we were forbidden to play Chopin. Yes, this nation understood the significance of music. "Let music disappear," I thought, "if it is to promote such objectives."

After a while, the Germans allowed us to reopen a few schools but on a considerably lower level. All theory was eliminated, and the teaching of history was prohibited. Only elementary and vocational schools received permits. A secondary school was unnecessary for a nation of servants. Algebra was forbidden in business schools, only liturgy was permitted in theological seminaries, and the composer's class was discontinued in music schools.

I am describing only fragments of what was going on.[9] This was a nation of poets and thinkers. I beheld the hideous caricature of a nation that had reached the heights and, having grasped the educational significance of science and art, had drafted them into the service of conquest. The scientist became, a state soldier. And they fully succeeded. That is what the German scientist became, fully and without reservation: not merely a soldier but a state mercenary. As a result, he lost the world. One can forgive the cruelty of the soldier on the battlefield, and one can forgive the hungry man who keeps his slice of bread for himself. One can even understand when the satisfied man who, thoughtful of the future, does not care for the hungry man. One cannot, however, forgive a scientist who does not want to share the products of his intellect with others and even robs them of what they possess. Nor can one forgive the artist who wants to reserve beauty for a few men only. Those wretched scientists and artists will always carry the stigma of traitors, because they have betrayed more than their homeland—they have betrayed the dream and hope of mankind that on the summits free spirits can discuss matters beyond intertribal conflicts.

That is how it was on the summits. Nor were the lowlands of life forgotten. Soldiers had the right to rob and make riots. Let them brawl. Thus, we saw awful scenes of robbery accompanied by uncanny hypocrisy. Non-Aryan medical offices, dental offices, roentgenological institutes, private libraries, instruments of all kinds, furniture, clothes, underwear, jewelry, and money— everything was taken away. I must describe a few specific scenes, because they reflect the reality far better than a terse report.

The Germans entered a Jewish apartment and literally took everything: pictures, carpets, furniture, shoes, and so on. The mother asked them to leave at least the child's bed. "A Jewish child does not need a bed," she was told. A few soldiers entered the apartment of Docent Brokman.[10] They were surprised by how few shirts he had. He told them that he was not rich. "Yes," one soldier said, "honesty does not pay," and he went on stealing. Two soldiers came to the apartment of a music teacher to steal her piano. They told her to play something, so she began to play a number by Beethoven. The Germans—generally musical people—were impressed. One of them was even inclined to leave the piano, but the other commented in anger: "That a Jew should have a Bechstein!" Really, what a pernicious nation! And so the piano was shoved out of the apartment. Jewish physicians' libraries were requisitioned. Nor were the scientists and physicians of the master nation embarrassed to use them. A young military physician walked into the study of Dr. Srebrny,[11] an elderly ear specialist of great merit, and unceremoniously began to pack his library. Dr. Srebrny had published valuable scientific papers in various languages, including German. One of his books, published in German was lying on his desk, and the following conversation took place. "Is this your book?" the German physician asked. "Yes, I wrote it when the Germans had respect for science." "Times have changed," the German replied and went on to say: "Haven't you ever heard of the Roman soldier who killed Archimedes?" Dr. Srebrny's singular response should scald every German with shame: "Indeed, I have. However, the name of Archimedes is known to you, to me, and to others, while no one knows the name of that Roman soldier." Those were prophetic words for German scientists, too. The world will probably never forgive them such degradation of science.

And here is another scene. A Jewish attorney who spoke beautiful German was being arrested. Did his perfect language impress the Germans? One was heard saying to the other, "What thieves those Jews are, they even stole our language."

Jews were not permitted to have more cash than 2,000 złotys. During an inspection of a Jewish attorney's apartment, it turned out that the German officer and his victim had been friends while studying at the same university. A conversation began, memories livened up, and the atmosphere of a youth spent together developed. "How much money do you have?" the German suddenly asked. The attorney, overcome by the pleasant atmosphere, told the truth, namely that he had several thousand. "Give me half," his friend said.

Refugees from Łódź began arriving. Everybody had been robbed and thrown out of the city, Poles and Jews alike. I know of cases where individuals who had been family friends for years and knew the value of the carpets and china belonging to specific families would now show up as *Volksdeutsch* [ethnically German] and take everything. A physician I knew told me about the tortures to which fleeing Jews were submitted: they were forced to run barefoot on broken glass; when they became thirsty, water was poured on the floor and they were ordered to lick it up. If I wished to describe such atrocities in detail, I would have to write entire volumes. However, my prime objective is to describe scientists, intellectuals, and leaders. When the stealing of furniture was at its height, and when people were being thrown out onto the street deprived of their last handkerchief, the daily *Völkischer Beobachter* printed the following commentary: "Horror propaganda maintains that we are taking furniture away from the Jews. The fact is that we loathe Jewish furniture."[12]

Let us turn to physicians. In Poznan and Łódź voivodships, the medical offices of Polish physicians were taken over by Baltic physicians. They presumably were told that these offices had been bought for them. I did not hear of a single comradely gesture of compassion. The Polish physicians were not permitted to keep a single needle or a syringe. In business relations the barbarity and cynicism were beyond description.

Dr. Vieweg was put in charge of the Social Security Health Center. He issued an order that under penalty of death all physicians must prove their Aryan descent for three generations. Non-Aryans and all women physicians were dismissed.

Professor Richter, author of some scientific book, became district physician. His assistants kept pestering the Polish physicians working under him

to buy his book. The lack of professionalism was striking. In 1942, posters o.
typhus were hung on poles telling everyone who found a louse on himself
to report to a physician. A decree was issued that every case of fever had to
be reported. And those were supposed to be professionals heading the sanitary department? Did they not realize how unrealistic such measures were?

Dr. Schrempf was the municipal physician, and a man characterized by
both energy and brutality.[13] He treated the physicians like orderlies, spoke
to them as if they were wagon drivers, and often threatened them with his
revolver. He was a tuberculosis specialist. I must admit that he effectively
introduced spa treatments, but he had not the slightest idea about our local
epidemiological conditions. If there was a case of typhus in an apartment
house, this house and the two neighboring houses were locked up for three
weeks. I will describe his measures in detail when I come to the discussion
of the Jewish district. The futility and uselessness of these measures greatly
contributed to the spread of typhus. He fired the most deserving hospital
directors—some were sent to camps—and in their place often hired very
young and completely inexperienced physicians.

I have already mentioned that all professional journals were discontinued. Medical journals were replaced by *Zdrowie i Życie* [Health and Life],
published in both languages as the government's official medical journal.[14]
They obviously had a poor idea of the standard of Polish physicians since
they dared to present them with such a product. Need I add that subscrip-
tion to the journal was obligatory; what splendid income this medical trash
must have brought! I will mention just one curio: tooth caries were supposedly caused by too small a jaw. In order to increase the jaw size, the author
advised throwing food morsels to the child so that, at an early age, the child
would learn to catch the morsels in the air. He concluded this nonsense with
a pathetic statement that if his method were followed, in fifty years there
would be no more dental caries in the General Government.[15] We wondered
whether ignoramuses were deliberately sent to Poland or whether the medical standards in Germany had actually fallen so low.

Such anecdotes could almost be amusing were it not for the great tragedy
they reflect. Articles appeared stating that Jews were carriers of germs that
did not induce diseases in them, but that in the interest of the Aryan population's health, Jews should be avoided. Presumably for this purpose, Jews
and Christians of Jewish descent were ordered to wear armbands.[16]
Disinfectors were sent to apartments from which Jews had been thrown out
in order to kill these dangerous germs. Separate street cars were allocated
for Jews to protect Polish people from infections. Jews were not permitted
to travel by train unless they had permits issued by state-employed German
physicians. This was quite a good source of income. Aryan physicians were
forbidden to treat Jewish patients and vice versa. And eventually, there was
talk of the necessity to establish a Jewish district in Warsaw, just as such ghet-

tos were being established throughout Poland. The Society for Aid to Jews [American Jewish Joint Distribution Committee, AJDC] asked me to write a memorandum to the German authorities on the subject of Jewish racial pathology.[17] The objective was to elucidate the question of whether Jews were indeed carriers of germs to which they were immune.

The memorandum was of no avail. A Jewish district was established in mid-October 1940.[18] I know that Polish physicians warned the German authorities that such a displacement of populations might well cause an epidemic. The answer they received was further proof of their tremendous ignorance: "Where Germans rule there is no typhus." Only later, when they set the whole country afire and destroyed the hospital system did they perhaps see how foolish their measures were. When I am able, I will describe the epidemic as it spread throughout the country. Today I will limit myself to the events I personally saw. The armbands isolated no germ carriers but caused a terrible persecution of Jewish people. On the streets I often saw old persons being beaten on their faces by whippersnappers. I remember one scene particularly vividly. Rosenberg, an artist musician, was returning from his father's burial. He had a beautiful, animated face. A young pilot approached him and, without reason, started to beat him on his face. There were hundreds of such cases. I know many battered and mutilated persons— young and old, simple and intelligent. I was told that a German professor who—prior to the war—had been repeatedly a guest of a Warsaw family, also visited the family now and stayed for tea. Upon leaving, he asked his host whether he would like to see him home and on the way share in the fun of Jew-hunting. Indeed, every nation has its fun and games. The English play football, the Spaniards have bullfights. This generation of Germans was trained in brutality to Jews so as to get rid of all pity, conscience, and human reflexes before setting off to conquer the world. I believe that this is the main source of German anti-Semitism: people were taught to have contempt and hate. And it worked.

I will come back to the details of the Jews' Golgotha, since I was personally confronted with the abyss of their suffering. For the moment, I will discuss my personal memories.

I was working hard with no income. My wife was the family breadwinner with her medical work. The friends of my youth extended a helping hand to me. (After the war I learned that a group of American scientists collected a fund to ransom me from the Germans.[19] This fund was liquidated when the Red Cross informed them that I had died). I received letters from Verzàr and mostly from my former chief in Zurich, Professor Silberschmidt. I knew that it was not an empty phrase when, in his youth, he often said that we were one family. He invited the three of us to Zurich. It was no trifle to assume the duty to support three persons during the war. It was difficult for us to decide to leave Poland. My wife's mother was without means for subsistence, while a person very close to us was in a concentration camp and his family was without protection.[20] Should we abandon them all? We accepted the invitation for our daughter only, because we saw that she was ill and only departure and a flight from the terrible events could save her. The Zurich University enrolled her immediately as the child of a former faculty member. Unfortunately, I did not receive a permit for her departure. My friends in Yugoslavia wrote to me and insisted that I should come. Without my asking for it, they sent me entry visas and dispatched to their embassy in Berlin a request to do everything to enable us to leave. Staff members at the Yugoslavian Embassy in Warsaw were very friendly and advised us to leave. For reasons I have just mentioned, we were unable to make a decision before August 1940. It was then that our friend was released from the camp, haggard and ill, but we were happy that he was alive. We were vastly relieved. His family arrived from the country, and we were able to leave my wife's mother in his care. Thus, when we were psychologically ready to leave, a letter arrived from a friend of mine in New York inquiring as to whether I would be inclined to accept a position there.[21] I cabled back that I was. It was as if the prison gates were opening in front of us, and that we might still save our daughter. After a while, I received a telegram from New York informing me that I should come, that my friends has signed an affidavit for us, and that Washington had informed all American consuls to help us. There awakened in me hope not only for freedom but also for the ability to fight. Over there, in America I would be able to tell everybody how our people were suffering. But how could we get out of here? During the first months of occupation, exit permits were extended *larga manu*. Subsequently, they demanded a ransom of seven hundred dollars per person from foreign countries. Colloquially, this is called "gangsterism": to maltreat a man and then demand money for his release. Political language has not yet devised the proper term for this. I went to various travel agencies. They showed me a decree stating that the last exit permits had been withdrawn. Exceptions were made only for those who had special support.

At that time, something completely unexpected happened. I received a summons from the Brühl Palace to report there on a certain day at a certain hour. What could they want from me? It was not impossible that they wanted to offer me a job in the laboratory or in the institute. I decided to refuse; I did not want to take advantage of any special considerations. I was met by a good-looking man aged about thirty. He introduced himself as Dr. Kohmann and said that my publications were known and highly valued in Germany, that racial interests were correlated with the problem of blood groups, that he was personally very interested in subgroups, and so on. He asked me where I was working and told me that, in view of my scientific achievements, they would like to help me. I thanked him, said that I did not need any help, that I was writing a textbook, and that in view of his interest in subgroups, I would inform him about my recent research. He inquired about my descent and religion, and expressed his regrets that I was unable to become a professor in Cracow because their legislation was based on the principle of racism. During the conversation he told me that a special district would soon be established. I said that this would be a hard blow for me, because I was living in my own house and that my sick child would not be able to take such a move. "Do call on us," he said in a friendly manner, "we will dispense you." The conversation had proceeded in the civilized manner to which I was accustomed during my former contact with Germans. He asked me whether I could obtain a certificate from a professor in Germany that I had never acted or said anything against Germans. I said that I could not. I did not know what they needed this certificate for, and whether they would not order me to work for them. I explained that I might cause my friends trouble with a request of this type. When we were parting, he asked me whether this summons had frightened me. I tried to behave like a man on an equal social status and said, "No, I thought this concerned the invitation the Yugoslavian Government might have sent to me." I left with a feeling of relief that I was able not to hate at least one German. To my consternation, I later learned that Dr. Kohmann was capable of kicking Jews. In this conversation with me, he had behaved like a gentleman.

I kept applying for a permit to leave. I was one of the few who could possibly carry the news abroad. I had high-standing friends in Yugoslavia and America, and they would believe my story of what was going on in Poland.

In view of this, I decided to ask Dr. Kohmann to support my application for an exit permit. I did not ask him for a job to work for the Germans. I asked for permission to leave. At stake was the life of our child and a mission to be accomplished: to inform the whole world about the behavior of the occupier. At the same time, I asked him orally to keep his promise, namely to obtain permission for me to stay in my own house till departure.

Dr. Kohmann lived up to his promise. I soon received a summons from Auerswald who was in charge of the resettlement project and later became the commissar of the Jewish district.[22] I was met by a courteous gentleman with European looks. I was amazed later when I saw how brutal he could be.

Two weeks later, I received a written permit signed by the sanitary authorities stating that in view of my imminent departure to Yugoslavia I was entitled to stay in my house. The text began as follows: "Concerns a Jew, Professor Hirszfeld. The professor is the greatest contemporary expert in hematology, therefore . . ."

I kept applying for the exit permit. I had to prepare a list of the items I wanted to take with me, state their value, and submit the list to the so-called foreign exchange commission. The commission would determine a certain—and rather considerable—payment to be made, and only then would

I be permitted to apply for the passport. If the passport was not issued, the payment would be lost.

So far, neither I nor any member of my family was wearing the armband. I even went to Dr. Kohmann and Commissar Auerswald without it. One could be imprisoned for such an affront. At that time, the Main Custodial Council[23] announced that it was intervening on behalf of Christians of non-Aryan descent who had made valuable contributions to the country as well as their families so as to be exempt from wearing armbands, from forced labor, and so on. We registered. Not only my and my wife's work but above all my uncle's name was inscribed in the history of Polish culture. We were granted the exemption without reservation. We had no idea at the time that this exemption would accelerate our misfortune. The exemptions issued by the Main Custodial Council proved to be a trap. Even though there was no ill will on the part of the members of the Main Custodial Council, they were guilty of recklessness. The occupying forces government seized the list of non-Aryan Christians who had been granted exemptions. Once their addresses were known, all these persons were forcefully sent to the district. As punishment for trusting the occupier, literally everything was taken away from them: furniture, books, underwear, and pictures. They were delivered to the district stripped of everything. We went through this experience as follows.

On the morning of February 26, 1941, the police arrived at our house with orders to deliver us to the district. I showed them the paper signed by the German authorities. The policemen were from the Saska Kępa; they had known us for many years, and their children were in my wife's medical care. They behaved with ideal delicacy. The senior policeman kissed my wife's hand and said that he would have preferred to apprehend bandits than cause us this trouble. I assured him that since I had the permit, there would be no trouble. He took the paper to show it to the authorities. We stayed under house arrest for two days, and were then informed that the police directorate did not recognize the power of the sanitary directorate, and that we had to be deported. I telephoned Dr. Kohmann; he would not believe me and said that he would clarify the misunderstanding immediately. When an hour later the matter was set on a knife's edge, I phoned him once more. My call was answered by his chief, Dr. Lambrecht, who had signed the permit for us to stay at home.[24] One would think that his own ambition would have not allowed him to disregard a decision he had himself made. Yet, he brutally told me that I was to obey police orders, that I would receive no permit to go abroad, and hung up.

I was furious that a mere puppy should dare talk to me in this tone. Our comfort was the dignified bearing of our daughter and the perfect conduct of the police. I will not mention the details, because this particular aspect of our last moments at home, namely the gallant demeanor of all those present, was our only alleviation: we were going into exile.

We quickly packed a few things and some food. In two coaches and with the assistance of the police, we proceeded toward the boundary of the Jewish State. It was raining, and the streets were covered with mud. The coaches were not entitled to enter the foreign state, and we had to change to a non-Aryan coach. The soldier on guard ordered us to place our suitcases on the muddy pavement and open them. He took everything he liked. He took the knapsack in which the food was placed, soap, and my wife's underwear to which he took a fancy. This was supposed to be a nation loving earth and blood. Love for someone else's property was supposed to be a Jewish characteristic. In one suitcase he came across my book published by the Manheim Academy when I was an assistant in Heidelberg. He asked what it was. I told him that I had written the book when I was an employee of the German State. His answer was "Jetzt bist du aber nur ein Jude" (Now you are only a Jew), and he went on stealing. The Jewish assistant policeman mentioned to me: "Don't say anything; this is a very decent German. Others take everything."

The following day, Germans went to my villa and took my furniture, pictures, carpets, and everything they liked. My books were requisitioned by the institute to save them from destruction. Later, the disinfecting team came to kill the dangerous germs that I presumably had been carrying.

Thus, everything was taken away from me: my position, property, house, furniture, pictures, and books. Finally, with consent from above, some nameless soldier grabbed the underwear and the little bit of food we had taken with us for the road. They took my position because they regarded Poland, the country where my family had lived for centuries, as their living space from which I had to be eliminated. They took the house and furniture we had acquired with our life-long work, because in their highest idealism they were dreaming of blood, earth, and iron, while I was only a parasite. They isolated me, because I was a carrier of dangerous germs. Finally, they decided to destroy me, because I had conspired against European culture and declared war on them.

That they stole all my possessions, the result of my life-long labor—I forgive them, because robbery is alas human, especially in wartime. But, that they did so in the name of an ideology, and that stripping of possessions was carried out not by mobs but by the elite—that I will not forgive them. Neither I, nor the world, nor history.

Chapter Twenty-Two

The City of Death

Once upon a time, the Turks decided to eliminate dogs from Constantinople. Their custom did not allow them to kill animals. Therefore, they placed the dogs on an empty island so that they would devour one another. This method of getting rid of dogs was later stigmatized as cruel and degrading.[1] That is how it was in those days. . . .

Nowadays, the Germans had decided to destroy the Jews. At first, like the dogs from Constantinople, they were supposed to die from starvation, lice, and dirt, and would devour one another.

A repulsive picture had to be presented to the world: here are the dirty, lice-infested, rejected, and starving parasites still fighting for their miserable lives.

With this objective in mind, ghettos were created.[2] At first there were many ghettos, then their number was reduced to tighten the loop around the condemned men's throats, and finally they ware jammed together in just a very few areas. Then, the ultimate blow was dealt.

In Warsaw, this plan developed as follows. There was talk about establishing a Jewish district from the very beginning of the occupation. This plan was repeatedly postponed. Epidemiologists warned the authorities about the danger of such moves; maybe my memorandum had also contributed to this delay.[3] But most probably this was the effect of Jewish money. Finally, in October 1940 the order was issued that under penalty of death all non-Aryans had to move to the district, leaving their furniture behind. Since this concerned about forty thousand persons, it was impossible to check whether the order to leave furniture behind was obeyed in full.[4] A portion of the city was surrounded with barbed wire and, at the expense of the Jewish Community, high walls were erected with just a few openings. Only small amounts of food could be brought in: up to two hundred calories per day per person, one-tenth of the amount necessary for life.[5]

Then, the Jewish Community was charged with all administrative duties without being granted appropriate rights. The district was headed by a German commissar whose duty it was to supervise, according to the orders received, that the Jews would die rapidly without provoking public outrage.[6] Then, various regulations were passed so that the condemned people could be efficiently robbed and so that the nation of parasites would be effectively exploited by the nation of idealists. For this purpose, which could be called parasitism in reverse, an eloquent definition was invented: the Jews must be taught to work so that they would finally cease to be parasites on foreign bodies.

Thus began a struggle for life and death. Whatever could be produced by Jewish intelligence, talent for improvising, hunger for life, and the ruthlessness of those fighting for their existence was tossed onto the scale. A nation unaccustomed to governing, without administrative experience, and with an overwhelming number of intellectual workers; a nation robbed, slandered, and wronged; a nation that had no internal discipline and which at best was used to expressing its social instincts through critique, opposition, and philanthropy; a nation with an intelligence higher than the average but with a defensive stance and ability lower than the average—this nation was forced from one day to another to organize administration, economic self-sufficiency, and even a sort of agriculture on rubble.

This resulted in a unique form of a fight for survival conducted by people condemned to death. Humanity still remembers the tragic: "Ave Ceasar, morituri te salutant." What were the perishing Jewish people calling out to their tormentor, to their countrymen, and to their—unfortunately very distant—friends? In this dance of death, how did they want to shape their lives and hearts? What did they dream about prior to extermination? Did they realize that they were doomed to be exterminated? Did they die with a proud bearing or in humility and despair? Oh, how I regret that I do not have the gift of a fiery pen to describe the last moments of a nation and, in a song, tell about its heroes and cowards, about those stricken by despair, and about those who to their last breath aspired to rise to the highest value. Unfortunately, I am unable to accomplish that. Yet, perhaps the fact that I am writing this book with my blood will substitute for my lack of a literary talent. Therefore, I will continue to weave the story of my suffering; because it also was the suffering of a nation condemned to death.

The gate was shut behind us. It was as if we had moved from a cold room into a crowded, stinking prison where we ceased to be human beings, where anyone was entitled to strike us, and where we were only part of a rejected, discarded mass. Actually, we were put into a concentration camp created in order that people devour one another and die from starvation, epidemics, and disgust. The population density in the Aryan sector was ten times less than here. There, mainly the intelligentsia was to be destroyed, here—everybody. There, people were allowed to live the life of servants; here, it was death in contempt. The streets were so crowded that it was difficult to walk; I had to push my way through a thick, compact mass. The people looked like ragamuffins; they were in rags, even without shirts. Hundreds of them, mainly women and children, were selling various things right on the streets: buttons, thread, old clothes, pretzels, cigarettes, some exotic sweets—anything. The street had a music of its own: an indescribable hubbub and buzz in which one could discern the thin, resigned voices of children: "Bagels, bagels for sale, cigarettes, candy."

It is impossible to forget those weak children's voices. There were mountains of mud and garbage on the streets. From time to time, a child in rags would steal a small package from a passerby and while running away, he would eat whatever was in it. People would run after the child with curses. If the child was caught, he was beaten, but he still continued to eat the meal. Frequently, I saw crowds of men, women, and children driven by the navy blue police[7] and assisted by the Jewish Monitor Service: deportees. They were carrying miserable remnants of their former possessions: a bundle, occasionally a pillow or a mattress. The luckier traveled by wagon. A moment's distraction meant losing one's last belongings in the crowd. I approached one such procession and asked them what their story was. They had been thrown out of their homes within five minutes and were not allowed to take anything with them. They had lived in small towns around Warsaw. Old, sick, and crippled persons were killed on the spot. On the way, those who lagged behind were killed. Once, a son stopped at the body of his dead father and was shot dead immediately. The faces of the deportees had tragic expressions: a ghastly fear or an ultimate stupor.

From time to time, I saw groups of ragamuffins led by the Jewish Monitor Service; among them, I occasionally noticed an intelligent face or a person dressed somewhat better. Those were inhabitants of houses where someone had fallen ill with typhus. One typhus case was enough to drive all tenants to bathhouses. The fact that they had to wait for hours and caught lice in so doing was not taken into consideration by the rulers. Every day I saw a typical scene: a hill covered with newspapers, with either very swollen or very emaciated legs protruding from beneath. Those were the bodies of persons who had died either of starvation or of typhus. Some homeless individuals died on the streets. Others were carried out of the house by relatives or other tenants to avoid funeral costs, which they were unable to pay, or to avoid the consequences of a typhus case in the apartment house. Those consequences were terrible; they could deprive people of roof and bread. This was the Germans' intelligent way of combating infectious diseases. And what about the lice crawling off the cadaver and spreading all over the street? The men sitting at the green executive desk did not want to be bothered with such unimportant details. The cadavers were picked up by the Monitor Service, but some lay for several days on the streets. They were carried off to a shack opposite the Community House and stored in charnel houses. There, they lay in layers piled one on top of the other, with deformed extremities, almost always naked, and often bitten by rats. Sad Jewish funerals frequently filed along the streets. There was a funeral parlor on every street. The Pinkiert Parlor was the most prosperous enterprise in the district.

The feeling of being enclosed in a prison was intensified by the fact that a pedestrian would frequently happen upon a wall or barbed wire. This was the way the authorities wanted to isolate the carriers of dangerous germs.[8] People who called themselves medical doctors supported these measures. Science had long ago abolished the medieval quarantines not only because they were cruel but also because they were useless. But since in this case the point was to liquidate the Jews and not the epidemic as such, quarantines turned out to be quite useful.

The walls had several gates guarded by sentries commonly called *wacha*.[9] Each sentry group was composed of several armed Germans who looked with contempt on the crowds, on the Polish policemen, and on serviceable Jewish policemen who, if they did not conduct themselves efficiently, were slapped in the face.

On the district side, the *wacha* was surrounded by swarms of ragged children; on the Aryan side, there were crowds of gapers watching the show. Those children were the district's breadwinners. When the Germans looked away, they would run across to the other side, buy some potatoes or bread, hide the purchase under their rags, and try to return the same way. The Polish police looked at this by and large with a condoning eye. The Jewish policemen were struggling with themselves: they had pity for the children, and they knew that they were nourishing the district. Without them, many would die of starvation. Those children not only supplied the district with food but often supported their families. The father was robbed of everything: his workshop, house, and often his last shirt. So, the child would bring in some potatoes. The German sentry behaved differently depending on the day. Occasionally, though very seldom, the guard would smile and encourage the child to run across. He too had children of his own, and those little Yid-kids resembled human beings, despite all. Not all Germans were hangmen. However, the guard often took his gun off his shoulder and shot right into those tiny children's bodies. They shot at children almost every day— can you believe it? Almost every day mortally wounded children were delivered to the hospital. All Jews were wearing armbands, but children were exempted from this order. This made it easier for them to smuggle. Imagine the following scene. A street car from the Aryan district passes by. While it is running, some children in the car throw bundles out onto the street and then jump off themselves. Other children who have been waiting grab these bundles and take off. If a German notices this, a chase begins. A policeman is running at full speed after a child. Sometimes, he takes only the contraband away. What does he care that the child and his family whom he supports will die of starvation. At other times, it is a chase for life and death, and the child is caught and killed by the German. Now you see children crawling up the wall. They must get over it quickly before the German turns around because if he notices them he will shoot. "Kriegsbefehl des Führers."[10] Is he having the ghastly vision of children being killed right now? This may be done as follows. The child is ordered to climb the wall again, and then they shoot at him to set an example to others. If the child's body falls on the Aryan side, it is tossed over to the ghetto like a ball. Such merchandise is not needed on that side.

These children prolonged the life of half a million district inhabitants for a year. If ever a monument is erected for the dead, those heroic children deserve it most. At this monument we should engrave the words: To the Unknown Child—the Contrabandist.

There were thousands of beggars pushing their way everywhere. The weird scenes resembled starving India: a mother nursing her infant with her desiccated breast, with the dead body of her elder child lying next to her. A man dying, or pretending so, stretches across the width of the sidewalk; his face is distorted from pain, and his legs are swollen from starvation and frostbite. I hear that every day they amputate frostbitten fingers, hands, and feet of those begging children. Are they great artists [sic]? Can life really look so horrible?

There were two distinct types of wretches. Some were balloon-shaped, had a pale, swollen face, almost invisible eyes, and legs like those of an elephant. Others were like skeletons covered with a pale yellow skin. The faces of children resembled the faces of old people. Those were the two faces of starvation: wet starvation and dry starvation.[11]

The walls of some houses had openings, and a bag was passed through them from time to time. Food was also tossed over the walls. Bribed policemen would turn away discreetly. But there were also motorcycles driven by armed Germans. At such times, smuggling was paid for with one's life. Prices were calculated so as to include death. In the district, food was many times more expensive than on the Aryan side. Yet, when you ate a roll, you often could not swallow it because it tasted of blood. There were other ways of smuggling, requiring not only courage but also imagination: calves and cows were transported in caravans, in coffins, and under hay or straw. All this required money. Some of it went to the Jewish policemen, some to the Polish policemen, and the lion's share to the German soldier. The cooperation was smooth, and the bribes were distributed according to the supremacy of races: the Jew got one part, the Pole two parts, and the German seven.

We had to pass bareheaded by the *wacha*. If someone failed to remove his hat, the soldier would shoot into the crowd. The first time I passed by the guard with my wife and daughter, I did not want to take off my hat. We were educated on the model of William Tell. Someone had to set an example of pride, I thought. Suddenly, I heard a voice from the crowd, "Don't put other people at risk!" I yielded. At first, we were obligated to greet passing Germans. Then, we were forbidden to. The result was that sometimes a Jew was beaten because he took his hat off, and another time because he did not. Persons for whom taking off the hat was a symbol of captivity did not wear hats. In the winter, people were extra cautious: when they saw a German from far away, they would cross to the other side of the street. In this way, an aura was formed around the Germans: they were circumvented like wild, biting animals. Occasionally, an automobile with Germans passed by. They slowed down to do some sightseeing. Indeed, these Jews did not resemble Europeans. They looked like wild animals. Usually, only zoological gardens post a sign reading "Do not tease the animals." The world also has societies for the prevention of cruelty to animals. But here, on the contrary, it was good to torture them. Occasionally, a young officer would pull out his revolver and shoot into the crowd. How amusing these scampering Jews were! Or a car pulled up, an officer or soldier got out and began to lash passersby with a long whip. Or else, he summoned a Jew and beat him: Is it not he who has provoked the war? Let Roosevelt defend him!

From time to time, large buses drove through the district with the curious faces of soldiers staring out the windows. The buses carried the sign "Kraft durch Freude."[12] Those were grand tours through the zoological gardens of "unified Europe." Goebbels wanted to show his soldiers the power of the German fist and teach them to despise anyone belonging to a foreign race. Indeed, those wretches did not resemble human beings. Goebbels had great knowledge of the German soul. Any other nation would have cried at the sight of these people resembling ghosts and would have shouted words of indignation in his face. But the German soldier laughed in obedience and despised in obedience. Thereupon, to kill a despised man was a mere trifle.

I found out the secret of how a man could be changed into a criminal. It suffices to introduce minor rearrangements in the human mind; deprive the future victim of human attributes and ascribe to him the features of an especially disgusting species: a bed bug, a rat, or a louse. This can be achieved by means of such journals as *Das Schwarze Korps*. They create automatic reflexes, which only minds accustomed to independent thinking can resist. But independent thinking can be suppressed with concentration camps and frequent marches lasting so long that thoughts follow like feet being put one in front of the other—to order. How else could one possibly explain the following facts.

A little girl tries to steal her way past the *wacha*. The soldier calls her and slowly takes his gun off. The child embraces his boot and begs for mercy. The soldier smiles and says: "You won't die, you will only quit smuggling," and he shoots into both her legs. The child's little legs are crushed. They must be amputated. Indeed, she will no longer smuggle. Here we have another scene: the soldier orders a child to walk forward, and he takes aim. The child comes back, kneels down, and begs for his life. Again, the soldier orders him to walk and with one shot into the back kills the little child. Or, a Jew buys a newspaper across the barbed wire from a boy in the Aryan district. Jews are forbidden to make purchases in that district. This crime deserves a death sentence. A shot is fired and the Jew drops dead. At that moment, a little dog comes along. The Germans like animals. This killer bends down and pats the little dog. This took place on Chłodna Street in Warsaw in the twentieth century after Christ was born.

Someone once asked a little Jewish girl what she would like to be. "I would like to be a dog, because the Germans like dogs, and I would not have to be afraid that they would kill me."

What I am recounting I observed personally or was told by totally trustworthy persons. Scenes were described to me for which the Germans will be ashamed for the next one hundred years. Jews who had passes went to work in the Aryan district—at least in the beginning. They had to run past the guards holding their hat in hand. Occasionally, the Germans stopped the crowd and ordered the people to hit the dirt or make knee bends, 100, 200 times, until the Jews fainted. At other times they ordered them to dance. The soldiers looked on and laughed their sides off. It was an "amusing sight," especially when an old, emaciated Jew had to dance. One Jew could no longer take it and said: "But I too am a human being." To which the soldier replied: "No, you are not a human being, you are not even an animal; you are just a Jew."

At first, we lived at our friends' on Grzybowska Street. The windows of our street gave onto a small square at the corner of Ciepła Street. At six in the morning, we were awakened by trucks, picking up slaves to carry them to work on the Aryan side. The Jews were a nation of parasites, and they had to be taught how to work. The Germans, brilliant educators, were able to educate even a Jew.

All Jews aged twelve to sixty were obligated to work. Some were working in camps. I will describe this "work" later. It was a martyrdom of which modern man had no idea. Only Satan could have invented it. Work in the Aryan sector, however, was sought after. There, one could buy potatoes and bread for somewhat less money and support one's family from the difference. The daily pay was about three złotys, which was not enough even for breakfast. People eager to work gathered on the square in crowds greater than the trucks could transport. The German looked on calmly as the slaves fought to get into the trucks. When the game bored him he pulled out his whip and began lashing; at other times, for the sake of variety, he kicked. This motion was evidently regarded as an attribute of power in Germany. I often saw Jewish physicians and attorneys being kicked; the buttocks were the same in all Jews. When the truck took off, the slaves who were clinging on the doors fell off; or their hats or bundles were knocked off, and then they would jump off the moving truck by themselves. And I thought as I looked on, "Both Jews and Germans have lost human traits, though for different reasons: the latter because they beat, the former because they are beaten. History will judge what will give rise to a better legacy for these two nations."

At around seven, the streets fill up with a scurrying black mob. We could hear the hum and the anxious voices of children singing the praises of their merchandise. Curfew was at about eight in the evening, and everything gradually calmed down. Occasionally, we went out somewhat earlier on Grzybowski or Rynkowy squares. Those were the only two places where one could see some sky and space. In the distance, I could see the trees in the Saxony Garden, where I had played as a child. Now I was one of this miserable mob; my wife and my child, too. Only shortly before I had been walking in the park of the Pasteur Institute in Garches, sitting on the bench where Pasteur used to rest. With pride I thought at that time that, as a result of my hard work, I had brought my child to the summits where the greatest human minds were dwelling. In fact, I had brought my child to Grzybowski Square. The poor child. I brought her up in Polish traditions and taught her to love her country. Now, someone else had come and said that we had no right to walk on this earth and chased us out of our homes. I could see that my daughter was unable to understand why she was here and what the world wanted from her.[13]

From the Grzybowski Square we could see the ruins of many houses and the Church of All Saints. Above was the beautiful sky they had been unable to wall up. We walked with our heads held high to see more of the sky and less of the earth.

I have described what met the eye. However, these scenes did not the least exhaust the abyss of the misery. Typhus was raging. An epidemic claiming victims unexpectedly always creates a special panic. The Germans too were afraid of typhus.[14] The rumor spread that if the disease were not combated the way the Germans wanted, they would evacuate the district and put persons responsible for public health into camps. No one dared oppose their orders. Yet, the orders were such that their implementation meant death, only of a different kind. Did not sealing houses for three weeks mean death by starvation for most people? And the disinfection was carried out in such a way that it destroyed everything. Did that not mean complete ruin? I soon noticed that the entire business of epidemic control consisted of issuing senseless orders and cleverly circumventing them. When this was noticed, severe penalties were imposed. Therefore, all those whose duty it was to carry out the disinfection stole regularly: Germans, Poles, and Jews. They were no longer human beings but hyenas. I will describe this later. I met many of my former pupils. Their eyes expressed hopelessness; they felt as if they had been put into an insane asylum. They could not believe that this was true. They were completely lost. I realized that if their thoughts did not latch on to something extraordinary, that would remind them of their former life, an important inner spring would break in them. One of them who loved books told me that he had only one desire: he wanted to die thinking of medical problems to which he had devoted his life, and not in ignominy. This kindled something in me that began to act as a pang of conscience. How could I help these people?

A priest does not help the body either, and yet unhappy, dying persons cling to him. He looks with his mild eyes and says that in heaven you will not be wronged but will find peace and friendliness. Who was to tell these Jews that life could still become beautiful for them? They just had to survive and not let the inner spring snap within them. I had always had the key to young hearts. Perhaps I could nourish them with lofty thoughts, and they would forget that they were . . . only Jews. I could talk to them about the serenity of a laboratory and the charm of scientific work. I could tell them about countries where Jews were not despised and where nobility was acquired through lofty thoughts. I was unable to help them, but I could offer them a moment of oblivion.

I would tell them that they were despised without justification. I would not tell them that they were a chosen nation. But I would tell them that they were not a cursed lot, so that they would take hold of their sore, humiliated souls and begin to shape them; so that they would not follow the hunger of life but would concentrate on life in a community that required mutual friendliness and compassion.

I was not familiar with Jews,[15] and I did not know how to reach their hearts. However, I had once succeeded in reaching the Serbian heart. Obviously, the attributes of the human soul were eternal. Race psychology? This rubbish could be left to the Germans. One could reach the unhappy ones with the religion of love. If the slogan "Unhappy ones of all lands— unite" resounded now, most of them would be Jews. Inevitably, someone would immediately say that the Jews pushed their way everywhere. Many months later, President Czerniaków[16] said that the presence of several scientists, mainly mine, brought comfort to many: they no longer felt such pariahs. Whenever I met with my former pupils I felt how much they needed the care of a priest.

The anti-epidemic orders issued by the Germans were worse than the epidemic itself. Yet no one dared oppose them. I had gained some recognition in Germany, I thought. Perhaps, if I placed the authority of my name on the scale, they might give the matter a second thought. I knew that criticism was forbidden. No one, least of all the Jews, was allowed to oppose the orders. At worst, I would be placed in a concentration camp. Even so, it would be in the service of a good cause.

I decided that if I got a chance to go abroad I would. There, I could fight for my people on a larger scale. If, however, this opportunity did not arise, I would stay in the district and would not hide. I did not want future generations, and my own conscience, to ask me the question: "Fate put you among several hundred thousand unhappy people. What did you do for them? Destiny gave you the key to young hearts. Did you try to open those poor hearts and soothe them?"

So, I stayed. I had nothing: no authority, no means, no laboratory. I was a stranger to these people; as a Christian,[17] I was rejected by the masses. I only had a great, boundless compassion. This was my capital and my weapon.

I told my wife: "I must stay, but you should leave this hell with our daughter. In the countryside she may regain her health."[18] I saw how she was torn between a mother's and a wife's instinct, and how the latter won. We told our daughter, "Darling, we have friends who want to save you. It will be better for us when you are on the other side." But our daughter was made of a noble substance: "Where you are, I shall be too."

So, the three of us stayed in that hell.

Only a few days later, our friends sent us forged papers and insisted that we should flee from the ghetto. On my end, I continued my applications for a permit to go abroad. We applied for a pass to Yugoslavia, hoping that this permit would be the easiest to obtain. In Yugoslavia our friends would help us. When they learned about our forced displacement, they reacted in a manner which touched us deeply: at their request, the young king granted me honorary Yugoslavian citizenship, and the Yugoslav Government requested our release on the basis that we were indispensable experts. We were informed that our citizenship papers were signed on March 24, 1941. But, the war broke out on March 27, and our hope collapsed.[19] Besides, at the same time we received a negative reply from Cracow. The money paid to the *Devisenstelle* was obviously lost. My attempts to get in touch with New York were in vain. Indirectly, I learned that by that time the Germans were releasing no one to America.

Only a short time before, I had spun the talc of a great world for my daughter and kept up her spirits by assuring her that soon she would get to know friendly people and regain her freedom, health, and joy of life. All that remained of those dreams were walls.

I had to get busy on something substantial. But how? How could I combat this misery? Two great powers had declared war on abused people: The Germans and typhus. I went to President Czerniaków and told him that I was not applying for a position, but that I was a bacteriologist, epidemiologist, and pedagogue and that, if I could be useful, I was offering my services. On the surface he gave the impression of a hard man. He was tall and stout and his protruding jaw resembled that of Mussolini. He wanted to be hard, but in fact he had the soft heart of a man who knew that he was doomed and wanted to sacrifice himself. He fought to the last moment. In his office in the Community House he had a large portrait of Piłsudski hanging on the wall.[20] It was the emblem and symbol of Polish statehood at the time. Once,

SS men came, beat him severely, and arrested him. A few days later, he was released from prison and went straight to his office, his face still bruised. A bombed-out house and an empty area were opposite his office. He converted it into a garden for children. When he was exhausted, he would watch their games, and this invigorated him. He supported science and art. He was reproached that he was spending the little bit of available money for nonessential things, while he should have purchased more food for the displaced persons. He had a firm attitude toward life, however. He would say that a nation lives by its spirit, that he was unable to save all, and that therefore he wanted to save those who were crucial for the nation's future. For this reason he took care of children, artists, and scientists. In his office I saw numerous pictures purchased by the Community House; he did not want artists to die of starvation. He was reproached for yielding to the Germans. Let those who made the reproaches remember that his resistance would have meant death for others. He was not afraid for himself. I personally witnessed how at a meeting with Commissar Auerswald concerning the epidemic he spoke so sharply that the commissar turned to his neighbor and said in a subdued voice, but so that everyone could hear: "Frecher Jude."[21] He fought like a lion for the life of those entrusted to his care. He was bold when only his life was endangered. He yielded when others were involved. That is why he supplied the Germans with the silver chamber pots, furs, and money they required. But, when the life of the Jews entrusted to him was demanded, he sacrificed himself. He believed that he had a mission to accomplish, and he liked me probably because he noticed that I was guided by no other motives than the desire to help. At the time I was opening a course for pharmacy students, he delivered an introductory address to the students. He said that in his youth he dreamt of the serenity of a laboratory (he was a chemist by education), that life forced him to a different type of work, but that he could only be happy in a laboratory. He spoke warm, sincere words about me, that I had a mission to accomplish here. He encouraged them to hang on. This man deserves a monument in human hearts.

Sanitary affairs were in the hands of Dr. Milejkowski.[22] Even before the war he was an active Zionist. By profession he was physician. He was community councilor and, as such, he was chairman of the Health Department.[23] In addition, the Germans appointed him chairman of the Ethnic Medical Chamber. This man had one ambition: he wanted to be the leader of Jews in his field. However, this was the ambition of a man who was fighting not for power, position or money, but for the love of his people. He was unable to take a firm stand against the Germans; nonetheless, he often got what he wanted from them.

Fighting the epidemic required my more active participation. It seemed to him at that time that I was competing with him for human hearts. This gave rise to conflicts that are still painful for me to recall. I soon saw that this small man was occasionally able to display a heroic strength. When Colonel Kon was arrested instead of him,[24] he was notified that he should hide. However, he was not one of those who would allow others to suffer in his place. He immediately went to his office and when the Germans came, he said: "I have been waiting for you." He could have saved himself easily; yet, I heard that he perished. He met his death voluntarily, as a leader. He did not want to abandon his people in misfortune. It is easy for one who feels the power of his country behind him to be a hero and say: "Civis Romanus sum." It is probably a greater heroism to meet one's death voluntarily while heading pariahs. A great heart was beating in his small body.

I visited Dr. Bielenki.[25] This physician and captain in the Polish Army enjoyed great popularity. He was tall, slim, and had a nice mild smile. His patients adored him. Later, when I walked with him to inspect disinfection operations, I noticed how the faces of the maltreated people would brighten at the sight of him. "If Dr. Bielenki is part of it," they would say, "we will not be wronged." I later learned that he had participated in Warsaw's defense and was in charge of a sanitary unit in the most dangerous sector. He was always full of initiative. During the epidemic he was in charge of the hospital.

He responded to my appeal during the typhus epidemic. He was one of those who have the heart of a dove and who can be tough, but only toward themselves. Whenever he came to the hospital, he was followed by a swarm of begging children to whom he regularly gave money. Smiling sadly, he said: "This will postpone their death by one month, but I cannot resist helping them now." He loved science, as many Jews do. I often recruited him as a lecturer, because he had enthusiasm and an ardent heart, even though he was not a good speaker. He was one of those physicians who not only cures the body but also soothes the soul.

Those were the men with whom I discussed the main points of my imminent activity. We agreed that I should proceed in three directions: organize lectures, combat the epidemic, and set up a laboratory in the infectious diseases hospital. Each task was a full-time job for one man. I had to cope with all three simultaneously.

Chapter Twenty-Three

Lectures and Courses

I had to be able to reach people even though meetings were forbidden. I had to invent something that could be justified in the eyes of the occupier. It turns out that antityphoid vaccinations were being performed by block physicians. The Germans ordered that they be summoned to periodic meetings during which they were threatened with the death penalty for various offenses. I suggested giving a lecture on vaccination as a cover-up for instruction for block physicians. Such meetings required no approvals.[1] Thus, I gave my first lecture to several hundred physicians in the Community House. The chairman was present, presumably to prevent demonstrations against me on the part of Jewish nationalists. At the entrance, a nationalist woman physician was actively encouraging everybody to boycott my lecture. Dr. Milejkowski, who was presiding, saw a national symbol in the fact that I had come there to speak. My first words were an appeal to prevail in dignity. The enemy wanted to take away from us, Poles and Jews, everything that represented science or art. Perhaps we would perish; but, if we must die, let us die with dignity. Our thoughts should not cling to petty troubles but should roam where they were permitted to think and create. I soon saw the surprised eyes of the listeners who were accustomed only to being scolded during such meetings.

Up to then, I had lectured to students of various nationalities. I had noticed that Jews did not like popular lectures. I often jested that Jews should be unable to follow a lecture for at least five minutes, otherwise they considered that it was too easy. So now I also delved into difficult matters, such as the complex structure of the bacterial cell and recent biochemical research. I saw eyes opening wider and wider in amazement: so there was still a world of ideas; the night has not yet devoured everything! When I finished, there was stormy applause, and the chairman expressed his hope that this was the beginning of a whole series of lectures.

I saw that I had struck the right chord, that these people were yearning for meetings, lectures, exchanges of ideas, and everything that resembled their normal prewar life. I suggested organizing lectures on infectious diseases, two sessions per week with each session lasting two hours: one hour of theory and one hour of clinic. Since the pretext was combating the epidemic, the occupier approved. This was to be a postgraduate course for physicians. I was warned that it would be a failure, because people's minds were absorbed with other matters and no one had any spare time. But my past had spoiled me. I believed that the course would be a success. I was coaxed into giving lectures by myself; however, I was not interested in solo appearances. I wanted to bring together those who were eager to make an effort, and those eager to listen, and the whole emulation [*sic*] that always sets the beginning of every intellectual activity. As lecturers, I recruited some hospital physicians and my former pupils whom I found here. After lectures, at which I always presided, I would start and maintain a discussion. I ordered beautiful charts to produce the impression of a real lecture. Each lecturer had to give me an outline of his talk. I discussed the lecture with him, made copies of the outline on a duplicating machine, and distributed the papers during the next lecture. This enabled me to control the material presented, and at the same time it offered the students a good compendium. People soon got used to the fact that twice a week they could find rest and recreation for their shattered thoughts and could forget the misery of daily life. People I didn't know came up to tell me that my lectures were their only bright moments. The course lasted two and a half months. I issued a table of contents with an appropriate cover page. On the program I wrote the motto: "Knowledge is my only hope and bliss; without it, I could not prevail." I sent those scripts to friendly people, including those on the Aryan side, so that they would serve as a proof and a stimulus preventing us from breaking down.

After the first series of lectures was over, dentists asked me to organize a similar course for them. The German Sanitary Board was headed by Dr. Hagen, supposedly a former social democrat, in any case a rather decent man.[2] Dr. Milejkowski obtained from him approval for a course in general pathology. Again we were giving lectures for several months, and again I was able to soothe those battered souls with science and to offer them moments of oblivion.

Meanwhile, Docent Zweibaum had obtained a permit to organize a course for the sanitary personnel. In actual fact, this was a clandestine first year of medical school.[3] We had several academic lecturers: Professors Centnerszwer and Lacks taught chemistry; Docent Zweibaum, histology; and I, bacteriology.[4] We set up a senate of sorts and acquired a few more lecturers. A fee had to be paid for these courses, which made them self-sufficient. The idea was warmly supported by President Czerniaków who helped us acquire a meeting place. Docent Zweibaum proved to be a very good organizer; he was efficiently assisted by his wife. We were eventually assigned facilities in the Labor Department at the corner of Żelazna and Leszna streets, right above the *wacha*. A peculiar mood prevailed during those lectures. They were given between five and eight in the evening, because all young persons had to work before dinner to earn their living. Frequently, there was no electricity, and the room was illuminated with candles or carbide lamps. From time to time we could hear shots: the *wacha* downstairs was killing a Jew. The young people were listening so attentively and quietly that one could hear a fly buzz.

What was I to talk about? Docent Zweibaum took a very formal approach to the school's purpose. He wished to establish a medical faculty and firmly believed that our students would eventually get credit for these lectures. I, on the other hand, saw the aura of a future martyrdom hovering over these young people. I felt that my main task was to keep up their morale. They often appeared to me as little, frightened chicks. I would look at their young faces and think how few of them would survive.[5] The power of hate was overwhelming. Should I, one day before their death, talk about pathogens and require condemned men to take an exam in bacteriology? No, I would capture their imagination with lofty ideas and alleviate their pain by satisfying a desire that is very strong in Jews: their hunger for knowledge. When I was discussing my favorite subject, blood groups, I led them to the banks of the Ganges River and said that it was there that I expected to find the answer to the question about the first mutation of blood. "Look," I said, "I am here with you behind the wall, and any old soldier can kill you or me to suit his whim. But no one can prevent me from roaming through distant countries, in my thoughts and this I can do because I have loved science above all." Thus, I gave them a vision of strong experiences, distant travels, and intense thoughts. I took those poor children by the hand and led them to the summits where the air was pure, where people prayed in ecstasy at sunrise, and no one despised them, and where, in the glory of mankind, they could build, think, and dream.

From below, I could hear shots and cries of victims. But the students sat looking at me and were completely absorbed in what I was saying. Indeed, the Word is a great thing when it is used to lift man's spirit. Never before did I speak with such plasticity and fire. I felt that I had to substitute for the life, youth, and love to which those children were entitled. Actually, it was not I who spoke. The spirit of compassion and love stood behind me and prompted me word by word, sentence by sentence.

There, I found myself as a gardener of human souls. I always wanted my words to create meaning out of chaos and to shape human souls. Now, moreover, I was giving them oblivion. Did I reach the Jewish soul? I do not know; I only know that I reached very, very unhappy people.

The subject matter presented in the lectures was bacteriology and immunology for medical and pharmacy students. This true objective had to be concealed under the pretext of combating the epidemic. Teaching on a graduate level had to be clandestine both in the district and on the Aryan side. When Docent Zweibaum opened the second course, I took only selected chapters to have a free hand in choosing discussion topics, because I decided to discuss, right under the hovering occupier, the most dangerous topic: racism. I wanted to take the curse of infamy off these young people, because it is easier to tolerate hate than infamy. I did not want them to die with the feeling of a deserved death. I realized that if someone denounced me, I would have to pay with my head for the lecture. But, by that time, I was relying on the spiritual power I had over people's soul. I met with signs of appreciation everywhere I went. When I was standing in a line, I was immediately called to the front, and public protests were rebuffed with just one sentence uttered in a mysterious voice: "This is Professor Hirszfeld; you try to lecture the way he does." I would walk into a restaurant, and soon the owners would approach me: "Professor Hirszfeld, our son is delighted with your lectures. Perhaps you would like some more preserves?" Or a student would come up to me after a lecture and say: "The war has taken everything away from me, my house and my family; but it has enabled me to listen to your lectures." Thus, I trusted that the firmly established ties would protect me even from the enemy. If not, were someone to denounce me—too bad. There are situations in a man's life when he cannot be motivated by caution. I will state the content of the lectures later when I describe the Jewish psyche in its bad and good aspects.

I had one more proof of my influence on the youth. Because of the daily victims in the district, it was necessary to organize blood donation. I went to see Colonel Szeryński, chief of the Monitor Service.[6] Later, I will say a few words about this man who could have become a hero but, instead, became a hangman. He supported my plan to use the Monitor Service as blood donors. We called a meeting, and I asked them to offer their blood, which was all the more necessary since their own colleagues were frequently dying of hemorrhages. The project failed. They dodged their turns for as long as they could and finally would faint after a few drops of blood were collected. The Monitor Service was made of poor stuff. I went to the head of the Organizational Department of the Monitor Service who was a lawyer by profession. "You want us to give blood for these people," he said. "Do they care for us? Do they pay us salaries? We got no earmuffs in the winter, and our ears were freezing." And I heard words of resentment and antipathy worthy of the greatest anti-Semite. I shrugged my shoulders and knew for sure that I would obtain no blood from the Monitor Service. In the past, I asked Serbs, Poles, and people of other nationalities for blood, and I never had difficulties in organizing a blood donor bank. Should the Jews really be an exception—willing to take but not to give? I appealed to my students, and they showed up in crowds. I gave each a certificate that "during the disaster, he (or she) saved the life of his fellowman with his own blood." Since most of them went hungry, I got additional food rations for the blood donors. Indeed, very many were giving their blood to save victims; some repeatedly. This was one more proof for me that the unwillingness of others was not due to some kind of an inherited cowardice but to a lack of tradition. The laws of heredity cannot be broken, but it is not difficult to change traditions in the name of higher values.

Thus we had organized a two-year course in medicine and pharmacy. We had now to think about the older students—there were about thirty of them. They generally worked in hospitals and were involved in the worst forms of medical practice, such as giving unnecessary injections and so on. They came irregularly to the hospital and did not study. The doctors' room in the hospital was dirty, cigarette buts littered the floor. I could see that things were going to seed. Was that the sign of the Jewish soul? I sat down and wrote a request for permission to organize a course of future study in view of the epidemic. Dr. Milejkowski obtained permission from Hagen. It was not the final permission, but we did not ask for it, not wishing to call too much attention to ourselves. I worked out a program, called the hospital department heads, attracted a few colleagues from town, and organized the fourth and fifth year of study. Next, I went to see Chairman Czerniaków and informed him that the lectures would materialize, but that it would be more appropriate if they were the expression of a collective effort and not merely that of the lecturers. Czerniaków understood this and ensured me with the preliminary budget. Contrary to the traditions of the Community, the money was paid regularly; the Jews did not spare expense when science and learning were involved.

I linked this activity with the organization of my laboratory in the hospital. When I arrived, I was given first a corner by the window and subsequently one small room. Meanwhile, the quarantine was liquidated and the hospital was given a new building on Stawki Street. I was allotted five rooms for my lab, next to which we established a lecture hall. We also created a laboratory for studying the effects of famine, directed by Dr. Fliederbaum and Dr. Apfelbaum.[7] I hired a draftsman, and we produced tables that would not have shamed the best universities in Europe. The walls were covered with tables and charts, the floors shined, and I forbade the wearing of rubber boots and the throwing of cigarette butts on the floor. Finally, I obtained the indispensable instruments.

These memories are dear to me because I could see that I was surrounded by the goodwill of others. There were no good heaters in the hospital, because they had been stolen or destroyed. I telephoned Mr. Kurowski, an engineer, and found out that he had a heater that cost 12,000 złotys. President Czerniaków promised to purchase it but in the end lacked the money. At a certain point, a young lady sent by Kurowski turned up to tell me that he did not wish to see our work hampered and was therefore sending me the heater for free, that I could pay him back later. At a time when showing friendship to prisoners behind the wall was punishable, this man did not hesitate to entrust me with no small fortune in order to allow me to work. The heater was beautiful. It was also the symbol of the selflessness of

Polish society. I very much wished for the Jewish philanthropic societies to pay Kurowski back after the war for his beautiful gesture, for which he had risked his life or deportation to a concentration camp.

I also had difficulty in finding microscopes; several physicians brought me theirs. I was able to buy one thanks to Chairman Gepner, a great philanthropist known for the social positions he took even before the war.[8] Upon entering my laboratory, one saw heads bent over microscopes, tables [charts and diagrams] hanging on the walls, and plants in the windows. Everything sparkled with cleanliness. Next door was the lecture hall and the emblems of school: a pulpit and a blackboard. The people who came were amazed and believed they were in Europe again. The lectures were scheduled daily and were obligatory for the students. Both the lecturers and the students felt the warmth of being cared for. Thoughts began to wander from every day worries and began to concentrate on subjects worthy of effort. It was not a question of who would earn more but of who would come up with the more interesting idea or draw to himself the friendship of young people.[9] The human soul is a mixture of good and bad. Every one wants to be good, one must just allow him to be so.

The students had no textbooks, the lecturers—no books. I invited President Czerniaków to attend a lecture promising him that he would breathe in the atmosphere of higher learning. He came with his coworkers, sat among the young people, and listened to the lecture.

What was I to speak about in the presence of the President of the Jewish Council? It would seem that everything separated us. But we were drawn to each other by a feeling of mission and a feeling of dignity. I decided to speak about the feeling of kinship. Are the Jews really a separate race; have they not mingled with the nations where they dwell? Have they always been foreign and hostile? I referred to my work on blood groups, which had proved that the serological structure of Jews' blood resembled the blood structure of the peoples among whom they lived.[10] I cited Czekanowski and Rénan as examples that the entire tale of the Jewish race was a myth invented in order to implement murder.[11] The President of the Jewish Council listened with attention and after the lecture came to ask me how he could help. "Mr. Czerniaków," I said, "these young people have no books. The Chairman gave me five hundred złotys for purchasing books. Only a few weeks later my laboratory was filled with lecturers and students who sat reading and studying. And when I bought some editions, they would open the book with emotion and look at it like at a holy icon. Because the book is the symbol of a certain way of life and the foretelling of its highest forms."

And that is how in what was hell there arose an oasis in which young medics could satisfy their spiritual needs. But how was one to satisfy the spiritual hunger of other doctors? After all, scientific meetings were forbidden. I proposed organizing a course on typhus. We obtained the permission, but no one showed up.[12] I couldn't understand such a lack of interest; I was told that Jewish physicians do not like to recognize the authority of colleagues. A professor is something else. Besides which, participation in such a course could give the impression that a given doctor was learning for the first time about exanthematous typhus, whereas typhus was already an important source of income. They suggested to me to keep the same program, but in the shape of medical conferences. We had permission to do so, but only for hospital physicians. Without asking for agreement, we linked the two projects and began the typhus lectures once a week. Crowds showed up. The lecture hall and corridors were full. I gave the first lecture on the etiology and epidemiology of typhus and then there were clinical lectures. I opened the lecture with a speech calling on a dignified posture and higher thoughts. These meetings became the basis of intellectual life for doctors living in the district. Unfortunately, I was unable to save the manuscripts; they would have been proof of the resilience of people's minds, as well as constituting a veritable scientific contribution.[13]

The conferences revealed a hunger for knowledge and good medical practice, but they also reflected the less positive sides of the Jewish mentality: the anarchy of thought and the nonrecognition of authority. The young assistants, sometimes with no experience of their own, would pronounce themselves authoritatively on the basis of what they had read. The lecturers' talks were sometimes too long. The community apparently followed the course of meetings, since a lecture or participation in a discussion was supposed to increase practice. Because of the type of permission the authorities had granted, nonhospital practitioners were not allowed to lecture. After finishing with the typhus cycle, without asking for permission, we tackled a new set of issues. Since the lecture hall was too small for all those who showed up, I founded a new association of block physicians. Under the cover of instruction sessions, we smuggled in lectures and discussions.

And that is what medical life looked like in the district: a full course of medical training, first year of pharmacy, two scientific laboratories—bacteriology and physio-chemistry—and two medical societies.

In the district, there were not many great physicians, but I must stress that among those physicians of Jewish origin in Poland, there were many outstanding ones of international acclaim. The anatomist Ludwik Hirszfeld; Samuel Goldflam, one of the best neurologists, well known in Poland and abroad; Edward Flatan, one of the best experts on the brain in Europe; Zygmunt Bydkowski, who the day before his death was still proofreading his works; and Maksymilian Rose, the best specialist of cytoarchitectonics, a wonderful neurologist, and pedagogue.[14] It is with emotion that I recall Dr. Zygmunt Srebrny, editor of the *Warsaw Medical Journal*.[15] I visited him in the district before his death. He was a fervent Polish patriot. He said to me with pain: "I love Poland, but I am under no illusions as to their wanting us. How painful it is to die with the consciousness of unrequited love." These doctors had founded before the war, together with a number of enlightened Poles, the Society for Preventive Medicine. The society has a beautiful record and contributed to raising the general level of medicine in Warsaw. I was happy to lecture there; I felt comfortable in that circle of people thirsting for knowledge. I am unhappy to recall that many doctors, even those of Jewish origin themselves, refused to give talks there lest—God forbid—they be accused of philo-Semitism.

I would like to say a few more words about my laboratory, since in the end it was the main component of my life. When I had arrived in the district, I had not formally founded anything more than a small laboratory devoted to urinary research. And in that large concentration of infectious diseases— more than two thousand patients—there were literally no bacteriological analyses being conducted. And I felt myself to be the protector of all bacteriologists. I was able to attract Róża Amzel and an excellent hematologist from Łódź, Dr. Kocen, and I named them my deputies.

I was able to profit from the services of an employee of the Institute of Hygiene, Tekla Epstein; an assistant from the Polish Free University, Zofia Judowicz; an assistant from the University of Warsaw, Dr. Landesman; several young girls who were students of Professor Szymanowski;[16] and a very talented biology student, [named] German. I reserved for myself the right to name, besides heads of departments, assistants and lab technicians and I could do so only with the support of President Czerniaków. I was accused even of autocracy.

My personnel was quite literally starving; Tekla Epstein once passed out from hunger in the lab. In order to help them, I produced an anti-dysentery vaccine and used the profits for extra food for the sick and income for the assistants; but even that did not suffice, and I had to find other ways of helping them. As it turned out, Professor Weigl from Lwów and the Institute of Hygiene sent me personally very large quantities of vaccines against typhus.[17] On the open market such vaccines cost around a thousand złotys. The vaccines I received I passed over to the Health Council, asking that those who were vaccinated pay a small fee. In this way, we opened a small administrative fund for the Health Council. We used it partly for a cleanliness competition and partly for people working on typhus. I passed over part of my income from lecturing to buy extra food for my assistants, and I tried to obtain reduced dinner cards. I also looked after two young people who were preparing their *matura*,[18] one of whom was the son of a very meritorious physician, Dr. Matecki. I tried to secure a grant from the Society for Scientific Work. Scientific work is well esteemed; Jews do not refuse money for it. Thanks to such a grant I was able to save from starvation talented people who were devoted to science. Such were the functions of a director in the district. I did not believe that in such conditions one would be able to carry out research. I handed out subjects like sacred images, so that each would have his own to pray to. To be honest, I did not believe that they would be heard.

But then a miracle occurred—maybe because I was so eager to assist the miserable. Our scientific work began yielding fruit, and abundantly at that. I would like to dwell on this a bit further.

Before the war, I had foreseen the existence of certain special forms of group O. For several years I had searched unsuccessfully for such forms. And then someone asked me to help him with an unclear case. I analyzed the sample and it turned out that this was the form I had long expected and looked for.[19] Being somewhat superstitious, I felt this as a sign from destiny that I should not break down but instead carry on. The main subject of research was naturally exanthematous typhus. During the Great War, Weil and Felix had published wonderful work on specific bacteria in patients' urine.[20] But for such research I needed a centrifuge, which I did not possess. So I came up with the following idea: I added patients' sera to the urine hoping that it would provoke greater absorption of the bacteria I was looking for; next, I dried the urine and filtered what was left onto a medium which, as I foresaw, would be optimal for such germs. The hypothesis turned out to be correct, and soon I was in possession of a greater number of strains than any labs in the world. I will not describe the many further investigations that will undoubtedly enrich diagnosis and also perhaps prevention. But that was not all: we ascertained the formation of precipitations in the urine under the influence of patients' sera. These observations opened up entirely new possibilities for exanthematous typhus. We were obliged to interrupt our investigations at the most interesting moment. I had never been jealous before, but I must admit that I discovered the feelings of jealousy and rage at the thought that German scientists were able to carry out research of which I had been deprived. I knew that I was on the brink of new discoveries and the thought that the Germans could beat me to it not through industriousness but through violence awakened in me negative feelings.

We made a whole series of observations about which I will not write, but there is one thing that does appear of practical importance. There were belts being sold in the district that were supposed to protect against lice and they had the approval of the Health Council. I involved German in the matter and we established that these were probably belts that had been used during the Great War by the Austrian armies. German guessed on the basis of his conversations with the dealer that these belts were impregnated with mercury. We tried them out on lice-infested displaced persons and they were indeed effective, but we wondered whether the dealer had not secretly made an agreement with the people we examined. We needed to find another way of carrying out the experiment. So we hired trustworthy people and stuck on them a small box with lice cultures. What happened resembled

Columbus's egg: you could count the lice as they fell, determine more closely the time of action, the changes in the lice, and so on. This method would, I believed, make possible quantitative experiments on lice. If we succeeded in replacing belts with injections of mercury, it would become one of the most important accomplishments in the fight against typhus and lice infestation.

Typhus was simply raging in the district and the vaccines used were those sent to us secretly by the wonderful scientist Professor Weigl as well as by the Institute of Hygiene.[21] In addition, very expensive vaccines were available on the open market. The International Red Cross sent vaccines made from the lungs of infected animals and demanded of the Germans that they be tested under my supervision.[22] Dr. Szejnman and I carried out immunizations and conducted a survey.[23] Up to then, the statistics had been fraught unfortunately with many errors. Working out the survey required a partly new statistical approach, which—I hope—will make the evaluation of immunization easier in times of epidemics. I should add that our survey did not produce evidence in favor of vaccine efficacy. It is possible that in conditions of mass production, the quality of the vaccines could not be sufficiently controlled. It is also possible that falsifications were involved. The value of our survey was in fact in the statistical methods it established. In any case, it can be said that the vaccine was of no avail in the district.

And so our scientific research proceeded smoothly. When I would enter the laboratory, I could see that confident expression—the one I had so often observed—on my colleagues' faces—faces radiating hope: a silent expression requesting further motivation, collaboration, recognition. I was giving those people condemned to death the possibility of the noblest of all lives: work with the hope of fruition.

Toward the end of my time at the Institute [of Hygiene], I had felt lonely. Here I once again experienced the warmth of people surrounding me. Was this because they were Jews? No, but because they knew longing. The cruel fact is that those who long are mostly the ones who carry out research. And I will say the same for myself: the enemy of creativity is satisfaction. Is that the price to pay for the kindling of the spark that many people have in them, but which dies out in most and burns on only in a few regardless of emotional stimulus?

I am writing this book in loneliness. Did I have to suffer in order to write it? Is there no other ink than the writer's sincere blood?

Chapter Twenty-Four

Typhus in the District

Typhus is the inseparable companion of war and famine.[1] After the Great War, there were thirty million typhus cases in Russia and over one-half million in Poland. This disease destroys more people than "the most brilliant" commander. It often decides the outcome of wars. It is transmitted by lice and possibly by other, still unknown agents. The germ grows in the louse, enters its feces, and when the louse bites, the contaminated feces may enter the microscopic wound. There follows a two-week long incubation period before the onset of disease, which is lethal in more than 20 percent of cases. An epidemic comes like an avalanche. In peacetime, it is hidden in separate endemic areas plagued by squalor and pediculosis.[2] For one implies the other. Man has freed himself of pediculosis not so much by deliberately combating it but as a result of the overall change in his living conditions. There was a time when the louse was a frequent guest at royal courts. When industry supplied us with enough shirts that most people could afford nightgowns and a change of clean underwear by washing them alternatively, pediculosis disappeared and so did typhus. Cleanliness and hygiene are the results of a higher standard of living. It is useless to tell a man to wash without giving him soap, and to change his underwear while taking his last shirt away from him. What has he got left to change? He cannot crawl out of his own skin.

In Poland, typhus lay hidden in rural villages of the East. There, when the peasant puts his sheepskin jacket on in the fall, he fastens it with a wire and does not take it off for six months. No wonder that lice find an El Dorado there. Infected children often do not succumb to the disease but transmit the germs and lice to others. Unfortunately, even public schools as well as spas and bath installations became foci of infection. During the Great War, Poland inherited hundreds of thousands of sick people from the German occupation. We had managed to decrease that number to two to three thousand cases per year. But the country was unable to cope with the remainder. We who were steering the fight against contagious diseases have guessed why. Typhus was being fought according to textbooks. Where typhus cases appeared, disinfection columns would arrive and delouse the population. By the time they reached the last hut, the first was infested with lice again. We should have come with soap, linen, and underwear instead of hunting for the last louse. It is not difficult to decrease the number of lice; however, it is difficult to eradicate them all. Martini observed correctly that combating pediculosis consists of killing lice at a faster rate than they reproduce themselves.[3] This can be attained without disinfection units; it suffices to have two day shirts and two nightgowns and to wash them regularly. Physicians, however, prefer to fight with complex equipment, police regulations, and circulars. Lenin had said: "Either the louse will conquer the revolution, or the revolution will conquer the louse." Only a physician knows how difficult it was for the revolution to gain victory.

The Germans were taking our laboratories away, depriving us of our last shirts, and displacing and chasing people from one place to another without asking whether they were coming from infested areas. Yet despite that, they felt that they were protecting people against typhus. Was it ignorance or ill will? Probably, it was administrative chaos, because it is characteristic of dictatorships that every *cacique* [local political boss] is a great Führer. A typhus epidemic broke out in the Ukrainian camp. But the camp cacique let the Ukrainians out, owing to political considerations. And no thought was given to the fact that they would spread the typhus. The Germans were chasing Poles from the Poznań area to the Congress Kingdom,[4] they were chasing Jews from one place to another and still another, and as a result they pauperized the whole population. After a few such displacements, even a fairly rich man became a wretch. And those lice-infested, shirtless, sore, and exhausted people without the strength to kill even a louse were pushed to Warsaw into the ghetto. Were those ragamuffins at least deloused? No, they were let in the way God made them, only with lice, germs, and despair. This was done under the pretext that the Jews were carriers of germs. And now I ask of the German physician: Have you ever been on the front? Have you ever seen soldiers infested with lice? With equal justification you could say that typhus is a German disease, because in our country it was remarkably correlated with German occupation. When one concentrates 400,000 wretches in one district, takes everything away from them, and gives them nothing, then one creates typhus. In this war, typhus is the doing of the Germans. History might forgive German medicine, because the latter cannot be responsible for political crimes. But history will not forgive the German doctors their hypocrisy: that they made Jews responsible for the epidemic they themselves had created, and that to all the faults ascribed to this unhappy nation, they added this guilt, too.[5] To make guilt out of squalor and malevolence out of misfortune, they proclaimed to the world that those alleged carriers of germs were seldom ill themselves but mainly infected the "goys" they hated. It was ignoble to slander those condemned to death. It was difficult to imagine that physicians and scientists could sign their names on such verdicts. Yet, their physicians did just that. I read about the German physicians' visit to Cracow Castle on July 11, 1943. They were given a reception by the governor who told them that the murder of three million Jews was necessary for reasons of public health. The head of the medical delegation bowed deeply to the governor and thanked him for having expanded the *Nebenabteilung* (subdivision) into a *Hauptabteilung* (main division), and promised him further loyal cooperation in accordance with the governor's instructions. This was Dr. Teitge, who even had the title of professor.[6]

In his description of English air raids, Goebbels wrote that when he heard the word gentleman, he would spit. What were we to do, pray tell?

Let us turn to the methods of combating typhus.

When cancer develops, a large portion of healthy tissue is cut out with the cancerous tissue so that all spread-out cancer cells are also removed. If several typhus cases develop in a house in a city with a population of one million and no record of that disease, then it is desirable to isolate this single focus as thoroughly as possible. It is feasible to close off the house and isolate the tenants for a definite period. It makes sense to isolate individual cases. When these cases become numerous, however, the measures must be changed; that is, sanitary police regulations must be replaced by comprehensive measures of preventive medicine. It is neither necessary nor possible to describe these measures here. In any case, the German chief of the Health Service knew how to cope with individual cases of such diseases as typhus or cholera, but he had no idea how to cope with an epidemic that was sliding like an avalanche or with an epidemic that was smoldering like embers. I will describe his orders and the results they produced when they were put into practice.

Whenever a typhus case was detected, the whole apartment house and two neighboring houses were closed off for three weeks. The tenants of the infected apartment were placed in quarantine for two weeks. A disinfection team was sent to the house, and it deloused all the tenants' belongings. The tenants too were forcefully bathed and disinfected. Quite logical, no? Mr. Schrempf demanded strict enforcement and threatened, with his revolver, the death penalty or a camp. In reality, the situation looked as follows.[7]

It was impossible to lock up all the inhabitants because they would have died of starvation. Nor was it enough to let only the cook out to shop; someone had to earn the money. Therefore, a tax was set up for a pass, and everybody who showed this "theater" ticket was let in or out by the police.

What about a bath for all tenants? But there were hundreds of such houses and tens of thousands of candidates for baths. The capacity of all bathhouses in the district was two thousand to three thousand per day. Thus, people were standing in lines for hours; persons with lice intermixed with those without lice. The result? The personnel in bathhouses were 100 percent ill, and a clean person was inevitably infested. Fortunately, bribes entered into operation and effectively counteracted the stupidity of orders issued by Dr. Schrempf. For a certain price, the bathhouses issued certificates stating a completed bath and disinfection. The income was divided among all: the disinfectors, physicians, and the police. As always, the division was done according to racial superiority.

What about disinfecting belongings? It was impossible to disinfect them in each apartment separately. The belongings were therefore heaped together in certain rooms. Obviously, the most infested apartment would be selected, and things from clean apartments would be brought down there. The doctors could not care less that an efficient delousing process also destroyed the last articles of clothing and underwear. Their owners cared very much. And in this way *not* carrying out delousing provided a new source of income for the disinfectors and physicians. Since, however, only the richest people were able to offer bribes, a trend developed to threaten mainly them with this procedure. The fact that a poor man was not deloused did not affect the prices.

This was the true implementation of the delousing, and official reports were sent to the government stating that so many persons had been bathed and so many houses deloused.

First thing in the morning, the Monitor Service would arrive at such an apartment house and lock it up. The apartment house committee would have been previously and unofficially warned and would already have prepared breakfast, including sausages and vodka. Polish and Jewish physicians would arrive, along with Jewish disinfectors and a German supervisor, whose presence would considerably raise the cost of the agreement. The Germans' capacity to kill was frightening, however, so the house committee would haggle less. What is not done in the name of business! The chairman of the house committee would discreetly hand over the required sum, the breakfast would proceed smoothly, and all race conflicts would be forgotten. A symbolic delousing pro forma would follow by placing several belongings into a few rooms. After all, regulations had to be obeyed. Here, I must disclose a secret of mine. I became interested in the delousing methods and gained the cooperation of a German who was incorruptible. We placed containers with lice into the bundles located in the rooms about to be deloused. If the lice died, it was probably from laughter. We found that all lice in the bundles survived the treatment. The disinfectors, those experts sent from Berlin and the Jewish team members trained in two-week long courses, were not aware of the trifling detail that sulfur did not penetrate a tightly bound bundle. Unfortunately, the lice did, because they penetrated all cracks. Therefore, the people were right when they said the disinfection only spread the lice. And therefore, I say: "The obligatory bathing of all tenants of a house made an apartment infection into a block infection, while the compulsory delousing converted apartment infestations into house infestations. If it were not for the Jewish bribery, the situation would have been much worse."

Such were Dr. Schrempf's orders.

Let us now consider the quarantine and the hospital. I visited the quarantine several times; it was a nightmare. Starving people lay on boards or on the floor. The washrooms were out of order, and they relieved themselves on the floor. They were properly louse-infested. For who could have deloused such masses? The quarantine was a family's economic ruin and the final blow.

The hospitals? There was an order to hospitalize all sick persons, as in Berlin or Hamburg, and thus isolate the source of infection. However, the fact that the hospital had been displaced three times, lost all linen, and had neither disinfection equipment nor beds—these were facts the Health Service was unable to absorb. A hospital is supposed to be a facility with a certain number of beds complete with mattresses, pillows, and linen. Here, people were lying two to three to a bed, dysentery was mixed in with typhus without proper nursing care, and patients were starving if their families were unable to bring them food. Occasionally, dead bodies lay for hours next to live patients. I heard one physician ask another: "Would you let your parents into a hospital like this?" Yet, every case had to be reported under penalty of death. Thus, a new procedure was developed. Some physicians would not report, but the risk they were assuming had to be adequately paid for. Often, it was impossible to call a good physician, because he, unfortunately, would report the case.

What was thus to be done when someone died of typhus at home? It meant disaster for all tenants. Two methods were developed to cope with the problem: either the body was put out on the street at night, or it was bought out—that is, for a certain payment it was placed in the house of poor people who, for several hundred or several thousand złotys, would agree to stand in a line or be placed in quarantine.

This is what the fight against the epidemic looked like in the district. Delousing supplies, which were partly being stolen, cost 5,000 złotys per day, while the ransom collected for not performing various treatments amounted to about 15,000 złotys. Together, this tragicomic nonsense cost about 20,000 złotys per day. For this money, one could have bought plenty of linen and fed many persons. But this would have meant sabotaging the orders of the occupiers. Furthermore, by that time there was a well-organized corporation of Germano-Polono-Jewish hyenas interested in maintaining the status quo. People were groaning and feared the delousing procedures more than the epidemic. The disease offered an 80 percent chance of survival, while the delousing could destroy all one's possessions.

This was the situation when I arrived in the district: terror of the plague and a still greater terror of the methods used against it.[8] By that time, Dr. Hagen had replaced Dr. Schrempf as chief of the German Health service.[9] He was basically confused, which was probably why we were able to influence him, since so-called active people mainly repeat what they have been taught. Drs. Milejkowski, Bielenki, and I wrote a letter to Dr. Hagen explaining that the situation there was unable to accommodate all sick persons, that the situation in the hospital was disastrous, and that it was therefore better to leave those sick persons at home if conditions were adequate. To determine whether they were, we suggested he appoint a paid commissar for carrying out the job. The quarantine was not as necessary as a hospital, and therefore it should be converted into one. It was not necessary to bathe and delouse persons who had no lice. We formulated all these matters, which were self-evident for a man with his head well screwed on, in a very professional style, and Dr. Milejkowski carried the letter over to Dr. Hagen. It was supposed to serve as evidence that the despised people had more common sense than the supercilious representatives of German science.

Hagen agreed, but the matter turned out to be not so simple. We asked him to let the Jews perform the delousing, not because they were less corrupt but because the people were less afraid of them. It was easier for us to suppress Jewish corruption than Aryan. Dr. Hagen agreed and promised to withdraw Polish columns.[10] On the same day, a physician from the Warsaw Health Department arrived at our Health Department. Whenever necessary, he would pretend to be a *Volksdeutsch*. He announced loudly that he had no intention of giving up this source of income. Indeed, soon we got a telephone call from Brühl Palace and were informed that the Aryan units would still remain for a certain period. Mr. Hagen made no money on this misery; however, others were ruling for him.

The epidemic was raging. It was a public secret that the actual number of typhus cases was greater than the reported one. President Czerniaków called us to a conference and proposed establishing a team that would continuously work on methods of combating the disease. As is usual at conferences, everyone expressed his view, and we left without making concrete decisions. Finally, the abscess broke open. Dr. Hagen was a type of person the French call "mouton enragé."[11] He mustered up a certain amount of energy, took his gendarmes, closed Krochmalna Street, and inspected all apartments. He detected a multitude of sick and unrecorded persons hidden in closets, toilets, and attics. Those proletarian buildings housed more than 14,000 poor persons. He ordered baths and delousing for all, only he forgot that there were not enough showers. When he was delousing the apartments, he did not consider where to accommodate the tenants. Thus, the poor wretches were lying on the bare ground; those who attempted to flee were shot. When he came back the next day to inspect the results of his energetic outburst, in each courtyard he found a dozen or so dead bodies of old and sick persons who had simply been unable to withstand a cold night on cold stones under the open sky. Dr. Hagen did what people all over the world do in such situations: he fired the chief of the Health Service, who was a woman,

Dr. Syrkin-Binstein. Later, when she was organizing school hygiene, I had occasion to see that she was a gallant woman. She was just not the right type for a Health Service chief, especially not under the circumstances that prevailed. Only a man able to arouse terror or enthusiasm could hold this position. No one was afraid of her, and by and large no one liked her. However, the stated cause of her dismissal was unjust, because Dr. Hagen's orders concerning Krochmalna Street were sheer nonsense. Apart from this, Dr. Hagen appointed Dr. Ganc to the post of the epidemic commissar, which made sense because Ganc was an energetic and honest man.[12]

President Czerniaków called on us again to say that we could not tolerate all this to be carried out right before our eyes and that if we were unable to cope with the epidemic, we would all be deported. He demanded that a Health Council be established. He asked me to formulate in writing the causes of the epidemic and feasible measures to be taken.

I felt that a crucial moment had come for me. The cause of the epidemic was the deportation of people from infested areas to the ghetto without prior delousing but after complete robbery. Someone had to tell it straight as it was. In the name of history one could not sustain the complaints that the Jews were spreading the typhus. I had warned them about this in my memorandum submitted through the Joint.[13] What were feasible measures? The only effective measure would have been to rebuild the entire social and sanitary structure. Sanitary—because it was necessary to prevent people from crowding in central bathhouses, standing in line, and so on. Social— because you had to let people earn enough money to purchase a new shirt or at least to wash the one they had.

When in the perspective of further events I now think about this memorandum, I see how naive it was. There was not enough coal to keep even the few public bathhouses in operation; nothing was available to set up a bath in each house, as I suggested. Furthermore, the political government was not interested in eradicating the epidemic but only in containing it. Let the Jews die, providing they not infect German soldiers. However, at that time, I heard only the voice of my own conscience. I knew that the time for settling accounts would come, and I would have to state before history whether I did not hesitate to tell the truth even at the price of my own life. In view of this, I wrote a memorandum about the causes of the epidemic.[14]

I read the memorandum to several persons: Dr. Milejkowski and hospital directors. Dr. Milejkowski was of the opinion that the text was written too boldly and that it was foolish to tease the occupier. I said that the roles had to be defined exactly. If what I had written was unsuitable, then my cooperation was not needed, because I had no intention of playing the role of an obsequious clerk. So I suggested that I alone would sign the memorandum as a university professor and member of Poland's Chief Health Council who, for many years, had expressed his opinions and presented his papers at various conferences held by that council, with regard to the epidemic situation in the country. After that, Dr. Wyszewiański got up and supported me saying the memorandum would save the honor of the district and the sanitary board. The memorandum was dispatched with my signature and my entire responsibility. I went home, sure that I would be arrested. Maybe I should not sleep at home? Then they would arrest my wife instead of me.

President Czerniaków came back. The memorandum was accepted. A conference to discuss the causes of the epidemics and the methods of combating it was to be held with Commissar Auerswald. The commissar did not regard the memorandum as a critique of the government. Perhaps he had not read it to the end. Or maybe it suited him well that not he but those who were ordering such crowds into the ghetto were guilty.

President Czerniaków suggested that I should become chairman of the Health Council. "What rights will the Health Council have?" I asked. "Because I have no time to give counsel to which no one listens." "You will have those rights which you take," he answered and went on to inquire, "Do you know how to fight an epidemic?" I replied: "Mr. President, this epidemic is like a volcano. It is difficult to catch lava with bare hands and direct its flow as we want it. Under the prevailing social and political conditions, I will not promise that I will defeat the epidemic. However, if I succeed in defeating the delousing, you will have good reason to be thankful to the Health Council." I promised to give him a reply soon as to whether I would accept the position.

I consulted with my friends. I learned that entire delegations had recently called on the president requesting that he appoint me chairman of the Health Council, and that a panic had spread among the people who believed that only I could save them. No wonder; Jews, when sick, always seek the advice of a professor. The delegation supposedly also included Councilor Eckerman, representative of orthodox Jews, who announced that it did not matter that I was not Jewish: "When you're sick, you consult a professor and don't ask him what his religious confession is." I went to my wife and asked her for advice. She was against it, being above all a loving woman. "The fight against the epidemic is nothing but a huge bribery," she said. "They will put filth on you and say that you have made money on someone else's misfortune. Jewish nationalists will say that you are pushing your way into their community, while Polish nationalists will say that you evidently considered yourself a Jew since your cooperation with them went so far. In the end you will be left alone while what is most important and essential in you, your scientific work, will suffer. Preserve yourself for the sake of that work and for our sake."

I answered: "By a strange coincidence, the hopes of tens of thousands of people have concentrated on me. If they learn that I am thinking of how to defeat the plague, maybe they will go to sleep tonight with peace in their minds. Maybe this will alleviate their panic that the nightmare is edging its way into their apartments and that death is descending on those they love. Furthermore, the Germans treat those unhappy people like pariahs, because they see a crowd of nameless wretches. It may just be that the relation will improve somewhat if the Germans begin to see among them persons whom they once respected and whom other countries still respect. Finally, maybe something can be done. If they slander me, I don't care. I'll tell you something, and please don't think I am conceited. Only the voice of my conscience is important to me, and I care only for your and our child's opinion. Yours, because you know the innermost secrets of my heart; our child's, because I wish to live on in her memory."

The next day, I informed the president that basically I was accepting the directorship of the Health Council if the statute would guarantee that the Health Council's resolutions, after approval by the chairman, would obligate all officials of the Community Council.

My wife was right, although differently than she anticipated. When I announced my decision, a woman I knew told me that she would regard this as collaboration with . . . the Germans. For those who thought that way, I have only one answer: "There are two categories of vile people. Those who blackmail and those who slander. The former at least gain some advantage from their action. The latter do it out of perverse impulses and internal filthiness. I do not care for their opinion."

Now I will describe how the Health Council was organized and what I saw while walking through the district and listening to groans not heard by many.

Chapter Twenty-Five

The Health Council

According to the initial setup, the Health Council had the right to make personnel recommendations. I heard many good things about a physician and a colonel in the active service of the Polish military forces: Dr. Mieczysław Kon, who lived in the district. I invited him privately to our apartment. He was a tall, slim man with a military posture. His beautiful face with distinctly designed features had a concentrated, powerful expression. I heard that he conducted himself bravely during the siege of Warsaw. He was one of our sanitary chiefs or inspectors and used to inspect the most dangerous sites. Under the battle fire, he learned that a bomb had killed his wife and daughter, but he did not interrupt his work. There was another candidate, Dr. Jerzy Landau, who was chief of a municipal health center prior to the war. His professional qualifications were very good, but Dr. Milejkowski did not want him, and I had to admit that a higher military person was more suitable in the presence of the occupier. The whole military posture of Colonel Kon indicated that he was able to impose his will. And in the district, it was essential to have someone who knew how to give orders. I remember our conversation at the time he was being appointed to the post of chief of the Health Department. When Dr. Milejkowski asked him whether he would accept this position, he answered: "Yes, but only until the Polish military forces call me." I liked his answer.

Up to then, the public health services in the district had been dispersed; the work was carried out partly by the Community Board and partly by a social charity: the Committee for the Aid to Jews.[1] Business conflicts were the order of the day. The simplest way was to make the representatives of the charity also members of the Health Council. Those members were Szymon Wyszewiański, a jurist and economist, and Dr. Chaim, a physician and cardiologist. Both were full of initiative and goodwill; Dr. Wyszewiański had good organizational experience. There were many things we thought through together. Other members of the council were the directors of both hospitals: Dr. Józef Stein and Dr. [Anna] Heller.[2] Dr. Stein was a typical scientist. He was a brilliant anatomist and pathologist. Before the war I persuaded him to work in cancer research and obtained a fellowship from the Potocki Foundation for him. At that time he held a minor position in the Health Department in Warsaw; while here he was in charge of several combined hospitals with a total of almost two thousand beds. He was hardly able to manage; he was too soft and mild. Any university could have been proud of his lectures; his scientific research was equally worthy. Dr. Heller had reorganized the hospital for children before the war. She was an excellent, thoughtful physician and the personification of energy. Her deep alto voice and her manner of speaking implied an emotional depth. Dr. Stein was a Christian and considered himself a Pole; Dr. Heller was a Jewess. After a while, we appointed two hospital curators to council membership: Maecenas Wacław Brockman and Michał Friedberg, both men of integrity and wisdom. Other council members were Dr. Bielenki; head of the Department of Health and of the Department of Hospitals, Colonel Kon; Dr. Wortman; and epidemic commissar, Major Dr. Ganc. In this way, the management of the Health Service was in the hands of two higher military persons.

By virtue of his office, Dr. Milejkowski was a member of the council as the sanitary chief. Initially, when the district was established. Jewish nationalism was so strong that Poles of Jewish origin were not allowed to hold any positions. The composition of the Health Department and Health Council was ample evidence that that period was soon to be over. One could find Christian Jews in other posts as well. For example, the post of the secretary general of the Community Board was held by an engineer by the name of Król who was a member of the Polish Socialist Party and had spent many years in exile in Siberia for his national liberation work.

We had conferences twice and then once per week. It was difficult to establish the statute. Dr. Milejkowski feared that I wanted to grab all the power; however, eventually, the commission found a compromise. Above all, the full representation before the occupying authorities was in the hands of Dr. Milejkowski. This was equally important for both of us—for him, because he wanted to represent his people in front of the occupier; for myself, because I wanted to have nothing to do with the Germans. The vicissitudes of that statute were quite curious; the interdepartmental approval in the Community Board was more bureaucratic than previous inter-ministerial procedures. Finally, when everything was adjusted, settled, and approved, the practice of mass murders began and the whole matter became irrelevant. However, I did not wait for approval but set to work immediately.

Our first concern was typhus. We began to teach people cleanliness and showed them how to get rid of pediculosis. We converted some delousing units into household service teams. We worked out such essential matters as the demand for hospital beds and linens. We saw to vaccinations. We modified all tedious and unrealistic orders, introduced administrative and professional control of disinfection, invited the engineer Szniolis[3] from the State Institute of Hygiene as the sanitary engineering expert, established a sanitary engineering expert, and established a sanitary engineering desk. We worked out a project for combating tuberculosis and venereal diseases. We organized school and camp hygiene, garbage removal from the city, and treatment of the terrible itch, which was spreading pervasively. After these matters were worked out in full, we began to investigate peace-time sanitary problems. To a few men I assigned the task of reporting on the control of food articles and on the nutritional state of the people. Instead of contagious diseases, we were facing a still more tragic problem: starvation. New diseases hitherto unknown to physicians began to proliferate, such as softening of the bones caused by certain vitamin deficiencies, a pathology discovered in the district and described by Dr. Jeleńkiewicz. Finally, at the request of Colonel Kon and with the cooperation of gifted builders and architects, we organized a typhus exhibition which, if it had been opened, would have convincingly demonstrated the lofty intellect and ingenuity of our community. One day prior to its opening, murders were initiated in the district. I do not even want to write about it. If those people had survived and the district had not been laid into ashes, the conference minutes and projects of the Health Council as well as of other councils organized there would have served as proof of an extraordinary fortitude and resilience. Our writings would have been cited by many to teach younger generations. In the light of subsequent events, however, our scientific achievements appear to me to be a bad historical joke. What does it matter now how the murdered ones intended to combat tuberculosis or venereal diseases? Yet, perhaps, there will be a few who will be touched by knowing how those condemned to death tried to live in dignity and to reassemble the norms of life. In my sector these efforts were represented by the conferences of the Health Council, which is why I wished to lecture in public. The council did not approve my paper in its initial form, perhaps because I also discussed things which I had created alone without the Health Council, or perhaps because I had not adequately emphasized the role of other council members, especially Dr. Milejkowski. I willingly pay honor to the memory of that man. He did what was possible during the occupation, and he was a prudent man. However, I did not want to change my report; I wanted it to represent not only the collective effort but also my personal endeavors.

Thus, the Health Council was composed of the elite and after a while became thoroughly familiar with the sanitary problems in the district, including not only the pathology of contagious diseases but also about starvation, food control, new diseases in the district, and so on. It could have played a greater role than that of Poland's Chief Health Council, because it not only acted as a consultative body but could also impose its will and enforce implementation. I believe that a similar statute is advisable for Poland's Chief Health Council to endow it with the rights of a parliamentary health commission. Nevertheless, the conferences were tedious, bore the stigma of all commissions, and the members talked too much. It took at least five minutes to find out that that someone agreed with the speaker's view, and everyone felt obligated to express his views on every point. I had frequently been chairman before the war, was used to directing discussions, and enjoyed the confidence of the participants. Here, I was often baffled, and I did not always understand the motive of opposition. Whenever I had to choose between the voice of my conscience and the council's decision, I went beyond the authority assigned to me. I was scolded vigorously, and I believe that if the council had existed for a longer period, some members would have considered me a despot and others would have regarded me as a blunderer. I would have acquired the reputation of an incompetent person mainly because I did not interfere in personnel matters since I saw that Colonel Kon was doing a perfect job. For my purposes, it made no difference whose relative would obtain a position. By and large, however, I have pleasant memories of that council. I also believe that its existence did some good for the district, even though I do not overestimate its role or the role of any institution of that type.

By a strange coincidence, the typhus epidemic began to subside soon after the council was established. Our efforts were concentrated on letting individual apartment and house infestation die out spontaneously. We simply did not spread the epidemic by issuing the wrong orders. I do not believe this was the only factor responsible for the improvement.; however, public opinion ascribed it to the new executive board and to the new spirit of the health service.

Now, I must discuss a matter which may seem superfluous to many readers but is necessary in order to understand the last moments of a perishing nation. For I cannot describe life without describing death. What means were used to exterminate Jews in a relatively peaceful way? How many died and of what? This problem is interesting not only from the historical but also from the epidemiological viewpoint. After the Great War, a scientific branch called experimental epidemiology developed in England and America. Animal populations were kept in cages, while the effects of various factors on the course of the epidemic were studied: hunger, population density, admission of newly infected or healthy animals, and so on. These animals were infected artificially, usually with typhus and typhoid pathogens. Central Europe could not afford such experiments because they were too costly. In this respect, the war implied a progress: the Jews proved to be good experimental material, considerably less expensive than mice and rabbits in America. Almost half a million people were squeezed into cages, all conditions that would facilitate the entry of pathogens were created, and all this was allowed to fry in its own juices. With the exception of a few grotesque decrees, no aid was given. When Mr. Kudicke received fifty million złotys to combat the epidemic in Poland, the only gift he sent to this greatest concentration of people and diseases in the world was a disinfector for eight thousand złotys.[4] For the rest, the occupier trusted in the hidden Jewish treasures. Most important, the motive was a deliberate desire to destroy. What proof do I have? Dr. Levy, director of the tuberculosis hospital for Jews in Otwock, went to Dr. Lambrecht, director of the medical district, to ask him for an allocation of food for the tuberculosis hospital. Dr. Lambrecht[5] spoke in a friendly way to him, even asked him to take a seat (Dr. Levy was once a practicing physician in Germany), phoned the supply department, and the astounded Dr. Levy heard the following question uttered by Dr. Lambrecht: "Do we actually want to destroy the Jews, or do we want them to live?" A question formulated in this way could have provoked only an answer unfavorable for the Jews. Dr. Levy received no allocation for his patients. When he was telling me this, I recalled Professor Sauerbruch's open letter about the conscience of German physicians. Now, this concept is probably regarded as a Jewish invention.

Let us now answer the question of what was the cause of death of the Jews who, in the twentieth century, had the honor of serving as experimental animals for a "united Europe."

The average death rate per month was about 5,000. This yielded an average mortality of 120 per 1,000. To illustrate the monstrosity of these figures, I will state that the prewar mortality among Jews was 10.5 per 1,000, while that among Poles was 11 per 1,000. The occupier's "concern" raised these figures to 22 per 1,000 for Poles and, as mentioned before, to 120 per 1,000 for Jews. The cause of death? During the worst months of the epidemic about 2,500 typhus cases that resulted in 700 deaths were recorded.

Thus, about 50,000 persons died each year. With this death rate, the ghetto's life span would have been eight years. Thus, the conditions created under the aegis of German science in order, as they said, to teach the Jews to work, meant extinction of a nation within eight years. In the wake of such facts, the triumphs of German organization became grotesque.

Of what did these 50,000 persons die? The number of those dying from typhus was 200 to 700 per month. Let us assume that the true number was three times greater. This still left 3,000 to 4,000 persons per month whose deaths were not caused by typhus.

The fact is that those people died from starvation. I knew houses where one-third of the tenants died within a few months. Mortality among deportees was so high that housing for successive groups was readily available. The dry reports of the Social Service make a more tragic impression than elaborate descriptions. I received these reports once every two weeks. Regularly once every two weeks I would sit over such a report in helpless agony, thinking that all our Health Councils, vaccinations, and conferences were only mental hygiene . . . for ourselves. No medical treatments could possibly help people who were dying from starvation.

I once invited Dr. Kerner, a statistician from the Supply Department, to a meeting of the Health Council and asked him to present a review of the state of nutrition in the district. The amount allocated by the government by means of ration cards was not the true amount consumed because smuggling was booming. I should mention that the rations officially allocated amounted to some 10 percent of the demand in the district, whereas on the Aryan side it ranged from 24 to 40 percent of the regular demand. Thus, Dr. Kerner conducted a poll in various social strata in the district, investigating the exact amount of food consumed. To orient the nonmedical reader, I will add that the amount of calories required by an adult person is about 3,000 per day. The highest aristocracy in this respect was represented by employees of the Community House, because they consumed as much as one-half of the amount of food required for life: 1,500 calories. All other strata, such as professional intelligentsia, workers, and craftsmen consumed fewer than 1,000 calories. And the deportees? They received about 300 calories. No wonder then that one could see thousands of beggars on the streets swollen from starvation. One could also see those hideous cadavers I mentioned before. I will describe the picture of this slow dying with 300 calories per day later when I will write about my trips to the abyss of squalor: the deportee houses and the child care houses. Right now, I will mention a man who devoted his life to children and who would readily have let his hand be cut off in order to feed his orphans. I am referring to Dr. Janusz Korczak.[6] For his orphanage he received such a poor allocation of food that if he had fed all the children with just that amount, they would have died within a few months. It was impossible to live with that amount of calories. In view of this, he decided not to feed infants in order to save older children who were aware of being alive. Janusz Korczak died a hero because, even though he could have, he did not abandon his orphans who were walking to their death. I am sure that when he faces the judgment of posterity, he will be forgiven for condemning the infants to death.[7]

Starvation has various effects. It changes the course of an infectious disease and induces specific pathological symptoms. I cannot avoid these purely medical problems because they are part of the picture.

When I read the district's death statistics, I was struck by the frequent occurrence of the term *exhaustion* and the rare occurrence of the term *tuberculosis*. Yet, it is well known how quickly tuberculosis spreads during famine.

Well, I have a secret to disclose. Tuberculosis is an infectious disease. According to the government's orders all who died of an infectious disease had to be buried in coffins. Most members of the "parasite nation" were unable to afford this luxury. Therefore, tuberculosis patients died of other diseases, because the payment for an appropriate certificate was much smaller than the price of a coffin. Presumably, a fairly profitable medical specialization developed during that period: issuing certificates for deceased men. To understand how quickly and how widely tuberculosis was spreading, one had to talk with physicians who were working free of charge in clinics and hospitals. X-rays of apparently healthy children from orphanages revealed that 12 to 15 percent of them had tubercular changes. My wife examined girls following the sanitary courses and X-rayed many of them. The percentage of positive findings was just about the same. Those were the apparently healthy students. Many had cavities in their lungs without knowing it. In hospital wards for tubercular children, the disease was growing and picking up speed like an avalanche, devastating those starved lungs that often had cavities the size of an apple. Mild forms were not seen at all. All cases were disintegrating, commonly called "galloping forms." These children were not even able to develop a fever. Absence of fever was striking in the wards. Normally, this is a good sign, but at that time it was a sign of approaching catastrophe. Attempts were made to separate the open, disintegrating tuberculosis from the initial stage, but children from the non-galloping ward were rapidly being transferred to the galloping ward. The poor little creatures knew that this was the end of their road, because it was impossible to separate the dying children; there simply were not enough wards. Thus, they watched as their friends died, several or even dozens a day.

Thus, a number of the starved people were dying of tuberculosis. However, hunger also killed without the intervention of germs. Children's hospitals were filled with two types of patients. Some were so edematous that they resembled drowned bodies: a pale blue face blown up like a balloon, eyes invisible, legs like that of an elephant, and the abdomen protruding like a mountain often contained ten or more liters of liquid. Others were skeletons covered with skin, hardly able to move the limbs, and lying powerless with legs pulled up. The senile faces had bushy brows and occasionally a mustache, beard, and sideburns emerging from parched skin. In the department headed by my wife, Drs. Jeleńkiewicz and Foman elucidated the cause of the senile appearance and hair growth. It was starvation damage to certain hormone-producing glands. Was it possible to save them? Most were in an irreversible condition, and the children were dying like flies. Yet, it occasionally happened that some poor souls improved even on the miserable hospital diet. Then, a relapse inevitably occurred. No sooner were they sent home after a marked improvement than they were brought back in considerably worse condition than the one in which they had originally been hospitalized. A number of research projects on the effects of starvation and the various forms of avitaminosis were begun. The findings would have made an important contribution to science; not all records were able to be saved.[8]

Starvation induced disturbances in menstruation: usually the disappearance of menstruation and barrenness. The number of newborn children was very small, and those who were born died after a short wandering through life. Curiously, people began to complain about pain in the limbs. This was usually treated as rheumatism or neuritis. After a certain period and in spite of treatment, the patients became crippled and all motion became impossible. According to Dr. Jeleńkiewicz's findings, this was softening of bones due to insufficiency of the fat-soluble vitamin D. Poland's Dr. Goldflam had been the first to describe this disease in Poland during the Great War.[9] Proper diet with ample amounts of vitamin D restored these individuals to health. Thus, apart from death by starvation, it was necessary to know the life or pathology of hunger. Most physicians were not familiar with these symptoms, especially the initial stages. Thus, I began to organize a course on starvation pathology while two special laboratories were investigating these problems scientifically. I believe that if the district had not been entirely murdered, science would have been enriched with very valuable research in this area. Jews have always respected science, and therefore it was not difficult to obtain credits for additional food when scientific experiments were stated as the objective. Generally speaking, the symptomatology of infectious diseases presented a specific picture in the district, and even typhus took a somewhat different turn: cerebral disturbances predominated. It is difficult to say whether this was due to pathogens with specific properties or to the route of infection. I had the impression that the disease was not being transmitted just by louse bites. It also appeared in very rich families for whom lousiness was out of the question and who never came into contact with the crowds because they simply never went out. I know of physicians who were infected by coughing patients. One more observation was astounding. The death curves in both districts ran in parallel. Even though the Jewish curve was many times higher, the weekly fluctuations were identical. The social conditions were different in the two districts, but the climatic conditions were identical. Possibly the louse feces got into the street dust, and a certain percentage of infections proceeded via the lungs, nose, or eyes.

There were relatively few intestinal diseases, especially typhoid fever and dysentery. This was partly due to typhoid fever vaccination; however, the small number of dysentery cases has remained a puzzle. In the few cases that did occur, our laboratory tests revealed typical Shiga-Kruse bacilli.

But the most curious were the pediatric diseases. Diphtheria and scarlet fever were very rare, measles somewhat more frequent, while whooping cough was relatively most frequent in the first year of life. This was probably due to the very low birth rate. Epidemics of childhood diseases are caused by pathogens present everywhere in a certain number. Their number increases when, as a result of natural growth, the number of susceptible individuals increases. Interestingly, diphtheria and scarlet fever developed mainly among Jewish children displaced from Germany. Local children were immune as a result of previous latent infections. For this reason, we performed no vaccination against diphtheria. I also feared this vaccination because of the strong prevalence of tuberculosis in the district. I realize that the recording of children's diseases left much to be desired.

What I am writing characterizes the course of infectious diseases in closed, starving districts. In such areas, typhus develops with a very great intensity but eventually it wanes spontaneously because survivors acquire immunity. Infectious diseases in children display similar trends, with the difference being that latent infections are more frequent than overt infections. In contrast, tuberculosis becomes very frequent and severe. The absence of dysentery in our district was probably accidental. I might mention though that during the great epidemic in Łódź, which also took place during the occupation, Jews were afflicted by dysentery less frequently than Poles. I do not believe that this was due to a racial difference. The underlying factor was probably different eating habits: more garlic and onion and less alcohol.

There is not much I can say about venereal diseases. By and large, poverty does not promote their spread. Rather they tend to spread during periods of economic booms. Prostitution existed in the district as it exists in all slums; but physical hunger does not enhance sexual hunger. We began combating venereal diseases by deliberately avoiding such drastic regulations as those specified by German legislation. Our regulations were adjusted to the ethical level of the district, that is, of its citizens and of the physicians. One should never issue decrees that cannot be implemented. I will mention one small curio. Even before the Jewish district was established, posters that read "Prostitution is Forbidden to Jewesses" were hanging in the health centers throughout the Aryan district. Basically, German legislation outlawed houses of prostitution. Military authorities, however, allowed such houses in occupied areas. On some streets, for example on Chmielna Street, long lines of German soldiers could be seen waiting in front of houses of prostitution. But order had to be maintained. The Polish police had to watch Polish prostitutes surrender to German soldiers. This honor was not granted to Jewish prostitutes.

That concludes the pathology of the ghetto.

The work of the Health Council was in full swing when the anticipated meeting with Commissar Auerswald on combating typhus took place. It was held in the State Institute of Hygiene, because Professor Kudicke, the institute's then director appointed by the Germans, had been granted much authority and fifty million złotys to fight the epidemic. By and large, Professor Kudicke conducted himself decently. He freed the institute's staff members who had been caught in raids and rented an apartment for himself instead of requisitioning it. My memory of him from Heidelberg was positive. I arrived in the State Institute of Hygiene with our community's president—Czerniaków, an engineer—but without the armband, and I was taking the risk of being arrested for six months. But I was simply unable to enter my institute as anybody but a Pole. We entered the late Dr. Szulc's office in which I had often officiated as deputy director. The place once occupied by me was now occupied by Professor Kudicke. Portraits of

German scientists Behring and Schaudinn were hanging on the walls.[10] Under normal circumstances I would have bowed to them with respect. On that day they were the symbol of oppression to me. Did the Germans not understand that they were destroying the legend that had gilded their science and art? When I entered the office, only Professor Kudicke was there. He rose, embraced me, and said: "How you have changed since I saw you last in Heidelberg." When we were parting other Germans were present. Kudicke raised his arm with pomp in the Nazi greeting but did not shake hands with me. By accident? Or through lack of civil courage?

During the conference, the conversation dealt with the usual subject: the Germans were complaining that the Jews were disseminating the plague and were doing nothing to prevent it; while the Jews kept explaining that they were doing everything they could, but that they had no means to combat the epidemic. The German's language has a good expression for such a discussion: *vorbeireden* [to skirt the issue]. Chairman Czerniaków spoke in a sharp and courageous manner. At that point, Commissar Auerswald whispered loudly to his neighbor: "Frecher Jude" [cheeky Jew]. Lambrecht, the district's sanitary chief, was also present. I had a strong feeling of aversion toward him since the time when, prior to my deportation to the district, I talked with him on the phone and he hung up on me. I was not accustomed to being treated that way. This was the first time I saw him, and he impressed me as such an insignificant zero that my feeling of hate vanished. My alleged protector, Dr. Kohmann, was also there and asked with acrimony why the epidemic persisted if the Jews had block physicians, disinfectors, and everything else that had just been mentioned by Dr. Milejkowski and Major Dr. Ganc. I then asked for permission to speak and said point-blank: "Because infected and louse-infested people were brought to the district after they had been deprived of all their possessions." I suggested that I should read to them our projects to convert a portion of the ineffective delousing units into household service units, to apply new methods of teaching people to observe personal hygiene and cleanliness, and to enable them to wash their laundry and themselves by installing local showers. After I finished reading, I handed all memoranda over to the chairman, including those which criticized decrees issued by German authorities. I thought: "It won't do a bit of good, but let them at least be aware of it." Indeed, it did not help a bit. Of the fifty million złotys earmarked for combating the epidemic, we received almost nothing. We never saw the 1,000 hospital beds we requested. Nonetheless, I was not sent to a concentration camp, probably because the commissar thought that I was insane since I dared to say the truth, and not in bad German at that.

Such was the conference of officials responsible for the greatest typhus epidemic during that period. A mountain gave birth to a mole hill.

I began to walk through the district, bestowed with the rights of the Health Council's chairman and with the undeserved aura that I possessed the power and the recipe to banish the epidemic, squalor, and crowdedness from the district. As an epidemiologist, I have always tried to familiarize myself with local conditions, because an epidemic is like a barrel that leaks in many places. One must personally find and analyze all the leaky spots.

I began to study the problem of the deportees. Only some of them found apartments and led a seemingly normal life with gainful employment. When I arrived in the district, there were still twelve thousand deportees alive, living in deportee houses. At first, they were cared for by the Community Board and the Committee for Aid to Jews.[11] The committee was issuing regular communications on the health, life, and death of its charges. Later, the Committee for Aid to Jews entirely took over the care of the deportees, while supervision remained in the hands of the Community Board.

There were three types of accommodations for the deportees: whole blocks of houses, larger five- to six-room apartments for several hundreds of persons, and smaller two- to three-room apartments for a dozen or so persons. In the courtyards, there were mountains of garbage and feces, often reaching up to the second floor. For reasons that I will explain later, the garbage collection service was not functioning in the district. The Supply Department often received and distributed rotten cabbage and frozen potatoes that were inedible, which improved neither the health nor the cleanliness of the district, because these products quickly reached the garbage piles. Because of the frost and unavailability of heating material, sewage pipes were damaged almost everywhere. As a result, there were piles of feces on the streets, in the courtyards, on the stairways, and even in apartments on the floors. I enter the stairway of a house and find no banister because it has long ago been used as fuel. I enter rooms and see people lying on sleeping boards and on the floor, actually not people but some kind of balloons or live cadavers. They lie naked or in some rags, without a blanket or a pillow. It is winter, but most rooms have no glass in the windows. What are those ghosts waiting for with their eyes fixed into empty space? For death. It comes with a merciless inevitability. The cause of death: starvation, typhus, or tuberculosis. The mortality in those houses was 60 to 100 percent. The children are in the courtyard. I do not write "play" but "are in the courtyard." These children do not play. Besides, they are not children; prematurely aged faces, a sad gaze, and their little legs like thin sticks. Some men apparently work but are unable to support their families. Those men get a bowl of soup so that they would not faint from hunger at work. But the families fall into the category of nonworking persons and are therefore classified by the master nation as parasites. The Community Board gives everybody a bowl of the notorious soup. Chairman Gepner, a philanthropist with a great heart, was right when he said that this kind of aid was wrong because it only prolonged the agony. I also believe that it would have been more humane to kill them at once. There were shoemakers, tailors, and washerwomen among them. It might have been better to give those bowls of soup for work within one's block. Some people had no strength to do any work, and they would have died faster. Others, upon getting up in the morning, would have a goal to work for instead of that hopeless emptiness. One could have used other opportunities and found employment for more tenants. Finally, it might have been possible to give not 300 but at least 1,000 calories to each. The Committee for Aid to Jews was making similar plans; it wanted to allocate 100,000 złotys, not for soup but for organizing labor. It was the eternal problem of aid to the unemployed. But here nothing was able to halt the avalanche of death.

I inspected dispensaries for deportees and talked with physicians and nurses; they made the impression of being honest and dedicated persons. The managers of deportee houses were not always up to the task though. Some were the personification of energy and selflessness, but there were many who stole from those unhappy wretches. I tried to cultivate within myself perceptiveness so as to distinguish selfless people devoted to ideas from thieves. I came up with the following formula: by and large, selfless persons had been working for social causes even before the war; one was a Zionist, another was a Bund member, while still another was orthodox—no matter what he was, he had to have faith. Those people worked for ideas. The thieves, by contrast, were always adroit, had good manners, and were eloquent. When they would say that they were proud of being Jewish, I knew that they were making good money.

The second type of facilities for the deportees were large 5 to 6 room apartments with grand rooms of the former bourgeoisie, or possibly school buildings. These accommodations isolated the deportees from society to the smallest degree. It was easier to supervise them sanitarily and to find jobs for the tenants. Men would set up some workshops, while women would go out to do laundry jobs. Small apartments located in basements were worse. It was impossible to supervise them, because there were too many of them and they were scattered. And there I again came across sights of extreme squalor and hopelessness.

The deportees were periodically sent to bathhouses. There they stood in lines for hours infecting themselves and others, while the delousing destroyed the last few things they still possessed. On our advice, a number of managers installed local showers and introduced the ironing of linens and underwear instead of delousing. Such a local delousing usually proceeded smoothly: the Jews gladly took showers to rid themselves of lice. Who cares whether this was due to a love for cleanliness or to fear of disease. The problem of personal hygiene applies not only to Jews but to peasants and workers also. Folk customs should be changed but not by talking at such length about personal hygiene that they cease to react. Washing and changing shirts should become a reflex, but first it must be made possible for them. No one is dirty out of love for dirt but through habit. Cleanliness determines a higher standard of living.

But the lowest depth of squalor I saw when inspecting the living quarters of the Jewish proletariat. There I saw a sick, feverish mother lying on a pallet and the body of her dead child next to her. I saw a seven-year-old child go begging and taking care of his two-year-old brother because the rest of the family had been struck dead by the disease. I saw a mother with insane eyes gnawing the dead body of her child. I walked through this pit of destitution overwhelmed by pity and disregarding the likelihood of catching lice and other diseases. This was life for millions of human beings in the twentieth century, and this was the revenge of the master nation. I recalled the words of Reduta Ordon:[12]

Where is he who is sending these masses to slaughter?
Is he sharing their courage, is he exposing his chest?

I had a vision of a castle suspended in the mountains: Berchtesgaden.[13] In it sits a madman who has brought the instinct of hate to ultimate perfection. He beholds the beautiful skyline of the mountains, sunsets and sunrises, and from time to time he opens Nietzsche and reads that compassion destroys life. If I could bring him here and show him this bottom of poverty, his heart just might burst. If I wanted to tell the details of my perambulations, I would have to fill entire volumes. I instructed our social hygienists to record individual pictures, just in case it would be my destiny to recount the last moments of a perishing nation. All records were lost during the destruction of Warsaw.

I also visited houses of deportees from Gdańsk. They consisted of an elite and spoke several languages, coming from the wealthier intelligentsia. One could verify Hitler's statement that Germans were veracious and Jews were mendacious, because there I saw the cruelty of the German nation combined with baseness. The Jews in Gdańsk had been told that they would go to Palestine, because the Führer was generously allowing them to emigrate to their country and take all their valuables and money. Some believed it, the more cautious ones sewed their things up. In Tczew, everything was taken away from them. The inspection was so exact that clothes were cut up into strips and buttons were torn off. They were ordered to exchange Gdańsk money for Polish currency. Very generous, was it not? However, when they arrived in Warsaw, they found out that they had been given unstamped Polish banknotes, which were invalid. They left their city as rich people but arrived in Warsaw as beggars. A very old woman with a still-beautiful face showed me a bundle hanging over her pallet: "This is my last possession: two dresses. When I exchange them for food and the food is gone, I will have to die."

I met a Jew from Mannheim where he used to have a tobacco shop. He still had German friends there, Social Democrats, but they were unable to help him because they might be sent to a camp. I asked him: "When the war is over and you have the opportunity to return to Mannheim, will you go back there or will you emigrate?" "I will go back there on my knees." Those German Jews were strange. They were beaten and despised, but they still wanted to see the Germans as a nation of poets and philosophers. In 1942, there came to the ghetto Jews from Germany. They were allowed to take their belongings, including valuables and money. All this was taken away from them on the Polish-German border or in Warsaw. The objective was to make Germans inside the country believe that the Jews were displaced in a humane way and were not robbed. One of those Jewesses called Hitler "unser Führer" [our leader] and proudly assured me that the Germans were unconquerable. My friend was right when he said that Germany was the Jews' hopeless and unrequited love. I met Dr. Hoche there. He was half French and had a beautiful, subtle face. His brother was even a high official in Germany. He lost all rights when he fell in love with a Jewess and wanted to share her and her two children's fate.

The only capital of the German Jews in the ghetto was their stock of drugs. The German authorities coaxed us to requisition all these drugs for hospitals. However, we did not want to rob beggars even for the sake of public benefit.

This was life in the district: extreme want and, to the other extreme, numerous restaurants and coffee shops. An abundance of pastry was displayed in the windows of coffee shops and bakeries. By that time, a new Jewish elite was formed who smuggled and traded with the Germans. It proved that the master nation had two types of specialists: those who hated and killed Jews and those who, jointly with the Jews, were doing good business and exploiting the paupers. I leave it to the reader to decide which type is more disgusting. Those mountains of pastry and well-supplied restaurants irritated the general public, but I personally was more indulgent. Those things are ugly, but they pop up wherever life is wasting away: a feast on pestilence.

I will confess my guilt: I too once had a good dinner in one of these restaurants. It happened as follows. I was bringing a graduate course for dentists to conclusion and the presidium decided to reciprocate. By giving me flowers? By donating money in my name to be used for the poor? In the end they came up with the idea of inviting me to dinner. For the first time in two and a half years, I felt the warmth of a full stomach and the pleasure of eating tasty food and drinking good vodka and even wine. And the mere bliss, the divine bliss of carrying on a carefree conversation about which is better, fried or cooked fish and what kinds of wine go with such and such fish. For a short while we forgot the war, the misery, and the death probably awaiting us.

I came home somewhat tipsy. There were several persons in our apartment, and they asked me what I had for dinner. I am not a bad talker, but I never saw people listening with such rapt attention. I began: "At first we had fried liver with onions and a herring with vodka." I saw everybody was delighted. "Then we had tomato soup." I heard a deep sigh. "Then Jewish style pike, then roast goose." The tension reached the zenith. "Finally, charlotte . . ." At this point someone cried out "Be quiet!" I assure you that even if the most heroic man were asked, after two or three years of starvation, how he would like to die, he would not choose death with heroic words on his lips but, depending on personal taste, with a piece of charlotte or fried liver in the mouth. This is so because when a man is dying, he first loses higher functions, while the so-called lower functions, including hunger, are still intact. I will even state a paradox: a man who still has appetite is not dying; when appetite is lost, this is the sign of an approaching end.

At that time, the Jews still felt a hunger for life. This was manifested in the existence not only of restaurants and coffee shops but also theaters and concerts. I did not go to the theater, but I went to concerts and cabarets. They were on a high level: the Jew was dying with witty humor on his lips. This humor was sometimes bitter and sometimes wistful. I remembered Gold's song about prewar Warsaw and about the charm of a big city, a summer evening, and carefree love.[14] Dante's words came back to my mind: "Nessun maggior dolore che ricordarsi del tempo felice nella miseria." [No greater pain than to recall in our misery a time when we were happy.][15] A strange atmosphere prevailed at concerts. Jews were not permitted to play any composers apart from Jewish ones. But those "nasty" Jews also loved Beethoven, Brahms, and Chopin, and they preferred the risk of being imprisoned than giving up this eternal beauty. I even remember Beethoven's Fifth Symphony. It was a hymn of joy played by condemned men. For it was inconceivable for those "wicked" Jews that condemned men could be forbidden to enjoy beauty, which was no one's property. To forbid listening to Beethoven was the same as forbidding one to cherish the sun. The sun would not become poorer by warming up a few more Jews. However, the master nation did not want to grant them this higher joy. The Germans outlawed concerts as a punishment for the fact that the Jews had dared to play Aryan composers.

In Schnitzler's drama "Professor Bernardi," the hero who has been done an injustice by various people makes the following profound statement: "I forgive all injustice and cruelty if they bring advantage to the evildoer. Baseness reaches its peak when it becomes useless." I believe history will sooner forgive Germans the robberies they committed than the fact that they took the ecstasy of the highest harmony away from men condemned to death.

In due course, they had to discredit Jews in the eyes of the whole world for the fact that the nouveaux riches dared to eat in the face of misery. For this purpose, a motion picture company arrived. A man eating in a restaurant is not very photogenic, however, so the order was issued to set up the table in President Czerniaków's apartment, complete with requisitioned bottles of liqueurs and wines. All this was photographed to show the world what a good life the president of the Jewish Council had.[16] People were caught on the street, delivered to requisitioned apartments, and ordered to eat meals requisitioned from restaurants. They were coaxed to eat greedily and messily, and were filmed doing so. Decently dressed ladies were caught on the street, brought to disgusting ritualistic bathhouses, and ordered to bathe naked and to wash ragged Jewish men so that the world could see how slovenly those Jews were. Or the circumcision ceremony was ordered to be performed for a motion picture, and if the mother was too neat, a dirty Jewess was caught on the street and put to bed instead of the mother.

Those were the methods used to arouse an ultimate repulsion among Germans against the condemned nation, in order to prepare them psychologically for a justification of the planned total murder—this time without survivors and without mercy.

Life was wasting away; 80 percent would have died anyhow, and 20 percent would have survived. But these 20 percent would have resembled human beings. In view of this, it was decided to make the living conditions even more severe so that more would die at a faster rate and survivors would suffer more painfully. The means were as follows: every once in a while, the district was made smaller, and 15,000 to 20,000 persons were shifted to the overcrowded remaining area. This was always done suddenly to impair orientation. That the craftsman was losing his workshop and the physician or dentist his medical office—this was precisely the objective. The fact that the Jewish mob stole mercilessly from their fellow men during such shifts only magnified the feeling of contempt for those who were perishing. There were many of those shifts, and actually no one was certain of the next day. But one day, posters appeared stating that Jews caught beyond the walls would be killed.

The decree was signed by Commissar Auerswald, a lawyer by profession. This was a man of consequence who would not allow a German word to be disbelieved. Soon, a list of persons killed was hung out. Somewhat later, a new list was posted. Then, they ceased to make these announcements, because it was a shame to waste paper. Very many were killed, and I can tell their story.

The prison director in the district was Lindenfeld, a doctor in law and judge of the Polish courts.[17] He was a good and noble man, and it was he who told me what follows.

The first person sentenced for trespassing the ghetto border was a glazier from Sienna Street who went over the barbed wire to the Aryan side of that street to buy a loaf of bread for two złotys or less. The second person was a beggar woman; she dared to beg on the other side. The third was a young mother of an infant a few months old. She went over to get some milk; she preferred risking her own life than that of her child. And so it went on endlessly. The shooting was ordered for seven o'clock in the morning. Stakes were set up in the prison's courtyard. The Jewish Monitor Service was granted the honor of binding the convicts to the stakes. Under penalty of death, the Polish police were forced to fire on command because the Germans wanted the disgrace of killing to fall on the Poles. The glazier, when he was being tied, exclaimed: "But they are people too!" The Jew who supposedly possessed Oriental cruelty could not understand why purchasing a loaf of bread should be punished by death. The young mother asked an orderly to deliver her infant to an orphanage. A young girl asked another orderly to kiss her before her death; she closed her eyes and probably imagined that her beloved was kissing her. The beggar woman took her watch, purchased or stolen, gave it to a policeman and asked him to carry it over to her family; her last thought was to help her relatives.

And so it went on without end. The German Attorney General was present. The commissar had not yet arrived, but it was decided not to wait for him. The commander gave the order. The bodies of the innocently killed drooped down. At that moment a car arrived and the commissar, fresh as a daisy and fragrant as a rose, jumped out. Was he pale from emotion? Did he realize that his name was associated with a crime? No, he said the following: "What a pity. Excuse me for being late. Das haben sie fein gemacht [you've done a fine job]."

Judge Lindenfeld is no longer alive. He was murdered under the following circumstances. He lived with his family in the prison building. They were having dinner: he, his wife, his twelve-year-old daughter, and their faithful housemaid, an Aryan, who did not want to abandon them during their misfortune. SS-men, spokesmen of the law, broke in and shot everyone, one after another. Now I am probably the only human being who knows the words spoken by the commissar during the first execution. I am communicating them to the world in the hope that on that day the commissar signed his own death sentence. Even my heart has become callous.

The death penalty was administered not only for crossing the district's border. At one time they needed fur coats for German soldiers; but probably for the hangmen, their wives, and lovers as well. The commissar passed the order that all fur coats be handed over under penalty of death. Some Jews burned their fur coats and jackets; they preferred destroying them to handing them over to the enemy. Others sent their furs beyond the wall. The smuggling operation was brilliantly organized. However, many turned their furs over because they had been seen wearing them and they feared blackmail from their fellow Jews. Again, there were long lines standing in front of the Community House in spite of the epidemic. And then the commissar came up with another great idea. He promised to free a number of Jews sentenced to death if he were supplied with 1,000 fur jackets. The following conversation supposedly took place between the Community President and the commissar. The president: "Mr. Commissar, the time has come for me to ignore the hierarchy's rules. I am speaking to you, not as to a commissar, but as to a lawyer; I could be your father. Do you know the name of General X? No, you don't. Well, he was the commandant of Saint Helena Island. He tortured Napoleon more than was necessary for reasons of security. However, when he returned to London, no one shook his hand. Remember this, sir." Thereafter, the commissar lifted the death sentence from several hundred Jews in exchange for 1,000 fur jackets. The chairman went to the prison to tell this to the convicts personally. The people on the street and the prisoners were all crying. I watched this and thought: "A bastille is conquered with blood and not with fur jackets."

During my wanderings in the pit of misery, I also visited the jail on Gęsia Street. As the chairman of the Health Council, I was entitled to enter every facility. Besides, there was typhus and dysentery in the jail. The jail, designed before the war for two hundred to three hundred people, now housed twelve hundred. I asked why they did not protest. "Because the Germans shoot the excess down. We are trying to hide them." I walked into the prison hall. On the left side I saw smaller cells for the intelligentsia; eight to ten persons per cell. Each window was barred, but the prisoners could see the courtyard, grass, and the sky. The cells for the proletariat had small windows right under the ceiling so that not even the sky was visible. I entered one such cell and noticed that everyone was standing. There was not enough room to sit down, let alone lie down. The air was so thick that after a few minutes I had to leave; I thought I would pass out.

Dr. Beiles, an excellent physician on the hospital staff, told me that he had visited a sick friend on the Aryan side, was caught, and put in this prison. He voluntarily assisted the prison physician. He told me that each day several persons were dying with the signs of carbon dioxide poisoning. The prisoners literally suffocated. I asked for what offenses they were arrested. It was always the same: the intelligentsia, because they had stayed outside the wall; the proletariat, because they had crossed the wall in search of work and bread. "How was it on the other side?" I asked. The reply was passionate calls uttered by adults and children expressing a yearning and gratitude: "Oh, the Poles are good; they gave me bread, soup, and even socks. I even spent the night there."

I had my armband on. Those poor creatures could not possibly know that my heart was growing at the thought that my nation, whom the world reproached for anti-Semitism, was good. Despite the death penalty imposed for aiding Jews and despite their inherited dislike for Jews, they helped them. "If Jehovah is keeping records of all the injustice done to Jews," I thought, "he will cross out insinuations made by Poles, scuffles at universities, and the separate benches." Because the Poles' antipathy for the Jews lasted only as long as the vision of the powerful Jew prevailed. It changed into pity when the Jew became a pauper. This was the Poles' stand during the martyr's death of the Jews. Of course, there were also other cases; I will recount them later.

I visited the execution site. It was in the back part of the prison compound and resembled a narrow garden. Green grass grew there. This place resembled a quiet little garden in a monastery. This impression vanished the moment I saw stakes and bloodstains on the walls. The jail was repeatedly visited by Swiss physicians who would come to help the Germans.[18] I do not know what fool brought them to that prison. They did not want to believe that the people were shot down for the offenses stated. I later read in newspapers that Switzerland was unable to comprehend the spirit of a "united Europe," and that hostile propaganda set the Swiss against the Germans. Do the Germans really believe that the Swiss physicians were not shocked by the terrible vision of people murdered for crossing the walls? This is a nation that honors William Tell as its hero. I would call this prison the epitome of German infamy if they had not surpassed themselves with the subsequent mass killings in gas chambers.

I also visited the cemetery. At first, the whole cemetery was in the district. Then, when it was filled it was expanded beyond the wall. It presented an awful sight. Among the monuments, armed soldiers were chasing smuggling women. Occasionally, an Aryan woman who was smuggling would put an armband on, join a funeral procession, and stealthily slip a piece of bacon into the hands of her Jewish associate during the funeral ceremony. The most horrible were the charnel house and the mass graves. The charnel house has been described previously. The dead bodies were grabbed by the arms and legs and thrown into a common pit. Ekerman, a beautiful man resembling Michelangelo's "Moses," was the cemetery councilor. He was a Jewish social worker, a deeply humanitarian man of letters, and a representative of orthodox Jews. He spoke beautiful Polish. He told me of shocking scenes, of which I will mention just one. When an infant was thrown into a common ditch it began to whimper. It was pulled out, brought to life, and sent to its mother. There, the dead body of her second child was waiting. "You don't want this one? Take the other." A few days later she brought the first one, too. Later, one had to get a pass to go to the cemetery on the Aryan side. Finally, people ceased to go to the funerals of even their close relatives. It was too easy to pass from the living to the dead state. Accompanying the deceased to the place of eternal rest was associated with the danger of losing one's life.

How were Christians of non-Aryan descent buried? Did the occupier bend to the pain of eternal parting? The mass was held in one of the two churches in the district. Then, relatives and friends accompanied the deceased up to the wall. The dead man was transported alone to the Aryan district that had been off-limits to him during his life. Occasionally, through high contacts and big money, a pass for one family member was obtained. On the other side of the wall or at the cemetery, Aryan friends were waiting. Most of them did not fear the cursed ones.

The street child. Swarms of lonely, begging, hungry children would catch one's eye every day. They made the most horrible impression in the district. It was a live reproach directed not only against the occupier but also against the Jews. Attempts were made to help them. Gallant efforts were made by Dr. Janusz Korczak, Romana Wilczyńska, President Czerniaków's wife, Attorney Pinkiertow's wife, Chairman Mayzlow's wife, Mr. Weitz, Dr. Rosenblum, Dr. Przedborski, Berman, and Dr. Zand; I am citing from memory and most probably I have forgotten a number of persons. What they accomplished was a drop compared with the sea of misery. Yet, it would be a sin of ingratitude not to appreciate their efforts.

Dr. Janusz Korczak's orphanage. Every Polish pedagogue knows and respects his name. I frequently visited that orphanage, because I felt that this was a higher world. Korczak did not even have a corner of his own. In one room were twelve small beds for boys, and Dr. Korczak slept in one of them. On a small night table he wrote his hymns to children. He taught them justice, friendliness, and a dignified stand. He would not admit children from the street, which gave rise to discussions at sessions. He did not want to save bodies alone, however; he wanted to sculpt human souls. He succeeded even in this hell, because children from the orphanage were known in the district as the personification of nobility. Mrs. Wilczyńska was his efficient and dedicated associate.[19] Together they organized concerts and lectures. Korczak had a special gift for discovering talent among the children. One industrialist named Grosman, an amateur philosopher, often visited the orphanage. He taught these children Greek philosophy. He told me that his only happy moments in that hell were when he was talking with the little Jews about the Dionysian intoxication with beauty. These children supposedly understood him, rejoiced whenever he came, and while discussing Plato would forget their predicament. One day, the Monitor Service arrived and demanded those children that were to be deported. For it was easier to carry off helpless children. When the order came to deliver seven thousand head each day the first to go were children, to the eternal, indelible disgrace of the Jewish Monitor Service. Korczak could have easily found a place to hide because he had many friends beyond the wall. I personally know that they got in touch with him. He, however, did not want the children to whom he had taught nobility, faith in mankind, and the victory of goodness to die looking at their hangman. He voluntarily went with them to their death, probably telling them: "Forgive them, because they know not what they do." This physician deserves to be considered as one of the martyrs of the idea of love.

Korczak evaded no work. There was a house for foundlings on Dzielna Street; it was hell on earth. A facility for several hundred was housing several thousand children. The smell of feces and urine struck one at the entrance. Infants were lying dirty, there were no diapers, urine froze solid in the winter, and frozen little cadavers were lying on that ice. Somewhat older children sat on the floors or benches the whole day long, they waggled monotonously and, like animals, lived from one meal to another waiting for miserable, very miserable food. Typhus and dysentery raged. The physicians were not bad men, but they were simply unable to cope with the shocking thievery of a personnel thriving on misery. Dr. Korczak decided to clean this Augean stable. With the assistance of Dr. Zand and Mrs. Henryka Mayzl[20] he brought this house of death to relative order within a few weeks.

I cannot possibly describe all the poorhouses.

However, I must mention the house for street children called "Good Will." Mrs. Pinkiert, a pedagogue by profession, was the director, while the institution was financed by Mr. Weitz. The establishment was criticized for having conditions too strikingly different from those prevailing in the district. The street children had clean clothes, each had his or her own clean bed with linen, his own towel and toothbrush, and never went hungry. As if by the touch of a magic wand, the gamins, beggars, and little street bandits bloomed into children who looked trustingly at people, without the hate of a pestered animal. The directress had a splendid idea: she photographed all the changes through which the children went, from the time they arrived naked or in rags, differing but little from animals, till they resembled sweet little cherubs. I brought my students there to show them what a heart can create. They had tears in their eyes. Who was this sorcerer, Mr. Weitz? Somewhat later, I came to know him a bit better. He came from simple Jewish stock in Małopolska [Cracow region and vicinity]. Upon arriving in the district, he set up a brush factory, began to sell his brushes to the Germans, and became a millionaire within a year. He considered it his duty to give back to society the money he had made off the Germans. He gave millions. In Otwock, he supported two hundred children. He promised to build a tuberculosis "preventorium" with his money. The Good Will House cost him a quarter of a million per month. He told me that he had enough money to live from dividends and that he could easily leave for Switzerland, but that he was employing 30,000 workers and that was dearer to him than life. He and his closest associates had strong unfulfilled paternal feelings; they took care of the orphans themselves, checked their weight, and procured mountains of fruit. The Jews are a strange people. Some looked indifferently at the death of street children, while others, overcome by some kind of frenzy, did everything to save them. I do not know whether Mr. Weitz is still alive or whether he was eventually classified as a parasite to be destroyed. In my memories he holds an honored place. I must forego writing about many, many others, especially the efforts of the Community Chairman and his wife. I will mention only that they fulfilled their duty. Thus, children were being saved, although not as many as should have been. But I must be frank: when I visited the two children's hospitals headed by Dr. Heller, a splendid human being, I had the impression that it would have been more humane not to prolong the life of those poor children. Not only because their bodies were in ruins but also because they were complete orphans, children of the street and misery, who had a terrible past and a still more terrible future. I remember one boy from Pustelnik. He looked as if taken straight from Murillo's pictures. The first time he was hospitalized for dysentery; he was saved from death. Several months later a nurse found him beyond the hospital walls dying from starvation, and she brought him back. This seven-year-old child was saved again due to the personal effort of physicians and nurses. Was he appreciative? He said in a sad voice: "Why did you save me the first time? If I had died I would not have seen how my daddy and mummy were killed." And there were thousands of such children.

I have described the life of a nation doomed to extermination. Yet, the master nation was boasting that it was teaching parasites how to work. Let us now take a closer look at the work offered and the methods used. There were street raids of the type staged on the Aryan side. But by and large the Germans loathed the Jews or else feared the epidemics, and therefore they charged the Community Board with the duty of organizing a supply of slaves. The Community Board established a Labor Office. This, as everything else organized by the Jews, immediately became a business that brought money mainly to the Germans and then to the Community Board and individual employees. There were two types of jobs: group work in the city on the Aryan side, and camp work in labor camps outside the city. Group work in the city was often a sinecure. The grouped Jews received some money, but their main income was trade. The main objective of the group work was to pay a salary to the German group supervisor and his Jewish assistant. The Germans earned good money, and only their greater greediness distinguished them from Jews. One could not see much Nordic pride and restraint. But the German reached its peak in the labor camps. I should like to point out that it was possible to ransom oneself from work in the camp. A person would be summoned to the Community House and a certain ransom was demanded, let us say two hundred złotys monthly. There would be a bargain and a final agreement on twenty złotys. Whoever did not have this amount was sent to the labor camp. Thus, helpless paupers were the victims, as usual. Most of these camps were torture chambers. Every camp had a physician from the district; most of the physicians were young, some were corrupt, a few were very dedicated. They told me confidentially what they had seen. On water constructions, people stood for hours in water up to their waists without protective boots. Depending on the season, rheumatism or frostbite was the order of the day. There were merciless beatings for everything. Once, a guard was called off his post because, while he was battering Jews with his gun, he broke it. And that was not permitted. The food was terrible, because absolutely everybody was stealing from the camp workers; Jews did not display more pity than the Germans. The workers slept in barracks on the bare ground. For trifles, they were shot to death before everybody's eyes, after they had been ordered to dig their own graves. The Ukrainian commander of one camp had a dog, a German shepherd, trained against Jews. This dog tore chunks of flesh from their bodies. I know of the following case. Two brothers, twins aged seventeen, ventured out of the camp into a village to buy some bread. The guard shot one dead and ordered the other to bury his brother. There were no court proceedings. The camp commander was the master of life and death. I know of a camp that had 120 workers, thirty of whom died within one week. The Labor Office was headed by Ziegler, a German from Mannheim. He was a good-hearted man and occasionally tried to ease the convicts' life. But what could a small clerk do in the face of the orders of the powerful Führer? People died in the camps from hunger, typhus, or sustained injuries, and often from all three factors at once. Frequently sick men were gotten rid of by being killed. Against this somber background, the stand of Polish peasants presented a bright contour. Most of them had compassion for the camp workers and gave or sold them bread, even though this was punished severely. I was told that priests preached that mercy be shown and aid be offered.

When I think of this Gehenna, I find only a few bright spots. One of them was the young people's drive to acquire knowledge. The Society for Supporting Agriculture, the so-called Toporol, has a beautiful record. It was operating even before the war on the basis of the concept that the Jewish community could he healed only by social restratification. The society did not discontinue its work even in such a concentration camp as the ghetto. Its members set up flower beds in courtyards, and they often converted house ruins and rubble into fantastic gardens where instructors taught children how to sow, harvest, and watch the earth's life cycle. Flowers or vegetables were grown on balconies, often under the tutelage of the Toporol youth. Courses in growing flowers and vegetables were organized. Finally, young persons were sent to the countryside, to farms and estates, to do agricultural work. Jewish youth highly valued this work. I read touching letters about work in the fields and about the goodness of the people they met. While camp workers described the worst of human wickedness, the letters written by the Toporol youth breathed courage, hope, and gratitude. A friend of mine, Hipolit Wohl, was an unknown patron of the Toporol. He was an ardent Polish patriot and nephew of a member of the National Government in the January Insurrection.[21] Hipolit Wohl was a former Polish official, and he had rendered the country good service by organizing the supply of provisions and cooperatives. Even though he considered himself a Pole, he wanted to emigrate, because he saw a degeneration of the young people dissociated from the earth. He dreamed not of Palestine but of Australia, and there he wanted to guide not only Jewish young people but all who were seeking other forms of life and who could no longer endure the hate prevailing in Europe. He believed that only work in the fields and contact with nature could regenerate man. He helped very many young persons. He was a rich man who squandered his money in his goodness. In all probability, he was murdered.

The Society for Supporting Crafts also has a beautiful record. It organized various courses in the locksmith's trade, carpentry, bookbinding, applied chemistry, and so on. Councilor Jaszuński made great achievements as an organizer. I was in personal touch with the building school whose heart and soul was the architect Jerzy Berliner. Eventually, the school set up an exhibition that was visited by many Germans with great interest. I do not know whether they suffered pangs of conscience because they had condemned to death a generation that was so gifted and so eager to work. The ideas and drawings of the young experts were often astounding. Berliner was in charge of the artistic aspects of our typhus exhibition. Professional directorship was in the hands of Róża Amzel. We hoped that the exhibition would survive the misfortunes of the wall-enclosed camp and would serve as eloquent proof of how difficult it is to break the spirit. Unfortunately, the exhibition too became the booty of the rabble. Because the master nation did not want the world to know that the Jews were not only usurers and paupers but also men who, while facing death, were capable of dreaming of beauty and implementing it.

I will say a few words about the Community Board as a whole. Jews do not respect their leaders. They comprise a nation that instinctively does not tolerate commanders. Thus, with a few exceptions, the community councilors did not enjoy a good reputation. It is not my concern to judge whether this reputation was deserved. I once spoke about this with the president of the Community Board, Engineer Czerniaków, who was never in the least suspected of being bent on gain. He told me that enlightened men broke down in this misfortune and that only people with nerves of steel retained their resilience. Only they were able to perform managerial functions; and tough people rarely possess a tender conscience. Besides which some of the employees were imposed on him by the Germans, and then he was helpless. Mainly, however, the nature itself of the district administration could not have been honest. The district's life depended on the so-called Transferstelle [Transfer Point] through which all agendas and transactions passed. The demands of the Transfer Point employees were diverse and fantastic. One demanded a silver chamber pot of definite dimensions, another required a horse-drawn coach to go hunting, still another ordered fabrics for dresses, furs, wallpaper, or cigars. I am citing facts. Those expenditures could not have been recorded in accounting books in the normal manner, and this alone was bound to corrupt any administration. What was this rich Community Board's income? The main cashier, an honest and quiet man, told me that often he had not even a dime in the cash register. Therefore, when money was needed, they often resorted to fantastic ideas. When it was urgent to collect one million złotys to purchase one thousand fur coats for the Germans, the Monitor Service arrested rich people one night and kept them until they were ransomed. Gangsterism? Yes, but it was not the Jews' fault.

I cannot describe everything in detail but only those things which I know and which I saw. Life is cruel. Thousands were dying from hunger and exhaustion, while at the same time and in the same desert, new life began to spring forth. Some new makeshift workshops arose, and some craftsmen began to manufacture various items, including oils and handkerchiefs. Eventually, the Jewish district began to believe that everything could be gotten there. Indeed, at the beginning, stocks of fabrics, overcoats, suits, stockings, and so on were hidden there. Evidently, the Jews had a special gift for concealing stocks and improvising new production. Unfortunately, an unjustified aura of an inexhaustible strength and boundless possibilities arose around the district. Soldiers standing on guard saw the smuggling of food articles that their own families did not have. I believe that this vision of the inexhaustible vitality of the Jew whom neither hunger nor an epidemic was able to eradicate was the reason why the master nation decided not to wait for a gradual extinction extending over several years but to take recourse to direct crime. In the parliamentary language, this was called "Die Juden werden aufhören zu lachen [The Jews will stop laughing]." This I will discuss later.

Chapter Twenty-Six

In the Shadow of the Church of All Saints

At first, we lived with our friends on Grzybowska Street. Not only the constant street noise but also the awful scenes we saw through the windows made our life there a torture. We did not have a moment of rest, and our daughter was falling into a progressively deeper depression. I was relatively lucky again: in September 1941, we received our own apartment in the rectory of the Church of All Saints on Grzybowski Square.[1] Unlike the church on Leszno Street where it was primarily only priests who lived in the parochial house, in the rectory of the Church of All Saints, all rooms and apartments were given to the parishioners except for the apartment of the Reverend Prelate Godlewski. Entering from Twarda Street and Grzybowski Square—a pit of hell, disgusting odors, and screaming people—one walked into the church courtyard, which resembled the courtyards of Italian monasteries. There was a narrow passage along the church wall, and in the courtyard stood a tall acacia. On the right side beyond the church, one could see the ruins of a fantastically huge but destroyed building. Between the church and that building, a small spot of earth was retrieved from under the ruins, and vegetables were grown. This was how life was reborn amid ruins. In the moonlight, all this looked like Pompeii destroyed, not by lava but by human hate. The windows of our apartment faced the small but beautiful garden of the church. Those church gardens surrounded by walls have a curious charm. We had the impression that we were in a secluded corner of reverie, calm, and friendliness, which had remained intact in the thick of hell. The priest of that secluded corner was Prelate Godlewski.[2]

Prelate Godlewski. When I utter his name, I am overcome with emotion. It was passion and love in one soul. He had once been a militant anti-Semite, a priest waging war by word and letter. But when fate brought him in touch with this abyss of misery, he discarded his attitude and devoted himself to the Jews with all the ardor of his priestly heart.

When his beautiful gray head appeared, reminiscent of Matejko's Piotr Skarga,[3] heads bowed down before him in love and humility. We all loved him: children and old people alike and we would snatch him from one another for a few moments' talk. Nor did he spare himself. He taught the children catechism, headed the Caritas movement in the district, and ordered that soup be handed out to all those who were hungry, regardless of whether they were Christians or Jews. He often came to see us, to comfort and hearten us.

We were not the only ones to think much of him. I would like to pass on for the records what the President of the Community, Czerniaków, thought of him. We got together at Docent Zweibaum's to mark the first year of our courses. The president told us how the priest had shed tears in his office when speaking of the Jews' misery and how he was trying to help and ease that misery. He told us how much heart this priest—a former anti-Semite—was showing the Jews.

Father Antoni Czarnecki was prelate Godlewski's assistant and deputy. He was a young priest and did not have the prelate's passionate relationship with life, but he possessed the sweetness and goodness of a clergyman. He was liked and respected by all. And his pleasant and sincere manner had a soothing affect.

It was a strange life. Never had I had such close contact with the church as when I lived in the Jewish district.[4] For a year, every day, morning and evening, I soaked up the atmosphere of quiet in the church. And I lived near people whose profession was a mission of goodness. On Sundays, all Christians—not only Catholics—went to mass. Every one participated: doctors, lawyers, those for whom baptism had been an act of faith, those for whom it was a symbol of national identity, as well as those who had at one point been baptized out of self-interest. But every one felt the need to assemble at least once a week in the church for worship. I observed many people who were not only believers but practicing: even daily mass attracted regular churchgoers. The penetrating cold of the partially destroyed church did not repel people in winter. Religious service in a church for people shut up behind the wall was the only kind of event available.

Gloria in exelcis Deo. Glory be to God, peace on earth to men of goodwill. And Grzybowski Square, Pańska and Twarda streets disappeared as well as the febrile, miserable crowd. We were surrounded by the coolness and atmosphere of a temple. A crowd absorbed in prayer. No longer did we see people struggling and hateful, we were in a group of engrossed people. We were bonded by a feeling of higher communion.

Agnus Dei. The personification of goodness and soothing contact with a Being of endless goodness. No need to be ashamed of one's ugliness. In fact, at that moment the soul ceased being ugly. A certain coolness of traces beyond the stars. Our suffering, the enemy, and the monstrous distaste for man who is no longer a man seem to vanish in a fog. The unending expanse is real, the light filters through and all is directed by the truth of timeless understanding and goodness.

"What is wrong, my son?"—"What should I love these monstrous people for?"—"For nothing. Love is a state of mind. It exists in all, though suffocated. It is an instinct like the hunger for life, like the joy of being. It is a delight, like drinking in the starry quiet, like the delight of beholding dancing stars. There are no small things, since every thing is evidence of the Spirit." Heavenly music resounds. And in its harmony, the soul bows down and sobs in humility and the world embraces in ecstasy and falls into oblivion. There are no more people nor ugly things, for every thing is the resonance of Great Harmony.

The end of the service and return to earth. But with a refreshed soul in cool and life-giving springs.

The sermon. The intellectualized public does not succumb to emotion when told of Christ's life as a child. Nor is such a public moved by the idea of reaching the Son through the mother, since such a request reminds them of earthly things. But they were moved when there was talk of Poland. Prelate Godlewski spoke beautifully and courageously. A homeland is like a mother: sometimes she crosses us, sometimes she is unfair in a moment of passion, but she is forgiven, because a mother wants only what is good for her children. Love for one's mother-homeland is powerful and that is what bonds a nation—more than common origins. A homeland is obtained through shared love. Suffering has meaning when it leads to greater ends.

That is more or less the way prelate Godlewski spoke and his words comforted spurned souls.

Many people were baptized in the district—both older and younger people, sometimes even entire families. Among them were some of my students, both men and women. Sometimes I was invited to be godfather. What motives could they have had? No profit was to be had: the change of faith in no way changed their legal status. No—they were drawn by the charm of religion—the religion of a nation to which they felt they belonged. A religion in which there is—or in any case is supposed to be—no room for hate. How tired these Jews were of the atmosphere of general ill-will. There stands before me one of my auditors, being baptized. She has a Semitic nose and thick lips, but in her eyes I see a deep yearning for human sympathy, which she wants to return with all her heart. Powerful people will come from the higher social spheres, the priests of a new religion. And they will take this little Jewish girl by the hand, protect her from hate, allow her to be good. For Christianity came to power because it gave the unhappy and despised the right to equality and human dignity. Equality before God—so maybe . . . before man as well. It is painful to live with the unmerited mark of Cain. And that stamp can—not only can, but must—be removed by the religion of love.

Such thoughts undoubtedly inspired this young girl when she was being baptized.

That is how I spent my Sunday mornings. Sometimes I walked beyond the walls since as a member of the Health Council, I had a pass. This was not the sign of a special favor on the part of the occupiers, since there were over a hundred such passes in order to ease the administration of the community. I could not force myself to walk in the Warsaw streets with my armband on and so I would take it off. I was risking prison; I was ready for it. The streets beyond the walls seemed to belong to another world. They were clean, neat, there were no crowds, there were instead serious and depressed walkers who were not, however, suffocated by a noisy crowd. Masses of Germans had totally changed the town's Polish character. In the Jewish district they were avoided and the crowd took off their hats. This gave the impression of heathenness since such walking little gods did not even deign react to such bows. Here, they were simply ignored. Over there—thousands of beggars and strewn cadavers. Here, Warsaw gave the impression of a neat little provincial town in which an army was stationed. I would meet with friends and acquaintances who greeted me with emotion. I saw that they cannot comprehend our experience. They had their own worries and pains. There was Auschwitz and Pawiak [Prison] and just about no family was left unscathed. But nevertheless there was hope that some would survive and ensure that Poland go on living. Over there, the shadow of death enveloped all: parents beheld their children and reproached themselves for having brought them into the world—to death and humiliation. In the district the young people worked themselves to distraction. For it was the only way to forget. Here there were official—though poorly attended—courses and also wonderful clandestine classes, which—as I saw—would be marked by the history of the occupation and bring forth a new elite. Some of the young people had given themselves over to trade and smuggling and, more worrisome, seemed to like it. And I thought to myself—may they not take on the bad sides of the trade they had inherited from the Jews. In the district, crime was rife: they killed openly on the street and when they wished to kill a Pole, they did so in the district. I didn't see why: surely they don't think they would be believed if they said the Jews did the killing? Here, they killed within prison walls and camps. Over there, clandestine life was relatively restrained: the Jews felt the threat of their situation as an elementary catastrophe that could not be opposed by any kind of human force—at best one could hide somewhere and individually escape annihilation. Here one felt the gathering storm in people's souls, the storm of protest and revenge. There, there was no leader to assemble and call to battle.

Here, they pronounced Sikorski's name with honor.[5] And yet I thought: if the war lasted, the same fate would await the Poles. For on the scale of the Führer's hate, the Poles are next in line after the Jews. And Hitler's pronouncements dictated by hate were fulfilled.

But visiting friends in *my* Warsaw, as if I had come from abroad, soon became psychologically impossible for me.

I would go back to the district on Sunday evenings. Before the checkpoint, I had to put my armband back on discretely in some doorway. It seemed to me as if I were putting on a leash. A German soldier would check my documents. I had to stand hatless, and I realized that I had gray hair. They would go through my briefcase. Sometimes I had a few candies given to me by friends for my daughter. I entered the abyss and immediately was enveloped in clamor, suffocation, and the specific odor of destitution. I went home by Twarda Street, Grzybowski Square, and I sank myself into the atmosphere of monastic quiet in our presbytery and entered the room where my wife and daughter were waiting for me. They were not allowed to go to the other side. They were condemned to this camp without rest or relief. My child was after all sick. I had asked Dr. Milejkowski to appeal to Hagen that my daughter might go once a week to her doctor on the other side. I had thought that maybe such impressions would distract her and allow her to forget for at least a moment. Hagen knew my works published in German. He had asked the chairman about me expressing some sort of compliments. But do you think he allowed my child to seek treatment on the other side? He was perhaps not a bad man, but like all of them he had no character and he refused. He claimed there were enough doctors in the district. Did he not understand that I wanted to give my child a moment's respite? I am not vengeful and so I do not wish on Mr. Hagen that he have to behold as his child fades away before his very eyes.

I stopped going to the other side even though I could. And in the end I couldn't any more, since Commissar Auerswald removed my pass, merely remarking "why should this professor hang around town?" And Dr. Kubliński, the medical commissar of the district who had graduated from a Polish University, removed Professor Centnerszwer's pass. The passes were removed from two Polish professors. But if my book is published, it will achieve what Commissar Auerswald did not wish to allow: the world will learn of his crimes and of the tears that were shed because of him.

Chapter Twenty-Seven

Race or Tradition

This may sound paradoxical, but it was only in the district that I came to know the Jewish soul.[1] I became acquainted with what was good as well as with what was bad. I met there a great number of people who were ethically and intellectually outstanding, toward whom it would be a great injustice if they were to fall victim to undifferentiated bad will or anti-Semitism. But at the same time, I saw that the average level of Jewish society in the district did not equal other societies, at least as long as they were in normal conditions. The hunger for life was too little tempered by sympathy. I am not speaking of capitalist exploitation—that is an international phenomenon. But I noted lack of pity for the destitute, bureaucracy taking no account of intellectualism, envy, and the tendency to deceive even among the intellectual elite. A block physician—having confirmed a case of typhus—was capable of forcing the patients to be treated by him. The quarantined Jew stole pitilessly from his comrades in misfortune. The [Jewish] Oder Service were generally recruited among intellectuals. They were monstrous. Buying people out and blackmail were common. To justify them it could be claimed that they had no resources and had to live off of something. And yet at such a crucial moment, when an entire people was being murdered, there should have been someone to gather young people and drive them to defense against rather than to participation in the crime. It seems to me a mistake to cover up the bad sides of the Jews, dwelling only on the high moral and intellectual level of many individuals. But there is a fundamental difference between what I say and what racists say. For me, the bad sides of the Jews can be explained by history and they can and should change; for the racists, they are inherited through the curse of the race. The Jewish mass was the most backward in all of Europe. But one of the reasons for this was that the Jewish nation was handing over its elites to the nations in which they lived. After all, even my uncle gave the Polish people his whole life's work and contribution to education.[2] That is what almost all those in our generation did, while the gray Jewish masses remained in darkness. And that is why I was offended by the assimilated Jewish intelligentsia's disdain for the Jewish masses. Generally, if a Pole, Serb, or Frenchman criticizes his country, it is so much proof that its flaws pain him. The best seek precisely to shake the consciousness of their nation. Among the Jews, it is different. The days of thundering prophets disappeared long ago. If a Jew speaks badly of the Jews, it is usually thought

that his disdain elevates him, that he ceases being a Jew by the mere fact of
his anti-Semitism. And he is surprised that the world has little respect for the
Jewish anti-Semite, or the anti-Semite of Jewish origin. I met in the district
an editor living in great poverty. Wishing to help him, I suggested he begin
writing news reports and I promised I would involve the President of the
Community. He wrote an article about how he was suffering being with Jews.
"Mr. X," I said, "is the worst part of this abyss of humiliation and enslavement
the mere fact that the Jews are left to themselves? It may be regrettable, but
surely this is not the most important?" Or else the wife of a baptized colleague told me that her husband wasn't looking for work in the district:
when he returns, she explained, the Poles will say that he wasn't hired
because he was Polish. A famous painter lived in the district, Roman
Kramsztyk[3]—a good Pole and European. He painted beautiful portraits. But
an architect acquaintance of mine said to me, "A ghetto painter shall not
paint my wife's portrait." I believe that if Poles had spoken in the same vein
about architects in the district, he would not have turned out very well. No,
these kinds of "auto-Semites" did not appeal to me. Destiny had brought me
into contact with suffering people and I was by profession a pedagogue. I
would try to tell them the truth.

I gave a series of lectures for auditors entitled "Selected Chapters." I
spoke about genetics, racism, and various issues of general biology. The
young people listened with enormous attention and with that kind of feeling
of trust, which is the condition for exerting influence. I decided, on the one
hand, to raise their spirits by explaining that they were not cursed, that the
bad sides of Jews had not been engraved over the centuries; and, on the
other hand, to demonstrate the necessity of inner renewal. My memoirs
would be incomplete if I did not record the main currents of thought that
drove these lectures.

Jews are accused of being a nation of parasites. This is a stamp of humiliation and one must ascertain whether such an accusation is deserved. Let us
take a look at this question first from the high viewpoint of the spirit, and
since it is the Germans who make the accusation, let us consider what the
Jews have given the Germans.

The names that I will cite below were taken from the Dutchman van
Miller's book, *The Germans and the Jews.*[4] Given the technique of lecturing, I
could not cite all the names for the students; I would like, however, to record
some of them in this book.

In the thirty-sixth edition of an anti-Semitic book by a German author—
Fritsch[5]—there are 264 names of Jewish authors, among whom are the great-
est, such as Heine, Boerne, Wassermann, Kraus, Altenberg, Hoffmannstal,
Schnitzler, Werfel, and Zweig. And also are there great composers, such as
Meyerbeer, Mendelssohn, Offenbach, Mahler, Schönberg, and Halévy. The
number of virtuosos is so great that one can only name a few, such as Thalberg,

Rubinstein, Moszkowski, Schnabel, Pachmann, Kreisler, Mischa Elman, Jascha Heifetz, and Joachim. Among the conductors: Herman Levy, Leo Blech, Oskar Fried, Felix Mottl, Bruno Walter, Siegfried Ochs, and thirty others. The Germans are indebted to Mahler for introducing Wagner and to Mendelssohn for introducing Bach. Referring to painters, this anti-Semitic lexicon cites forty-three names, among which are those of Liebermann and Uri. Among the scientists, nine Nobel Prize winners: Baranyi, Meierhoff, Ehrlich, Michelson, Lippmann, Frank, Haber, Wildstätter, Einstein, and two half-Jews: Beyer and Warburg. The Germans are proud of a series of discoveries made by Jews: Neisser discovered the microbe causing gonorrhea; Fraenkel, pneumonia; Wassermann, the serodiagnosis of syphilis; Weil and Felix, the serodiagnosis of exanthematous typhus; Ehrlich, salvarsan and chemotherapy for syphilis; and Morgenroth, chemotherapy for pneumococcal infections. The number of Jewish mathematicians in Germany is quite simply staggering, among whom are George Cantor, the creator of the theory of multiplicity; Kronecker, whose entire works were published by the Prussian academy; and Minkowski, who established the mathematical basis for the theory of relativity. Germany is famous for its chemistry and owes the Jews the following discoveries: Liebermann created alizarin red thus laying the grounds for the dye industry. Caro was the discoverer of aniline red and methylene blue; Nikodem, Caro, and Frank laid the grounds for the production of artificial fertilizers; and Frank was the creator of the potassium industry. Haber, the creator of the synthesis of ammonia with air [sic]; and Wildstaetter, the scientist of chlorophyll. As for the physicians, there was Henryk Hertz, who proved that light waves and electricity waves were identical; Michelson, and Frank. As for the serologists: Sachs, Schiff, Witebsky, and the mathematician Bernstein. I will stop there. Van Miller's book contains twenty pages with such names.

And so the Jews are parasites? They gave more to German culture than those who despise them.

I have cited those names unwillingly in that genius and talent is a God-given grace. It is not a question of dissimulating negative traits of a people by holding forth the talents of their kin. The fact that the Germans produced Beethoven does not in the least entitle them to steal from their neighbors, and the fact that the Jews produced twenty Nobel Prize winners for Germany does not justify ugly things. It is only a question of determining what is ugly.

The Jews are accused of being social parasites. Apparently a quarter of the apartment buildings in Warsaw belonged to Jews. Parasites? This is a misunderstanding. If a quarter of such houses belonged to Jews, then the Jews had invested their money wisely. For they built or lent out on credit in order to build a quarter of the city.

The publishers of Polish medical books were Jews. Should we therefore boycott these publishers and destroy Polish medical literature? Were it not for these Jewish publishers, such books would have come out a few or a good many years later.

If the exportation of eggs or bacon was mostly in the hands of Jews, that means simply that Poland profited from the international relations and the trading talents of the Jews who knew how to organize their exportation. The Jews were not in any way parasites, but on the contrary, an important motor in economic life.

Social "parasitism" occurs only when unnecessary or harmful social functions are performed. Jews were creditors in the country. This was "parasitism" because agricultural credit ought to be in the hands of the Agricultural Bank and not in the hands of private citizens earning high percentages on it.

Jews once ran roadhouses in the country and in small towns. This was harmful. Contemporary hotels should be run according to other principles, and country social life should be based on public meeting places, not on roadhouses.

But the fact that roadhouses are not socially beneficial, does not mean one should stop helping the Jews export eggs and bacon. For the money earned by Jews stays in the country and creates new branches of production, just as any other money. For a Jew does not eat the money he earns, but uses it to build a factory or a house, or puts it in the bank. He "parasites" only the money lent abroad, since the percentages earned through work are transferred outside the country. Consequently, one must avoid parasitical professions. And this should be said in the first place to the Germans: the thesis of *Lebensraum* (living space) is a parasitical thesis. For they wish to take objects and land created and worked by others. A country that imports human force in the form of agricultural workers does not have the right to claim that it has too little living space, since in fact it wants to fulfill the function of a master toward his slaves. And that is what parasitism is.

Bringing in commissars, confiscating Polish and Jewish goods—that is parasitism.

If in the institute that I had a part in molding there now works Mr. Nauck and Mr. Kudicke,[6] whereas I—expelled—pine for my workplace: Who is the parasite, I or they? And who is profiting from someone else's work?

This record must be neat and clear. In world opinion, the Germans are less condemnable for the fact that they have robbed than for the fact that they wished to justify such robbery by their superiority. One can forget stealing by force and even violence, but one cannot forgive the scientist who seeks to substantiate it. For a scientist is bound by another perspective and another approach.

So much for the question of Jewish parasitization. It is partly a question of misunderstanding and even ungratefulness, and it is partly true. Just as some reproaches are partly justified; indeed the Jews in the district knew very well of what their negative sides consisted.

Too much hunger for life and lack of sympathy is ugly. The feeling of being the chosen people and yet feeling disdain for one's own country is ugly. The difficulty in recognizing in others motives other than self interest is ugly. The lack of heroic ideology is ugly: the last heroes were Maccabean. A good chess player impresses more than a good soldier. A lack of understanding for the pathos of heroism. The conflict of Wilhelm Tell is thought to be of no consequence.

And yet: the admiration for the intellect and the readiness to devote one's self to intellectual work is itself admirable. The faith that talent comes from nobility and not birth is admirable. The Jews wish and know how to be grateful. They sense the international character of science and art. When they found schools or fund scholarships—it is with great willingness and regardless of confession. And I do not believe that they compose an international mafia striving for some kind of secret power. Just like every other nation, they wish to be rich. Neither more nor less.

A swindler, on the one hand, and a great philanthropist, on the other. But does great capital generally have its source in hard work or in conjecturing and theft? The Jews remind one of an untended garden: many weeds, and among them beautiful covered-up flowers. The average ethical level may not be high, but numerous individuals are above that level. Whence the misunderstanding. Those who fight anti-Semitism point to the elite of people with exceptional ethics and intellect. Anti-Semites draw attention to the generally low moral level. Who is right?

This reminds me of the story about the difference between a pessimist and an optimist: the pessimist sees in Swiss cheese only the holes, the optimist—only substance. The issue must be tackled in another way and the question asked: Are the bad sides of the Jews inbred and therefore inalterable—are they just as engraved as the criminals of Lombroso who were condemned throughout time to deceive and be parasites? Or else can one take the same mental ingredients and mold a different soul? And I see here two issues: Are Jewish traits inherited, constitutional traits or do they depend on external circumstances? The same question is raised with regard to numerous traits: What is of the realm of the constitution and what are the external influences? Only for the Jewish soul did the Germans *a priori* simplify the question. The psychology of races is discussed there not as a part of *ethnology* but as part of genetics. Their answer could not be correct, because their thesis was wrong. In the name of objectivity, I will not for the moment go into their psychological motives, but instead take a closer look at their arguments and on that basis consider their demonstrative value.

Let us try to embrace the racist viewpoint in the form of a few theses. The first one is the principle one: the soul reveals the race as it looks on the inside and, likewise, the race is the outer side of the soul (Alfred Rosenberg).[7] And therefore it was necessary to create a corresponding racial component to the mystical soul. As it turns out, three such components were invented. Hitler propagates the thesis of the Aryan race; Günther, of the Nordic race; and Clauss, of the Nordic soul.[8] Hitler writes: "There exists only one holy truth which is also an obligation and that is to keep human elements in order to enable noble evolution."—Hitler believes that there exists some sort of homogenous Aryan race and continues—"As the winner, the Aryan organized those people below him and regulated their activities according to his will and his goals. . . . As long as he heedlessly minded his position as master, he was not only master but also the guardian and creator of culture, which was based solely on his own talents. But when the defeated began to raise their heads, the separation between the master and the servant broke down. The Aryan betrayed the purity of his blood and lost his place in paradise. He sank into the chaos of races and slowly lost his creative talents and increasingly resembled the defeated instead of his victorious ancestors." I will only briefly mention the protest of anthropologists, that there is no such thing as a Semitic race, or an Aryan race, that there exist only Aryan and Semitic languages. To speak about an Aryan race is the same kind of nonsense as speaking about a long-haired language. I am not surprised that a house painter does not study anthropology, but I am surprised when the Minister of Justice, Frank, says: "Every scientific theory must ask itself the question: Does it support National Socialism?"

The second dogma is faith in Nordic-ness. And its apostle is an anthropologist by the name of Günther. Mercilessly mocked by German anthropologists, he was subsequently praised to the skies. According to Günther, at most 10 percent of Nordics can be considered part of the German nation. But even this author believes that specific spiritual traits are closely linked to distinct anthropological ones that are fairly Nordic.

Finally Kraus stressed the dematerialized image of the Nordic soul: "The Nordic human being possesses a soul which speaks through silence. He is lonely, closed up in himself and lives only by what he does. That is why the Nordic is afraid lest he reveal too much. He speaks not through what he says, but through what he keeps silent. To pause while speaking is characteristic. The noblest way for him to express himself is silence. . . . Fidelity. The Nordic soul stands alone before God. . . . The Nordic does not abase himself before the crowd, nor does he go to the fair with his prayers, but keeps to his quiet little room. He would be ashamed to be overheard."

Am I to criticize this image? I shall only ask: Do Hitler and Frank speak through silence? And does this mob of bought-out commissars and torturers in concentration camps have a Nordic soul? It is not worth writing about this in Polish. It is so disgusting that the mouth goes dry. But one thing is interesting. Hitler's "Aryan" viewpoint is a transfer of the struggle to live on social issues, it is a conscious opposition to Christian culture, it is the sanctification of violence. The vanquisher is better. *Nota bene*: the beautiful Aryan gifted both with a powerful spirit and a beautiful body came to Europe and found there various locals. And his sin was that he had intercourse with them and committed a racial vice. This had to be undone. Should one therefore forbid the descendants of Aryan winners to unite with locals? It would appear that they should have forbidden the long-headed Nordic under penalty of death to have children with an Alpine or Dinaric type. That would at least have been consequential. But in that case where on earth do the Jews come in? Are they then the ones who are the beaten sub-humans, to whom Europe once belonged before the enlightened Nordics came along?

Just to think that this rubbish cost millions of people their lives. I will not write about this at greater length; I will only mention that according to the investigations of Czekanowski and his school, there are many Nordics among Poles and Polish Jews.

Now I will elucidate the problem of racial purity from the viewpoint not of anthropological traits but serological properties. It is incorrect to monopolize Nordic features for Germans alone. The proof or probability evidence of an intermixture of races is supplied by serology. It is a disputable question whether the shape of the skull can change. The permanence of blood groups is beyond dispute. At one time, I established that the B group was characteristic of Eastern nations, and at that time it seemed to me that Jews belonged to those Oriental nations. When my research was expanded, however, the results showed that the blood group picture of Jews resembles the blood group picture of the people among whom they live. Dutch Jews have less B than German Jews and these have less B than Russian Jews. Furthermore, Dutch Jews have less B than Germans in Königsberg or in Saxony. The reader may be interested in the tabulation of data presented below and its explanation.[9]

How can this be explained? It cannot be ruled out that Jews have a greater predisposition to give rise to mutations. The many hereditary diseases among Jews can support this thesis. Yet, I believe that we must seek an explanation of the similarity between the blood of Jews and the blood of people among whom they live in historical factors. In 1883, Rénan published a book on the subject, which is very topical now: *Judaism as a Race and Religion.*[10] The following excerpts are from that book.

> We analyze Judaism as a manifestation of a race and assume that the Jewish people who created the Judaic religion have kept that religion for themselves and have preserved their ethnic composition in an unchanged state. This is not so. . . . Settlers brought by the Assyrians incorporated many completely alien elements into the Israeli masses. . . . In Syria, the number of converts was so large that Syria was Judaicized to a considerable degree. . . . During the Roman era, Judaism had no ethnographic significance but was something universal and had its adherents everywhere. Most

Table 1. Blood group distribution of Jews and the populations where they live

Origin	Groups			
	O Percentage (%)	A Percentage (%)	B Percentage (%)	AB Percentage (%)
Dutch Jews	42.6	39.4	13.4	4.5
Dutchmen	45.7	41.2	9.6	3.5
German Jews	42.1	41.1	11.9	4.9
Berliners	36.5	42.5	14.5	6.5
Polish Jews	33.1	41.4	17.4	8.1
Warsaw Inhabitants	33.7	38.4	19.4	8.5
Jews in Kharkov	28.6	42.3	23.5	3.6
Russians in Kharkov	27.3	40.0	24.1	8.6
Jews in Samarkand	28.9	31.4	32.7	7.0
Persians	30.6	31.8	31.6	6.0

Italic and Gallic Jews were such converts, and the synagogue was established next to the church as a religious minority. . . . Gaul had many people who accepted Judaism through conversion and never had been in Palestine. . . . In the Arab lands, Judaism made great conquests.

In the eleventh century, Jews practiced polygamy and often took slave women for wives. Many of the prisoners of war bought by Jews were circumcised and later sold as Jews.

Czekanowski says the following in his comparison of Arabs and Jews:

The oriental component is more frequent in Arabs than in Jews, especially European Jews who have little of that component, at most 20 percent, while this factor is almost twice as frequent as it is in Arabs. Thus, we have here a very archaic formation represented by Arabs from Iraq. This formation is represented by Jews from Yemen. In this formation, the Mediterranean component is more frequent than it is in the oriental. A younger formation is represented by Arabs from Yemen, the oriental Jews called Spanioles, and Samaritans. They present an Armenoidal admixture. Jews from Central Europe and the Caucasus must be regarded as young formations. They are the result of the Semitic formation being imbued with the blood of autochthonous people.[11]

Thus, you can see that the desire to replace the concept of "nation" with the concept of "race," as well as "the call of blood," are empty phrases. A nation is a coherent group of people inculcated with the same culture, melted in the fire of history, and feeling love for the same country. Everything else is just cant created to facilitate conquests.

Blood group research indicates that the various races are intermixed—no less and no more. It is improper to regard, as some German researchers do, the presence of the A group as a proof of the influence of the Nordic race. The concept of a nation is based on love for one's country and joint culture and traditions, and not on the call of blood or race. The Polish nation contains Nordic elements as well as Armenoid, Mediterranean, and many other elements. An attempt to ascribe the attributes of power to one race would destroy all statehood. It would also have destroyed German statehood if the Germans had taken racism seriously. As it was, they needed racism to destroy the Jews and to pull the northern countries into the orbit of the German imperialistic plans. Even though these premises are quite clear, let us continue to make an objective analysis of the theses of racism. They contain the following errors.

Race or national psychology is an exponent of complex psychophysical processes transmitted by writing, words, song, or picture: thus, transmitted from one generation to the other as tradition. There are no genes of loyalty to the English king, there is only a tradition of loyalty. There are no genes of swindling or trade in Jews, there is only a commercial tradition. There may be special preferences arising from a constitutional basis. The body build of a smith is different from that of a tailor. Yet, there are no smith or tailor genes, there are only complex states of mental reactivity. The athlete who must burn off his physical strength prefers to become a smith, while the weaker man chooses to be a tailor. Modern genetics does not believe that genes always become manifest; they condition one's reactivity, which then becomes manifest under the influence of external factors. Even the Germans write: "Es vererben sich Reaktions-normen" [reaction norms are inherited]. When we analyze mentality, we must realize that certain impressions, examples from history, first sensations, and folk songs shape one's psyche and give it a certain content and direction. But even when the reactivity is inherited, it does not necessarily condition every specific trait. Most traits are inherited on the basis of several genes. For example, the red hair color is determined by several genes that are inherited independently. A red-haired parent may transmit to his child a gene which in combination with the maternal gene creates not red but a different color. The inheritance of organs or facial features, such as the nose or eyes, is all the more complex. Random association of various genes may result in a feature quite different from that of the parents. This is clearly seen in the heredity of mental diseases. Specific diseases are not inherited, but rather a mental instability which conditions creativity in one child and insanity in another. Kretschmer has demonstrated it completely unequivocally.[12] Racists analyze race psychology on the same plane as some mental diseases—not as a reactivity responding to social imperatives, but only as an

internal compulsion. Mentality is only a reflection of the race. A chasm separates the scientific principles and the practical consequences of racism. The
analysis of the heredity of mental diseases, even though it is only in its beginnings, is based on certain studies. But, no one has investigated or knows anything about the heredity of those complex mental junctions that determ ine
race psychology. A priori we may assume that the transmission of psychic ⋅ raits
depends on joint action of many genes. Racism is based on analogies with
applied eugenics. We can breed certain properties, such as milk productivity
in cows or acute smell in dogs. In nature, this is accomplished through natural
selection, which preserves types that are best adjusted to the surroundings. In
a naïve manner, racism has introduced this analogy into sociology. The
Nordic mentality, the racists say, is best adjusted to the conditions prevailing
in the central European area, while Slavs or Jews are fit only to be serfs or
cadavers. All these arguments are plucked out of the wind, because traits that
form the basis of race psychology cannot yet be used for genetic analysis. Let
us attempt to make a hypothetical analysis of the so-called race traits. The
Nordic supposedly has the mentality of a soldier, the Jew the mentality of a
swindler. What is the possible psychological basis of these traits?

Courage? It may be the result of a lack of imagination in anticipating
danger, or it may arise from the rage of battle that is typical of cruel people.
Sadists usually make good soldiers.

Heroism? It may signify lack of compassion for the enemy or susceptibility
to persuasion. See, for example, "Bartek zwyciężca" (Bartek the Conqueror).[13]

Subordination to the ideas of the leader? It may be the result of a lack of
criticism or civil courage. For example, the German professor who agrees to
follow the directives of national socialists.

Now, let us analyze the mental components of a swindler. He must penetrate the mentality and the needs of the buyer, and he must have a well-
developed imagination and organizational talent. Lack of conscience? What
is better—to swindle or to rob and kill?

Thus, it should be obvious that certain mental elements can be put
together in a socially positive or negative way, exactly as individual bricks can
be used to build a beautiful or ugly edifice. But even these psychic elements
are certainly not inherited as such. We do not know the heredity paths of
imagination or civil courage, because these are complex processes.

Most important, we do not know whether race psychology is subject to
heredity at all, or whether it is only the manifestation of traditions and customs. Granted, we do know that race psychology changes through the ages.
The psyche of Germans at the beginning of the nineteenth century during
their struggle for independence was different from in 1848, which eventually changed under the influence of Bismarck's conquests, and is still different today.

The same applies to Jews. Fate deprived them of land. To preserve the nation, a number of regulations were worked out to preserve the Jewish identity. These regulations have not always preserved rational traits, and often they fixed on purely external features. Yet indeed, these regulations enable the Jews to prevail. However, the mere desire to prevail is typical of weeds. Noble entities want to bring forth fruits. Therefore, the mentality of a nation must be held by firm yet loving hands, and it must be sculptured so as to correspond to the contemporary concept of beauty and dignity.

The verdict condemning the Jewish "race" is unjustified. This "race" has the right and the possibility to rise. One must not close one's eyes to one's own faults, but stride forward with persistence and courage. It may well be that the immense suffering of the current Jewish generation will eliminate the negative traits and sculpture a new mentality.

Those were approximately the words I spoke to the young people. They listened to me with suspended breath. After the lecture, several of them came up to me and told me with overflowing emotions: "We thank you. We feel that you have taken the curse from us."

When I was speaking to them, it seemed to me that I was fulfilling the duty of a teacher who was showing new roads to his pupils, roads beset with difficulties but also offering a hope for a better future. Unfortunately, I was speaking to human beings sentenced to extermination.

Chapter Twenty-Eight

The Beginning of the End

In the first half of July 1942, persons who had contact with the Germans warned the Community Board that the Jews were in danger of being deported from the ghetto.[1] A person sent by certain well-meaning Poles whom I did not know personally came to me suggesting that I should register myself and my family in a foreseen workshop, because those not employed in such workshops would be forcefully deported. I thanked him for his concern but did not accept the offer, telling him that I was aware of that circumstance but had other plans. President Czerniaków heard this rumor and immediately inquired with Commissar Auerswald and with the Gestapo. He was told that those rumors were groundless and that they were downright *quatsch* [rubbish]. Presumably, at the beginning, the local government knew nothing about this ordinance, because it was passed by Hitler himself. This is supported by the following evidence. I was told by a delegation member sent by the Main Custodial Council to the local government to inquire about the plans for destroying the ghetto. "Keep your hands off," they were told. "It's the leader's military order." It was probably in connection with these plans that the order was issued to register non-Polish Jews who were interned in Pawiak.[2] Refugees from Lublin warned us that the deportation of the Lublin Jews to their death had had an identical beginning. At the same time, the frequency of killings sharply increased. Persons involved in social and political activities, members of the Polish Socialist Party and of the Bund,[3] and chairmen of trade unions were hauled out of their homes at night by the Gestapo and were shot in courtyards or on streets. Bakers and smugglers were also killed in large numbers. After those nightmarish nights, people would recount to one another gruesome scenes: a wife who did not want to leave her husband and died with him, or a child leaving home in the morning stumbling across his father's body in the entrance hall. We could feel the gathering storm but were unable to determine against whom the action was directed, because people belonging to the most diverse social groups and strata were being murdered. Almost every day, men and women, Poles and Jews, were brought from the Aryan side and killed, usually in the door of houses on Orła Street or in ruins.

This was being done in broad daylight. The killing technique was as follows. An automobile would halt in front of a house, an elegant officer would get out and point with his hand to the man or woman sitting in the automobile to get out and walk to the door. I do not know what he was saying, but once the person started walking toward the door, he would shoot the victim from behind. The Monitor Service had the standing order to remove these bodies and bury them in the Jewish cemetery. Occasionally the procedure was different. A friend of mine, Dr. Miński, was so battered that it was hard to recognize him at the cemetery. Another friend of mine, the young Wolski, was also severely battered. His father ransomed the body: there was no unbruised area on it. The poor man told me that he thanked God that his one and only son had died as quickly as one hour after being arrested. Such a death befell Earl Raczyński, Henryk Teplitz who was the brother of the well-known banker in Milan, and many others. On July 20, the murders began to multiply in number. On July 21, so many murders were committed that on Chłodna Street, for example, there were twenty-six bodies and, to quote the inhabitants, the air smelled of butchery. One stairway was entirely covered with human blood. Professor Raszeja[4] was killed on that day. He was summoned to a sick person's house, and he had a legal pass. His former assistant, Dr. Kazimierz Polak, a nurse, and relatives were all present. SS men broke into the apartment and murdered them all. Shortly thereafter, the order was given to close both churches and take the keys and the passes from the priests.

I remember the moment when Rev. Czarnecki, pale from emotion, came and communicated this dismal news to us. We had the impression that an abyss was opening before us. A friend of mine, the young lawyer Tadeusz Endelman, had long had the intention of being baptized. In view of the approaching death threat, he asked the priest to perform the ritual immediately. Rev. Czarnecki did not refuse him this last comfort. After the baptism, the priest went up to the altar and began to pray. A handful of parishioners were present. Everyone felt that the moment was a farewell to life; everyone cried. Later, Rev. Czarnecki visited all tenants of the parochial house to say good-bye and to encourage them. I came back from the city at that moment. He said good-bye to me with tears in his eyes, made the sign of the cross, and left.

In an avalanche of events, it is difficult to keep the individual staves apart. I should like to move back somewhat. As if in a prophetic vision that life in the district was coming to an end, on July 1, 1942, President Czerniaków summoned a great meeting to which the main representatives of the community were invited. Several hundred persons were present. The meeting resembled a reception: tea and pastries were served. When the meeting was in progress, a Gestapo man walked in. A modest reception, a table set with tea and cakes and sugar for several hundred people—all this might have given the impression of a feast. Various persons presented reports on the work done in the district, the people's great ability to improvise, and the Jewish craftsman's love for his workshop, which was compared with the Polish peasant's love for his land. It was reported that 80,000 persons were working and that perhaps their work would pay for their right to live. The president delivered a speech, reading excerpts from his diary. The laconism of the text was shocking: on such and such a day, so many persons were shot to death; on the second day fur jackets were demanded; on the third day so many persons were ransomed; on the fourth day there were so many funerals. Among other things, he said: "Let it be our comfort that there are scientists and artists among us who suffer with us. Therefore, we must not feel like a nameless crowd of parasites." When enumerating the scientists, he mentioned me in the first place. I thought that if my suffering could really bring comfort and wipe tears away, then it was not in vain. He spoke about the tragedy of Jews who, contrary to other peoples, had to demonstrate their right to live—a right to which every creature was entitled. This did not exhaust the agenda of the meeting. During the break—this was an unforgettable moment—piano music resounded and we heard Chopin's preludes interwoven with chords from "Jeszcze Polska nie zginęła" [As Long As We Live, Poland Is Not Lost; Polish national anthem]. Jews were forbidden to play non-Jewish composers. The fact that Chopin was played at an official meeting had a special meaning. My main desire is to tell future generations that during the last official meeting of the Community Board, the Jews played "Jeszcze Polska nie zginęła." For this song, the piano player, the president, and most of those present could have been sent to a concentration camp. Yet, I can assure that in no one's eyes did I see fear; on the contrary, that song was their expression of hope and gratitude.

The president knew that one period of the Jewish Golgotha was coming to an end and that they had to present evidence of their constructive power. He did not realize, however, that this was to be the swan song and that the end was so near.

The actual "action" began on July 22.[5] In the morning, three vehicles halted at the Community House: two automobiles with SS officers and an open bus with armed soldiers. An immediate conference with the councilors was ordered. During that time, an automobile was driving through the city, and the SS men were arresting more prominent persons, according to a list. They did not adhere to the list strictly but frequently took wives instead of husbands or neighbors. They were looking for, among other persons, the president's wife, announcing that she, the patroness of many children's institutions, would be shot to death if she would not supply a large number of children for deportation. Fortunately, they did not find her. In the afternoon, the president was called to the Gestapo office. There he was informed of the decision to resettle Jews to the East, and he was told that the Jewish government had to deliver seven thousand to ten thousand persons each day; children in the first place.

I previously mentioned the internal struggle of that man—how he yielded when money or fur was demanded and how he did not yield when human life was at stake. Now he was facing a great decision, one to which only men of great spirit could rise: to prove with his voluntary death that his people had the right to live. Indeed, with his death he proved that the Jewish nation has sons gifted with heroism. He came back from the meeting and wrote a few words to his wife: "They demand that I kill my people's children with my own hands. I have no choice but to die."

He was right. The legend of a perished nation will not be stained with the memory of a leader who valued his life more than his duty. He knew that only sacrifice and suffering could create a legend that would prevail; that it was necessary to die in order to live eternally. If he had died by the hands of the hangmen, he would have been one of many. He had to die so that a legend could arise around the suffering of his people. He will go down in history as one of those pure minds who understand that the ideas and nations for whom one dies have the right to live.

People in the district were terribly shocked by his death. Everybody understood that the fate of Jews had now been sealed. And the Germans? They expressed their supposed surprise that he could have done such a thing. They could not understand that a Jew could be a hero, because they had been taught that only the Nordic mind was capable of heroism.

A number of conferences were held to establish the procedure for delivering the victims. The Community Board was not told how many persons in total were to be deported; there was only a haggle for daily contingents. Simultaneously, they ordered the immediate evacuation of the infectious diseases hospital in Stawki, because the building was to be converted into a reloading station. At first, one day was given for the evacuation of a hospital with some 1,500 patients; then, one more day was given. At conferences with the Community Board, they established which categories of people would not be deported: employees of the Community House, the Supplies Department, the Monitor Service, and the Medical Chamber. These employees deluded themselves that their parents would also be spared. At first, the announcements were signed by the Community Board, but soon the honor of murdering one's own nation was given in full to the Monitor Service. Colonel Szeryński was appointed to the position of chief hangman. Unfortunately, I must immortalize the name of that modern Herod. He was an officer in the prewar police, a Jew by origin; a tough and strong man with a thunderous voice and the conduct of a commander. He resembled a *condottiere* [a leader or a member of a troop of mercenaries]. Not without personal charm, he made the impression of being capable of brutality. He introduced a certain discipline in the Monitor Service. To a certain degree, he satisfied the need for a uniform which had arisen in all modern societies, including among Jews. During the collection of fur jackets, he made a mistake: he sent a fur coat, one of little value, beyond the wall. This was disclosed, and he was arrested and placed in the jail in Pawiak. He was threatened with the death sentence. Then he was pardoned on the condition that he would assume the charge of conveying his compatriots to slaughter. Did he know they would be sent to slaughter? At the beginning maybe not, but later it was beyond doubt.

He could have headed the only organized combat unit and died so that his name would be uttered with respect by his people and by his enemies. He was not equal to the task. Therefore, he died a Judas. Toward the end of the action, he was wounded by his compatriots' bullet, and later he supposedly took his life. Did he do it because of pangs of conscience, or did he know that he would be spared neither by the occupiers nor by his own people?

A new period in the district's life began. Commissar Auerswald was recalled; and the governing was taken over by the party, especially the *Umsiedlungskommando* [Resettlement Squad], which presumably was headed by General Groener.[6] The action was carried out by the satan of malevolence, well acquainted with the abyss of human baseness. The historical figures of evil, such as Nero or Caligula, turn pale in comparison with that man. He decided not to soil German hands with Jewish blood but to create circumstances such that the Jews themselves would supply their own victims. He invented the following procedure. For supplying the contingents of human flesh, the members of the Monitor Service were to be rewarded with the nondeportation of their parents. Other employees protected only their wives and children; the suppliers of human flesh also protected their mothers and fathers. Just think, many a man would probably have done what the president did: would have killed himself rather than, with his own hands, deliver children to murderers. But at that moment, he might have visualized his old mother, and thought: they will take the children anyhow, but I can at least save my one and only mother from this inevitable death. I ask you now does the history of the world know such infamy? Will the German nation ever be able to cleanse itself of this shame? To convert people, many of them decent, into hangmen of their own nation for such a price as one's mother's life! And they did not even keep their promise.

This was the way in which people, beginning with the employees of the Community House, were forced to participate in the crime. Only later was this left to the Monitor Service. I personally witnessed how physicians, disinfectors, and clerks were used against the people. At first, the people could gain no orientation as to the methods used to supply the quota of live flesh. The evacuation of children's houses and deportee points was the simplest. Streets began to fill with processions of ghosts in rags, beggars swollen from starvation, and children emaciated into skeletons.

The sight was ghastly. Some were walking driven by the police, others were shoved onto trucks. The culmination point was reached when they started catching people on the streets. I will never forget this hell. Mothers

with children and children walking alone were attacked by police function-
aries and the resettlement squad and thrown into trucks. When a child succeeded in jumping off the truck, a chase would begin that could freeze a
normal man's blood in his veins. Screams of parents, cries of children, and
finally the sobbing of the catchers themselves mixed with the relentless rumble of shots. . . . The Germans were enforcing order. They requisitioned a
number of rickshaws and drove around the city. Some enforcers were armed
with light machine guns and without reason would shoot into the terror-
stricken crowds. Others would lash passersby with long horsewhips. Shots
were fired into windows and balconies; to show oneself on the balcony
meant becoming a target for deadly bullets.

All exits from the ghetto but one were sealed with bricks. The Jews were
trapped. It was dangerous to ride in a streetcar, because the soldiers stopped
them, killed some passengers, and carried the others, without checking
their documents, off to the reloading point in Stawki. During that period,
documents were of no use because they were not being checked. Without
exception, all who were caught were delivered to the reloading point or
directly into railroad cars. The streets began to resemble an insane asylum.
At a certain point, I was informed that an assistant from the State Institute
of Hygiene, Miss Helena Rabinowicz, had been caught and was at the reloading station. I had to save her. I took a rickshaw and went to that reloading
place. The driver warned me that we could both die, because bullets were
flying in all directions there. I told him frankly: "I have to save my assistant."
He was visibly moved and took me there.

The hospital in Stawki, currently the reloading station where people
caught on streets were placed, presented a sight compared to which Dante's
hell was nothing. The entry was blocked by a cordon of the Monitor Service.
In front of the hospital building, crowds of despairing people were calling
out the names of their relatives and friends; in the window, I saw desperate
faces gone insane from terror. Inside the building, people were lying
crowded on the floors in halls and stairways. Bathrooms could not be used,
because no one was permitted to leave his place. Thus, they were lying in
their own urine and feces. They were kept this way without water and food,
without linens and beds, and without the possibility of using the bathroom.
They were worse off than animals going to slaughter. I went through all
rooms, looking for my assistant; through the wards where not so long ago
lectures had been given and through my laboratory where new ideas and a
hope for discoveries had been born. Finally, I learned that the assistant I was
looking for had been released, but that another assistant, Tekla Epstein, had
been deported. I left this hell. A man from the Monitor Service whom I
knew would not let me into the railroad cars. "There is sure death there," he
said, "and your head can still be useful." I left the place. Suddenly, a young
boy who had seen me go in and out without difficulty ran up to me. "Sir," he

said, "I implore you, take this to my mother. For three days she has not seen a drop of water." He handed me a bottle with water. The last gift for a mother. "Sir, I am her only son." I took the bottle and tried to cross the cordon, but I was not let through. Unfortunately, I was unable to bring a mother the last gift from her only son: a bottle of water.

The Germans were raising their requirements: they demanded not seven thousand but nine thousand persons each day. At the same time, people began to realize that it was not a resettlement but death. I learned that the cripples and elderly were being shot at the cemetery nearby and thrown into mass graves. Other people were being loaded into railroad cars, one hundred instead of forty per car. The German soldiers were saying that the trains were going toward Bobruysk, but some persons wrote down the car registration numbers and noted that the cars were coming back within six hours. They went only as far as Malkinia next to which the death camp with gas chambers was located.

There is a special technique for leading cattle to slaughter. The animals are trained and are led as if to a pasture. The Monitor Service accomplished this in the following manner: they distributed a circular saying that those who would come voluntarily would be given bread and marmalade and that their families would not be separated. It was a small price for luring hungry people into a trap: a little bit of ersatz marmalade. The scenes varied: some wept as they walked, some families rode on wagons, and younger persons walked with bundles on their shoulder, hoping that they would be allowed to live in exchange for work. Even catching people can be done methodologically. Each day, new districts and houses were raided. In the morning, they were surrounded by the Monitor Service with the assistance of several Germans. To create a proper atmosphere of obedience, they would shoot into windows and kill a few persons. Next, they would issue the order for everybody to come down to the courtyard, and then they would search apartments and attics. Those who were found hiding were killed on the spot. In this way, thousands were dying each day: women, the elderly, children, sick and healthy persons alike.

All stores were closed and smuggling stopped. Most people had to choose between dying from starvation or submitting to the order. There was always the hope that perhaps they would not be killed, only deported for labor.

The Germans kept raising their demands for human contingents. The Monitor Service was unable to keep up with them. Finally, they were threatened that hostages would be shot, that the chief officers of the Jewish police would be shot, that one hundred orderlies would be shot, and that the right to protect their parents would be withdrawn from them. I spoke with an orderly, a doctor of social science, an intelligent and noble young man. He was crying like a child and said that he was unable to sacrifice his mother. Yet, not the slightest mercy should have been expected from the tormentors.

In the jail, they killed most prisoners and deported the rest. They were thrown into trucks like cattle. Women were pulled by their hair. Any orderly who attempted to free persons who were officially exempted from deportation was shot immediately. Besides, killing the orderlies was the order of the day. As long as the people were still at the reloading station, occasionally their documents were checked and some were released. Once the railroad cars were pulled up though, all were loaded on them indiscriminately and their fate was sealed.

Szaulises [Lithuanian soldiers] and Ukrainian armed boys were taking part in this action of killing. They were the cruelest.

A poster announced that only those working in German workshops for the army would be permitted to stay. They would live together in factories and would receive soup and a few złotys. The workday would have twelve hours to break their habit of living like parasites. There were several of these workshops: Toebbens, Schultz, and Felix. Others sprang up spontaneously. Generally, the capital was Jewish, the machines were Jewish, the workers were Jewish, but the firm was German; to separate the parasites. Some shops had textile machines, but most were to function as a cottage industry, and the Jews were to supply their own tools. A satanic cunning was behind it. Within a few days, all sewing machines were concentrated in a few places where men and women, oldsters and youngsters were sitting and sewing buttons onto some kind of uniform; they received a bowl of soup and imagined that they were in paradise. It was better to sew buttons twelve hours a day than to die. A profitable business immediately sprang up around those shops: several thousand złotys were charged for enrollment, and some mysterious Germans were bribed. A nation that supposedly despised gold and loved blood, earth, and iron did not loathe that Jewish money. Indeed, it was easier to earn by means of iron than through swindling. At the government's order, whole blocks were requisitioned and the tenants were thrown out of the houses into the street within a few minutes. On this occasion, hundreds were killed. When Nowolipe Street was being evacuated, two hundred persons were killed. The blocks were surrounded with wire, and a parade began: twelve hours per day for six złotys, soup, eighty grams of bread, and two cups of coffee. And there was the hope that one might survive. One "humane" shop owner, a German, said: "My Jews will be the last to die."

The reloading station was a kind of purgatory on the road to hell. It looked like this. Each day, from seven thousand to nine thousand people from the city were brought there, partly on wagons but mostly on foot. The square was surrounded by a cordon. Whoever got behind the cordon ceased to be a human being and became an animal assigned to work or to death. Sometimes the Germans would segregate the people gathered on the square. Those who had certificates that they were working in the shops were directed to the right side, while the others were loaded into cars. They had to pass by a row of armed Germans who pitilessly battered them with sticks and whips and almost continuously shot into them without reason. One could also hear the relentless report of guns and machine guns. The dead bodies of orderlies who had been inefficiently performing the role of hangmen were lying next to the railroad cars. Eyewitnesses told me the details of the killings, which made my hair stand on end and which no report can render accurately.

The Community Board received no information as to how many were to be deported and how many could stay. The people, thrown out of their homes, robbed of their belongings, and caught like wild animals, were stricken by panic and despair. A decree stated that after barracking, any Jew found outside shop areas would be killed on spot.

Thus proceeded one of the most tragic slaughters known in history, a slaughter of almost half a million people. The cynicism of that operation consisted of the fact that Jews were forced to catch Jews for the reward of being entitled to defend their mothers and fathers from deportation. The feeling of helplessness created an atmosphere of despair and panic that was more horrible than death. Henchmen are often interested in the psychological reactions of their victims. Here, for example, they would tear a child from its mother, load it into the car, and watch, with a grin, to see whether the mother would follow the child into the car or stay out. It was gay at 103 Żelazna Street where SS men directing the operation lived. Stolen food was abundant, and there was music and singing: "Kraft durch Freude" [strength through joy]. They were not aware that they were killing people. They thought they were freeing the world from lice.

What sense was there in that monstrous slaughter? German newspapers published a commentary that the Jews were again making "Greuelpropaganda" [horror propaganda]. The truth was, the commentary stated, that the Jewish Community Board had requested the nonproductive individual to be taken away.

In the next chapter, I will tell what happened to the deportees. By a stroke of fate, I lived for a while among those condemned to death. Several weeks later, in a completely different role, I spoke with people who had been forced to participate in the murder. When I was still in the district, I thought that the Germans wanted to introduce slavery, and that young persons would be spared from death because they were needed as a labor force. Unfortunately, even this was excessive optimism: the Germans' hate was greater than the mandates of national egoism.

Jointly with our hospital, my laboratory was transferred to Żelazna Street to the children's hospital for infectious diseases. The laboratory was not set up, it was not functioning, and I went there rather irregularly. All my efforts were concentrated on saving persons I knew and obtaining appropriate certificates for my laboratory workers. They too came irregularly to the laboratory. More and more were missing, because they had been caught and deported. Those on the list of community employees at least had some hope of saving their lives. However, hiding one's parents and bringing them some bread was enough of a task to absorb one's entire energy. It was known that the moment would come when most patients would have to be relinquished to their fate. Then what were the physicians to do? Their obligation was to protect their patients. It was impossible to protect all of them from their cruel fate. But a dozen or so patients were the mothers of the physicians. On behalf of professional courtesy, it was decided not to let them be murdered. It was the last service of one colleague to another: those mothers should not die looking at their henchmen; they should fall asleep in peace. When the order arrived that the human contingents from the hospital had to be presented the next day, fifteen elderly mothers of physicians were put to sleep with morphine injections. They had an easy, quiet death. They were buried in a little square in a joint grave. May their deaths lie on the conscience of German physicians.

I have described the first act of the tragedy when there was still hope that half of the people would be left in the district, and that among the deportees at least young persons would survive. During the next few months, the fate of all was sealed. I will recount later how this took place.

I do not know what attitude will be dictated to German intellectuals by their reason of state when they learn that their compatriots have murdered whole nations. They may say that it was only a few bestial degenerates who committed this uniquely atrocious act, that German intellectuals knew nothing about these plans, and that it is improper to blame all Germans for what has happened. Or they may arrogantly say: "Es ist nicht wahr"—it is not true—as they said thirty years ago.[7]

In my mind, I see Germans, including the just ones, who will not be able to believe that their nation could have been so base. We Poles know this and do not need to cite names. However, it is my desire to cite the names of my associates who, with their life's work, have gained the world's recognition and have proved their ethical and intellectual value.

Róża Amzel, assistant in the State Institute of Hygiene, was a subtle and well-educated woman.[8] She spoke and wrote seven languages and had the gift of presenting enrapturing lectures. She was an exceptionally good human being and helped everybody in the institute in writing papers for publication. She completed several research projects for the League of Nations Health Committee. When the war broke out, she was in Paris at the Pasteur Institute. She took the last train to Warsaw to be in the country during the war. During the siege of Warsaw, she displayed great courage and helped me in organizing the blood donor bank. She was my deputy in the district and initially planned to stay there. She was caught several times to be deported, but finally she succeeded in fleeing to the Aryan side where she hid for almost a year. Eventually she was arrested by the Germans and killed with her mother. She was my closest associate. We published jointly many papers. She was a beautiful, good, and courageous human being.

Tekla Epstein, assistant at the State Institute of Hygiene, was the archetype of a humble laboratory worker. She was very conscientious and perceptive and published valuable papers on whooping cough, bacteriophages, and other problems. I conducted my last work on typhus in the district with her. She found her fulfillment in work and science. In the district, she was quite simply starving. She was among the first ones to perish, because she was caught in a streetcar on the first day. She was the incarnation of diligence and loyalty.

Dr. Bronisława Fejgin was for many years an assistant at the State Institute of Hygiene. Full of ideas, she established her name abroad due to her excellent research on bacteriophages. She worked in the United States, Paris, and Monte Carlo for a long time. In the district, she was chief of the laboratory of the Community Board.

Helena Rabinowicz, assistant at the State Institute of Hygiene, was a woman of exceptional goodness and had a great pedagogical talent, which she put to work at the courses given at the institute. She did not want to part with her sisters and perished with them at a time when employees of the Community House were not yet being deported.

Dr. Maurycy Landesman, assistant at the Warsaw University, was a researcher full of character who published valuable papers on diphtheria.

Dr. Mieczyslaw Kocen, head of the hospital laboratory in Łódź, was a brilliant hematologist and the archetype of a quiet scientist.

Dr. Kazimierz Polak was sent to me by the military authorities prior to the war but subsequently became a surgeon. He was esteemed and liked by his colleagues. With courage and dedication he gave medical care to people in labor camps. In the district, I entrusted him with directing blood donor centers. He was killed at a patient's bed, together with Professor Raszeja.

Dr. Kaiser had worked under me since before the war. He was a conscientious physician and had a loyal heart. With his fiancée, he fled the district to join the partisans. On the way, his fiancée was wounded. He did not abandon her; first he killed her and then himself.

Dr. Mieczysław Szeynman, a pediatrician, for years participated in our vaccination actions. In the district, I entrusted him with the preparation of the questionnaires on typhus. He was a good, intelligent physician.

Józef Stein, MD and PhD, was an anatomist-pathologist with an exceptionally broad education. Before the war, he had studied cancer in my laboratory. His typhus preparations are exhibited in many German museums. He did not want to save himself because he did not want to abandon his patients.

Dr. Antoni Lande, was a man of exceptional ingenuity; his doctoral dissertation received an award from the Medical Department. He returned to Poland during the war to save his elderly mother.

Róża Zaydel, author of valuable papers on diphtheria vaccines, left Warsaw before the siege and worked in Kowel. When a ghetto was established there, she could not endure the humiliations and committed suicide.

From among my Aryan associates, Zbigniew Kostuch was my assistant before the war. He was a man with unusual abilities. He was killed by Ukrainians when he attempted to defend his father who was being beaten by soldiers.

Dr. Olgierd Sokołowski was Poland's best expert in tuberculosis. Together we organized blood transfusion during Warsaw's siege. He was a man of integrity with an ardent heart. He was shot to death by the Germans on one of the first days of the uprising when he was performing his medical functions in the Wolski Hospital. It is an irreparable loss for Poland. Together with him, all other physicians who were in the hospital at that time were killed.

Dr. Zbigniew Krajewski, head of the Łódź Subsidiary of the State Institute of Hygiene, was an excellent physician and organizer. Of all those mentioned above, he was the only one allowed to die a soldier's death.

I am citing the names only of those whom I know have perished. I do not know how many were killed by the Germans in Warsaw during the uprising. May they forgive me that I am not mentioning them here.

Except for Dr. Krajewski, *all* the others were murdered, some with their wives and children. Since death alone was evidently not enough for the good-hearted Germans, they magnified the pain of death with the sight of dying relatives. Some were shot to death with machine guns or torn to death with grenades; those who were more lucky died in gas chambers. German sentimentalism had nothing against an easier death if it was easier and cheaper.

I sit here in my solitude and think of them all. I cannot even go to their graves to place a few flowers and say: "Since you have lost all your relatives, I will remember to tend your grave." I am unable to do even this for them, because they died nameless I know not where or when, and were either thrown into a mass grave or cremated with many others. Perhaps their ashes are now enriching German soil. Therefore, I have only one way to build them a small monument: to recount in this book their aspirations and tears.

They were all good human beings; an antithesis to the artificially created image of Jews. They were full of dedication and enthusiasm. They did not see a great future ahead of them, but they were pleased to be able to add a building stone to the progress of the science they adored. They were content with being a bridge that was leading into the future. And they were so grateful. Yet I was able to give them so little, only an opportunity to spend their last year of life—which was to be a year of humiliation and disgrace—in a dignified posture, busy with their beloved work.

Perhaps it was my destiny to survive them in order to tell about them. About those people who yearned for goodness and noble work. Who never did or wished wrong to anybody. They did not have the nature of masters for whom the world is an object of conquests and robbery. They wanted to give the world more than they took from it. Thus, they possessed the only mark of nobility worthy of this term.

I remember these people with deep pain and wistful affection. Now, I will have to try alone to rebuild our edifice of science. If I myself do not perish.

Chapter Twenty-Nine

Leap into the Unknown

What was I to do? It has never been my habit to abandon unhappy people. But I wanted to save my wife and daughter. Previously, when I still had the pass, I had gone to the other side and made arrangements with friends so that my wife and daughter would be able to flee the district. However, the story of one-and-a-half years ago happened all over again: my wife did not want to leave me, and our daughter did not want to leave us. If we were to die, then together it should be, or we should save ourselves together. I was also considering how we could save the young people who kept coming and asking me for my plans and advice. We discussed this problem with Professors Centnerszwer, Lachs,[1] and Zweibaum and decided to set up a workshop for the students; however, nothing came of that. The power and methods used to wage destruction were ruthless. It was absolutely impossible to save anybody. Should we all die together? How willingly I would have stood in rank and file if it had been possible to oppose the Germans with arms and to die in order to document the protest. But there were no weapons, there was no resilience and, most important, there was no certainty whether the Germans planned to kill everybody. At the same time, any indication of armed resistance would have condemned the whole district to destruction. I believe that this was the reason why even brave individuals did not rise to resistance at that time. One can risk one's own life and even the lives of one's pupils, going with them and encouraging them to fight even though the fight may be hopeless, but it is very difficult to decide to expose everyone to a cruel death. Therefore, I decided to leave. At one time, I had faithfully stayed with Serbs when it seemed that they were perishing, because they were dying under the flying banner of combat. I stayed in Warsaw during its siege because we were fighting. I decided to stay in the district even though there was no armed combat, but there was a fight against the epidemic and the foolish sanitary methods, and there was a fight for the souls of the young people who had to be comforted and supported. But now there was no fight; there was only panic, humiliation, and slaughter. For the first time in my life, I decided to leave. I must say frankly: the idea of dying in humiliation and disgrace was unbearable to me. I could not bear the thought that such a death was awaiting not only me but also my wife and child. Yet they refused to flee without me.

The Polish people frequently displayed goodwill and friendliness toward me. I was touched most when this was displayed by persons whom I did not know personally. I remember how gallant Polish policemen were when we were being displaced from our home, how they encouraged us, apologized, and emphasized: "You will come back here." When I once asked the city government to disinfect my apartment, the disinfectors did not want to take a tip, saying: "One does not take money from Professor Hirszfeld." I remember a worker from the telephone company who repaired the telephone in our apartment in the presbytery. He asked my daughter whether I was a relative of that professor. When he learned that he was in that professor's apartment, he refused to accept payment. I previously mentioned the touching story of how we acquired an incubator for our laboratory. Such incidents were becoming more frequent and so impressive that I will never forget them. When the "action" was in progress, a man whom I did not know personally, Engineer Albinowski, came up to me on the street and told me that my friends had instructed him to save me and my family, and that the only rescue was to assume work in the shop.[2] Everything had been prepared: papers and even an apartment next to the shop. Since this could temporarily save us from deportation, I accepted the offer.

The greatest surprise, however, was expecting me at home. Engineer D. was waiting for me and announced that a friendly Pole, *Magister* Potocki, had instructed him to save me and my family, regardless of the cost. Everything was ready, and an automobile would be waiting for us on the other side and would take us to the country. That helpful friend had also asked Engineer D. to tell me that in due time, he wanted to see me dancing at a ball. Evidently, my sincere love for students was alive in the memory of a friendly man. Touched to the bottom of my heart, I accepted the offer.

Thus, we decided to leave the district. I told this only to my closest friends, because the matter required secrecy. I called Miss Róża Amzel to our apartment, I had set her and her mother up in the shop, I gave her a work certificate from the Department in Health, and I said: "We have decided to leave this place. Soon I will be an obscure man and will disappear from the surface. On that other side, I won't be able to do anything, I will simply not exist. In some aspects, such a civil death is just as radical as a physical death. What do you intend to do? I have ensured you here as best I could. I have also made arrangements for others of my former employees to the best of my ability. I got them some money and I will try to get them appropriate certificates." She told me that she had neither papers nor means of subsistence on the other side. Should circumstances force her to leave, she was counting on aid from Professor Szymanowski and her friend, Julka Seydel.[3]

I parted from her with the thought that it was possibly for good.

We were to leave disguised as workers, joining a group of slave laborers who were working outside the walls. It was impossible for all three of us to leave on the same day. Our daughter did not want to go first, fearing that she would be saved while we would perish. I on the other hand, did not want to leave our daughter behind. Thus, we decided that my wife would go first. She left the apartment at seven in the morning, disguised as a worker woman. The meeting point was on Ogrodowa Street. A group formed, and they started going. From a distance, I saw my wife moving amid the drab mob. A German sentry was counting the slaves as they were passing by the checkpoint. He did not like some woman in the first group. Unceremoniously, he shot at her and killed her. Maybe he would not like my wife either? No, he paid no attention. My wife disappeared in the crowd behind the wall. I went home. My daughter was not well: she had a fever of 40 °C [104 °F]. We had to leave the next day. However, this was impossible on that day, and we had to postpone it to the following day. My daughter could hardly stand straight. We were both in worker's clothes and mixed with the crowd at the checkpoint. Again, there was the same game of life and death. I walked in front of my daughter; should they shoot, I might protect her. We passed by the guard with the swinging gait of workers going to work. Several hundred meters farther on, the group spread apart, and we were free. Our friends were waiting for us. I walked past the cemetery in Powązki and entered fields. There was space and space around me. I never knew that space could be so beautiful. I felt the scent of fields and greenery. The feeling of being a beaten dog yielded place to the feeling of belonging to a nation that was fighting not only under the motto of life but also under the motto of freedom.

This was the last opportunity to leave. The next day, the henchmen broke into the presbytery and deported everybody, except for a few who had committed suicide.

Several months prior to our flight from the district, I had met Dr. Maria Wierzbowska,[4] a friend of my wife. She is an exceptional woman, a personification of bravery, intelligence, and social consciousness; hard and good at the same time. In her youth, she had fought in the defense of Lwów. An excellent physician and scientist, she gave up her medical practice to take care of homeless children. We talked about the probability of the ghetto people being deported to the East, and I said that in that event, I had decided to hide my wife and daughter. As simply and frankly as could be, she said "Naturally, Hanka [Hanna's nickname] will stay with me," even though she would be threatened with the death penalty. It was at her house that my wife found her first shelter. My daughter and I were awaited by a friend of mine, Stanisław Kiełbasinski,[5] and his brother Jan. At first I went to Jan's apartment in Żoliborz. I was so weak that I fell down like a log and remained prostrate till the afternoon. Then a barber shaved me down to the scalp to change my looks as much as possible. In the evening, they took me to Maria Wierzbowska's, because I was half unconscious. I knew that any traitor could reveal me to the Germans. I was not afraid of death. My psychic breakdown was caused mainly by a loss of faith in man. It was difficult to get used to the idea that one could die for nothing. Our daughter was ill. With a fever of up to 40 °C still, she had found shelter at the apartment of our friends, the worthy Mr. and Mrs. Voit.

My wife and I spent three weeks at Mrs. Wierzbowska's. Engineer D.'s story about a pension that was supposedly waiting for us proved to be a fairy story invented by him to facilitate our decision to leave the district. That was important to him, because he obtained from Mr. Potocki much more money than the normal amount. It was impossible for me personally to thank Mr. Potocki for his initiative and selflessness, and therefore I sent a friend of mine to forward my words of appreciation and return the money expended. Every morning my wife would leave to stay the whole day with our sick daughter who lived nearby. Those who have never experienced such moments will be unable to understand that crossing from one street to another meant exposing oneself to the danger of losing one's life. Thousands of spies were roaming the city, Volksdeutsch, Poles, and Jews whose lives were temporarily spared in exchange for denouncing their countrymen in hiding. Supposedly, the price was not very high: a few hundred złotys per head. Blackmail was the order of the day. Even some dark-complexioned core-Aryans were halted on the street and inspected somewhere in a doorway to see whether they were circumcised.[6] Every day we heard of the death of someone we knew.

I spent entire days inside. It was difficult to suppress normal reflexes and not to answer a ringing phone, not to open the door when someone was knocking on it, and not to approach the window, but to hide in a corner when our hosts were visited by their friends. We had to give the impression that we were no longer alive. Our friends who knew about our flight did not deny such rumors. The news spread that my wife, my daughter, and I had committed suicide. In the evening when it was dark, I would walk over the plots nearby. It was a strange feeling to walk through a city in which one had no right to live. On my way, I would see couples in love: German soldiers walking hand in hand with their girlfriends. I would walk with the brisk stride of a busy man. After half an hour, I would go back to the apartment. I did not attempt to see any of my colleagues at the institute, because I did not want them to be reminded of my existence and thus feel obliged to be in touch with me. I had been too long a donor there to be a receiver even for a short moment. After a few days, the news reached me that my assistant, Miss Amzel, who initially had not wanted to leave the district, had arrived at the conclusion that a further stay there meant sure death and had fled with her family. When we were leaving the district, it was not certain who was risking more, because fleeing to the Aryan side was a leap into the unknown. She stayed with a friend of hers. It was impossible for me to see her, because her friend lived in the same apartment house in which the institute's employees also lived, and she feared my visit. Miss Amzel sent me desperate letters, telling me how horrid people were and that she no longer wanted to work in the institute after the war. She was filled with bitter resentment. I asked one of the pediatricians who were helping me whether he could not also help Miss Amzel. He said that he was trying to save his own colleagues and that bacteriologists should help her. I personally was powerless. The people who were helping me were risking their heads. My house in Saska Kępa was under observation, and the friends of mine who were living there had had to flee. Therefore, I was deeply hurt by the words spoken by one of Amzel's friends on whom she was counting: "The Hirszfelds should stand upon their heads [idiomatic expression meaning should do their damnedest] to help you." "On whose head, Miss J.?" Because as far as my head was concerned, I could not even stand upon it myself. At that time, my head had the value of one hundred złotys; this was the official price for denouncing me. I had no apartment of my own, and I had no right to endanger someone else's head. Another friend on whom Amzel was counting said: "I am appalled at Hirszfeld, he has abandoned Róża Amzel." Here is what that powerful Hirszfeld was going through at that time.

For three weeks, my daughter had fever and was literally exhausted. I wondered whether she was going to die. My wife fell in the apartment and broke her hand below the wrist. For a while we believed that if she did not go to the hospital she would remain crippled. If she went to the hospital, she might be recognized and betrayed. Finally, she told me: "It seems to me that we cannot keep up the struggle and must give it up." I encouraged her: "We have a sick child, we have no right to leave her." My wife agreed: "Whatever happens to one of us, the other must live to save our child." While she was being examined by a surgeon, I was hiding in another room.

A few weeks later, our daughter's health improved, and I took her out of Warsaw to Miłosna. We found accommodations in a little house in a forest, jointly with our faithful Zosia Ossowska who had lived with us on Saska Kępa since the beginning of the war. For the time being, my wife stayed in Warsaw.

In the meantime, Docent Popowski wrote to Dr. Szymon Starkiewicz,[7] telling him that I had to hide with my daughter and asking him to find a shelter for us. A few days later, we were informed by wire to come. My wife and I decided to part, because it was impossible for the three of us to hide together. Besides, her hand was in a cast, and she still needed medical attention. We said good-bye, not knowing whether it was for good. I went with our daughter to Wiślica in Kielce Voivodship, while my wife took our place in Miłosna and stayed there in Zosia's care.

The trip was nerve-racking. Every so often, inspections were performed in search of smuggled bacon and fleeing Jews. A person with Semitic facial features would not have gotten away. We were helped not only by our looks but also by our completely unconcerned conduct. Finally, we arrived in

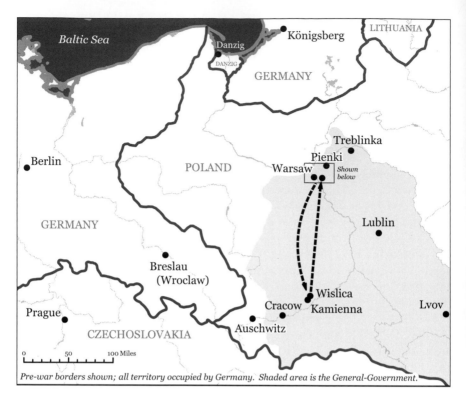

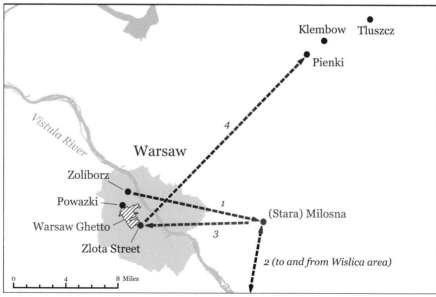

Map 2. The Hirszfelds' travels after escape from the Warsaw Ghetto, 1944–45. Drawn by O. T. Ford

Wiślica where we were expected by Dr. Szymon Starkiewicz and his wife, Wanda. The same day, we went to a remote village on the Nida River.[8] At first we lived in a peasant's house. Contact with rural people did us good; they surrounded us with warmth and friendliness. I was an unknown, obscure man whom the vicissitudes of war had driven to the countryside with his sick daughter.

Chapter Thirty

The Life of an Obscure Man

One warm September day, my daughter and I were lying on the bank of the Nida River. The river was flowing quietly. From afar, we saw the park in Czarków and trees bending under the weight of fruits. The sun was shining. With us, bathing in the water, were a young country schoolteacher and his very nice wife, both brave, young, and devoted to the national cause. We were basking in the sun, watching the calm current and little white clouds sailing over the sky, and we were trying to forget. Perhaps this hell was only a bad dream.

A young boy of twenty came out of the water. He had a simple and fairly nice face. He sat down next to us and told us his story.

He had been mobilized as a *Junak* to work in the Baudienst [Construction Service].[1] He could not avoid this, because they would have taken his parents. The Germans put these young men into barracks and drilled them for several days with lectures on how the Jews were the source of all evil in this world. They were shown pictures in which the Jew was presented as a louse or bedbug. Later, they were given a large amount of vodka to drink. When they were half unconscious, they were ordered to surround the town and catch the Jews. Then they took the poor Jews out of the town where several armed Germans were waiting for them. The Jews were ordered to dig a ditch. Then the *Junaks* were ordered to grab one Jew after another by both arms and lead him toward the German. The German fired a shot, and the *Junaks* had to throw the Jew into the ditch. If the Jew was killed, it was not so bad. But some were only wounded. It was unpleasant to throw a screaming Jewess into the ditch. The worst was to throw children into the pit. One *Junak* broke down in tears. These young boys were actually still children who had been forced to play the role of hangmen. The teacher asked: "Were you not ashamed?" "Yes, at first I was ashamed and very sorry, but then I got used to it. Besides, what could I do? One fellow refused, and immediately he got a bullet in his head." The teacher's wife had tears in her eyes, and the teacher was looking at the boy with sadness and was crying inside not only because of the Jews' death but also because of the moral death of the youth of his own nation. At the age of twenty, no one can carry out the tasks of a hangman with impunity; even if he is poisoned with alcohol.

I was looking into the calm waves of the Nida River and had a vision. I was leading my daughter by her hand. Walking along with us was that very same boy with his honest Slavic face. Such boys used to be the friends of my Marysia: blond hair, blue eyes. But this time, they were holding her by her arms and were leading her to the killer. When she began to cry and tried to tear herself loose, one such fair boy hit her with a spade on her head. On her fair head. I was not even permitted to die voluntarily, because I had to hold my child by her hand till the last moment.

The Nida was flowing quietly. Little clouds were sailing over the sky. Farther downstream, trees were bending under their fruits. Above us was the sun whose joy it is to shine. Next to us—this young boy, a murderer.

Those were the circumstances of our freedom.

Because of my daughter's poor health, it was difficult for us to live in a peasant's house. Dr. Starkiewicz[2] advised me to go to the manor where the living conditions were better and where a man from the city was less conspicuous. The manor was in the vicinity, and it belonged to a friend of Dr. Starkiewicz, Mr. Adam Grabkowski. The owners had the reputation of being hospitable and good-hearted people. We decided to ask them for accommodations, and Dr. Starkiewicz and I went there. My daughter stayed with Wanda Starkiewicz. I cannot write about Dr. and Mrs. Starkiewicz without emotion. The thought that such people exist is a comfort to me. It is incomprehensible how people like them and the German criminals can belong to the same species. Starkiewicz reminded me of Judym;[3] he was all goodness and nobility.

When I saw this Polish mansion in a beautiful park by a pond and met the landlady who greeted me cordially, I thought that I had entered paradise. The house was bright and radiant. The windows were facing the park with beautiful poplars and willows at the pond. But I was impressed most by the people. Mrs. Grabkowska had an expression of goodness and distinction on her beautiful face. Her movements were young, and her whole posture was radiating what we call class. A moment later, the landlord walked in: he had a tall figure, kind, blue eyes, and the same expression of dignity and kindness. Dr. Starkiewicz presented my case: a Pole, a scientist, forced to hide with his ill daughter. They agreed immediately. I apologized that I had come without invitation, but I was sure they would want to see me before they would let me into their house. I said that under the conditions, no payment could match their hospitality, but I asked them the favor of allowing me to pay for my room and board. They did not want to hear of it. Quietly they indicated that it was their duty to help the intelligentsia and said that they would send horses for us the following day.

The next day, the manor horses came to fetch us. We cordially bade farewell to our hosts and the teacher and his wife who had shown us so much friendliness, and left for the mansion. Mrs. Grabkowska led us into a beautiful, large, south-facing room with a charming view. My daughter looked around as if she had entered a different world. I thought: perhaps here this poor soul will find comfort and health.

We began a new life, that of unknown people. I was no longer myself, but no one asked me anything. Many people were hiding at various estates. Everybody knew that they were hiding under assumed names and that no one must be asked any questions.

The house was full of people: relatives who had lost roofs over their heads, and strangers. There was a Jesuit father and a physician with his wife. About fifteen people sat down to table. Mrs. Grabkowska was like a good spirit: she surrounded everyone with loving care, quickly learned who liked what, and tried to please everyone as best she could. All this was done with simplicity and great culture. During the six months I stayed there, I did not see her commit the slightest social gaucheness. It is not difficult to maintain social culture on neutral terrain, but it is not easy to maintain an appropriate tone with respect to homeless and wounded people who are not accustomed to taking advantage of the hospitality of others and to living in someone else's house. Our hosts created an atmosphere in which we felt at home. I remember the Christmas Eve that also was the host's name day; we spent the day and the evening like one family. The hosts' kindness radiated on us all. Ludwika Grabkowska was the good spirit of the house and helped everyone she could. She always personally wrapped packages for her wartime godchildren, and she had an improbable number of them. I do not know of anyone who requested a favor of the Grabkowskis and left with empty hands. They were individuals of high interior culture. They were musical, knew literature and history, and loved art—these people were near to my heart. They were also close to me because of their kindness, their desire to help others, their attenuated hunger for life, and their general attitude, which was non-intrusive but faithful. When they were sitting on the terrace gazing at trees shimmering gold in the setting sun and were absorbing the beauty of this world with the thought of the Goodness that is eternal and survives battles and hate, I felt the thread of community uniting me with them. Only they were made of a different crystal. They had the harmony of souls that for generations had been absorbing beauty, had seen no suffering, and had not been fighting for life. But in addition I carried in me the world of suffering masses and a dissatisfaction with life that required justification and sanctification through a goal. I was not intoxicated with life as such. Had I been born among them, I would probably have been happier and, like they, I would have sat peacefully at sunset and drunk in the beauty of the world. I would not have had in me this unrest that had made me leave my native country and then, on its behalf, rush to the rescue of the perishing Serbs and finally descend to hell. I realize that my creativity is the result of this disharmony and of the gnashing pain, which is the cause of dissatisfaction and the motor of action. To embrace various worlds, one must have them within one. And this hurts.

Mr. and Mrs. Grabkowski were very religious. They had their own chapel in the manor. One Jesuit father from Cracow was always there and said Mass every day. All who lived in the manor attended Mass. However, I had the impression that our hosts rose higher in the ecstasy of prayer. Moreover, for them Christianity was not only a belief in a Superior Being but also an essential motor of their actions. These people were incapable of hate. They had a kind attitude toward all. I believe that given the opportunity they would forgive their enemies in the name of Christ's dictates.

Yet, these through and through good and just people despised Jews and ascribed all evil to them. They were convinced that Jews were the eternal enemies of the church. In their opinion, even Protestantism was the result of Jewish conspiracy. They did not differentiate Jews; evidently, they had never met Poles of Jewish descent. Nevertheless, they had pity for Jews and severely condemned German cruelty. They, as well as many other people from the same sphere, were a good example of the ennobling influence of the Christian culture. In spite of their antipathy for Jews and their desire to rid Poland of them, they condemned the murders committed by the occupiers. It was obvious that Poland was incapable of such actions. Here, in the house of these good and noble people, for the first time I gazed into the abyss separating Poles from Jews.

In the city, one talks about art, literature, and a hundred other things that represent the common interest of intellectuals. In the country, one feels the reality of the memories of childhood, the first prayers, and the charm of a little village church. In rural areas, one feels the power and formative influence of folk customs, and one hears the call of the earth that does not penetrate urban pavements. This forces one to take a different approach to the problem of assimilation; this process must be deeper than just a formal acceptance of the culture of the highest spheres. I will talk about this later when I consider the overall aspects of Polish-Jewish relations that have emerged from this war.

It was not easy to shape my life when I was writing a textbook in Saska Kępa with my faithful assistant and, with the power of my imagination, I had to create the recipients of my ideas. But the effort was immeasurably greater now that I was alone in the remote countryside and had to carry on without the rhythm of life to which I was accustomed. It is difficult not to be oneself, and it is difficult to live without the capital we call the past. I arranged my external life as follows. In the morning, I studied statistics as applied to medicine and worked on the various chapters of my book. I read Shakespeare in the original and took delight in his endless beauty. After lunch, I worked with my daughter on her English and German. In the late afternoon, I went for a walk alone or in company.

The manor was frequently visited by neighbors, priests or landowners, who told gruesome stories about hunts for Jews in villages and towns. The place would be surrounded by German gendarmes assisted by Polish police and the Jewish Monitor Service. The peasants had previously been ordered to supply a wagon. The game thus caught would be transported to the execution place. For a while they would keep the victims behind wire, often naked. Then they would shoot at them with revolvers or machine guns. Mothers would perish with children in their arms. The descriptions indicated that the tormentors were fanciful. One execution took place along a road. German drivers stopped for a while and shot at the Jews the way one shoots at ducks. When they had had enough of the amusement, they continued on their way. Occasionally, such a killing was accompanied by a small reception: a table with snacks and vodka was standing next to the execution place. The German killer would eat and drink with his hands stained with blood. Occasionally, passersby were allowed to watch how the Jewish problem was being liquidated; at other times, no strangers were permitted to stand by. Tarnów had public executions. I was told that the steps of the church were flooded with blood. In another town, a truck carrying soldiers with machine guns was driving along the streets. The residents were ordered to come out of their apartments. One such truck left mountains of dead bodies behind. Somewhat later, better methods of killing were established: gas chambers in Treblinka, Majdanek, and Oświęcim (Auschwitz). Once, I remember Polish policemen came to the manor; they were invited to the table and offered vodka and sausages. They had simple, honest faces; most were young. Between the first and second glass of vodka, they told us how they were being forced to catch Jews and deliver them to designated places. One of them told us how he once carried off one thousand Jews and then left to wash and have his breakfast. Later, this simple man wanted to see what was happening to them because he felt pity for them. He found that nothing was happening any longer because they were no longer alive. When the prisoners were being conducted by the Polish police or the *junaks*, many a Jew escaped. Sometimes they were allowed to escape for money, sometimes out of pity. The Germans were hard and often tied the Jews so that they would not escape; to be sure, they used barbed wire. Clothes had their value, so the prisoners were ordered to undress, even in winter. What did it matter that they froze, before death. In one neighboring town, the sub-prefect— supposedly German sub-prefects are individuals with university education— amused himself by shooting at children in their mothers' arms. One Jew could not endure it, and he assaulted the sub-prefect and cut his throat. However, he did not do a thorough job. The Jew was killed, while the sub-prefect recovered and swore vengeance. What else could he have done? There are ways and means to encourage or force people to commit murder. Several policemen were killed for having had pity on Jews. The peasants often hid Jews. Once, a young Jewess was caught and was forced to confess under torture where she had been hiding. The peasants who had been hiding her were killed, while the Polish police were informed that Jews were denouncing the peasants and that, in order to save their countrymen, they should kill Jews at once. Or, those who would kill a Jew would be allowed to keep his clothes. To document what benefactors of Poles the Germans were and how smartly they were taking care of the Jewish problem to the country's advantage, they arranged a public auction of the furniture of the dead Jews. Shabby peasants from the neighborhoods were seen carrying various traps, pots, and pans. It was a queer sight. Even objects have a soul and they will scream about the blood shed. This Jewish pot reminding one of tears, blood, and lack of conscience, could it have brought happiness? Could the peasants' standard of living be raised that way? Together with the Jewish problem, it was human conscience itself that was being liquidated.

One young policeman once said that he had been participating in a search for bandits and was killing Jews along the way: "As for the Jews, I am not sorry for them." Mrs. Grabkowska was not made of the stuff to condone the liquidation of the Christian idea. Without hesitation she said: "Are you not ashamed?" It was not always possible to intervene in such a manner, however. Once, when we were sitting at the table, one man was telling about his household troubles: "A son of a gun of a Jew sold me bad seeds. But I got my satisfaction: he was soon killed." An embarrassing silence fell at the table.

I discussed these matters with many people: landed gentry, clergy, peasants, and policemen. Almost all were anti-Semites, but no one approved of these methods of liquidating the problem. The longer the tension lasted, the more unendurable the atmosphere was becoming. When a sick man suffers very much, others wait for his death as a relief. Thus, in spite of their internal protest, many people eventually began to sigh: "If it only would end soon," because they were unable to endure this atmosphere of pain, suffering, and cruelty.

During these talks, my daughter and I would sit at the table with indifferent faces. But in my child's eyes I saw how the weak threads which were still tying her to life were breaking. When she heard about these killings, she had a more real picture of it than the others. She felt that the loop was closing on her and that she had to flee into a world where there was no torture. At that time, it seemed to her that there was no such place in the world.

Chapter Thirty-One

My Evening Song

Having spent one-and-a-half years in the district, I now perceived my contact with nature twice as intensely. Every morning I looked at the sunrise, and at sunset I would go for a walk to absorb the richness of colors for which I had been yearning for so long. I walked as a stranger though, while life was passing by at a distance. I felt that a loop was closing in on me, and that my strength would not last for long. And I felt that it was time to gradually bid farewell to everything that was dear to me. I once read a play by the Swiss poet Widmann:[1] *Die Maikäfer-Komödie*—The Maybug Comedy. How those who have dug into the ground yearn for the sun. Ecstasy of flight and joy of life. They throw themselves on young leaves in inebriation. But the leaves feel and suffer. The song of perishing leaves resounds. They had the same right to live as the May bugs, but the eternal law of nature had destined them to be the joy of May bugs. At the end, the May bug also dies, pierced by the pin of a playing child. The bug dies with words of delight: "God, I forgive the world as a man forgives a woman's lie for her beauty."

My end was also approaching. Thus, I recalled that I had not only been a passive link, a minute part of a regenerating life, a leaf made to be devoured, but that I had also had the pride and pleasure of a giver. I was using my will-power to recall old memories. I saw the eyes of young students fixed on me who did not even know that it was through them that my longing for the far-off yonder and eternity was extending. I would also tell them—possibly from beyond my grave—that I wanted to teach them how to behold. And that while beholding, they should not unlearn how to see and admire. While in appearance training them for a profession, I wanted to breathe into them a feeling for the beauty and harmony of the world. Now, like that dying May bug, I wanted to forgive the world as one forgives a woman because she is beautiful.

I will give my feelings the form of a preface. But actually, this is my evening song:

I am sad, Savior! For me in western skies
You poured out a radiant rainbow array. . .[2]

Introduction to the Textbook of General Immunology[3]
Writing this textbook may well draw to a close my pedagogical and scientific activity. Lectures have never meant for me sterile professional training

because I believe that institutions of higher education must be schools of the spirit. The student must learn not only those facts necessary to pursue a profession but also the philosophy and charm of the given field of knowledge, and he must gain insight into the craft of thoughts and the technique of creative research. The objective of higher education must be to give more than what is required by professional life; it must teach to think, take hold of issues, wonder, and admire. A man who looks thus at the world becomes a better man. And that is why science must not only educate the intellect but also carve the soul. Teaching in this sense has been my passion and, like everything that is the fruit of a passion, my lectures have inflamed young people, in this—I will say with no false modesty—I was successful.

However, the teacher also has another task: to attract pupils and establish a school. A pedagogue seeks a young soul to sculpt, a researcher seeks resonance. Not an echo, but rather friction because only resistance manifests a young eagle that is searching its own path and flight. At the root of this aspiration is the desire to live on through the young—the immortality of an inner drive of which the reality can be felt. *Non omnis moriar . . .*[4] Attracting students and creating a school has been my second passion. I have put much energy into this, and I have had fairly many pupils. Of this I was proud. I often felt that lecturing was like an expedition to gather young souls. Not many of my pupils have remained faithful to the work I initiated. Some were led by life to similar but more descriptive work than the experimental research which is closer to my heart, while others were forced to do organizational or field work. Many had to abandon laboratory work. Still others died in the prime of their work or were killed in a cruel manner before they could learn to know the charm of creative work. Only a few have continued in my direction and have carried on in my field. I do not know whether this is to be blamed on my scientific premises lacking dynamics or on the period and circumstances. I do believe, however, that great dynamics would have overcome the obstacles of the era, and therefore I am inclined to assume the guilt—a sad confession after so many years.

Still, some problems born or developed in my laboratory have become everyday topics of contemporary science. I might mention constitutional serology and blood groups, the specificity of regressive and productive processes, the serological specificity of tissues, the application of some aspects of coagulation in immunology, the matter of latent infections, and a certain interpretation of vaccination. Nonetheless, these problems have not attracted many researchers in our country. I am writing this even though I gave several persons the stimulus to work and an opportunity to walk on summits where the air is pure and where ideas can be projected into the far future. Nevertheless, it has become empty around me. When I think of my laboratory and my deceased or murdered associates, my thoughts roam through cemeteries. It is not good to survive one's children. I feel surrounded by the frosty air of loneliness.

Having lost my laboratory, I have implemented my dream of many years' and written a textbook of immunology. I have written it in the form of lectures, because this best corresponded to my inner rhythm, although these lectures are markedly expanded. When lecturing, the researcher is hampered by a certain bashfulness that the judgment of his concepts should be left to students who are conditioned to his viewpoints and are biased in favoring him. By and large, students are not enraptured by the correctness of the lecturer's concepts but by their own dynamics and the fact that they are witnessing the birth of ideas. A textbook—if it is not narrowed down into a compendium, can give the necessary facts, but only after they have been processed in the smithy of experimentation and criticism exercised by the lecturer. Such a textbook offers a synthesis of knowledge which is adjusted to nothing else but the absorption capacity of young minds. It speaks to the young people and to the experts, and thus to the pupils and judges all at once. What the researcher has thought through and has perceived in his favored scientific branch—this he can express in an appropriate selection of facts and their illumination. His thoughts take a broader perspective and radiate over the span of several generations. The fact that those near and far away can exercise control gives the author a feeling of intellectual freedom without the fear of unwittingly projecting into aborted flight a young person who trusted him. In such a textbook, one can express not only facts and specific concepts, but also a basic intuition for natural phenomena.

I have attempted to do just that. I do not know whether I have succeeded. Future researchers will write better books. Several contemporary books, such as those of Nicolle, Jeans, Zinsser, or Natanson,[5] prove that a new style of writing is being born which ensues from an understanding of the beauty lying at the basis of natural phenomena. Beauty must be found. If it is understood, hymns arise. Similar phenomena are revealing themselves in science. One feels that science has begun to reject mere accumulations of facts and is gaining increasingly more understanding of the whole, the indivisibility, and the beauty of nature. One cannot present strict facts by means of hymns: however, a feeling for the beauty and meaningfulness of natural phenomena endows a description with a special melodiousness, facilitates bifurcation of thoughts, and discloses unanticipated associations of ideas. At that point, the difference between science and art fades away. Bacteriology is in an especially fortunate position, because microbiology presents an excellent picture of struggling life, while the branch making for the subject of this book presents the mechanisms of this struggle. Immunology is the basis of bacteriology, epidemiology, and the pathogenesis of contagious diseases. The science of resistance is the story of a cell which, once it has been struck, retains in all subsequent globuliogenic processes—and perhaps in others as well—the mark of the initial stimulus, of the original chemical rhythm. Antibodies are a manifestation of the general law of the memory of live matter. The organism defends its integrity regardless of whether it is penetrated by germs or by inorganic particles suspended in the air. Climatic and food allergies prove that the body resonates together with the outside world and reacts in a basically similar manner to a number of stimuli, regardless of their nature and destructive power.

Let us consider the basic phenomenon of microbiology: the formation of various strains of pathogenic bacteria. Unknown causes condition the formation of susceptible types of macroorganisms and pathogenic types of microorganisms. Coincidence brings them together and forces them to stay together. Thus, we witness the birth of the infectious disease. A race develops between the defensive ability of the one and the destructive power of the other. This struggle gives rise to new means of defense and attack and to new races and strains of microorganisms. Often the germs change their chemical structure, they actually assume new forms of existence, and in this way they escape the specific reactions of the macroorganisms. Mutual adaptation develops, giving rise to unsusceptible species and races which do not and need not defend themselves because they do not suffer. Animal species and races are formed which, without becoming ill, transmit germs, and complex biocenoses are formed which include several species in a sphere of common and interrelated interests. The world is filled with billions of microorganisms which induce no diseases. This gives rise to the carrier state and latent infections—mankind's greatest plague which blocks our victory over infectious diseases and, at the same time, the greatest triumph of nature which tends to preserve the greatest number of species in spite of interspecies conflicts. In filterable viruses, especially in bacteriophages, one can experimentally produce new unsusceptible strains, even new races. How can we fail to admire the flexibility of matter and its will to endure which, through the greatest calamity—dissolution of microorganisms—create new resistant races and sing the immortal hymn of life. The chapter on the formation of microorganisms or on filterable bacteria could actually be written in the style of "Genesis from the Spirit." The same is true for the chapter on bacterial filtration.

Experimentally, by means of certain chemical or mechanical injuries, we can produce diseases in animals and, from the diseased focus, we can raise bacteria which cause the animal's death, can be transmitted and cultivated, and possess all signs of a separate life. Are we dealing here with the secret of the creation of life, the eternal dream of mankind? Or have we only disturbed the balance between the asymptomatically infected cell and the pathogen, and thus created false appearances of new life? A definitive answer to this question is crucial for the epidemiological vision of the world, the formation of new infectious chains, and consequently the future of preventive medicine. Or, let us consider the phenomenon of latent infections. We must abandon the theory that after millions of years of a joint journey, macroorganisms and microorganisms are two separate worlds capable of independent life and, upon contact, always hostile and mutually destructive. The thesis of a struggle for life and death does not adequately consider the phenomenon of symbiosis and does not hear the harmony which is the basic melody of nature. Macro- and microorganisms represent two interwoven worlds which tend to develop forms of peaceful coexistence. . . . Fight for survival is not the only slogan of life. There may be a law which could be called the Law of Least Suffering. The gnawing which we call suffering and disease is a rather rare breakdown in this coexistence; pathology is a minute deviation, a sort of tribute to the matter of preserving the largest number of species and the most buoyant world possible. Therefore, this book cites many examples of the positive, life-promoting effects of infection. Infection lies at the origin of evolution because it indirectly conditions the formation of new organs. I regard all these phenomena as the philology of infection, superimposed on which are detrimental phenomena classified as the pathology of infection. A similar illumination of infection is suggested by a broader view of the phenomenon of an epidemic. An epidemic involves not only the formation of new infectious chains but also a disturbance of the balance between germs and the population of fauna and flora. Our struggle against the so-called civilized diseases has gone astray because asymptomatic infections and the ubiquitous presence of germs have not been considered.

When we look at immunology in detail, we again see impressive defensive processes. We would narrow down our research methods if we did not emphasize the purpose of these phenomena. They display such beautiful dynamics and such a clear purpose that it is impossible not to see and not to emphasize their great, mysterious will to live.

What richness of ideas is offered by modern serology. Immunochemical investigations have demonstrated that inorganic particles produce essentially the same defensive reactions as those produced by cellular agents. Modern serological methods have even touched upon the secret of the formation of matter. The greatest puzzle in serology—what the formation of bonds consists of—has been elucidated. The science of allergies tells us that there are struggles and conflicts between macro- and microorganisms which are not the consequence of the initial effect of the microorganisms but result from a defense which goes beyond the goal. Finally, what enrapturing prospects are opened by the chapter on blood groups. Unknown stimuli induce mutations in all parts of the world in human beings, animals, and microorganisms alike. Serological races are formed which, driven by the eternal unrest characteristic of man, flow in all directions and produce assemblies known as blood groups. By studying a drop of blood, we can investigate the prehistoric and historic journey of peoples. Thanks to investigative methods of tissue groups, it has been possible to check the blood groups of mummies and determine the serological properties of races long extinct. Here we are facing a problem which has recently been raised in my laboratory and which has not yet been touched by world literature, the problem of the uneven distribution of group properties. In my opinion, this is the result of incomplete mutations. Serology is facing a question which, I hope, can be resolved experimentally by investigating the distribution of group properties in various regions of the world: Has the human race become ossified in its ability to produce new serological strains, or is it still able to form and magnify new properties and thus to create new races? Other biological sciences can envy serology the wide scope of these problems.

I have discussed certain aspects of immunology in the light of biology. But the resistance phenomena also have their social ramifications. Just as a chemical analysis of stars cannot explain their motion, so bacteriology and immunology cannot explain epidemic developments. Yet we must base our considerations on these two scientific pillars, because an epidemic develops when social susceptibility is great, and becomes extinct when resistance builds up in a large number of individuals. In these phenomena, resistance induced by latent infections in one generation is interwoven with the resistance acquired by former generations, selected, and sculptured in germinal plasma. Without familiarity with these facts, we cannot understand protective vaccination. Therefore, this book contains the main data from genetics and states the main methods used in medical statistics.

Not all chapters in a textbook can carry the mark of the author's own experimental work and subjective illumination. In this textbook, two chapters dealing with new branches are an objective compilation drawn from other sources. These are the chapters on filterable viruses and immunochemistry, probably the two most grandiose branches of modern microbiology.

This was the content and tendency of my lectures, and these are the psychological motives of this book. Perhaps it will inspire or encourage someone to my manner of thinking. The term "thinking" narrows the essence of the approach. The investigator has a certain specific mental rhythm, a nostalgia for the unknown and for creating new ideas, and a desire to participate in this process. The source of love is the need for love. The source of an idea is a yearning for it. The intellect supplies the content, but the yearning creates the foundation. Research activity is the result not only of definite intellectual interests but also of an even directionless passion for thinking. To awaken this passion and make it conscious in those who possess it in a dormant form is the objective of this book. I cast it at the hearts of the young people, even though I am not sure that it will be granted to me to enjoy its resonance.

Chapter Thirty-Two

My Greatest Defeat

A calm, sunny September was lingering on, as it usually does in the autumn in Poland. Orchards were aplenty with fruits. The park in Czarków was a kind of paradise, and we often walked there. But what my daughter liked most was to visit her favorite priest. This is one of my most painful memories. Once we went to a neighboring village in search of fruits; we were directed to a presbytery. A pretty little rural church was standing on a small hill, and next to it was the presbytery—a small white house surrounded by a small but charming orchard full of apple and plum trees. Visible from afar was the outline of the park in Czarków, and nearby on a hill was the village cemetery. All this was imbued with a special charm of quietude and bliss.

In this orchard, we met Father Teofil Zawartka.[1] He was an old man of seventy years and resembled those trees that have roots in that earth of ours and bring forth fruits on which human souls regale. My child came to love this old man. She was longing for kindness, friendliness, and the quietude of that orchard, and she wanted to forget human malevolence. He also liked her; he was touched by this young plant so helpless and weak, and he felt that she had been uprooted from her soil and was longing for warmth. On every sunny day, she would go to see him. They took walks together in the orchard and understood each other without saying much. He always had a nice surprise for her. With a smile, he would pull out of the pocket of his habit a beautiful apple or a handful of nuts; or they would go together to a tree, and he would let her pick fruits of a special sort known only to him. They would bask in the sun and chat. My daughter returned very much comforted and warmed up from these walks. She drew from them courage and hope that perhaps not everything in her life had been forfeited and that maybe she would succeed in once more growing roots into life. The priest's niece and nephew also lived in the presbytery. They too grew fond of Marysia. She would tell them about her study in far-off Paris, and then she would recall the world, her past, and her happy childhood.

On All Souls' Day, I went with her to the cemetery to visit the graves of fallen soldiers. The autumn sun was mild, from afar we saw hills covered with stubble, and everything breathed calm. At one point, my daughter said to me: "Daddy, this is where I would like to be buried." This was, as it subsequently turned out, the only of my daughter's wishes I was able to fulfill.

When the late autumn rains came, the impossibility of going out, and the evenings stretching out, increased the intensity of our empty life and separation from the world. Her health deteriorated and her longing for not being became more acute.

She came up to me and said: "I heard the servants in the kitchen say that there were no leftovers from dinner because everything had been eaten up by those beggars. The beggars they were referring to were we and the others taking advantage of the hosts' hospitality." And my child continued: "It hurts me so much, Daddy, that you have lost your luster here and are such a nobody. How different it was to see you radiant and giving to others in Warsaw and in Paris. At that time, I felt that as your daughter I had different rights."

I then decided to attempt the most difficult game of all: to create, with the power of my words, the vision of a great world, of academic life, and of tension and hope. During the long winter evenings, while the others were sorting beans, I read novels aloud to them. Then, I suggested that instead of reading novels, I would give lectures. True enough, I wanted to reciprocate to the Grabkowskis for their kindness by offering them some intellectual entertainment, but my main objective was to do something so that my daughter would not feel like the child of a nobody and so that she could again look into the future with confidence. Thus, I began my lectures. I tried to discuss matters that were interesting to all, and so I talked about the mystery of the formation of the world, about modern attempts to explain the formation of species, and similar problems. Where had all the students gone who used to come from different departments to listen to my lectures? How remote and unreal was my lecture in Paris when among the students I saw the glowing face of my daughter! Here, as an added attraction to bean sorting, I was talking to elderly persons with the idea—no, with painful hope— that the ardor of my words might warm up my one and only freezing child.

Mrs. Grabkowska later asked me why I had unmasked myself to such an extent. "Dear Madam, I would give my life to breathe it into my child." When I was talking, I had the impression that my daughter would gradually forget the tormenting thoughts, the tragedy of being in hiding, the memory of dying people, and her longing for her own home. I felt that I was leading her into a great, radiant world and bringing back to her the carefree times of her study in Paris.

I offered her moments of oblivion. Sometimes she would tell me after a lecture: "How beautifully you spoke, Daddy, how you are fighting for us."

But I lost this battle. I did not succeed in arousing a hunger for life in my child. She developed a fever, and a physician diagnosed pneumonia and warned me that in view of her general exhaustion, this was dangerous. He advised me to summon her mother.

This was in January 1943. Persecution had reached its climax. Non-Aryans were killed on the spot the moment they were detected. I was facing the gravest decision in my life: Should I risk my wife's life? Finally, I decided that if the worst had to come, let the last gaze of my daughter fall on her mother. Even for the price of her life—no, our life—because I had no intention of surviving those who were dearest to me. I summoned my wife by wire, and I knew that she would come in spite of the danger.

Twilight fell in Kamienna. My daughter was lying in fever. On the first floor, Germans were having a feast. They arrived for a hunt and ordered a dinner for themselves. The nation of knights would not deign to ask for hospitality. We could hear calls of "hoch" and the chatter of gay and well-fed people. My daughter told me: "I wish so much I were at home, Daddy, but don't write to Mommy. She would worry too much."

I knew that my wife was already on her way to see her dying child. With a pain that made all my senses more acute, I saw all the dangers to which my wife was being exposed. In my imagination, I saw her being caught by a soldier, bending down, kissing his hands, and imploring him: "My child is dying, let me go to my child. I swear I will come back, but I beg of you let me go to my child."

A newspaper was delivered with Goebbels's speech printed in it: "How those wicked Jews are hiding." In my mind's eye, I again saw my wife who had wiped so many tears but never caused a single tear to be shed. Before me was my dying child, asking me: "Daddy, why can't I go home? I want to see my friends. And I want to see my mother. Why can't I go back home?"

From below, I could hear the strong, juicy male voices.

I felt hate and disgust rising up to my throat and suffocating me. Hate for juicy, complacent voices. Hate for German hypocrisy, which combined bestial cruelty with sentimentality, and for this disgusting behavior, which they called culture. In that moment I felt that German words, German thought, and German writing would forever induce in me a feeling of disgust and hate of which I would never be able to free myself. That disgusting voice of self-assurance of soldiers who saw nothing and understood nothing and were only able humbly to obey orders.

I knew that millions of people would have the same associations.

My wife arrived. She spent a week with her child. She closed her eyes. Our daughter was led over to eternity by her favored priest.

She rests in peace in the rural cemetery where she wanted to be buried. The bells of the village church played for her the tune of eternal sleep. Polish peasants came and sang. The winter sun was shining.

My whole life's work and all the human kindness I had assembled enabled me only to let my twenty-three-year-old daughter die in a bed, surrounded by good and kind people, and to bury her in her own grave, albeit under a foreign name. For under her own name, she had the right neither to live nor to die.

Chapter Thirty-Three

The Origins of This Book

My wife and I are lonely now in this world. Our love and hope rests under a birch cross in a cemetery in a remote village. We stand over the grave with an open wound and with the memory of the withering of a weak plant that could not withstand the evil of this world.

Our daughter grew up in an atmosphere of warmth, kindness, and friendliness. She always saw her parents in the aureola of giving people. She never felt handicapped, and with her radiant blue eyes she looked confidently into the world. Until the Germans came and made her feel the whole power of their hate.

As long as my child was alive, I felt like a link in a chain leading into the future. Now this chain is broken. Instead of the eternity that used to stand before me, I see a few months or years filled with longing and torture. How rich I was before, and how poor I am now. Loss of the feeling of permanency is tragic. I remember the delight I felt when I was showing my child the world and saw how she was absorbing its beauty in delight. I wanted to travel with her and draw human friendliness to her so that she would not be lonely. I myself would live in her heart. I no longer yearned for beauty or emotional experiences; all this was meant not for me but for her: human love and the pleasure of life.

After our daughter's death, we stayed only a few more weeks in the hospitable house of Mr. and Mrs. Grabkowski. Our daughter's death made a grave impression on the neighborhood. This was incompatible with our necessity to remain hidden. Besides, staying on would have been torture: every object reminded me of my daughter.

A Jesuit, Father Mokrzycki, tried to console me. He gave me a holy picture with the inscription: "He who has loved and understood many, and has suffered and forborne much can make many people happy."

No, I can make no one happy. To make someone happy, one must be happy oneself. I belong to those millions whose souls have been so wounded that they can radiate only suffering. My thoughts roam now among graves: my child's grave, the graves of murdered victims, and the grave of a now-massacred hope that human beings were something more than dogs in who, if you train them properly, can develop the reflex of biting everybody they see. What point was there of appealing to people's conscience? I must flee, flee from this abyss of tears.

Yet—I say to myself—perhaps this terrible suffering into which stupidity and greediness, clad in the robe of a world-outlook, have thrust mankind will still arouse public conscience. Maybe at least those who have suffered will understand.

I decided to be the voice of those who suffered, to speak to those who were still able to feel. Out of my pain, I decided to forge a weapon to move the public conscience. I would show the endless pain of parents whose children have died. No, not who died, but who were murdered before their parents' very eyes. Death in combat can be an ecstasy, a vision of victory, and a promise of the nation's or the idea's future. It may be that some parents imagine that their children have died such a death and are unaware of their pain and fear of a premature death. I, on the other hand, will write about the death of the victim whose last impression is befouled with the sight of a murderer or a murdering robot, and is burdened with the loss of faith in man, with the complaint of senseless torture, and with the question cast in the last moment: "For what sins?"

No, I will not make anybody happy. Happiness is a luxury inaccessible to our generation. Who would dare to speak of happiness in the ocean of suffering? But perhaps I will succeed in awakening the consciences at least of those who suffer and of those fools who are proud of their children's death. I will tell them about the boundless pain of people who lived in a space claimed by Germans as their own. I will tell them about a mad extermination of people as if they had been lice or bedbugs.

Still, humanity is tired of the sight of pain and tears. Those who for years looked death in the eyes now want to see a child's smile and the joyous awakening of springtime. At most, they want to see how such suffering could have been caused and how it can be prevented in the future. If I speak only of tears, they will reject this book like all White and Blue Books.[1] They have seen enough misfortune. Why think of it again?

Thus, I decided to write the story of my life. How my spirit was shaped in those remote times when the air was not imbued with hate. My dreams and my aspirations. How I believed that creators of science could sculpt human souls. And about their great guilt and betrayal of the Truth.

May this book be a monument to those who departed prematurely— among them my daughter. And to my associates for whom I wished to be a father.

I had the impression that those who perished and those who are still suffering were standing behind me and bidding me to write the tale of their torture and the guilt of their contemporaries.

Chapter Thirty-Four

Extermination Camps

Staying on in Kamienna after our child's death was impossible for reasons of safety and owing to our emotional state. Our daughter's decease had shocked the neighborhood, and more attention was being paid to us than was advisable. Furthermore, a stay among rooms where every corner reminded us of the life and suffering of our child was one continuous path of painful memories. We received a letter from Mrs. Laura Kenig, with whom my wife and Zosia had spent half a year, inviting us both to come and stay with her in Stara Miłosna. We decided to go back to her. Again, during the trip and in railroad terminals and streetcars, we had that awful feeling that at any moment someone might recognize me or my wife and spread the news that we were alive and where we had been seen. This would have meant death for us as well as for our friends. Laura was not afraid. This good, deeply religious, and extremely diligent woman loved her lot of land which she cultivated; while her husband was in German captivity, she had a hard time making ends meet. Much has been said about the greediness of those who were hiding others from the Germans. We were dealing with completely different people. We had to implore and coax Laura to accept payment for the apartment. Equally worthy and disinterested was her sister, Irena Przedpelska, in whose little house nearby we spent the summer, while we spent the autumn and winter in Mrs. Kenig's hospitable house. We were not the only ones whom she was hiding. Before our arrival, three ladies had lived in her home. They had moved out and were subsequently detected as non-Aryan and killed by the Germans. On many evenings, we would hear a tap on the window. Mrs. Kenig or her daughters, who certainly were not rolling in riches, would carry out hot food and cigarettes. The man who tapped on the window was a Jewish butcher who was hiding in the forests and whose whole family had been murdered. Mrs. Kenig did not hesitate to help him. Toward the end, he was hiding on her property. In our misfortune we evidently had good fortune: we met only worthy persons.

In Miłosna, I learned about the fate of those I had left in the district. In the first place, I should like to mention a few names. Physicians from the Department of Health, including Dr. Milejkowski and Dr. Syrkin-Binstein[1] were packed into a railroad car and sent to slaughter. Dr. Syrkin-Binstein proved to be a wise woman: she took morphine with her and injected everyone with a fatal dose so that while dying, they would not have to gaze upon monsters with a human face. Dr. Fliederbaum—the personification of subtlety and kindness and an excellent scientist and physician—pushed his little son and his wife off the fourth floor before himself leaping to his death. Dr. Józef Stein and Dr. Heller, the chief physicians of the hospitals, died at their posts; they refused to abandon their patients.

It was after the war that I learned about the fate of the chief nurse, Mrs. Grynberg. Her child was with her friends on the Aryan side, and she already had Aryan papers. She had orphans in her charge, however, and she did not want to abandon them. She struggled between the instinct of a mother and that of a nurse and social worker. The latter won—she stayed with the orphans. She wrote to her friend, "I am entrusting my child to you; raise him as your own." She gave her documents to others. She perished voluntarily in the service of a great cause.

My family, including my old aunts who were worthy women, was murdered. I have an especially touching memory of one of them, Mrs. Ruta Kon. She was all human kindness and nobility.

I was told the story of Dr. Wiener from Lwów who was hiding with his father in the basement of one of the university buildings. His friends brought him food at night. One day, his old father lost his mind and started screaming. To save himself and others from a cruel death, the son was forced to poison his father, cut his body into pieces, and carry them out at night one by one to bury in the ground. I believe that human history does not know and will never know more monstrous tragedies. If I wanted to write down everything I heard, I would have to fill this book with stories about facts each of which could be the subject of a shocking drama. I cannot do this. I will only make a general outline of the lives of those condemned to death.

All Jews in the ghetto were put into barracks. After the mass deportation to the execution camps, the initial district population of half a million was reduced to 30,000 or 50,000. All were put into the so-called [work]shops. They were not permitted to leave under penalty of death. Their "reward" was initially six złotys per day and subsequently nothing. The whole sum of about a quarter million złotys per day was paid to the SS men; for this money, the nation of masters ate, drank, and enjoyed life. It sometimes cost thousands of złotys to get admitted to such a shop. My secretary, a woman with higher education, described to me living conditions in such a shop. It was headed by a young German who was almost always drunk. He would walk through the shop with a whip and would see to it that the slaves did not get even a moment's rest. One time he became sober, looked around, and was surprised that the Jews resembled human beings, because he had been taught that they were "sub-humans," something between an ape and an Aryan. Even his beatings though were not as bad as the constant visits by other Germans. They would come to the shop to take a certain number of them to their death. The victims humbly awaited their turn. Thus, they lived under the whip, in hunger, and in constant expectation of death. I will quote below the terse report written without pathos by my secretary; it is all the more eloquent.

Shops existed even before the deportation action. As a rule, the owner was a uniformed German, and the directorship was Jewish. Prior to the action, it was not difficult to get admitted to such a shop. The pay was low, a few złotys per day (at that time, 1 kilogram of bread cost 15 złotys). The workers received a breakfast consisting of 1 liter of coffee and 250 grams of bread with marmalade, and a dinner consisting of about 1 liter of soup. This was the way it was in large shops, such as Schultz's furriery and Toebbens' dress factory. Toebbens had retail stores in Germany, and here elegant dresses were sewn for his stores. Schultz was working for the military.[2]

Things changed the moment the action began. Fictitious firms sprang up and charged tremendous amounts of money and sewing machines for admittance to work; later, it proved that certificates issued by such firms were of no value. Those who found employment in real factories were forced, from the middle of August, not to go back home after work, because chances were that they would not return at all. As a result, buildings designed for 150 persons were inhabited by over 3,000 family members, including small children. No wonder pediculosis was spreading. Then an ordinance was passed prohibiting Jews to receive money. The factories had to organize a food supply. The daily meals consisted of 250 grams of bread, black coffee, and a soup that was so disgusting that it was inedible in spite of the hunger. Immediately smuggling arose, but food products were very expensive. In our factory, there actually was no work to do, while in other factories people worked in two shifts. The main objective of the shops was actually to extract money for the German owners and the Jewish managers. Under the firm's cover, items bought for pennies from desperate people were sent to the other side and sold there for thousands. The German owners and Jewish managers became rich within a few weeks. This excellent business began after 7 September—that is, at the time of the apparently ultimate selection for death or for slavery. That was when a food supply was organized in factories, and food products were sold to the workers for huge sums of money. Some products were allocated by the Community Board. But everything had to be paid for, while salaries were not paid. Toward the end of August, the workers were ordered to occupy houses in the vicinity of factories. Almost no one went out on the street; besides, there was no reason to, because all the stores were closed and the streets were completely deserted. A stranger could get into a shop only through bribery. The firm in which we worked occupied four buildings: two apartment houses and two factory buildings. The four of us lived in one room, and we shared the kitchen with another family. Frequently, soldiers would come and take whatever they liked and could put their hands on. It often seemed to us that our last hour had come. As of October, we were given only 125 grams of bread and half a liter of a very poor soup. To buy anything at all on the free market, we had to sell our clothes. We were paid very little: 100 złotys for shoes, 200 for an overcoat. Our work would begin at 7 in the morning and end at 5 in the afternoon, including a half hour break for breakfast and dinner. Penalties were imposed for walking in the courtyard during work hours.

In other cities, the situation was no better. Everywhere, Jews were concentrated in a few ghettos in which the same work system was introduced: hunger, beating, twelve-hour working day, and the constant waiting for death which would come with inexorable certainty. At first, the Jews would flee from one town to another, but the orb was gradually contracting. The most courageous and lucky ones would join the partisans, while others were hiding in forests. Ghosts would sometimes frighten travelers on highways: these starved people, overgrown with hair, and with an expression of deadly fear in their eyes, no longer resembled human beings. When it was announced that whoever killed a Jew could keep his clothes, there were some from among the mob who would go after Jews as if they were going hunting. Trains traveling to Treblinka or Majdanek were especially interesting. Jews who jumped off the train usually had foreign money or valuables, which they had received from fellow travelers who no longer had a shade of hope for survival. Sometimes, such ghastly persons would show up around settlements, and groan: "Poles, help us." Occasionally they were given aid, but by and large, this only prolonged the agony because those who helped them were also risking their lives. I was told shocking incidents when whole families were killed because they had offered shelter to a Jew. Jewish children were hidden and fed in Warsaw's market halls, and some superintendents hid them in the garrets of apartment houses. The children were in danger if they were noticed by a *Volksdeutsch* woman. Those women were implacable. A patrol composed of a Polish policeman and an armed German was walking along Nowy Świat Street. They stopped a Jewish child. The child asked that his life be spared. The Polish policeman begged for mercy. It did not help; the child was killed on the spot. A crowd of no-longer beggars but ragged or naked phantoms was chased through Grochów and watched by hounds. They were walking to their death.

The liquidation in smaller towns—for example, Piotrków—took place in the following manner. At first the order was given for all Jews to remain in their apartments. In order to torture them, water, electricity, and gas were disconnected. Next, all Jews were ordered to show up on the square, so that the fittest could be selected for work. At the same time, robbery began. Women—the wives of SS men—came, packed Jewish belongings, and carried off full suitcases. Those who were unfit for work were chased into trains. Killing Jews was not enough; they had to be tortured. They were ordered to take off their shoes, and then they were put into cars sprayed with quicklime. The cars filled with Jews were left on the side tracks for several days—without water or food. Why should condemned men be fed? Finally, they were transported to Tomaszów and from there to shooting places or to gas chambers. Similar news was coming from all regions in Poland. In Kołomyja, all Jewish physicians were summoned to the SS office; most of them committed suicide. Other Jews were chased outside the town and ordered to dig a ditch. Then they were ordered to lie down in the ditch, one next to the other. Boards were placed on them, and a German walked along the boards and killed one after the other. This first layer was covered with a second and third.

The world was being informed of all these crimes by radio. But the world would not believe. By radio and through the press we learned that some people in the outside world were worrying lest the natural growth of the Jews should decrease under the occupation. From time to time railroad cars arrived from Germany, Czechoslovakia, and Holland. Elegant ladies in furs and with suitcases would inquire whether there were hotels in Treblinka, because they were going there for a longer period.

Were there hotels in Treblinka? I will tell you what Treblinka was like because there are people who had been to hell, seen it, and come back.

There were rows of barracks and the death chamber was in the middle. A modern gas chamber worthy of German science. People were chased naked into it, the chamber was locked, and a small window was opened through which gas was piped in. Those near the window would fall first. In five to fifteen minutes all would be dead. It was a very humane death provided it was not prolonged, whether in jest or for reasons of economy: the occupier would pipe less gas in. Then groans were heard for hours. What I am recounting here was written by a man prior to his death. The paper was found by his friends who were forced to play the role of aides. Then the bodies were put into large furnaces and cremated.

I am describing all this the way it was stated in the Polish underground press on the basis of authentic documents. Several years later, the historic Nemesis arrived: one of these death camps was captured by Soviet and Polish military troops. A commission of experts was established under the chairmanship of Professor Siengalewicz, a just and wise man who would not have stained his person with a lie or exaggeration. Even though I read this document and visited Majdanek two years later, I should like to state at this point the dry but therefore all the more shocking report of the special commission established to investigate the crimes committed in Majdanek.

Excerpt from the report issued by the Special Polish-Soviet Commission:[3]

The Hitlerites have established a chain of concentration camps on Poland's territory: in Lublin, Dębina, Oświęcim, Chełm, Sobibór, Biała Podlaska, and in other localities. They brought hundreds of thousands of people from occupied European countries to these camps for the purpose of exterminating them. In these camps, they established a system of mass extermination of undesirable populations, mainly the intelligentsia from occupied countries, Soviet and Polish prisoners of war, and Jews. A huge death combine was established in Majdanek. The Germans called it *Vernichtungslager* which means destruction camp. The camp occupies an area of 270 hectares. Its construction was begun at the end of 1940 and was completed in 1943. There were 144 barracks, each accommodating 300 persons.

The camp was surrounded by two rows of barbed wire and by special guardhouses. High-voltage current flowed through the wires. The camp was guarded by SS troops and 200 German sheep dogs. There were continuously from 25,000 to 45,000 persons in the camp. The prisoners were systematically murdered, and their places were taken by new transports. The citizenship of the inmates was established on the basis of later discovered passports and identification cards belonging to the murdered persons, on the basis of logbooks found in the camp, and on the basis of the testimonies of witnesses, both Poles and Germans.

The prisoners were systematically starved. For the slightest offense, a prisoner was deprived of even that minute portion. Most prisoners resembled skeletons covered with skin; Soviet prisoners received almost no food at all.

The workday would begin at four o'clock in the morning. The Germans would break into the barracks and chase the people off their sleeping boards with whips. The rally would last three or more hours, and the prisoners were beaten and tortured during that time. If a prisoner fainted, he was given the finishing blows with sticks. Those who worked were beaten at every step. The squads returning for the so-called dinner would bring back battered and wounded persons as well as the bodies of murdered persons. During the evening rally, those who had worked poorly were beaten with cowhide whips and rods. Frequently they were beaten to death. Professors, physicians, and other professionals totaling 1,200 persons were forced to carry heavy boulders. Those who fell from exhaustion were tortured to death. A group of Greek scientists was annihilated within five weeks. The torture system was diverse. Some methods had the nature of jests, as for example stunning a victim by hitting him with a board on his head, sham-drowning him, and the like. Some were killed by being hit with a stick on the occiput [back of the head], by being hit with a boot in the abdomen or in the groin, or by having their heads submerged in dirty water and pressed with a boot until the victim died. Some prisoners were hung by their arms which were tied together in the back.

One barrack had crossbeams under the ceiling; a large number of persons were hung on these beams at once. Witnesses state that frequently an infant would be taken away from its mother's breast and would be killed by being hit against the barrack wall. Or, right before their mothers' eyes, children were caught by one leg and arm, then by the second leg and arm, and would be torn apart.

Prisoners of war, soldiers of the former Polish army and Russian prisoners of war, were murdered in masses.

Persons of various nationalities and ages were brought to Krępicki Forest in trucks, all their belongings and valuables were taken away, and they were shot to death.

On 3 November 1943, 18,400 persons were shot to death in the camp. Completely unclad prisoners were led by SS men to ditches in groups of 50 to 100, ordered to lie down on the ground face down, and shot to death with machineguns. To drown the shots and the victims' screams, powerful loudspeakers were installed and lively music was played.

The most popular method of performing mass murder was suffocation with Cyclon B2 gas and carbon monoxide.

The selection of persons to be killed in gas chambers was made by German physicians: Blanke and Rindfleisch.[4] During the initial period of operation of the camp in Majdanek, the Germans buried the remains. As of 1943, they began to cremate them. Two furnaces for the cremation of bodies had been built in 1942, and five more furnaces were built later. One furnace could hold four bodies with cut-off limbs; in 24 hours, it was possible to cremate 1,920 bodies.

The commission has established that 60,000 bodies were cremated in the crematorium. Over 300,000 were burned on huge stakes in Krępicki Forest. Another 80,000 bodies were cremated in two old furnaces. And at least 400,000 bodies were burned on stakes inside the camp.

Robbing the prisoners and the dead was organized into a system. Majdanek had 820,000 pairs of children's, men's and ladies' shoes. The factory brands were from all European cities.

Prior to entering the gas chambers, all victims were ordered to undress and bathe. The objective was to facilitate robbery. One woman refused to undress. She was thrown alive into the furnace. A German witness testified that he saw how her hair caught fire and then how her body began to burn and shrink.

It is impossible to present all the details of this crime; it goes beyond human imagination. I will cite excerpts from the speech by Attorney Sawicki, prosecutor of Majdanek criminals.

Words are a human invention, but what has happened there is inhuman: degradation to the extent that a father would hang his own son in order to survive one more hour in the camp; or the inexplicable heroism of a mother who, having had one breast cut off, nursed her infant with the other; or the raping of women in the presence of their husbands and children.

There is a barrack, and in it are 300 women tortured during the day, debased and beaten. It is evening. Women, cringing on their sleeping boards, are waiting in fear and are tormented by relentless apprehension: Whose turn will it be today? In what condition will he arrive, and what pleasures will he demand? Those long hours of waiting prior to being raped. Now, the rapist has gone away choking with desire, drunkenness, and that irritating spiteful smile. The disgraced, crushed woman is looking for privacy, but around her are 300 pairs of wide-open eyes and 300 whispering lips. There is no escape, there is only barbed wire all around the high-voltage current. One road is left: into nonentity through the crematorium. But she must patiently wait her turn assigned by the German.

Now there is a procession of several hundred small, naked children. You see their plump, warm, little feet. Three hundred children are marching on snow, hardly able to raise their bare feet bleeding from wounds. Three hundred children with sad eyes. Do you know those sad eyes of Jewish children?

The terrible cry of women awaiting their death behind the Cordon of Roses next to which this nightmarish procession of children was moving toward the gas chambers. Do you hear the whimper of a child feeling death, do you see how this starved girl is being encouraged to speed up her steps? Is it possible to comprehend what was happening to the mothers behind the wire who had to watch all this before they were allowed to follow their children? Could anything still have been terrible to them except the torture of waiting for death?

Visualize a blank between night and day, when the moon has gone but the sun has not yet come up and the world is covered with a gray, early fog and cold, sticky dew. There are frozen, cringing bodies—and nothing else. Only blue smoke from the crematorium, barbed wire, and groans from the gas chamber. That is how it is day after day, week after week, night after night. You are awaiting your turn. Look at the line of trucks filled with women and children, look how they are bidding farewell to their companions in misfortune. The departure looks like a trip, but beyond the turn there is nothingness. Nothingness is preceded by death, and is preceded by monstrous torture, by rupturing one's lungs from pain in the suffocation chambers. There is awareness of this agony which will come in a moment, and the most monstrous awareness that one's own children will go through this agony. A mother's awareness that she is powerless while her child is being tortured. That is Majdanek!

In your bedrooms you have seen children with large, blue eyes and dilated pupils, when they say nothing, do not cry, and only look at you with courage and faith. Now imagine the same blue eyes, the same extended little arms, and the same darkness—but in gas chambers. The same gaze of a mother, but with the breath of blood from ruptured lungs. This is Majdanek.

There is a slate half written with words, and a slate pencil. Perhaps the chatter of a child in this barrack has succeeded in conjuring up a smile on starvation-swollen faces. The writing on the slate breaks off with the words: Dear Mommy . . . Do you feel that the mother's heart must have burst before she went to the gas chamber?

Before my eyes stood a five-year-old boy who was holding his three-year-old brother by the hand. He stood on the square when shots were being fired, and he was comforting his little brother with these words: "Don't worry, it doesn't hurt much and it won't last long." I saw how he wiped the tears on the little boy's cheeks and I saw how, a few minutes later, these tears mixed with a thin streak of warm blood. Forgive me, my child, my helplessness, the helplessness of human good will. There were hundreds of such heroes in Majdanek. Who will count them, who will pay tribute to them?

Allow me yet to recall the tortures undergone by those whose necks were bent and whose heads were held by spiked boots under water. Do you feel the gurgle of the last breath and the tension of a man who wants to live and is constantly being pushed into water? Do you see the cut-off fingers of Soviet prisoners'-of-war clinging to the frames of trucks from which these prisoners were being tossed into death pits?

There is still another perspective on Majdanek.

Many thousands of German bookkeepers, financiers, cashiers, warehousemen, railroad men, mailmen, telephone operators, engineers, physicians, jurists, agronomists, chemists, and pharmacists—they all were organized into one big machine bringing death to their defenseless fellowmen. Do you feel the monstrosity of the aggregate of people possessed by the idea of destroying others in the fastest, cheapest, and most efficient manner? This is Majdanek!

And yet another perspective on Majdanek.

Something the Germans have brought to perfection and can be modeled on is their talent for being systematic, organized, and thrifty. These three greatest German virtues were put in the service of the greatest evil. Can you comprehend the crushing ruthlessness and monstrous efficiency of this sinister machine? This is Majdanek!

There is yet one more monstrosity of Majdanek. During the day, there is beating, torturing, gashing, and cremating, and at night this German goes back to his wife. Outside the shadow of death, there is a brightly illuminated barrack, and inside there is a table covered with a white cloth, a phonograph, and a piano. The German washes the blood off his hands and takes home little shoes pulled off a child while it was still convulsing in agony. Or it is Christmas; under the Christmas tree lie the dresses which belonged to gassed victims, medallions torn from the necks of murdered persons and into which the German now places his photographs.

Now I must speak of the penalty for the accused?

Are you surprised that from Stalingrad to the Pyrenees and from Italy to Norway a voice is resounding calling for gallows for the traitors? Something unusual must have happened if there is no peace in the European consciousness, if people cannot return to their normal business until a forest of gallows springs up.

If the accused could understand the language in which I speak, I would wish them that at the moment when the loop tightens around their necks, when they are pulled up the gallows, when their bodies stretch in the last motion, when they breathe their last breath, and when their eyes become fixed—that the glaze of these eyes should reflect the image of that mother in the crematorium who snuggled her child to her very last moment, that they should see the relentlessly flowing tears of that woman. That they should see her in the instant they are hanged and in the instant they die. That they should see that mother's tears when they are put in their graves and when they decompose into dust for eternity.

The destruction of Poles occurred not only in camps. Now and again, hundreds of persons were shot to death in Pawiak; people were caught on streets and arrested indiscriminately. Whole villages were surrounded, and all residents were deported and killed. Entire trainloads of Polish children were deported. The catchword was that where the sword of the German soldier has reached, there the plow of the German farmers should plow.

This is how the Germans were acquiring their living space.

Two years later, I stood by the gallows when the culprits were being hanged. Their faces were so degenerate and base that one could feel no humanness in them. Many people stood by and watched how their bodies were suspended, and each had only one thought: that death was too small a punishment for them. I was thinking that in a normal era, such people would have killed a few persons and raped a few women, and we would have learned about it from a newspaper report. The crime was that these degenerate gangsters were given, by the German people, the right to decide the life and death of entire nations. And when I sought to discern who was guilty of these hideous crimes, I had unfortunately to come to the conclusion to which all others who saw such crimes have also arrived:

A nation capable of idolizing criminals and giving them power over itself and the rest of Europe is accomplice to the crime.

Chapter Thirty-Five

The Last Upsurge of a Perishing Nation

Editors' and translator's note: This chapter was based on a brochure entitled *Na Oczach Świata* [Before the Eyes of the World]. It has not been included in this translation because of the circumstances as described below, which was in an editor's note from the Polish edition of Ludwik Hirszfeld's *The Story of One Life* (Warsaw: Czytelnik, 2000), 453:

Professor Ludwik Hirszfeld did not have accurate information as to the author of this brochure and ascribed it to the Jewish resistance. Understandably, while he was writing his memoirs, he did not have the means—nor did he probably see the necessity—to verify his sources. It was only in 1950 that the question was elucidated thanks to the Society of Authors, Composers and Publishers as well as the Jewish Historical Institute. The author of this brochure, which was written for the Propaganda Commission of the Warsaw Division of the Home Army, is Maria Kann. This fact was further confirmed by the author herself as well as Dr. Aleksander Kamiński, who, as director of the Propaganda Commission, had requested her to write the brochure.

Chapter Thirty-Six

A Chased Animal

Toward the end of February 1943, we moved from Kamienna to Miłosna to live in Mrs. Kenig's little house. Her property consisted of a garden and a small pine forest totaling some fourteen acres. We were free to move about within this small property. Sometimes, we would go to the nearby forest or walk over the surrounding meadow, but it was usually in the evening so as not to attract the neighbors' attention. We played the role of an elderly couple who, due to illness, had to retire to the countryside. Except for our closest relatives no one knew the place of our residence which, incidentally, was not safe at all. The three ladies who lived at Mrs. Kenig's prior to us were denounced by a seventeen-year-old boy, and the son of a judge, and they were killed. In Wiśniowa Góra, a village nearby, the German police happened upon a man and, having checked that he was Jewish, killed him on the spot. On the highway, several Polish women who seemed suspicious to the Germans were killed. Thus, while we kept the appearance of a peaceful life, we actually lived on a volcano. In our private life, though, we felt no distress due to Mrs. Kenig's tactfulness and kindness.

I immediately engrossed myself in work. I wrote several chapters of my textbook and translated them into English. While translating, I had the vision of a free America to which I had been invited, and the awareness that I had something to say to the world alleviated my painful feeling of being a chased animal. I would work from eight to twelve in the morning. Then I would chop wood for an hour, and in the afternoon I would read, mainly in English. In June 1943, my wife gave me the idea of writing down what we and Poland had been going through. I decided to go a bit further back and describe the breakdown of science and scientists under the influence of fascism. The first chapter I wrote dealt with the 1935 congresses in Rome. Those closest to me liked this chapter, and so I decided to expand my reminiscences and write a memoir of my whole life. I began to write in mid-June. Sometimes I rose at four or five in the morning, wrote a few pages, and then went back to bed. Later, I dictated to my wife who typed, and in the afternoon I outlined the content of subsequent chapters; thus, I kept writing in fierce engrossment. The whole book was written within two months, with the exception of the chapter that deals with the period after we left Miłosna. On Sundays, my friends and nephews would come, and I would read them what I had written during the week. They were my critics. When they were shocked and tears came to their eyes, I knew that I succeeded in presenting the atmosphere of despair which is the actual content of this book. While walking in the garden, my wife and I sometimes pondered whether it was good that we had not gone to America and why, for what sins we have been so terribly punished. We would contemplate that a higher destiny bid us to partake of the suffering so that we could later tell it to mankind, "You—they will believe," my wife would say. The idea that my anguish had meaning and that it was my duty and my mission to recount the suffering of people who lived in the "life space" claimed by the Germans as their own kept up my spirits. This was the reason I did not break down. Writing this book enabled me to prevail in yet another way. The thought that I was far removed from the battle against the enemy depressed me. I did not feel physically fit to go to the woods and participate in active combat. It was impossible for me to participate in underground teaching, because I might call attention to my person and imperil many others. At the same time, writing this book was a fight against the enemy, and it was a dangerous fight. I was writing this book in a little house, while in the garden around it German soldiers, SS men, and Vlasov men[1] were doing their exercises. The tapping of the typewriter could have called their attention, and one glance at the text would have immediately given us away. How often Zosia had to stand guard and watch whether the Germans were coming too close; if they did, we hid the typewriter and the manuscript.

Except for our closest relatives and friends, no one visited us. I did not give my address to my colleagues for the same reason I did not ask them for help. The idea that they could subconsciously feel superiority with respect to or pity for a banned man was unbearable. I only sent them my scientific manuscript for safekeeping because I considered this their professional duty and not a personal favor. I did not send them the manuscript of this memoir in order not to expose them to risk. The peril of transporting and hiding several copies of this manuscript was assumed by those closest to me. Still, it was very hard to live so lonely a life without the exchange of ideas and without dramatic tensions, seemingly dissociated from life.

It was during that period that Róża Amzel was killed. At first she hid in Kazimierz. But she had to flee from there and went to live with my close friends in our house in Saska Kępa. One day, instructed by the Germans, the villa commissar accused my friend of hiding me and told him that he was going to occupy the house for himself. Amzel and her mother had to leave the house and, after various ups and downs, were arrested in a lodge and killed. I asked several of my colleagues to help her; however, the burden of care fell not on them, but on Dr. Kiełbasinski.[2] When she was arrested I tried to save her. At that time, a certain Sobkowski was prowling in Warsaw; he was a *Volksdeutsch* and later a *Reichsdeutsch* who freed prisoners for money. Through a friend I forwarded him the money he required, an amount which was rather large even for that time. He did not lift a finger and even stopped answering the phone after he received the money. I hope that the hand of justice will not miss him. Amzel's death was a great blow to me.

The year 1943 passed. I would have broken down after my daughter's death if it had not been for this book. In March 1944, our relatively peaceful stay in Miłosna came to an end. Rumors were being spread in the neighborhood that Mrs. Kenig was hiding Jews, and such rumors usually provoked an arrest. We had to leave. It would have been suspicious if we went to another rural area, therefore, we decided to hide in Warsaw and live there under house arrest. We found room and board at 39 Złota Street; this time we paid big money as did all Jews in hiding. The risk of hiding Jews was included in the price. Thus, I was not spared this stretch of Golgotha either: as a nonliving entity to live in hiding among living men, to see no sun but at best illuminated walls, and to experience the condition of a man chased like an animal. Our closest friends tried to alleviate our loneliness. At dusk, my wife, accompanied by Zosia, would go out, and I was never sure whether she would come back. I would go out only once in a great while and would roam through the ghetto gravesite; a few times I went near the Saxony Garden. These walks were unbearable: I felt like a cadaver walking among living beings.

Periodically I would learn of the tragic death of someone I knew. Each case involved a special tragedy. Blackmailers came to the engineer Heyman's apartment. He asked them politely to be seated and said: "I will not give you such satisfaction." Before they could catch him, he had swallowed poison. Dr. Frenkiel, chairman of the Łódź Medical Society, died in a similar way. Professor Centnerszwer, my brother-in-law Professor Władysław Sterling,[3] and his wife, my sister, were killed in the following manner. At first, neighbors began to whisper that Jews were hiding in the house endangering all the tenants. Then, armed bandits came, robbed the apartment, and shot at the tenants killing my sister and my brother-in-law. Upon noticing the robbers, the neighbors called the German police. The gendarmes came, checked the bandits' identification cards, and let them go. It turned out that they were bandits serving the Gestapo and they later shared the booty with the German police. Something similar happened to Professor Centnerszwer. His Aryan wife in whose apartment he was hiding was summoned to the Gestapo. She was threatened with the death sentence for hiding a Jew, even if he was her husband. She tried to explain that her husband had arrived only three days earlier. The German officer was courteous: "Madam, you could have hidden him even for six days." They robbed her completely. To characterize the intellectual level of these murdered victims, I will mention that a few days prior to his death Professor Centnerszwer handed the manuscript of his latest scientific treatise over to a friend of his, and that to the last moment of his life Professor Sterling was translating poems by Verlaine and Baudelaire. The Polish underground was combating spies, blackmailers, and informers, and whoever was proved guilty was punished by death. The case of Professor Sterling was also handled by the court of underground Poland; still, this will bring neither him nor my sister back to life.

Life in Warsaw was a nightmare. The underground movement was growing from day to day. Hate for the occupier was manifested in assaults on the German thugs. They in turn would close off a street, catch the pedestrians, and sentence them to death. One could see processions of victims clad in white paper clothes. Their mouths were probably plastered, because they made no noise. They were shot to death on the streets. At first their names were announced publicly, later they died nameless. One could smell terror and the stench of blood in the air, and one walked over places wet with the victims' blood. Every once in awhile we heard that all Pawiak prisoners had been deported or killed, and that large-scale arrests and executions were going on throughout Poland; the names of murdered persons were forwarded to their friends. This Gehenna lasted for years and grew worse toward the end of the occupation. If I do not write at length about these matters, this is because there is not a spot in Poland that did not go through the same Gehenna. It is not necessary to write about something everybody knows. This must be assessed in the form of facts and figures, and this duty rests not with me but with the historians of the era.[4] Therefore, I will not describe the martyrdom of the Lublin region, the deportation of whole villages, and the shooting of whole families. My duty is to describe the tragedy of scientists.

Lwów went through the most horrid tragedy.[5] Several days after the occupation of Lwów, the Germans arrested all professors of medicine proceeding on the basis of a list of names which, as I have heard, had been prepared by Ukrainians. Almost all professors in the medical school were killed, some with their families and servants. They include Professor Progulski, Professor Nowicki with wife and son, Professor Grek with wife, Professor Rencki, Professor Ostrowski with wife and accidental house tenants, Professor Dobrzaniecki, Professor Hilarowicz, Professor Grzedzielski, Professor Maczewski, the eighty-year-old Professor Orlowy, Boy-Żelenski, Longchamps with three sons, Professor Piwarski, and Professor Pieszyński. From the Technical Institute: Professors Bartel, Witkiewicz, Kasper, Weigel, Stozek with two sons, and Łomniski. I have held in my own hand the list of professors who were either killed or kept in concentration camps. Their number was stated by the underground press and, if I remember correctly, it was 240. This was the implementation of the Hitlerites' most terrible and cynical plan: the plan to exterminate the entire Polish intelligentsia. They did not succeed in exterminating it in full, but the wounds they inflicted are so terrible that, unfortunately, they will not be able to heal. No one of us doubted that after the war, even in case of victory, we would have to rebuild Polish culture from scratch. This deliberate destruction of the intelligentsia, done in cold blood in accordance with a mad concept, is the German nation's greatest sin which can never be forgiven.

There is another fact which proves the self-degradation of German scientists. Performing tests on human subjects is basically unavoidable, because every new medicine and every new operation must be tested on someone for the first time. This is normally done with great caution in clinics or hospitals under strict control of the best experts. Furthermore, these tests are done for the purpose of introducing new medical treatments. I believe that this was the first time in medical history that experiments were done on captured women belonging to other nations; experiments which were done with the objective of solving abstract problems and which had no direct connection with treating the woman in question. I had in my possession an underground report from Auschwitz stating the operations performed on women in the camp—with a degree of precision allowing the expert to understand the nature of the problem. The objective of some experiments was to find morphological changes in the chromosomes of reproductive cells; the ovaries were systematically removed in various periods of the cycle. The names of all physicians performing these operations were stated. I copied them all, as well as descriptions of other operations performed on Polish women. Unfortunately, these papers were lost during the insurrection in Warsaw. I can cite other facts. Professor Supniewski,[6] dean of the Medical School in Cracow, told me that one professor who was in the Oranienburg camp had an abscess on his buttock and went to see the physician to have it cut out. However, the treatment proved to be simpler: the physician kicked him in the rear and the abscess burst. In Majdanek and in a number of other concentration camps, weak persons were gassed. It was the physician's task to select persons for death. Is not the physician who, even under compulsion, was performing such a function as accomplice? I will not elaborate on the abuse of the Red Cross symbol. Treblinka supposedly had a dispensary over which the Red Cross flag was flying. Whoever went there was struck dead behind the curtain with a blow on the head.

I am unable to judge whether these horrible facts that prove the downfall of medical ethics were known in Germany. We can judge the German nation on the basis of what we experienced. The hideous crimes should have incensed the German physicians who were in Poland to some kind of active expiation or desire to aid. Very few such facts are known. Dr. Zelch in Wilno supposedly conducted himself decently; I even heard of a letter he received from the famous pediatrician Pfaundler with thanks that he was trying to alleviate the wrongs done to Poles. Warsaw's municipal physician Dr. Hagen presumably protested against the deportation of Polish children for which he was sent to the front. Professor Kudicke at the State Institute of Hygiene in Warsaw maintained a proper attitude toward the institute's members. But all this was a drop in the sea of infamy. I personally must say that when the Germans decided to kill everyone in the district, none of the German scientists and physicians who were in Warsaw, not even those who knew exactly who I was, warned me or offered me the slightest help. Among them were Kudicke, Wohlrab, and Laves who visited me in the district hospital and talked with me about science. Even Kohmann, who offered me assistance at the beginning of the occupation, did not lift a finger when death was staring me in the eyes. I do not want to say that Germans are incapable of friendship with respect to people of other nationalities. I even have evidence that some of them are capable of such friendship, but for understandable reasons I still cannot yet write about this. I am only making a general statement that those German scientists who were in Poland did nothing to alleviate the fate if not of all then at least of Polish scientists. Our account with them is crystal clear. We owe them nothing, absolutely nothing.

Only the Dungerns gave us from afar proof of their friendship and assistance, and their luminous personalities have remained immaculate.

I will return to our everyday life. As mentioned before, we went out very seldom and usually at dusk. I know persons who did not open their door for several years. This was so because the general situation gave rise to two professions in the service of evil: spies and blackmailers. Spies included both Poles and Jews in the service of the Gestapo. The worst were the blackmailers. They could wring the last penny out of their victims. The work of the National-Radical Camp was not wasted either; I was told facts about denunciations of socialists and Jews. I know the case of a young boy who committed suicide after he learned what services were demanded of him by the party. Again we found this mixture of nobility and wickedness, which is typical of modern society; children chasing a small Jewish child shouting "Yid," and a woman, a complete stranger, who takes this child by the hand and saves it. I know facts about denouncing diseased Jews who were hospitalized for treatment. It would take volumes to enumerate all instances of this lurid mixture of the noblest and basest impulses.

While writing about the fate of scientists, I must mention Professor Kopeć who was killed as early as 1940. He had a brilliant and subtle mind and was a scientist of great measure. I should like to pay tribute to Kazimierz Pelczar, a professor from Wilno;[7] he and I jointly worked out the statute of the Anticancer Society. I should also like to pay tribute to the memory of Docent Rajchman,[8] a pure heart incapable of compromise. They were shot. When I think about these pure souls, people who kept working till their very last breath, and then about German scientists who at best did not agree with the regime but did nothing to fight it—a gulf looms before me.

Minister Dr. Chodźko had been Poland's delegate to the [International] Office of Hygiene.[9] During the German occupation of France, Professor Reiter, formerly Germany's delegate to that office, was the German commis-

sar in France. When the director of the Paris Office of Hygiene asked him
to take care of Dr. Chodźko in Warsaw, he received the following answer:
"The Poles are condemned to death. I do not deal with the fate of condemned men."

Reiter was chairman of the Reichsgesundheitsamt and a professor at
Berlin University!

It may not have been easier for unknown persons to hide. Unfortunately,
I was well known from lectures and my wife from her medical practice. Our
landlady, for example, told us that she used to go to Professor Hirszfeld's
lectures and loved them. She did not recognize me, however. Suffering had
evidently changed me. When I went out, I met acquaintances. We would
take a walk at dusk. I always went out with my briefcase in hand, pretending
to be a working man. Life was moving forward somewhere far away from us.
I would walk through the district to see the presbytery where we once lived,
on Grzybkowski Square, and to Leszno Street where the sessions of the
Health Council had taken place. On the surface life was going on as before.
At dusk, people moved like shadows. There were the same stores, a butcher
here, a shoemaker there, and people were performing the same functions,
had the same feelings, and were just as attached to their workshops as the
others had been, those who lived here previously but had been exterminated. No one spoke of those others. They disappeared from the earth's
surface so radically that there was no one left to weep for them and remind
the current owners that they had not inherited shops but tears and suffering
so immense that they could never crush them. These people lived on a cemetery. For me, every store, every stone, and every street corner bore traces of
tears and blood. What constitutes a wailing wall for a homeland lost? The
loss of their country condemned Jews to wandering. Now, they had all but
disappeared from the surface of the earth. When one portion of a nation
survives, it builds a legend around those who have died. Then, they live on
in poems, songs, and memories. I realize that I am unable to be the trouba-
dour of their life or of their anguish and death, but when I walked with my
wife through the district, it seemed to us that the shadows of the dead were
filling the streets. And those shadows were closer to us than living men,
because these living people did not feel the anguish they had experienced
during the nightmarish past.

They experienced it later. The anguish of the premature death of those
dear to their hearts, and the anguish of a ruined workshop. Now that I am
writing these words, the whole of Warsaw is one huge cemetery. This is the
manifestation of the eternal law that evil spreads all over. There are laws of
living together which require goodwill toward all and which are defined by
the Christian ethic as the maxim of love for one's fellow man. Without this
law of living together, no society can exist. Some believed that it was possible
to murder Jews without simultaneously killing the great idea of love for one's

fellow man. The example of Warsaw, a cemetery of people and ideas, proves that this is impossible.

After a while, I gave up my evening walks: I could stand neither the sight of life nor the memory of death. I tried to work, but it was difficult to muster mental vigor. In this atmosphere of horror and panic, with the feeling of a chased animal, and torn away from nature and everything that gives life a certain charm, I lived for three months. With the greatest effort, I tried to set up some forms of everyday life to direct my attention on something else and give myself an injection of the will to live. I began to dictate lectures in English to my wife. While dictating, I would imagine that I had succeeded in regaining my freedom, that I was in America, and that I was telling the Americans about the suffering of people and summoning them to retaliation, to boycott German scientists. I imagined people with eyes fixed on me listening with attention, and thus I was forcing myself to expel the terrible reality out of my mind. I could not keep my English lectures because they would have betrayed me in the case of an inspection. After they were typed, I burned them immediately.

Those deliberate hallucinations and the immobility of house arrest undermined my health. The two exercises which I could afford—several hours of polishing the floor of the entire apartment and running around the table—did not help. I lost all vigor and felt that my body and mind were withering. I do not know what the end of my stay in Warsaw would have been if friendly people had not invited us to the countryside and helped us find an apartment there.

Some thirty-odd kilometers outside of Warsaw, next to Tłuszcz, there is a place called Klembów which is surrounded by beautiful forests. Nearby is the village of Lipka where my wife and I lodged in the house of Stanisław Kaflik. I introduced myself as a disinfection clerk, and at first I was coming for Sundays, and later when people got used to us, I said that I had gotten a leave, and we stayed there permanently. This place offered me the opportunity to get to know another social stratum. I had spent one-and-a-half years in the Jewish district where I learned misery and the highlights and shadows of Jewish mentality. I spent half a year on the estate of nobility, one year at Mrs. Kenig's in a suburb of Warsaw, and finally three months in the city where I learned the feeling of being a chased animal. Now in the countryside, I had the opportunity to learn the attitude of the Polish peasant toward the current events and the tragedy of Jews hiding in rural areas.

Leaving the stifling air of the city and house arrest, we came to ourselves in the true Polish countryside. In front of our house, grain fields, spotted here and there with islands of trees, were swaying in the breeze. Nearby stood a dense leafy forest which farther on gradually changed into an evergreen forest. Beautiful 100-year-old oaks, hornbeams, and ashes shimmered with all kinds of autumn leaves. The most delightful thing in the forest was the singing of birds, which was so vigorous and frantic with joy that we walked into the forest to draw a vitality that was alien to modern man.

We had to modify our daily routine. The dictating of English lectures, the writing of books, and even excessive reading did not correspond to the modest personality of a disinfection clerk. I played the role of a tired elderly gentleman who was helping his wife in household chores and was not used to mental work. Only when I noticed that people got accustomed to us and no one had any suspicions as to our true position were we able to afford more intensive work. Keeping up with the role of an elderly gentleman tired of life, I chatted about various things and approached the local population trying to learn its attitude toward political and social problems and attempting to get to know the psyche of the Polish peasant during occupation. Again I had a chance to see the whole wide span of the Polish mentality from lofty, disinterested impulses to an abyss of greediness and blackmail.

Our host, Stanisław Kaflik, had for a dozen years or so been a runner at the Ministry of Treasury, saved some money, and built himself a neat little house. I still remember this man with the feeling of deepest respect. He deserved admiration not only because of his humane attitude toward the world and complete honesty but also lack of any pettiness. This man did not have a trace of the so-called peasant greediness. He was filled with love for work and his family and a subtlety and refinement of feelings worthy of social summits. The Kafliks were very upset when they told me about the killing of Jews: "They were people too and often good people. We went to see them like our own people. They often gave us things on credit; even lent us a few bucks. Some of their children were like little angels. To think that these little children were killed like ducks. Where is the German culture?" This was the spirit in which these decent people talked, and they did not know how my heart was filling with inner warmth that this was the voice of the simple Polish man. When we talk about the tragedy of Jews in hiding who found no shelter, we forget about those millions of Poles who had no opportunity to manifest their attitude toward the crime. The overall picture that the persecuted Jews must have formed was quite different. The Klembowski forests are the first woods outside Warsaw in the direction of Treblinka. Through them led the Jews' torturous road from Warsaw to the end of their road. When trains went by, ten or more Jews would jump off. Those packed in the cars knew that they were going to their death and that neither gold nor dollars would help them. The few who decided to jump off often received a fortune from the others to help them survive. When the bravadoes fell on the tracks, they often broke their legs or were hit by bullets fired by the escorting guards. As a rule, the train would not stop because of such a trifle as the flight of a dozen convicts; this merchandise was aplenty and it was not counted. These men roamed the forests, and many of the wounded died there. Some begged the local population for aid. This was no trifle: the death penalty was being imposed for hiding or giving aid to Jews. Nevertheless, some people hid those Jews or would give them a plate of food. Yet, these manifestations of a good heart cannot conceal another picture: the fugitives were also prey for many local inhabitants. They would go hunting for Jews to rob them of dollars and valuables. In a ringing frost, they were often undressed bare, or coats and suits were pulled off dying men. I know of a case where a man with a good heart asked those hyenas to wait with the fleecing of clothes at least until the dying victim was dead. I was told about a young woman whose legs had been cut off by the train. She was dying on the tracks from loss of blood and from cold when the hyenas showed up and stripped her bare. Local Jews were also hiding in the Klembowski Forests (one survived for many months), and their Polish friends would bring them food. Eventually, someone would bring German gendarmes, and the latter would kill the victims with one grenade. The price for delivering a Jew was small: his clothes or some vodka. Some Jews were hiding under assumed names. It was not difficult to get forged documents. As a rule, the government of underground Poland would give them to any Jew without charge. One Mr. D., a well-known local drunkard, and his comrades, first blackmailed the Jews and then denounced them. Once, the German gendarmes came and found only the wife and the child, unfortunately circumcised. The mother conducted herself heroically before her death. Her mind was made up: to protect the good people who had been hiding her. Thus she assured her persecutors that the peasants knew nothing about her background. She took her little son by his hand and said: "Come, Stefanek, don't be afraid. It does not hurt at all. These gentlemen are good, they don't want a child to cry." Two shots were fired, and gone were the mother and Stefanek. One Jew from Wołomin told me that his barber friend had been hiding him for two years. Nevertheless, he did not think well of the local people: "For them, this was a game and laughter. Many Jews would have gone unnoticed if it were not for the local scum." I saw that the Jews in hiding were looking at the stand of the people from a perspective that was understandable but incorrect. A goodhearted man could save at best one or a few Jews and that not for long, especially when he had no means. One evil man, a blackmailer or informer, could destroy thousands. Therefore, I believe that we must not generalize the fact and we must not rely on the stories told by victims. Out of the 3.5 million Polish Jews, only several tens of thousands survived. Those survivors will, unfortunately, have a false picture of the whole when they think about the mob that persecuted them and robbed them of their last shirt. Warsaw residents who had to hide in rural areas after the uprising went through a similar Golgotha. The Polish mob has always and everywhere indulged in plunder. It was in a certain historical moment that Jews were their only victims. It is obvious that this inhuman attitude was facilitated by their worldview. But we must not condemn all society. Many of these blackmailers were later killed by the government of underground Poland and even by local people. The guilt rests not with Polish anti-Semitism but with the German Government, which placed a certain category of people outside the law. I was told the story of a gendarme in Tłuszcz who boasted that he had killed 150 Jews with his own hands. Once, he noticed two children holding each other by the hand: a little girl and her little brother. He aimed—then he lowered his gun. He gave it to his aide. He too was unable to shoot. This was finally done by a third German. Local people told me this story with tears in their eyes. They showed me the nameless graves of people who were killed without guilt or need, were wept for by no one, and yet were pitied by the local people.

Thus we went on living in contact with nature. From afar, news reached us about an approaching storm and about attempts to shape one's life in spite of the ghastly violence. The Germans still represented a power against which Poland could hoist only the maxims of humanitarianism, the right to live, and the idea of good. The fight waged by the Polish underground could do the Germans some harm but it could not defeat them. Still, cracks began to show on the edifice of German power. More and more frequently we heard about victorious fights by the Red Army. The front was approaching from month to month, and finally it rolled over our village and brought us back to life.

By the end of July 1944, we heard about the great victories of the Soviet Army which was approaching Lublin. Wilno, Białystok, and Lublin were recaptured from the Germans. The troops were moving on like a thunderstorm. On Sunday morning July 30, we were awakened by the rattle of tanks on the highways. We ran out and to our great surprise we saw Russian tanks proceeding in the direction of Wołomin. The tanks were moving on as if on review. The soldiers' faces were poignant as if aware of a great historical moment. We greeted them with joy. We believed that in this simple and non-dramatic manner we were finally gaining our freedom. Three tanks drove into our village, and the officer entered a hut to wash. Then a trifle reminded us that the war was still on: German planes appeared and bombed the village. The tanks withdrew, half of the village was burned, and a dozen or so persons were killed. Not only we were confused by the tanks. Everybody was counting on the imminent arrival of Soviet troops, and Warsaw broke out in an uprising which was one of the war's greatest calamities. The foremost squad of tanks was partly crushed. During the following few days, German troops retreating from the east were rolling through our village: at first it was Hungarian and Croatian divisions which were all in rags, and finally German divisions. As chance would have it, I even established direct contact with the German soldiers, because they would come to the peasants, and I and my wife had to play the role of interpreters. Conversation with us, elderly persons who had good mastery of the German language, evidently reminded them of their family for whom they were longing, because they would come to our house and talk with us for hours. It was strange that I, a man condemned by them, was an object of their interest and even of certain friendly feelings.

Some soldiers ruthlessly plundered the village, others asked only for milk and eggs, but still others would bring their own products and ask only to cook the meal, reciprocating with cigarettes and chocolate. Their longing for their families was striking; even after a brief conversation, they would show photographs of their wives and children. Even though they had been fleeing for months, they were not broken but firmly believed in German power and a reversal of the events. They told me that they had a new weapon ready and that they would drive the Russians back, even if the war had to last for five more years. There was a strange difference in their view on the events: they said nothing about the atrocities committed by Germans but spoke for hours about Russian atrocities, including alleged plucking of eyes. Once, three soldiers arrived to take cattle from the peasants, obviously without compensation. Their commander, a native of Vienna, told me how difficult it was for him to play this role. He displayed human feelings and complained about the atrocities of war. At a certain point he asked me: "But you certainly must appreciate that we have knocked out your Jews. The Jews are guilty of everything." I looked at the face of this young boy who had pity for the peasants and did not want to take their cows but spoke in cold blood about the murder of millions of Jews as if he were talking about millions of lice. And I saw the terrible crime committed by Hitlerism with its inhuman propaganda. All German soldiers were warning us that the Russians would come soon; they were telling us ghastly stories about Russian cruelty, and advised us to flee with them to the West.

In the meantime, the uprising broke out in Warsaw.[10] At night, we saw the glow of conflagration and heard cannon shots. In helpless pain we were looking at the death of the city and of persons dear to our hearts. The German soldiers told me: "Warsaw has been captured by bandits, but we will manage them. They will all be . . ." and they made a significant motion with their hands at their throats. Right now, I am not analyzing whether or not the uprising was necessary, I am interested only in the mentality of the German soldier and of the one and only nation that could muster contempt and hate on command. For these soldiers, the insurgent was not a hero but only a bandit. This term deprived them of the compassion and respect they had for the enemy-soldier. I wondered whether this was a national trait or a respect for power and the printed word. If it is veneration for the printed word that can cause such a disappearance of one's own sound judgment, then it would be better if no one could read and were forced to think independently about what is going on.

On August 10, the German authorities ordered our village to be evacuated. They ordered us to go westward in front of the army. No one thought for a moment of surrendering to this order. The Polish peasant does not abandon his land so easily. We decided to hide in the forests until both armies had rolled away. This was a very risky undertaking not only because of the death penalty for hiding but also because at a certain moment we would find ourselves right on the front line. We formed a camp composed of twenty-two persons: our hosts, their close relatives, and a few friends, including Professor Blaton with wife and family. We took with us only the most necessary food and buried our personal belongings in the garden. We had one wagon at our disposal, put our things on it, drove up to the tracks, turned off into the forest thicket, and burrowed ourselves. We spent six days in this shelter. However, within two days we were discovered by a German tank formation located in the vicinity. The officer took all our young men to work. We were all upset because, in spite of the officer's reassurance, we were not sure whether or not they would detain our men. At their request, I went as their interpreter and supervisor. My work squad was composed of a physics professor, a graduate student in physics, a college graduate, a gardener, and an organ player. On the way, I conversed with the officer who was a merchant from Hamburg. I was aware of the danger threatening our young boys, and I decided to impress the German. This maneuver might save them. So I told him that I personally knew Professor Nocht,[11] [former] director of the Hamburg Institute of Tropical Medicine and honorary citizen of Hamburg. I realized that with this statement I was partially betraying my identity and assuming a risk which seemed less important than the imminent peril for our whole camp. The Germans ordered ditches dug for the officers. This was the first time I had seen a German military field camp from such a close range. Commanders were sitting at a table, and every once in a while officers with reports arrived by cars. I learned that this was the front SS formation which was always directed into places exposed to special danger. Indeed, the officers and some soldiers looked like young robber knights: they had sprightly movements and hard, racy faces. A radio on the table was playing "Tosca." I heard a radio for the first time in five years, and it had to be in this atmosphere of combat trumpets. Suddenly, Russian planes appeared above us, and we heard shots and the resounding of falling bombs. No officer moved, they continued to listen to a beautiful aria. We took it as a point of honor, and did not hide either. After completing work, the officer released my boys. I uttered a brief thank you to which he replied: "Gentlemen, we should like to thank you." We went back to our camp with the feeling that the German military elite had better external conduct than the formations we had met so far. In our camp we learned that during our absence German soldiers had come from that formation and, threatening a nineteen-year-old girl with revolvers, took her watch away from her. Was it a coincidence or a principle? We will see later.

We thought that our services for the German camp were completed. Not so. The next day, the same officer came again and took six men, this time for the whole day. The work, to be sure, was not hard. Our men worked with the German soldiers who treated them in a humane way, offered them cigarettes, and gave them a good dinner. The German soldiers took pigs, cows, and poultry from the neighboring peasants; no wonder then that the casserole was excellent. The whole situation reminded me of the robber knights whose external culture certainly was higher than that of the merchants and peasants they robbed. This tradition evidently facilitated their way of living. However, the work for German soldiers did not suit us, all the more since they began to approach our women with unequivocal propositions. It took a great diplomatic skill on the part of my wife and myself to save the situation. Here again, the soldiers would come for hour-long conversations with me. I remember one miner from Saarbrücken and his friend, a German from Romania. They brought candy for our children and told us that immediately after the war they would quit military service, because they did not want to be gendarmes in conquered countries. I asked how they could explain Germany's defeat in the East. They said: "Treason by generals. But now that . . ."—they made a hand sweep around their necks—"these generals no longer exist, everything will be all right again. Hitler knows what he is doing."

We decided to leave our shelter, because of the Germans being too close by. The next morning at four o'clock we descended to the village of Pieńka and continued to Pieńki Górne. The road led through swamps and flooded rivers. We got into a barnyard and found a large number of refugees there. All around, German cannons were deployed facing east. Every once in a while guards would come and take people to work: felling trees and digging ditches; for women: washing underwear and peeling potatoes. The situation became severe, because the soldiers began to select mainly young and pretty girls. One mother succeeded in going to work in place of her daughter. Once, two soldiers came, holding a grenade in one hand and a revolver in the other: "Open all packages. There will be an inspection for the presence of weapons." One stood holding the revolver, while the other inspected. The inspection consisted of a ruthless robbing of suitcases, briefcases, fountain pens, and watches. They left with their booty, but immediately returned and addressed one lady: "You old woman, give us the watch you are hiding immediately." The lady protested that she had no watch. The German swung out to hit her. I could not bear this sight, so I intervened and said in my good German: "If you absolutely want to have a watch, I will give you mine, but leave this lady alone." "Give it," was his reply. I gave him my watch, but that did not help. Under the threat of battery, this lady still had to relinquish her watch. These robbers were members of the famous SS division called Viking. One of our women who had to peel potatoes for these soldiers saw how they showed the robbed objects to one another: silver spoons and watches. She heard how they considered how they should send these objects with the field mail so that they would not be noticed. Several weeks later I was walking through the freed Mińsk Mazowiceki and, for lack of my own watch, I had to ask other pedestrians the time. Without exception, I always got the same answer: "A German stole my watch." This was the real aspect of the chivalrous legend of the famous Viking Division. We had had enough and decided to leave Pieńki and again hide in the woods.

In the evening of August 16, we arrived in Pieńki Dólne. The village was deserted, and we saw only a few refugees. We heard cannon shots from afar. Suddenly, a platoon of soldiers came up from behind the river and, threatening with revolvers, they started catching people for work. The same thing happened all over again. I was serving as an interpreter. A young noncom started a conversation with me and, only a few moments later, told me how hard it was to perform the function of a hangman, that he had lost his mother and sister in Hamburg during air raids, and that he was trying to be tough. My wife asked him for some favors for our group, and they were given immediately. Again I saw this queer picture of people who individually were not bad but were performing the monstrous functions of catchers. The noncom left, and another one came up from the tank formation we had met in the forest. He told me that I reminded him of his grandfather. He was a Dutchman who ten years ago had immigrated to Germany and joined the SS. He was completely drunk, owing to which fact—perhaps—he had some personal charm. Swaying from leg to leg, he told us that he believed only his wife and Hitler. He showed us a photograph and began to philosophize on the matter that Hitler wanted the very best. When I asked him what would happen to the Polish people who had been captured, he told me frankly that they would not come back and that tomorrow at five in the morning Germans would come to drive all the remaining people westward. I blessed his drunkenness and when he left, I told everyone about the imminent danger. We all fled the next morning at four o'clock. This saved the lives of many of us. We proceeded toward the village Zamość and, before reaching it, we turned off into a thicket and were so well hidden that we were invisible even from the nearest road. We dug shelters out and decided to stay there until everything was over. A thunderstorm was coming, so we went to the nearby village for the night and returned to our burrow at dawn. Several elderly women, including my wife, stayed in the village to prepare a meal; I went with the men and children to the forest.

That day, August 18, was the turning point in my life. Two armies rolled over us, and a turnabout took place that brought me back to my former self. At first, the forest was filled with German soldiers and tanks. Our people in the village brought the news that the Germans would shoot anybody they found in the forest. "They must first find us," we said. Occasionally, German tanks would drive by, sometimes two steps away from our hideout. In the air, Russian planes were circling and dropping bombs on the valley. The forest could be attacked at any moment. We clearly saw the German artillery, which was standing in Górne Pieńki and shooting over our heads. The Russian artillery was giving increasingly louder replies. From time to time we heard the thud of wounded and falling trees. The missiles from Russian tanks and planes began to hit the nearby German tanks. We felt no fear whatsoever; we watched the battle at arm's reach, a battle that was to bring us freedom. At one moment, the Russian artillery started its play. It was neither thunder exploding from time to time nor the whistling of individual bullets, but a continuous roar. We felt as if an avalanche of iron was rolling over us. Trees were crashing, and grenades exploding all around us. We hid in our very shallow shelters; my foster son and my wife's niece were with me.

The roar was coming from the *katyushas*. This hurricane lasted four hours. At about four in the afternoon, the cannonade suddenly broke off. Soviet soldiers broke into our thicket and feverishly began to dig a shelter for themselves. There was a short conversation by phone and the squad disappeared. After a while, a German soldier with a revolver broke in and asked where the Bolsheviks were. I was overcome by a strange feeling of gratification, and I replied with an irony the taste of which I thoroughly enjoyed: "Just about everywhere." The German took off, and from afar I heard the calls of the German commanders. Some fifteen minutes later, I saw Soviet troops pouring by, no longer individual soldiers but tens and hundreds of them. Their eyes were full of eagerness and enthusiasm. I did not see a trace of alcoholic intoxication, contrary to what the Germans had told me. This was the first time during the war that I saw soldiers intoxicated with victory and shouting: "Victory!" In that moment I forgot my personal anguish and troubles and with the whole power of my heart I felt that a dawn was coming for our country and that the German might was breaking down. That I would again be myself and that the period of living like a chased animal was ending. Overcome with joy, I turned to a Soviet officer and said: "Thank you, sir."

With this I will conclude my memories of the war. I will not describe the reception I was given by my friends. I will just mention that I was overcome by a feeling of warmth which is felt by a person returning to his native country. I see a symbol of this return in the fact that within several weeks after we had assumed our regular identities both my wife and I had been appointed professors in the newly established Maria Curie-Skłodowska University in Lublin, that my colleagues had elected me as the university's assistant rector, and that I again had been ordained to rebuild the country, remove the cinders, and cultivate human souls. I will not write about this, because once again my time and my heart are filled with action. Still, the moment may come when I will want to look back in concentration on what I have gone through and to glance at the edifice on whose construction it was my destiny to participate. I will describe this when I am able to entitle the second volume of my memoirs *W Poszukiwaniu Wiernych* [In Search of the Faithful].[12]

I did not write this book with the intention of writing a novel. It never occurred to me to draw entertainment from my pain and that of an entire nation. I wrote this book with my heart's blood in order to fight for certain maxims and ideas. I am fighting against extreme nationalism, and I am fighting for the proper role of scientists—the role of promoters and champions of truth. After I had written the story of my life, the role of a man of letters became remote and alien to me, and I again felt the desire to be a scientist who—for better or for worse—formulates questions. The question of nationalism reached its climax and its most tragic solution of the Jewish problem. The breakdown of modern science and the subordination of science, which is a search of truth, to the goals of a political state bent on conquests culminating in the German theories of racism. I should like to devote the last two chapters of this book to these two issues.

Lublin, November 1944

Chapter Thirty-Seven

The Turning Point for the Jewish Nation

My destiny has been curious. I came from a completely assimilated family and I lived among progressive Poles and assimilated Jews without feeling essential differences among them. Then I lived one-and-one-half years in the world's greatest concentration of Jews, during a period when they were forced to create something like their own state. Next, as a native Aryan, I lived in an aristocratic, conservative estate. Problems previously alien to me and which I had underestimated struck me with full force. Before that it had seemed to me that the problem of Jews in Eastern Europe was similar to that of Jews in Western Europe. Western Jews wanted to fuse with the nations in which they lived; they did not want to form a separate nation, but only a religious community. Indeed, in the West, the process of assimilation had progressed at a rapid rate. However, it took the French Revolution to free the Jews from ghettos and make them citizens with equal rights. In Prussia, there were only 4.5 percent mixed marriages in 1876, whereas out of 103,000 marriages among Jews, 33,000 [or approximately 33 percent] were mixed in 1900–25. During the past thirty years, 100,000 German Jews—or 20 percent of the total—left the Jewish community. If we consider that Jews belonging to the richer urban classes have a smaller natural growth, it becomes clear that Western Jews were on the road to complete fusion with the Germans. The French and American power of assimilation was so great that hardly anybody belonging to the Jewish community would apprehend that this compassion for the fate of Eastern Jews could be interpreted as Jewish patriotism. Besides, only American and English Jews displayed this compassion. The German Jews rather disliked Eastern Jews, fearing that their influx might hold back the assimilation process.

In the district, however, I became convinced that Jewish patriotism existed. And, where there is patriotism there is a nation. It would be contrary to the concept of justice and freedom to forbid a certain community to consider itself a nation. The assertion that it was only the persecution that converted the feelings of religious unity into the feelings of national unity, and that the Jews are not a nation because they have neither a language nor a country of their own, is incorrect. Persecution means a common fate and destiny, and destiny does not have to, but it may create a nation. Western

Jews did not want to be a nation and were not. Eastern Jews wanted to be and were. The masses regarded themselves as a nation with a separate destiny and separate properties. The fact that they relinquished their elite to the nations in which they lived has no bearing on the problem. The lower strata, those primeval, surging, rapidly multiplying, longing, and persecuted strata felt that they constituted a separate nation and, moreover, the Poles too regarded them as a separate nation. What was the origin of this recently growing feeling of mutual strangeness?

Jews came to Poland in large numbers during the fourteenth century. They came from Germany and brought with them a special kind of denatured German language, a jargon. In Poland, they were granted autonomy for their cultural and religious affairs; their occupations were crafts, trade, usury, and keeping inns in rural areas; and they lived in a complete social isolation. In cities, they lived in open ghettos, which were quite unlike the districts established by the Germans during the occupation. It was the freedom they enjoyed, the right to observe their customs and religious mandates that enabled them to develop independently of their surroundings. The Jewish religious regulations had been created for the purpose of making them culturally distinct from others and enabling them to survive as Jews in spite of the fact that they had no country of their own. Special holidays, kosher kitchens, ritual slaughters, different clothes, circumcision, different architecture of shrines, and dissimilar religious customs created a clear-cut boundary between Jews and Poles. Most Jews lived in cities. The bourgeois German culture was closer to them than the aristocratic Polish culture. They liked to study and hold offices in Germany, they frequented German spas, and consulted German physicians, so that one can even say today: the Germans have been the Jews' unhappy love.

Toward the end of the nineteenth century, masses of Russian Jews banned by the Tsar, the so-called Litvaks, started coming to Poland. These Jews were imbued with Russian culture. They were indifferent toward the Poles' struggle for their own Polish culture. The assimilated Polish Jews, however, participated in the Polish underground struggle and in the building of Polish science and literature, and they disliked the Russophile attitude of the Litvaks. At that time, progressive Polish spheres knew no anti-Semitism. This was manifested in belletristic literature: *Jankiel* by Mickiewicz, *Meir Ezofowicz* and *Eli Makower* by Orzeszkowa, or *Chana Rubin* by Świętochowski.[1] The assimilated Jews' contribution to Polish culture was considerable. At the same time, Poland housed masses of fanatics who paid tribute to [*sic*] the worst prejudices, dressed differently, ate differently, and spoke differently from the people around them, and for whom Polish culture was alien; or at best they were indifferent to it. Jews living in eastern voivodships preserved their affection for the Russian culture from which they had parted relatively recently. For understandable reasons, the Jewish proletariat sympathized with Communism. No other patch of earth had such a great concentration of Jews as Poland: they constituted 10 percent of the total population, but made up about 90 percent of the population in many small towns, and about 30 percent of the population in Warsaw. Some performed useful social functions: they were professionals, craftsmen, industrialists, merchants, and workers. Others, however, dealt in trade on such a low level that they eventually prevented trade from being organized on a sound basis. When not so long ago an improved trend arose to organize trade on the basis of cooperatives, conflicts ensued with private trade and with Jewish tradesman and, moreover, not only social but also racial arguments were frequently used [by anti-Semites].

The Peace Treaty of Versailles recognized the Jews as a national minority, gave them special rights to establish their own schools, and put this national minority in the care of the League of Nations. This was a mistake. The minority rights guarded by international institutions did not help the Jews at all. On the contrary, they deepened the chasm between the Jews and the Poles and, what was even more detrimental for the Europeanization of the lower Jewish strata, such rights encouraged others to associate the concept of a Jew with cultural backwardness. The Tsarist Government admitted only a small percentage of Jews into universities. For the Jew, the Polish Constitution opened the road to professions that had been inaccessible to them during the Russian period. At the same time, enjoying the rights of a minority, the Jews opened schools where subjects were taught in Yiddish or Hebrew and where the children did not learn to speak proper Polish. It would not have occurred to a Jew in America or France to send his children to such schools. The Jewish religious schools, the so-called heders, were alien to modern pedagogical trends and yet, on the basis of their minority rights, they were entitled to exist. The assimilated Jewish intelligentsia drew the proper conclusions from this state of affairs: realizing that the concept of a Pole of Jewish faith had become an empty concept, they would espouse baptism. The young Polish nationalist movement did not want to accept the Jewish masses which, for their part, deliberately struggled to preserve their distinctiveness. Only left-wing parties, such as the Polish Socialist Party and the Bund, kept in touch in the name of joint social ideals. Jewish schools that were under the influence of the Bund emphasized the Polish statehood idea. Alternatively, schools supported by counties, especially religious schools, emphasized Jewish nationalism.

Both Polish and Jewish nationalism were strengthened by the flourishing of national ideas in Europe, which was associated with the lower classes gaining public voice in the nineteenth century. The rate of social and economic life intensified in Poland, and democracy began to break the barriers between the various social strata. Previously, with the exception of the assimilated intelligentsia, Poles and Jews had no social contact. Now, Jews had their own representatives in the House of Congress and could exert some control over the country's fate. There were many Jewish industrialists, physicians, lawyers, and journalists. Some were carriers of Polish culture, while to others this culture was alien or indifferent. This alienation had tragic consequences during the occupation. When Jews were forced to hide to escape being murdered, it proved that many Jews, even those from the intelligentsia, had no contact with Polish spheres and had no personal friends among the Poles. Many from the lower strata could not even speak Polish. The process of assimilation had certainly been moved forward, but it could not perform miracles. While discussing the program of private Jewish schools, one well-known Bund leader had said: "With this program, the feeling of belonging to the Jewish nation will disappear after three generations and will change into the feeling of community with the Polish nation. But let us allow this process to proceed spontaneously, without compulsion."

As I mentioned earlier, modern life began breaking down the barriers. At the same time, however, as a result of greater contact, people began to perceive the strangeness more acutely. One can say that some of the former "pro-Semites," people for whom the Jewish problem had been of no significance, became anti-Semites. Racist ideas took root in this soil. These ideas were accepted mainly by the petite bourgeoisie to which professionals belonged, not only in Poland but everywhere else as well. It was ugly and grotesque for associations of engineers, who had no notion of race problems, to introduce the principle of searching for non-Aryan grandmothers.[2] It was worst when Rydz-Śmigly attempted to combine the right and left wings but actually accepted the right-wing program for domestic purposes, at a moment when Poland's very existence depended on keeping up with the world's great democracies. The Jews responded by strengthening their isolation. Separate Polish and Jewish associations of physicians, medical students, merchants, and so on were formed in the country. To illustrate how acute the mutual antagonism was I will mention that Jewish teachers' associations in the ghetto refused to accept baptized Jews, regarding them in spite of the common misery as national traitors.

A qui la faute? [Whose fault is it?] In matters of this type it is best not to talk about guilt. Two historically dissimilar nations lived on one soil. As a result of competition and differing customs among the lower strata, joint ties and tasks were forgotten. The sound and fruitful work done by Jewish scientists, physicians, and industrialists was regarded as penetration by alien elements. Rural areas did not have enough physicians, but Jewish physicians did not like to go to the countryside. Medical schools began to introduce a *numerus clausus* to equalize the percentage of Poles and Jews coming from cities and thus ensure medical service in rural areas. In some schools, however, this trend changed into a *numerus nullus*, which was lawlessness and injustice. This gave rise to monstrous events in academic life: beating Jewish tudents; introducing separate benches, on which Jewish students refused to sit, and justly so; and assaulting democratic professors. Academic life in Poland became nightmarish and unworthy of higher schools. Professors did not always rise to the task and often subordinated themselves to the rioting youth.

That is how it was immediately prior to the war: ethnic fights fed by German propaganda and agitation. Even at that time the only way out was for some Jews to emigrate and for others to change their social structure and absorb Polish culture. That process would have required several generations and a higher emigration rate which, incidentally, was being promoted by the Polish Government. This problem was tragically solved during the war. The number of Jews in Poland who have survived cannot be determined right now, but it is very small. Out of the 3.5 million, maybe 10,000 or at most 100,000 have survived. Almost all Jewish property was taken by the Germans. Now, with the Germans gone, these factories and stores will be transferred to Poles. There probably is no power that could take these assets away from the Polish owners. Thus, nothing is economically binding the Jews to their previous existence, which they must now rebuild anew, possibly with the aid of foreign countries. Another, unanticipated factor enters the picture: the survivors include assimilated and Europeanized Jews who have Polish documents and most of whom will not want to return to their past; and the lowest strata who due to their primitivism saved themselves by hiding in the woods. There are a few Jewish children who must be brought back to health and raised. Here, a question arises that has never before existed in the current scope: Where should their lives be rebuilt, in Poland or in a new homeland? And in what spirit should we educate the surviving Jewish children who were hiding in peasants' huts, monasteries, or with their Polish friends? Should the Jews leave them where they are now and where they can be readily absorbed and assimilated, or should these children be saved for Jewry? When solving these problems, one must primarily consider psychological factors in order to prevent forever another tragedy of the type we have just witnessed. We must reckon with the grave psychological burden lying heavily on both sides. Poles will be especially sensitive to anthropological and traditional differences: the vision of a beggar who should be happy that he is allowed to live and should present no claims on equal rights. Jews will feel gratitude toward their Polish friends who saved them and their children, often at the risk of their own lives. At the same time, the Jews will also remember the Polish police who were forced to kill, the blackmailers, and that Polish man who did not return the property entrusted to him for safekeeping. We must understand the bitter resentment of a Jew who sees a stranger in his apartment or factory. I am talking about the spiritual burdens of Poles and Polish Jews. Western Jews were also murdered, but I am not familiar with the reactions of Western societies. I do not know whether

German Jews will regard the whole German nation as morally decayed, o
whether there were some Germans who, like the numerous Poles, paid no
heed to the danger and offered aid and comfort to the Jews. I do not know
whether German Jews will be able to forget the exile, insults, and murders.
I am writing only about what I saw and experienced in Poland.

The problem of minority is not just a Jewish problem. The Turks and
Greeks understood this and solved this problem by exchanging their minorities. This was difficult at first, but now all are pleased. Such a solution is
possible, however, only where neighboring countries have mutual minorities. The Jews have either complete assimilation or emigration as their
choice, because the vicissitudes of the eternal Jewish wanderer have been
too tragic to be allowed to continue.

So far I have discussed the decisions that can be deliberately made by the
Jews. However, what will be the stand of the Poles? Do the Poles wish to, and
will they, absorb the remaining Jews?

I discussed the problem of racism previously. Racism was nothing but a
tool to facilitate the conquest of neighbors by arousing contempt for them.
When the Germans created the concept of racism and racial isolation, they
poured just as much disdain on Poles as on Jews. The only difference was the
rate of destruction. One must wonder at the shortsightedness of Polish right-
wing parties who did not understand the role of anti-Semitism in the
Germans' fight for a greater Germany. Even so, one must admit the existence of a certain aversion for an alien tribe, regardless of political considerations. This is grounded in the given tribe's desire to prevail. This desire
was especially strong among Jews as a result of the fact that they did not have
a country of their own. Traditional and biological exclusiveness was their
only weapon. This way of defending one's existence is justified only among
nations that are oppressed or have no country of their own. In the days of
the partitions, it was understandable when a Polish woman rejected marriage with a Russian or a German. A nation living in its own country has
other means of self-defense than exclusiveness in selecting a spouse.
Currently, not one of us will protest when our soldiers bring their English or
Russian wives to Poland.

Every individual is surrounded by an aura of association with his origin.
When we meet an Englishman, we involuntarily think of the power of his
nation and of the English ships that proudly sail the seas. Such associations
affect our relations with each individual Englishman and exert an attractive
effect. The power or nobility of birth play the role of secondary sex characteristics. When an Englishwoman meets a Pole, she associates him with
memories of heroic battles for freedom, national suffering, and the like.
This may sound paradoxical; nevertheless, in our relations with the opposite
sex, we profit from the beauty or power of his or her nation. What aura surrounds the Jews? The first childhood memories concern the story of Christ's

crucifixion, and then we see the image of an alien noisy Jewish crowd and the fact of their social inferiority. Even the most beautiful type of Jew will be burdened by these associations. Only very independent minds can see the individual value of a person, and therefore, only they extend their hand to foreign nations and races. Such people are pioneers in the area of choice. When the Jews cease to be the despised minority mass, the spiderweb of allegedly infamous associations enveloping valuable persons owing to their origin will burst and vanish.

Finally, let us not forget that racism has not only been unmasked as scientific nonsense but has also been militarily defeated. Racism has been compromised not only scientifically but also politically. Poland must strive to find axioms that bring people together. It should not and, let us hope, it will not profess outdated slogans that have been rejected by its great allies.

Therefore, I believe that when the vision of a minority indifferent or hostile to the Polish society disappears, the feeling of alienation or ethnic inferiority will vanish among the Jews—in the name of the great precepts of equality and love for the fellow man.

What then should the remaining Polish Jews do?

When they had to hide, they cursed their traditional and cultural alienation. Many a man perished because he spoke poor Polish, did not know folk customs, and was circumcised. The Jews are tired of the antipathy surrounding them and their passive resistance to alien cultures. They no longer want to be different through such nonessential matters as avoiding pork, ritual slaughter, and circumcision. They want finally to feel at home; not by force of minority rights controlled by foreign powers but due to a common culture, common customs, and a common national feeling. Thus, a portion of Polish Jews no longer want to be a separate nation. Yet, in others, the suffering has intensified national solidarity, and they want to remain Jews and become a free nation. Both aspirations arise, not from a calm reasoning but from endless suffering, tragically accompanied by murder, passion, and yearning.

In the ghetto, I came to understand the pride and aspiration of some Jews to develop without the need to adjust to others but rather by creating their own new forms of life. Yet, I also came to understand the desire of Poles, which is shared by most surviving Jews, that the citizens of one country should be bound by a community of culture and tradition. Having accompanied the Jews on their last road and later—in the role of an Aryan—learned the source of antipathy toward Jews, I would like to tell the world what I think about the problem of Polish and perhaps other Jews as well.

The greatest tragedy of Jews is not that the anti-Semite hates them, but that the good, gentle people say: "He is a decent man even though he is Jewish." The tragedy is not that their religion is forbidden, because this is not so; but that good, peaceful people perceive the swaying at prayer, to which the Jews have the same right as the Christians to kneel, as a queer magic. No one forbids them to speak Yiddish, but they end up not knowing the language of the country they have inhabited for centuries. No one forbids them to eat garlic, but they end up smelling differently than their surroundings.

In places where various races fuse to give rise to new nations, every race contributes its customs to the whole. The Italian eats macaroni, the Englishman roast beef, and the Jew garlic. No one sees any evil in this because neither the macaroni nor the garlic is a symbol of life; they are simply macaroni and garlic. Here, however, in this disgusting Europe, in this cauldron where nations are roasted, even completely insignificant customs become national symbols. Here, not tolerance but idiosyncrasy prevails. Countries are paranoid about their minorities in general, not just the Jewish minority. Here, the neighbor is hated in general. One could withstand hate, but the worst is the fact that nations do not tolerate each other and fray each others nerves. Thus arises the tragic feeling of loneliness and separateness. How painful it is to die with the feeling of unrequited love. Are the children, grandchildren, and great grandchildren of these homeless wanderers doomed to suffer the way they do?

The Jews have the following choice: either they will pray differently from their contemporaries, speak and eat differently, and always feel foreigners, or else, if they decide to become this country's children, they must conform completely to other citizens. Or, if they want to preserve their separateness but want to free their future generations from the curse of unrequited love, then they must acquire a country of their own whose soil they will cultivate and defend with their own blood if need be.

I understand that it hurts one's pride when, in spite of the indisposition of the surroundings, one has to try to fuse with a given society and reject one's own traditions and customs. But this phenomenon prevails everywhere that barriers among various nations, races, or castes are being lifted. At such a time, one must give up the traditions of one's nation, tribe, or class. This must be done for the sake of the happiness of future generations.

This is the current state of affairs. Jews, or actually a handful of Jewish survivors, are facing an alternative: to conform in full or to emigrate and create their own land. Both decisions are equally respectful and both require the work of several generations. Those, however, who are now making a choice must know: noble and happy is not he who endures—this is the happiness of weeds—but he who aspires. Everything that is alive wants to be the beginning of a new life; but that which is noble wants to bear fruit. Let everyone look into his heart and make his decision:

Homeland here or homeland there.

Because this European hell has no room for national minorities.

It may be that new shocks, new fights, and new loves will come and create a new man transcending the boundaries of national cultures. Differences in customs may become essential in the wake of great new goals and tasks. I am speaking about the tragedy of my contemporaries who have the right to regard the country in which they were born as their homeland.

Chapter Thirty-Eight

The Great Guilt

The auditorium of Heidelberg University, July 11, 1943. The Heidelberg I dreamt of in my youth and which attracted delighted and appreciative youth from all over the world. The university is sanctified with the names of the greatest minds. Here arose the idea of a chemical analysis of stars. Here arose the purest inspirations; here was the shrine of pure science; the sacred legacy of a nation of poets and thinkers.

The professors are entering. They are cool and festive and bear the expression of earnestness that is typical of persons who, accustomed to thinking in solitude, believe that they are participating in important state undertakings. Mr. Goebbels mounts the podium and speaks:

> By intellectualism we understand a degeneration of sound reason, or rather the presence of intellect in excess of will power. . . . Men of pure science are inclined toward a certain reserve with respect to public life. The usual objective is to escape contact with daily and material needs and to make pure science independent. Who can resent this, the scientists? The German nation is rich enough to be able to afford such a luxury. Still, the history of German universities has many examples of men of the spirit speaking out loudly in moments of national peril.

And Mr. Goebbels shouted: "Men of science and inventors—to the front!"

My thoughts were probably shared by scientists all over the world:

Mr. Goebbels did not mention his worry of how to explain to the world that six million Poles and Jews, defenseless women, children, elderly people, workers, and scientists, have been caught, arrested, robbed, tied, and killed. Mr. Goebbels spits when he thinks of the English air raids on the civilian population who could after all hide in shelters and leave the cities, thus disorganizing the smooth running of the state, which was the Englishmen's very objective. But he does not spit when he thinks of the senseless murder of defenseless people. Instead he mobilizes the company of absent-minded professors gazing at and still believing in the forgotten legend of a nation of poets and philosophers. Priests of pure science—that is a luxury for a nation. During normal times, one can afford such a luxury. In moments of danger they can bend themselves to declare with internal conviction and with pathos: "Es is nicht wahr . . ." that is, it is not true that several million defenseless people have been murdered. Even the sparrows chirping on the roofs know! Only an absent-minded specialist in pure science could not have noticed what is by now common knowledge. But even if he has noticed, what are orders for? And the one about the welfare of the nation being more important than the truth? How convenient those millions of murdered defenseless Poles and Jews from the East and elsewhere, and now at just before noon it has become obvious that the history of the world will not be written by Germans but by the Americans, the British, and the Russians: we have to account to the world for those useless murders. And pay. Therefore, gentlemen scientists, to the front, march!

Mr. Rector of Heidelberg University: What was the sense of converting the university—five minutes before the end—into a department of the Ministry of Propaganda and defiling a sacrosanct place with the presence of a man who has never stained his speeches with a word of truth?

Koenigsberg, July 12, 1943—The memory of the great Kant is still alive. Every year philosophers from all over the world come to honor the memory of a man whose only passion was pure thinking. This time, soldiers and mobilized scientists arrive. Major Dr. Wilhelm Ehmer[1] delivers a speech about "The Spiritual Waging of a War and Its Ethical Significance." The major and philosopher declare in one voice:

England has no writings about soldierly ethics. We are a land-bound nation, a nation of soldiers; by tradition we are soldierly and chivalrous. England rules its empire at a distance and conducts its military showdowns with the smallest possible involvement of its troops. It does not respect the profession of a soldier and, according to its slogan "right or wrong," it prefers the unchivalrous means of starvation, slander, and lies.

We, a young nation, are facing an experienced enemy devoid of conscience. Our objectivity—our pride in times of peace—which places our self-criticism above the criticism of the enemy as well as our onward looking manner of thinking differ from the English "wait and see." The Great War left us with a spiritual burden of tragedy and disappointment of which our enemy is well aware and which he subtly tries to exploit. It is difficult for us, men of science, to assume an uncompromised party stand; nevertheless, we must. One moment of doubt wreaks more havoc than outside attacks. Our intelligentsia should remember that in some situations there is nothing more stupid than such a metropolitan intelligentsia. He who wants to judge and foresee should not forget that no one is able to foresee the outcome of the war. Responsible forecasting is the business of the few who have the basis to make judgments.

United spiritual power is greater than a spiritual double-edged sword. An absolute faith in our right to live and absolute loyalty are the core of our soldierly Germans and thus all-German ethics, and only they have meaning in the face of the mysterious and initially overwhelming head of the war Gorgon.

I should like to ask the major who so easily assumed the party stand: Do you know that the Germans have killed more innocent people than they have killed soldiers in combat? The nation of masters has no right to talk about chivalrous ethics and not accord it to other nations until it washes itself of the infamy of murdering defenseless people.

In July 1943, when at least 90 percent of the Jewish population had been killed, lectures on the Jewish problem were given in Jena University. The main lecture was given by Professor Heberer,[2] a well-known expert on chromosomes. I do not know whether Mr. Heberer considered it advisable to inspect the objects of his studies. On the basis of his judgment of Jews, I am inclined to assume that he talked about matters of which he knew nothing, since he characterized Jews in the following way (*Warschauer Zeitung*, October 14, 1943):

Intolerance of persons confessing other religions, lack of charity, vindictiveness, a calculating mentality . . . Great eloquence . . . Desire to rule and exploit others exercised with cold cruelty . . . Their laws require dissolving all mixed marriages and punishing all who object to this.

One would think that Professor Heberer was presenting an objective description of the Germans. But he believed that this was a characterization of the Jews, and he continued: "We have solved the Jewish problem for Germany and we will solve it for Europe. This solution states: eliminating in full these parasites out of the bodies of healthy nations."

Did the professor realize that he was blessing the murder of millions and that he will be made responsible for inciting to murder? Did the professor realize that in spite of his study of chromosomes, he should pay for this speech with his head on the gallows jointly with the murderers from Majdanek?

In Brunswick, Professor Six[3] spoke about the tasks of the German cultural policy in foreign countries (*Deutsche Allgemeine Zeitung*, September 17, 1943):

Implementing our cultural policy means carrying through the idea of leadership. The reason there for this is to acquire control over the spiritual leaders in other countries. The means in cultural policy are different from the means of regular policy: in the first case, we must attract and convince. By inviting foreign artists and scientists to Germany, we establish a foundation of trust.

Thus, we learn that relations with foreign scientists do not arise from a community of interests but from the aspiration to gain leadership over other nations. This is what has been going on in the present war. Not only human bodies but also human souls were mobilized. And how was it before the war? Did the sacrosanct nature of science, objectivity, and respect for human ideas exist? Germany had a Society for Blood Group Research. Its secretary general, Dr. Steffan, wrote a textbook on blood groups, which he concluded with the following words: "There is no doubt that the study of blood groups has established a new foundation for our knowledge of nations and races. Familiarity with the past and the present allows us to draw conclusions about the future and gives us guidelines for our population policy." In an official journal edited by Steffan (*Zeitschrift fur Rassenphysiologie* 7 [1933], 116) I read the following article written by an unknown author:

> Credit is due to Mr. Stephan [*sic*] for recognizing the significance of blood groups for race problems at a time when the Jews have begun to be silent about and conceal their race characteristics. . . . It is well known how cunning Mongols are. It is striking that once the correlation between races and blood groups has been established, blood group protocols in Asia have begun to change their appearance. Asian researchers are trying to show that, contrary to what is characteristic of colored races, their compatriots do not have mainly B and N groups but O, A, and M groups.

I would not cite such rubbish written by an unknown author if it had not been published in an official journal that usually published serious scientific papers by German and foreign authors.

Professor Jordan,[4] who is considered an authority on theoretical physics, lives in Germany. This scientist observed that X-rays either kill bacteria or leave them unscathed. From this he concluded that each cell must have a control center and that injury to that center kills the cell. On the basis of this falsely interpreted experiment (we do not know whether or not a bacterium becomes ill, because we do not know its pathology; we can only distinguish live from dead bacteria), the author proposes that societies too must have their control centers. That in a society this control center is represented by the leader, and that Germany should be proud to have Hitler as its leader. Such a publication would have been impossible in any other country for the simple reason that it would have provoked such amusement that the author would have been killed by public opinion.

Books were burned, authors ordered not to be cited, scientists ousted from their posts, "German physics" and "German mathematics" were created, and finally came institutes—imposed from above—supposed to provide the answers, such as institutes for the study of Jews, Eastern institutes, and so on.[5]

I see blood on the hands of German scientists, on those who wrote about race hygiene, the Nordic soul, living space, a mission in the East, and in whatever other ways violence was anticipated and motivated. I see blood also on those even more numerous scientists who knew that this was nonsense but kept silent and on the street did not even greet their colleagues who had fallen into disfavor. There are moments in the life of a nation when a man must not keep quiet lest he become an accomplice.

If a murderer comes and wants to murder my relatives because their eyes are not blue enough and their skulls are not long enough, should I point out to him that their skulls are long enough and that one must not murder? Is it worthwhile devoting entire books to whether there are 20 or 50 percent Nordics among Poles or Jews, or rather should we write—no, knock it into people's heads so that they should know it down to the tenth generation—that one must not murder, even non-Nordics. We must put an end to that well-bred style—the style of scientists—because we will lose sight of the essential matter, the psychological motives of mass murders. I no longer want to be one of those well-bred individuals who walk on summits where they hear pure thoughts but do not hear pain or passion. I want to find an expression for the pain experienced by parents whose children are being murdered. They are not being killed in a war, they are being murdered in cold blood.

When the first German papers appeared in which the science of blood groups was used to justify population policy, I protested in a book I published in French entitled *Les groupes sanguins*.[6]

I wrote that I wanted to be separated from those who associated blood groups with the mystique of races, and that I created the concept of a serological race, which had nothing in common with the anthropological race. I called a nation a group of people who loved one country, were attached to its traditions, and were imbued with its culture. I wrote that a misinterpretation of a piece of research could be detrimental not only by possibly ruining scientific objectivity—the desire to find the truth—but also because it could destroy what has been and always will be the source of creative effort in science and art, namely inspiration. Thus, I highlighted (in a general perspective only) the danger of changing incentives for scientific work when inspiration is displaced by motives of a secondary nature. A war came and showed a danger incomparably greater than what I anticipated, a danger reaching the foundation of intellectual life. On the basis of the thesis of race inequality or the higher value of the Nordic race or some mystic Aryan race, the conclusion was drawn that eugenics required weeding inferior races the way one weeds crabgrass or kills vermin. This was the stand taken by the Germans with respect to the conquered. We witnessed murders unprecedented in history. Every human being in Poland is haunted by the indelible visions of people being caught like mad dogs, people being deported and killed, the ghastly visions of millions of defenseless people being chased to the place of mass extermination. These acts were made possible by a belief in the superior value of one race, based on trust in science. The survival of just one race was presumed to condition progress. For German authors, race purity is a matter of race hygiene. In this way science became an accomplice in the greatest murder in history. More than that, the conscience of the world must accuse German science not only of collaboration but also of instigation to murder. It is no longer a matter of inspiration—who would contemplate roses when forests are burning—it is a matter of the most essential and profound foundations of a scientist's work. "La science sans conscience est vide" [science without conscience is null and void]. A science in the service of planned depredation is bound to be followed by a liquidation of scientific thinking and a loss of trust in scientists as spokesmen of truth. We can no longer speak of a danger, because the crime has been committed. We can only try to elucidate the causes of self-delusions and consider how we can weed out certain categories of mental aberration and how we can insulate ourselves from mental fomenters.

Should I write about the value of the various races and analyze the arguments that brought about the mass murders? Instead I shall analyze the German way of thinking. German science has facilitated the confirmation of its demise through its own cynical formulation of the essence of the law and science as we see them. A law is not an assignment to bring an act into agreement with home policy—even if it has a historical basis—but should be an interior mandate developed according to what is felt to be right. The law— to my mind above individual or collective life—has been bent to fit the needs of a race and its expression through statehood. In Germany, a law is something that is advantageous to the nation. Do the German scientists not understand that this formulation of a law is only a legalization of lawlessness, and that if a society wants to respect lawful order, it must endow the law with the dignity of the absolute? In this formulation or actual distortion of the concept of the law, the world can see for itself that all crimes were committed in the name of the German nation's right to live. With this concept of law, were the Germans able to unify European nations?

Let us turn to the other principle concerning the role of science—the concept of the scientist as the state's soldier. I believe that most German scientists would frown at this formulation. However, I suspect that given the very peculiar atmosphere that has spread over their country, few German scientists realize what distaste this concept causes all over the world, how much it isolates German science, and how it kills all desire to conduct any kind of serious discussion with them. If military analogies must be used, only one formulation can be made, namely, that the scientist is a soldier of truth. This thesis is correct even though the concept of truth is variable and subject to evolution. If discussions about the scientist's duty with regard to truth have elicited little conflict up to now, this is because scientific thought and views have not always reached the level of a world outlook. But a true scientist is guided by the immortal example of a researcher ready to die with the words "Eppur si muove" on his lips.[7] German scientists probably still feel insulted by the words of the "great Fritz," that king-cynic who said that he could find a professor who would prove as correct what was politically advantageous on every corner like a paid trollop. Do the German researchers not comprehend that the Hitlerite formulation of a scientist's role differs only in the pathos but not in the contempt for the scientist? No, the scientist is neither a soldier of the state nor of the nation, despite the fact that he is subjected to historically developed traditions and despite the fact that his thoughts are affected, though to a lesser degree than those of the average man, by the impression and even the needs of the moment. But he ought to strive to make his thoughts independent. Just as religion is comprehensible only as acknowledgment of the One Supreme—the concept of the German god is a distortion of the idea of God—so science can be only the spokesman of the truth. Or it will not exist at all. Because science is not for supplying the means of governing. The individual who used the latest technological achievements to design modern gas chambers in which the Germans murdered the Jews was no scientist. He was a technologically trained hangman. The same is true of the ethnologist who was ordered to find traces of Germanism on Polish land. A scientist does not take such orders. A man of this type is a mercenary in the service of the Reich and operates with scientific weapons.

What then is truth? To what service is the scientist called and in what shrine is he to be a priest? Again we are facing the need to prevent the subconscious relativism of our mental impressions from being displaced by a conscious subordination to the needs of the moment. In the assessment of the value of races by German science, we find a formulation which, projecting the theory of the fight for survival upon social relations, distorts the concept of the truth for the needs of the war. Truth is what is supposed to be advantageous for the nation. Thus, victory is advantageous. The concept of value acquires a utilitarian nature. Here is what is said about the concept of total war: "Total war requires such a superiority in all fields that victory in itself is proof that our cause is just." Thus, justice is victory, because victory proves righteousness . . . by the mere fact of superiority. The principle of righteousness is consciously eliminated. Justice, compassion, equality, love for one's fellow man, or at least respect for life—all these maxims created by humanity as the foundation of living together are, for the Germans, an illusion concealing the inferior species' desire to live.

By the time these words are published, the proud victorious theses will probably be rebutted in Germany as a moral sanctioning of defeat. Those of justice, the right to live, and independence will be advanced as the only possible foundation for nations to live together. Since this will not restore the life of the murdered millions, however, we cannot dismiss their former thesis as no longer valid and concentrate on the order of the day. German science, or at least one of its branches, must, in full awareness of its guilt, begin to comprehend of what the error of its premises consisted and how it brought about mass murders.

The error to which their current concept is paying tribute consists of using the idea of the fight for survival as the only exponent of social problems, eliminating the concept of justice from international or interracial relations, and replacing the concept of value with the concept of biological adaptation to certain living spaces. According to their thesis, the fight for survival—and only this fight—shapes social life and is the source of progress. As the fight for survival selects traits which guarantee survival, so—they believe—the fight of nations selects types which are best adjusted, are the strongest, and thus are the best. According to them, the victor is better because he is better adjusted, and the right to live is granted only to the best. "The earth should belong to the best," said Hitler.

They will turn their back on these theses and this judgment when the sword they raised turns against them and when they, as the defeated ones, will wish to be admitted to the community of free nations. Then, they will feel close to the following ideas which have guided us, small—according to them inferior—nations from our cradle.

The species' fight for survival and the marvelous usefulness of our organs have nothing in common with a military victory. The victor is not the one who is biologically adjusted to the given living space and is the best with respect to that space. A war is not a combat of more or less biologically adapted entities. A war is a combat of more or less technologically advanced societies whose energy is to a greater or lesser degree directed on despoiling foreign lands. As a result of the defeat, certain social shifts may take place and cause intellectual and social limitations among the defeated and degrade them to lower functions.

This was the Germans' goal and not the victory of the biologically best. In Poland all secondary schools and universities were shut down. [The contents of] scientific institutes were carted off: German scientists themselves carried out this pillage. Cracow professors were invited to a lecture on Hitlerism's attitude toward science, and when they gathered, they were deported to Oranienburg and executed. In Poland, it was forbidden to sell books with a cosmopolitan outlook to Poles. It was forbidden to play Chopin. Is this an equal start guaranteeing victory to the best, or exploitation of momentary supremacy in order to rule out the possibility of equals fighting with equals? For the Germans it was "strength through joy" and the pleasure of creative work; for the Poles—the life of servants. As for the Jews, even the life of serfs was not granted them. The valuables and libraries acquired by the Jews, the factories they built, the musical instruments that brought them comfort—everything was taken by the Germans. Was it in the name of the victory of the biologically better species, or was it in the name of a more exuberant life for the Germans as the nation of masters? We who have been treated as sub-humans by the Germans should be allowed to state how we see the superman of their cult. The masses of armed plunderers, the commissars robbing our property, the corrupt officials, the men throwing us out of our homes, the persecutors exterminating us in camps, the university-educated mercenaries killing on command—those are the creatures we are supposed to regard as biologically superior? They may say: new territories are not conquered with ethics and culture; the conqueror is a brute, a blood-thirsty mercenary. But, they are followed by divisions of great spirits. Therefore I ask: In the future, will the road to proliferation be closed to these beasts of prey who indeed deserve the name of sub-humans? Or should we acknowledge them as supermen?

The same day that the Dresden Philharmonic Orchestra arrived in Warsaw (a memorable date for the Poles: October 17, 1939), the first round-up took place on Warsaw's streets. Society was stunned to see for the first time that people could be caught like animals and later killed, defenseless. Shall we forgive them because the Germans are musical and constitute an elite? What do we care about an elite whose moral backbone has been broken. For many years to come the nation of poets and philosophers will live in Europe's memory as the nation of murderers, blunderers, and thieves. He who saw the German's deed in Poland will be overcome by the utmost disgust at the idea of scientists justifying victory as the guarantee of progress that opens new space for superior human beings.

I wonder whether the Germans are aware of the blow Hitlerism has dealt to German science and its radiation? I know that they still include great, enlightened minds to whom injustice will be done if humanity, upon learning about the crimes committed on defenseless people, will turn away from the Germans. We can forgive death inflicted in combat where man fights man face to face and where readiness to die absolves one of the sin of killing. But we cannot forgive murders. In cold blood the Germans murdered millions of civilians. There must have existed in Germany a moral climate that facilitated murders publicly announced by the Führer. His statement that Jews would cease to laugh meant that all Jews would be exterminated and that no witness would remain to tell the story of these crimes. Even for the sake of one's own homeland, certain moral norms may not be transgressed, and so much the less for personal comfort. Many German scientists felt repulsion and even contempt for Hitlerism, but personal advantage was more important to them, and that lack of color and of one's own opinion, which the Germans call internal discipline. Those murders were committed in the name of *Germanism* and for the sake of their nation's benefit. Did they have the right to keep silent?

These murderers will tell us that they relied on the characterization of races supplied by German science. That they were exterminating inferior, parasitic races and that in the perspective of history these murders will be recognized as a lever for progress. That the respect for human life should not hinder mankind's aspiration to develop upward.

Let us analyze this reasoning.

Certain character traits regarded by the Germans as good and worthy are supposed to be inherited according to the great biological laws. Because one or another anatomical property of a mental illness is inherited, German scientists conclude that certain national psychological traits are inherited. The mentality of a nation is supposed to be the expression of race, and race is the mentality of the nation seen from without. The arbitrariness of this interpretation went unnoticed probably because racism was not a scientific concept but a new faith, the objective of which was to knock down the "infidels" and to absolve the planned violence with which new living space was to be conquered. The errors of these premises are so obvious that it was even unpleasant to discuss them as long as the tragedy of the murdered people did not force us to analyze the reasoning behind it.

Simple laws of heredity govern relatively simple anatomic and physiological traits. However, most traits, especially mental ones, depend not on one gene but on groups of genes whose joint effect makes implementation of the trait possible. The genes constituting such groups are generally inherited independently of one another, and the change of just one component in the group can change the given trait completely. Thus, a stray Jewish gene in an Aryan embryo does not necessarily produce a characteriological [sic] feature so disliked by the Germans. In the light of genetics, tracking Jewish great grandmothers is ridiculous.

Mental traits are not hereditary units. In fact, one inherits not traits but a certain reactivity. Education then allows social factors to extract and develop those psychological traits that are considered positive. The principle of every genetic experiment is to let identical external stimuli exert their forces on the experimental subjects. Have the various nations lived under identical social conditions?

No, the soul of a nation is not the sole expression of race. National tradition cannot be identified with racial constitution. No one, absolutely no one, knows where tradition and history end and where heredity and race begin. Teaching race psychology in connection with genetics in the way the Germans have done, and not with ethnology and history, is aiming at a preset answer and not an objective investigation.

Let us now turn from the pseudoscientific formulation of race psychology to the ethics advanced by racists. Europe is dominated by Christian ethics partly based on the Old Testament, which proclaims: "Thou shalt not kill!" The Christian ethic goes further, requiring universal kindness and love for one's fellow man. Even during the darkest moment in history, these maxims have guided humanity as an ideal which is not always strictly followed but which at least imposes certain moral bonds. Racists have deliberately proceeded against these theses, relying to a large degree on the anti-Christian stand of Nietzsche. I could cite many excerpts against Christianity from Hitler's and his adherents' writings. I will cite an article by Wachler, which I happened upon in the *Warschauer Zeitung* and which is representative of that trend.

> Nietzsche identified the man of worth in the Icelandic saga that is, in a literature not contaminated by Christianity—and on this basis he built his aristocratic world outlook in a passionate revolt against religion and morality brought from the East. In a sense, this is a volcanic eruption of northern heathenism, the Fatherland's rebellion against aliens. . . . His Teutonic soul can be explained by his aversion for the Jewish concept of sin. . . . He rejects Christianity for the same motives: the essence of life is eternal resistance, eternal fighting thanks to which progress and grandeur are possible. A fight based on honor, valor, and courage, thus not on the traits of sore weaklings and merchants, but on the properties of strong and noble natures. . . . The morality of masters and slaves . . . Awareness of human inequality.[8]

We see here a tremendous overestimation of the concept of fighting and a deliberate underestimation of other impulses. My intention in this book is not to investigate how these concepts gave rise to an idea of leadership devoid of human conscience. The instinct of solidarity is inborn, only its scope and reach are historically conditioned. Humanity was on the best road to expand this concept. The German science of races has narrowed it and halted progress for decades and perhaps even centuries. Let us now consider how the concept of solidarity and love for one's fellow man arose.

In nature we see a fight for survival. This fight shapes and maintains every living thing on a certain level. If it did not exist, the world would be populated with nonadapted individuals. Out of necessity, the fight for survival tolerates no compassion and is as ruthless as any other fight. Religions, especially Christianity, teach compassion. Nietzsche, who regarded Christianity as a religion favoring weak and maladapted individuals, stated a generally accepted idea. And here we have a paradox. When Darwinism advanced the theory of evolution, it was combated by the Church, but the thesis of a fight as the motor of life was not subjected to theological critique. It was assumed *a priori* that it was correct, and it was not discussed whether there are phenomena in nature which correspond to our concepts of ethics.

What is the essence of the fight for survival and of its mental counterpart—lack of compassion or hate? By creating the ethics of love, did religion really make a mistake? Is there really no place for kindness in the household of the universe? Wherever a biological fight is necessary in nature, we see the formation of a combative instinct, an enragement which turns that struggle into one of the periods of greatest tension in life. Within a species, the period of fighting is associated primarily with the period of acquiring a mate. Therefore, the period of sexual maturation is a period of the greatest dynamism and militancy. Here, the fight guarantees biological progress, because victory guarantees to the offspring traits required for living. Understanding this phenomenon was a triumph for biology. But the weakness of sociology has been projecting such a concept onto other phenomena. The need to have compassion arises in the life of a community when it comes to taking care of the weak. In the final analysis, every individual living in a community is weak, because in one way or another he depends on the services of the whole. Thus arises a feeling of solidarity with as its highest sublimation the thesis of love for one's fellow man. Such love arises when it is socially justified and vanishes from the consciousness as the motor of action when it is biologically harmful. The appearance of seemingly contrary instincts depending on the period and the biological and social needs is one of the most interesting phenomena in psychology. In my opinion, there is no contradiction between the thesis of love for one's fellow man and the purposefulness of a fight. The pleasure of a fight which displaces compassion on one hand, and love or kindness for one's fellow man on the other are psychological states which are adjusted to the various periods of life, which are biologically and socially equally necessary, and supplement each other.

The requirements of community life create by way of spiritual sublimation the idea of kindness and solidarity, which, in our consciousness, reaches its maximum manifestation in the idea of love for one's fellow man. The relation of love for one's fellow man to the simple feeling of solidarity of interests is similar to that which exists between the feeling of love and the sex instinct. At one time, Russian nihilism attempted to separate the feeling of love from the sex instinct. This was a mistake. The feeling of love is a subjective perception of the sex instinct by individuals endowed with imagination. Such sublimation is quite frequent. We must only differentiate impulses associated with individual life and the life of posterity from impulses associated with community life. I would like to illustrate this idea with the following examples.

Racism has become a religion seemingly based on biological theses. This was expressed probably most emphatically by General Morch who, in his paper *Das Reich*, wrote that total war was the manifestation of superiority on so many fields that victory itself was proof of righteousness. The thesis of victory as the exponent of value has become the credo of the religion of the racists. In their opinion, the Germanic nation is adapted to the living space in central Europe, and therefore it is the best nation for central Europe. Racism wants to find a permanent basis in the instinct of fight. However, this

Table 2. Biological and social impulses of the individual and community life

	Direct sensation	Spiritual sublimation
Impulses of individual life (biological)	• Sex instinct	• Love
	• Fight instinct	• Hate
	• Pleasure	• Happiness
	• Pain	• Suffering
	• Disgust	• Concept of esthetics
	• Feeling of rhythm and harmony	• Concept of beauty, music, art
	• Fancy for secondary sex characteristics	• Love fetishism
Impulses of community life (social)	• Dislike of moves harmful to community	• Conscientiousness
		• Love for fellow man
	• Solidarity	• Religion, ethics, law

instinct is justified only for certain biological and historical periods, while the Christian ethic offers basic directives for life in a community.

The instinct of solidarity and its sublimation in the form of love for one's fellow man represent mainly social requirements, while the instinct to fight represents mainly biological requirements. Both are necessary for the survival of the species. However, it is naïve to reject the idea of compassion, kindness, and love for one's fellow man as the motors of social life just because on a different plane or in a different sector of life this idea is not applicable.

This is not the only area in which we find errors in the racist religion and its inferiority as compared with Christianity. Christianity professes free will and personal responsibility. Racism professes that everything is hereditary and irreversible and that a certain race psyche is engraved in the chromosomes, just as mental disease is. Thus, there is no need to spare races that are considered bad. Murdering them means implementing the laws of nature and accelerating natural selection.

We are dealing here with two religions based on different motivations and creeds. Racism takes pride in being based on scientific theses and rejects love for one's fellow man and respect for life. Christianity relies on God's word without attempting to justify love as a historically developed social need, and it endows the feeling of love with the dignity of God's command. When we ask which ethics is more life-giving, we can have no doubt about the answer. One ethic creates a striving toward good as the Absolute. The other deifies the race, rejects the concept of good as the Absolute, and sanctions murder.

I have sought to analyze somewhat more deeply the concept of inequality of people and the rejection of the Christian ethic. I have done so because the consequences of a deliberate rejection of these laws of conduct, which for millennia have been a shining ideal toward which humanity must strive, have been disastrous. Arbeitsführer Ley[9] did not hesitate to express an idea which evidently has been avowed by the Germans as a whole: "We are a nation of masters and as such we need more nourishment and more space."

I have no doubt that the Germans will be ashamed of these views, as they will be of having searched for lost great grandmothers. But unfortunately not on the basis of their scientists' protests but as a consequence of imminent defeat. We cannot build the life of societies on such principles, much less implement the bold idea of unifying Europe. Since the German intelligentsia was unable ideologically to oppose the theses of contempt for other races and to impede the murder of defenseless people—no one, not even the greatest pacifist, and not even the greatest admirer of German science and art, will want to spare the Germans the school of suffering. Not in order to take revenge, but in order to break forever their myth as a nation of masters.

German scientists may claim that they wanted to separate their nation from races alien to their living space, but that they did not want to murder defenseless people. That even the loftiest ideas have been defiled in the hands of the common people. That the religion of love was also spread by the sword and the ideas of equality and fraternity by the guillotine.

While I am writing these words, I am not only isolated from my laboratory but also deprived of the right to live. In my eyes I have the vision of people being murdered, and my heart is burning with the memory of my dying child who was not allowed to return to her own home and was not permitted to die under her own name. It is not my concern to find attenuating circumstances for the Germans.

Nevertheless, I say: perhaps. Perhaps those scientists did not want to murder and plunder our culture. Maybe their sin is shallowness, boastfulness, and arrogance, but not instigation to murder. Then, for God's sake, why did they not reject crime while the voice of conscience could still resound as a cry of protest? Why did they allow the development of that climate of contempt and hate and that self-adoration of their nation?

It will be too late to express remorse after the defeat. Humanity will liquidate their blindness. But those of you on the summits will be embraced with its punitive contempt.

Miłosna near Warsaw, June–August 1943

Chapter Thirty-Nine

Afterword

The Polish edition of my book ended at the point I was able to come back into my own. "Once again action fills my time and soul," I wrote at the time. "But maybe one day I shall again choose to look back and describe the edifice of which the collaborative building turned out to be my destiny." I had planned to do so in the distant future, when I would be able to look anew with a certain distance on the history of the world and the fate of my country and determine whether the stance toward life and thoughts which I expressed in my book can still be brought to life or whether they would belong to the large family of the last Mohicans longing for a time that will never return.

That was what I thought. But then there came a letter from Mr. J. Brown,[1] who demanded that the story continue. Mr. J. Brown, well acquainted with the mentality of the American reader, is of the opinion that without the continuation of my book one cannot satisfy the curiosity of the reader who has become interested in my destiny.

I am unable to do so in full. Not only because I lack time, but because I still lack the perspective and I do not yet see the direction which the world is taking. And yet I cannot refuse in the face of an editor who took on the difficult task of introducing an author from a far-off land to the American readership. I shall therefore present the further events of my life in summary, without going into any depth or seeking to synthesize, while all the while trying to bring out the issues that are currently equally affecting America and Europe.

My memoirs speak of the Russian-Polish-German battles that freed me and gave me back my identity. We came back to the village with the fellow countrymen with whom we had hidden in the woods only to find there funeral piles and ruins. Quite literally there was not a single settlement left untouched. And we saw the pain of the peasant who discovers his entire patrimony burned and must start all over again with his bare hands. At first we lived in a barn and became acquainted with a new world, the mentality of the Soviet soldiers and the different reactions and conceptions of social and national issues. Battles were still being waged nearby and we were in fact never sure whether the coin might not flip back, whether the German armies might not threaten us once again. Every night we could hear the roar of the cannons destroying Warsaw. And above Warsaw, which we could see from a distance, there rose a constant steam of smoke and we saw the sky in flames.

It was around then that I found out that a government had been installed in Lublin. My students were organizing a health service. So I betook myself to Lublin in October 1944. What curious times those were. Most of Poland was still in German hands, while in Lublin a life was being sketched out and organized supposedly under the aegis of Poland.[2] I moved in with a colleague, the director of the State Institute of Hygiene.[3] I kept bumping into people I knew who greeted me with the words: "My goodness, you're still alive? We were all sure you had perished." One constantly came across people, physicians, professors hidden hither and thither in villages, country estates, or who had fled burning Warsaw and Praga,[4] stripped of literally everything, clothed in shreds, destroyed, often without even a change of shirts and on all their lips hovered the same question: What should be begun, how should we organize life in a country that was still burning? I soon received the mandate to see to the organization of the State Institute of Hygiene in liberated areas. But I must admit that resuming the work I had so liked before the war did not attract me.

At that time Henryk Raabe (future Polish ambassador in Moscow) began organizing a university in Lublin.[5] The idea of creating a university on ruins while the crematoriums were still smoking and most of Poland was still occupied enthralled me. I accepted Raabe's proposal to participate in establishing a Medical Faculty. I did not, however, wish to become dean for I felt that society must wait and see whether years of suffering had not broken me—I did not wish to take trust on credit. I therefore occupied myself with what I believed to be the most important: selecting young people to be accepted as students. Several of us professors were in Lublin at that time. We needed to seek out older assistants who could assume their functions, find buildings, collaborators, draughtsmen. My wife took over the Chair of Paediatrics. At that time there were all sorts of people, and from all walks of life—political, literary, and scientific. Soon there came to Lublin one of the docents from the Medical Faculty of the University of Warsaw who had been by chance in Praga on the other side of the Vistula when the uprising broke out. Praga had been taken after a while by the Russian and Polish armies and those who were there found themselves cut off from their families and homes [in Warsaw]. They could see the town burning on the other side where their wives and children were. Nevertheless, they decided to organize life and came to Lublin proposing to create a Medical Faculty in Praga. And they are the ones whom I promised to work with. The entire faculty was housed in the hospital, while the lecture hall or rather the hospital room was divided into two by way of screens: on one side we installed the students' dormitory and the other served as a lecture hall. Far off we could hear the roar of cannons. But the students listened totally absorbed by the subject matter. I could not lecture there regularly and went only several times.

My unwillingness to accept a position on nomination was soon bypassed since the academic senate chose me to be "prorector."[6] It was too early to think about scientific research. Besides organizational activities I worked on editing my books. I read the war memoirs described in this book to many people who had been shattered as we were. I saw their emotion when they realized that the boundlessness of their grief had been materialized in a book. I saw how important it was for them that someone tell of their tragedy and that the information be passed on to the world.

Young people were flowing in from everywhere—the Polish youth of the German occupation, the youth who had hidden in forests, persecuted and tortured and only sometimes been able to find refuge in some hospital. Every day they arrived in crowds, their youth having been deceived by the world, harassed and penniless. They were often without papers or else with false ones and it seemed that for them there was no tomorrow. I personally saw to it that they were taken in. How easy it had seemed to me to evaluate a young person beginning his or her studies after the *matura*,[7] when the knowledge proffered by school contributes to his quickness, whereas a certain exterior attitude contributes to his grasp of things and general upbringing. But how difficult it was now to evaluate a young boy having escaped from prison or having hidden for years, or whose family had been killed and whose eyes expressed protest and pain. And finally, how was one to organize his life? Before being able to give him a book to read, one had to secure him some bread and a roof over his head.

We somehow managed. The military authorities and medical faculties agreed to mobilize all the young people on the condition that they keep to their studies. And thus the army clothed and shod them, gave them a roof over their heads as well as straw mattresses and cots. And when after a few weeks I looked at these children who not long ago had walked around in disbelief and with the words "what will they do with us?" trembling on their lips, I saw people who were young and enthusiastic and in whose eyes one could even see a gleam of hope. And finally there came the unforgettable day of the opening of the Maria Skłodowska-Curie University in 1944.[8] The university was housed to begin with in the Staszic secondary school. Laboratories were installed in various classrooms, and the opening was held in the modest gymnastics room converted into a lecture hall. And the minister of education could recall with pride that one of the most important discoveries of the scientist whose name was borne by the university took place precisely in a modest wooden hut. She knew how to work in hunger and poverty. And it was under such an aegis and with such a motto that the University of Lublin was opened.

I shall not write about the organizational details, which are of interest only to Poles, nor about my colleagues who set themselves passionately to the job of creating a new scientific life. I will only mention that once again my optimism led me to dream. I wished to make out of Lublin . . . a Polish Oxford. We started drawing up plans for the buildings; we wished to create a university town and a handful of enthusiasts from the Medical Faculty participated in the undertaking.

That is what the last months of 1944 looked like for us. In January, the offensive began, Warsaw was liberated, and the occupiers fled Poland. With a trembling heart I betook myself to Warsaw but what I found there was indescribable. Except for a street where the Gestapo had their headquarters and a very few streets in the southern areas of the town, there was literally not one building left unscathed. I will not go into the history of the insurrection.[9] It was a heroic but crazy attempt that was stamped out in two months. For the few weeks that the insurrection lasted, the population lived in cellars and the conquest of each street by the Germans was accompanied by the murder of people [men, women, and children]. It was then that almost all the doctors in the Wola Hospital were shot, including Dr. Sokołowski, about whom I have already written. It was there that the Ukrainian Army, which was fighting with the Germans, killed my student Dr. Kostuch when he tried to protect his elderly father. It was there also that they burned to death my colleague Prof. Lawrynowicz. Fischer's trial and the confessions of von dem Bach displayed the boundlessness of German cruelty, since Hitler had ordered the murder of the entire population for the "crime" of the insurrection. Part of the civilian population who gave themselves over was evacuated in dreadful conditions and some were sent on to forced labor in Germany. And then house after house, and then entire towns that had already surrendered, were destroyed for no apparent strategic reason. The Germans defend themselves in their conscience by the fact that the Allies bombed their cities. I experienced from the very beginning of the war the bombing of defenseless Polish cities, and I saw what every soldier ought to consider a criminal act: the killing sometimes even of those who had given themselves up, and the pointless destruction of towns, out of pure revenge. I will not describe the wanderings of peoples, of millions of prisoners, who returned from the German camps, nor the marching of the armies toward the east. Europe was then hell and an insane asylum combined. The armies driving down on to the west were filled with such frightful hate, the millions coming home from Germany bore such an awful grudge, that it will take decades for these feelings to abate. And what about the state of the Poles and the handful of Jews who had survived? They could not have been happier. Just like a string drawn out tight does not curl up, there was no place with greater joy than in this corner of the earth.

While Warsaw was still occupied, and Praga was being fought for, I went to Wesola to try to find the manuscript of this book. The entire area was still occupied by the armies, and the woods had been cut down to such an extent that I could not recognize the area. However, Mrs. Kenig's house remained intact together with her library and the manuscript of the textbook I had written. I had buried the present book in 1944 in a solitary house with no floor just before fleeing from Wesoła one dark night. But now I found soldiers living in that house. The officer pleasantly agreed to open up the floorboards that had been laid down in the meantime and to my utmost joy the manuscript lay unharmed in glass jars. In Lublin I reworked the text very slightly and added the chapter on the liberation. I submitted it to be published in February 1944, and it took twenty months to come out—such were the arduous publishing conditions in Poland.

And so I lived in Lublin preparing working conditions for myself and others. I could not and in fact did not wish to return to Warsaw. The National Institute of Hygiene and School of Hygiene founded by the Rockefeller Foundation had been seized by the occupier and were only partially destroyed. I could have returned to my former position. The vision of Warsaw in ruins, of my destroyed little house where I had been so happy, and especially the thought of my assistants who had died so tragically were so painful that staying in the institute would have been psychologically impossible for me. When I walked the Warsaw streets, I yet saw before my eyes the hundreds of executed people, women pursued by tanks, the roundups of innocent people. It was beyond my strength to remain there. My colleague Dr. Przesmycki and I therefore agreed that he should become director of the Institute of Hygiene, whereas I would chair the Scientific Board that we decided to establish.

I turned to research I had initiated during the occupation. The organization of a medical faculty and chair absorbed much of my energy. But fairly soon this relatively peaceful life was interrupted. It began by the invitation of several Polish scientists, among whom myself, by the Academy of Science in Moscow to its 220th anniversary.[10] The Soviet Government wished to demonstrate that the conditions for scientific work had not died out during the war and were being looked after by the government. This impressed me very much.

I saw in my mind's eye the Polish ruins and felt still the despair of occupation, and in Moscow one could feel the scale of reconstruction and the pride of victory. The road to Moscow led through such deserted areas that one traveled for hours through emptiness, without a sign of life. In Moscow though, the universities, libraries, museums, everything was humming with life. We were in Moscow for the victory celebrations and the military review on the Red Square made a great impression. We were welcomed with great hospitality, served banquets the exquisiteness of which surpassed most scientific congresses abroad. It was in Moscow that for the first time after the years of the occupation I met with scientists who had come from all over the world: Mr. and Mrs. Joliot, Huxley, Szent-Györgyi, and many others.[11] It was there that after so many years of isolation I participated in a conference at which scientists from the United States, England, and France spoke about the discoveries they had made during the war.

I will not describe the laboratories and high level of science in Moscow and Leningrad, the psychology of Soviet scientists, their fervour for work or the organization of higher education—this would require a separate chapter. Indeed it would be inappropriate to be tempted into writing a synthesis of what we saw without studying it a bit further. But one thing was particularly striking: the care the government was taking to raise the level of culture. And it seems to me that this is the secret for assimilating the many peoples living in that country. The Soviets not only allow but even favor the separate cultures of the peoples making up the Soviet state. They look after the languages of even small tribes, in relatively uncivilized areas they organize universities and academies, but often the sometimes primitive languages cannot express all the new concepts present in modern culture. Life then requires the use of a fairly rich Russian language in order to be understood. A lady ethnologist from Leningrad told me that every year she spent several months among the relatively primitive tribes of Central Asia. The parents often uphold all the old prejudices, the mothers live in harems, whereas the children live the life of Western European students, become assistants, professors, and so on. And I believe that in this very rapid modernization of culture lies the main secret of Russian progress and victory. People who during the Great War had at best known how to carry a rifle had now grown up with the motor and could fulfil the requirements of modern war.

I returned from Moscow in July 1945. The Soviet Government let us dispose of a luxurious former tsarist train wagon. A surprise was awaiting me in Warsaw: a summons from the Ministry of Education to go to Wrocław and organize a Medical Faculty. And thus there stood before me a new task reaching beyond what I had planned in Lublin. The point was to use some of the buildings and laboratories owned by the university for Polish young people. After all our universities were partly in ruins, we had lost two universities and one Polytechnic (Lwów and Wilno),[12] the universities in Cracow and Poznań had been so carefully evacuated that there did not remain a single microscope. The new universities were lodged partly in private apartments and had literally nothing. Poland had to provide higher education for over 60,000 young people who had been deprived of education during the occupation and were now flowing into every school, yet there were not the conditions for teaching. And Poland was threatened by the fact that she might not be able to create her intelligentsia, which was essential for the country's cultural life. The fight of the Germans with the Polish intelligentsia had ended with the formers' victory. I have before me the list of more than two thousand names of professors, architects, and physicians murdered in concentration camps or shot. Who was to rebuild the lost or destroyed schools? How could one require of such a consciously destroyed nation, bled white, to reconstruct buildings on its own, to organize laboratories, to write and print burnt manuscripts?

And thus I decided to take upon myself the difficult task of organizing a Medical Faculty at the University of Wrocław, I did not wish for Poland to be an eternal beggar especially where the most difficult values to create were concerned—the values of culture.

I arrived in Wrocław on August 1, 1945, to begin work alone, without my wife.

Wrocław had been acquired in May 1945. The scientific institutes lay partially in ruins. The costly instruments and books had been partly hidden in bunkers, partly evacuated and—mostly left with no owners—become the prey of robbers. The university had incurred enormous losses during the bombings. The Physics, Chemistry, Anthropology, and Pharmacy institutes had all been burned as well as the Pharmacology Institute in part, whereas the University Library had been reduced to ashes and half of the main university building was also burned. The Institute of Anatomy and Legal Medicine had also been seriously damaged. The more valuable apparatus and an important part of the library contents had been evacuated and the remaining scientific objects had been damaged as a result of blasts and quakes. That is how things looked when a group of scientists led by the dean, Professor Kulczyński,[13] alighted in burning Wrocław. One of the professors and a few doctors saw to saving what they could for the Medical Faculty. They had to engage in essential reconstruction, insure instruments, books, and so forth. In mid-October a few more professors arrived together with my wife who had been summoned from Lublin to chair the pediatrics department. The work was hard-going but we could see how young people were yearning to learn. The academic senate therefore decided to call the university to life even though the buildings were not ready and the halls and dormitories did not all have glass fitted in the windows. Scientific societies soon followed: in medicine, mathematics, archaeology, the Copernicus Society, and so on.

The Institute of Hygiene—which I had taken over—had in fact been less destroyed, and what is more, it still possessed a series of apparatus that the Germans had not had the time to take away. True, there were almost no modern microscopes. A laboratory technician of many years [who had remained in Wrocław], of Alsatian extraction, produced for me the laboratory inventory. He was a sort of . . . laboratory patriot and was behaving loyally since he had seen that I intended and knew how to look after an institute. I asked him where the rest of the good microscopes were. "Die hat uns der Blumenberg weggeklaut." [Blumenberg stole them from us.] *Uns*— that meant him and me. Blumenberg was the foreigner, for he was no longer director of the laboratory.

With few exceptions the German professors had already left; some of those who had remained visited me in my capacity as dean, asking my permission to live in the clinics or for other favors. It was a curious change about: I had recently been pushed out of my work, hidden under a foreign name, and now I was dean of a faculty and the master of these buildings. Despite all my feelings of resentment, I tried as much as I could to help them. I know that Dean Kulczyński did likewise as well as the other deans. We still have the letters of thanks indicating that the Polish university officials did what they could to help the local scholars. The one thing I could not prevent myself from doing was reading them a description of Majdanek. Each and every one of them insisted that they knew about nothing and that only in the last year of the war did the rumors begin. I must admit that I have a hard time believing them, murder and looting in Poland were too widespread. I rather think that they had erased from their consciousness news that indicted the German nation with such responsibility. I was struck by the total lack of understanding and lack of perception of the abyss that had been dug between German scientists and scholars and those of the rest of the world. One of them had asked whether he might lecture to Polish doctors on the subject of the future man. Summoned to the anatomy chair of some German town he asked whether he might first work in our dissecting room. I told him that I would gladly let him have a study and let him use the library (incidentally, scholars of Jewish origin were forbidden from using German libraries), and that only during vacation could he have access to the dissecting room, since there was almost no family in Poland who did not have some close relation murdered by the Germans and that in such circumstances I could not guarantee that the students might not act very disagreeably toward him. As far as I know, despite the Polish scholars' decision to break with German science, the majority of Polish scientists tried to behave civilly toward their German counterparts. Before the armies marched in to Wrocław, Professor Kudicke and Dr. Wohlrab were here and they had brought with them the inventories of the National Institute of Hygiene from Warsaw and treated one of the institute's members, the sanitary engineer Dr. Szniolis, like a prisoner, not even hesitating to use the informal "Du" with him. Wohlrab was imprisoned by the Soviets and the prosecution officer in charge of his case asked Dr. Szniolis how he had behaved in Warsaw. And Szniolis acted the way most of the Polish intelligentsia did: he defended Wohlrab thus saving him from execution. But when Wohlrab later extended his hand in order to thank him, Szniolis would not shake it.

Those initial days of establishing the laboratory were unforgettable. There were still in Wrocław literally thousands of young people wandering listlessly around and hovering about the school gates. One began to see the well-known disagreeable sides of postwar relaxation. When homes are abandoned and great goods lie on the streets, how easy it is for the borderline to disappear between what are ownerless and what are private possessions. One had thus to remove the bad elements and support the good ones. I began my lectures before the universities had been officially reopened in order to provide the young people with some spiritual nourishment. They participated in the reconstruction, covering roofs, clearing the rubble, straightening up the laboratories and finally in January 1946 the Faculty Board held its first session. We admitted more than four hundred students. As I am writing now the university and polytechnic comprise eight thousand students, 160 percent more than before the war.

And once again my most essential chord began to resonate in me: longing for scientific work. It is difficult to combine the tasks of dean and faculty organizer with those of researcher and pedagogue. We were not receiving any scientific journals and the occasional visits of Anglo-Saxon colleagues sent by UNRRA[14] awakened in us a hunger for knowledge that could not be satisfied. I feared that the issues which most concerned me were not those attracting contemporary science. Just like after World War I when I had the disagreeable feeling that I was on the peripheries of scientific life, left out of some bountiful development happening elsewhere. And once again I longed for when I would be able to quit writing about tears and suffering and about the longing for those who would never return and instead begin planning and delving into scientific questions, evaluating them, and rejoicing in uncommon relationship with people for whom the axis in life is science. In other words, I wished to be revived in a country unburdened by pain and memories, without ruins, and with the full dimensions of life. That country was the United States. I asked my friends to let the Rockefeller Foundation know that I was alive, and that my wife and I wished to visit the United States.

There soon came a telegram from the Rockefeller Foundation inviting us for a visit of several months to the United States.[15]

In May 1946 we left on the UNRRA ship, the *Harvard Victory*. For the first time in years I experienced that feeling of freedom, carelessness, and adventure when we felt the sea breeze and were surrounded by the boundless ocean. Only he who has been imprisoned or lived through an occupation and been pursued like a hunted animal can understand such a feeling. Our ship was supposed to sail to New York, but on our way an order came to go to Houston on the Mexican Gulf. We spent seventeen days aboard as the only passengers. We befriended the captain and officers but quickly realized how difficult it was for Americans to enter into the mentality of tired and . . . sad people. The captain looked at the photographs of the camps as at drawings of some inferno. Horrible, but unreal. There was a nice library on board and I happened upon Hans Zinsser's book *As I Remember Him*.

I had great admiration for Zinsser as a scientist and writer. His books have been translated into Polish and have many admirers. *As I Remember Him* is his autobiography. After a scientific expedition to China in 1942,[16] the author diagnosed himself with an incurable disease of the blood—leukemia. He therefore set out to write an autobiography reflecting not only his personal life but his scientific efforts and the fantastic development of science in the United States. He ends his book with a beautiful sonnet in honor of life:

> Now is death merciful. He calls me hence
> Gently, with friendly soothing of my fears
> Of ugly age and feeble impotence
> And cruel disintegration of slow years.
> Nor does he leap upon me unaware
> Like some wild beast that hungers for its prey,
> But gives me kindly warning to prepare:
> Before I go, to kiss your tears away.
>
> How sweet the summer! And the autumn shone
> Late warmth within our hearts as in the sky,
> Ripening rich harvests that our love had sown.
> How good that 'ere the winter comes, I die!
> Then, ageless, in your heart I'll come to rest
> Serene and proud, as when you loved me best.

I too went through a period when it seemed that I must perish and bid farewell to the world. I had written a letter of thanks to life. The fundamental melody is similar to Zinsser's sonata.

> And the outer limit comes my way. And so I remember that I was not only a passive hearth, some small trifle of burgeoning life, a leaf destined to be torn off. No, I also knew the pride and the delight of he who gives. . . . I wish to express forgiveness of the world. God, I forgive you this world, just as one forgives a woman her lies for her beauty.

Zinsser also describes certain experiences and life situations reminding me of my own: the typhus epidemic in Serbia, contacts with some of the same scientists, as, for instance, with Nicolle. And I then thought that Little Brown & Co Publishers who had brought out Zinsser's memoirs would be the best choice for mine. The contrasted destinies of two scholars: one living in a country of good fortune, constructiveness, and great scale and the other in a country of rubble, ruins, and dejection.

My wish was to come true. Mr. Brown showed interest in my manuscript, and our first conversation already gave me the feeling that the mentality of a scholar from ruined Europe was not foreign to him. I had a defender in the person of a subtle connoisseur of Polish literature, Dr. Guterman,[17] who began translating my book.

Waiting for us in Houston we found telegrams from Dr. Lambert[18] and our friend Arthur Coca. We traveled to New York from Houston by motor car under the care of our young friend Olsen, a medical student whom we had met on the boat.

When I recall Coca, I am filled with great sadness and tenderness. Our friendship had lasted from the days of our Heidelberg assistantships all through the years. Coca had twice visited us in Warsaw and the first time had given beautiful lectures on his scientific research. Only when I arrived in the United States did I discover that it was on his initiative that several outstanding American serologists had raised funds in order to pay the occupiers a ransom for me, my wife, and our daughter. I have already mentioned that the Germans had organized this kind of gangsterism in order to attract hard currency. In 1943, Coca had been officially informed by the Red Cross that we had as a family committed collective suicide. Ella Coca told us how her husband had come home and shown her the Red Cross's letter without a word and then broken down in tears.[19] And those tears sealed our friendship forever. Not only was Coca the personification of warm hospitality but also his charming wife and collaborator, Ella Grove-Coca, whose work in allergy has given her an outstanding place in science.

I recall also with gratefulness Dr. R. A. Lambert of the Rockefeller Foundation—his council full of understanding and his friendly, sincere attitude.

I wished to become acquainted with the teaching of medicine, on the one hand, and with the developments in microbiology—and especially virology—on the other. I visited an entire series of institutes and attended many conferences given by deans of various medical schools. I remember them all with respect and warmth. Contrary to the European system, deans in the United States do not change but are named permanently and are thus lasting organizers who have a clear conception of educational methods and of the type of physician-social worker whom they wish to train. But I soon realized that one could not hope to transport the American teaching methods to ruined Europe. In the good American schools (of course not all universities are so well equipped) teaching is so precise, and there are so many lecturers that students can delve into scientific detail at firsthand. This type of teaching is for the moment unattainable for us and for most European countries. How far off it seems when Welch looked jealously at the European laboratories and together with the great organizer of American science, Flexner, introduced them into the United States.[20] Teaching in Europe and Poland must be adapted to our present possibilities. Only in a few years will we perhaps be able to teach on such a level.

I nevertheless decided to devote my greatest efforts to acquainting myself with the scientific developments of the last years in my own area. It will be of no interest to the American reader to find out in which laboratories I worked and the subjects I discussed with various scientists. And yet I cannot resist the temptation to recall several studies that I feel to be of particular importance. Maybe the reader will be interested by which of the discoveries in my area—made in his country—are of greatest importance to world science.

I shall begin with the studies that are particularly close to me, which I link to my own efforts. In 1926 I worked on the question of whether a pregnancy proceeds normally when mother and fetus belong to differing blood groups. I did not succeed at that time in defining the pathological states that would be the indication of such an incompatibility. But I stressed that some group combinations provoke miscarriage and still birth more often than others. I considered, however, the question of how the mother and fetus live together despite serological differences as inconclusive. For twenty years I was isolated in my viewpoint—further, I was opposed in my efforts since many researchers considered the question to be of no consequence, artificial. How pleasant it was for me to hear Levine—the co-discoverer of the new blood trait called Rh—say to me during our first meeting in New York: "We should ask you to forgive us that for so many years we did not believe in the reality of your views."[21] For, in fact, the incompatibility of blood between mother and fetus, as it turned out, plays an enormous role, but does not affect first and foremost the incompatibility of groups A, B and O but rather, the new Rh factor, discovered by Landsteiner and his collaborators Levine and Wiener. A series of phenomena in miscarriage also can be explained by this serological conflict between mother and fetus. While I was still in New York, I gave a paper on the transitional forms of blood groups in which I presented my research little known in America before and during the war: my study of Pleiades, which I considered my scientific testament and published in the *Journal of Immunology.*[22] It seems to me that it can contribute to the understanding also of questions regarding the new Rh factor.

Coca's theory regarding a new form of allergy that he discovered and which expresses itself by an acceleration of the pulse in response to a harmful agent[23] appears to me as highly interesting and fertile. Coca explains various kinds of pathological phenomena by the hypersensitivity to various foods. Coca shared with us uncommonly interesting stories and considerations.

Craigie's work on so-called bacteriophages represents a great progress.[24] Bacteriophages are small entities that attack bacteria. Craigie managed to train them in such a way that they react only to certain types of germs of typhoid fever. This is of enormous epidemiological importance: Craigie's bacteriophages are like hunting dogs trained to catch only certain types of microorganisms. I went to visit Craigie in Toronto. Craigie likes to work in a small laboratory in a beautiful park. I spent several unforgettable days with him: his nervous face trembled with emotion when he told of his experiments. I was so taken by my stay with him, that I ended up not going to the Niagara Falls. On my way back I visited my young friend Witebski, who is professor of bacteriology at Buffalo and who has produced some beautiful studies on blood groups.[25]

In Boston I met Gordon, the chief epidemiologist of the U.S. Army.[26] He told me that thanks to DDT, Cox's vaccine against exanthematous typhus, and excellent sanitary organization, the American Army had identified only five cases of exanthematous typhus and thirty of typhoid fever throughout the entire war. This ought to be considered the greatest sanitary revolution in the world. In recent wars, millions perished from these diseases, and during the last war a terrible typhus epidemic occurred in Poland as a result of the poor German health system.

I will yet mention the wonderful development of the science of viruses, thanks to culture methods worked out by the American scientist Goodpasture as well as Stanley's marvelous discoveries of the crystallization of viruses[27] that build a bridge between life and lifeless material, Cox's interesting research on the culture of exanthematous typhus, and Hirst's on influenza infection. The works of Avery and Dubos of the Rockefeller Institute on the mechanisms of change of bacteria type also deserve a special place.[28] Dubos is a Frenchman by origin, a man with an enchanting smile, subtle and of volatile intelligence. His book *The Bacterial Cell* reads like a novel. Its motto is "paradoxes are better than prejudices."[29] Finally the wonderful development of immunochemistry represented above all by Heidelberg[er].[30] His laboratory is within the Columbia Medical School. His office is a closet of a few square feet, the rest of space being given over to laboratories. It is to him that we owe the memorable work on the chemical structure of the aureola of pneumococci. He is personally the personification of modesty and charm, he unites wonder with courage. I was deeply moved by meeting with my former teacher and friend Ulrich Friedemann.[31] He had managed to escape Hitler's hell. His works on the meningeal blood barrier are full of unexpected observations and open up new scientific horizons.

I am giving these few details because meeting these people and the discussions I had with them were of deep importance to me. I could finally speak about thoughts stretching toward the future. How easy the recipe for happiness seems when one speaks with such people—it's Goethe's old recipe: look every day at something beautiful and speak with a wise man. And in the face of such an easy recipe for happiness, how savage, how morbid the European conflicts seem. But unfortunately there are new conflicts and new issues. The first is the question of how to organize scientific work and the second—more tragic—how to ensure the freedom of science and responsibility of scientists in the face of history. Both issues are linked to the wonderful developments of physics and the discovery of the atomic bomb. Thousands of scientists were drawn into work, each one received a specific task often without realizing what its finality was. But the war is over and scientists wish to return to their own subjects. And now the question arises as to whether society can give up the collective work of these scientists. On the one hand, its supporters claim that only collective work was able to draw from an isolated discovery an enormous source of energy. The supporters of pure science will stress, on the other hand, that the free and nonobligatory work of scientists provide the mass of knowledge into which technology was able to delve as into a treasure trove of unending riches. A discussion then arises in which most scientists defend independent creativity. They point out that the activity of a scientist—if it is to be based on preestablished tasks—will quickly become sterile. There is too much talk about great factories and not enough about the first flash of thought—too much about the music of machines and the poetry of collective work, and too little about the student and assistant in love with science, where on a half-ruined workbench they had the first vision of an atom.

Related to the second question is whether the scientist has the obligation to consider the effects his discoveries may have. During the war in the face of the dangers of Nazism, the question was simple, but now emerges a deep conflict. Raymond B. Fosdick, the director of the Rockefeller Foundation, expressed in great depth the inner conflict of scientists. I will cite a passage from the activities report of the foundation for 1945: "The pursuit of truth has at last led us to the tools by which we can ourselves become the destroyers of our own institutions and all the bright hopes of the race. In this situation, what are we to do—curb our science, or cling to the pursuit of truth and run the risk of returning our society to barbarism?"[32]

If the activity of a researcher is not associated with humanitarian and constructive goals, should it not be limited? Can scientists allow themselves to limit their work only in view of specific objectives without asking what the social worth of their work is and what is its goal? Can we say to scientists: we must differentiate among truths, those which give scope to human happiness and those which threaten that happiness? Does science not bear responsibility and must the scientist go ahead with searching for the truth even if that truth will transform the earth into ruins and ashes? A certain English journalist exclaimed: "Will no one hold back these damned scientists?"[33] But how is one to hold back scientists, how is one to foresee what will come of our discoveries? After all, every discovery can be used for or against society. Man's enemy is not technology, but irrationality. Science reflects the social forces that surround it. And the author concludes: "Do we still have time? The feeling of fear and unrest accompanies this generation like so many threatening shadows. There is no escape from them. But if we are incapable of building normal foundations for the world, fear itself will not suffice to find solutions to these issues."[34]

My life has shown me what an unworthy role scientists can play when they do not feel themselves the militants of truth.

And with these words I wish to end this book.

Appendix: Biographical Annex of Frequently Cited Names

Editors' note: The following people are mentioned in more than one chapter by Hirszfeld in his autobiography. For more information, see the notes accompanying the initial reference in the text.

A

Stanisława Adamowicz (1888–1965) was born in Latvia and studied medicine and statistics in St. Petersburg and Geneva. She worked for the National Institute of Hygiene in the industrial city of Łódź and was an associate of Hirszfeld's in Warsaw, assisting him with blood transfusion efforts during the German siege of the city. (chapters 7, 20)

Róża Amzel (n.d.–1943/1944) worked in Hirszfeld's laboratory in Warsaw between the wars and was a frequent coauthor with him on blood group research. Hirszfeld arranged a scholarship for her to study in Paris before World War II, which she was unable to use. During the German occupation, Hirszfeld appointed her a deputy of his laboratory in the ghetto. She was subsequently captured and shot by the Germans. (chapters 7, 8, 16, 20, 21, 23, 28, 29, 36)

Ludwik Anigstein (1891–1975) studied philosophy in Heidelberg and medicine in Dorpat and became Poland's leading parasitologist between the wars, working with the National Institute of Hygiene from 1919 to 1939. After his dismissal from the Institute of Hygiene in 1939, he emigrated to the United States and became professor at the University of Texas Medical Branch in Galveston where he was visited by the Hirszfelds in 1946. (introduction, chapters 7, 8)

Heinz Auerswald (1908–70) was a lawyer and SS member in charge of the Warsaw Ghetto from May 1941 to November 1942. He was never convicted of war crimes. (chapters 21, 22, 24, 25, 26, 28)

B

Joseph Jules François Félix Babiński (1857–1932) was the son of Polish *émigrés* who fled Warsaw for Paris after the Revolution of 1848. He studied with

Charcot and became one of France's best-known neuropsychiatrists. (chapters 6, 14)

Alexandre Besredka (1870–1940) succeeded Mechnikoff at the Pasteur Institute where he continued to work in immunology. (chapters 14, 16)

Anna Braude-Heller (n.d.–1943) attended medical school in Switzerland before World War I. She headed the Berson and Bauman Children's Hospital (established 1878 and now Warsaw Children's Hospital) in the 1930s. The hospital was within the boundaries of the ghetto, but she refused to leave and died in the Warsaw Ghetto Uprising in 1943. (chapter 25)

Henryk Brokman (1886–1976) studied in Warsaw, Berlin, and Heidelberg and was one of Poland's most prominent pediatricians.(introduction, chapters 8, 21)

C

Józef Celarek (1888–n.d.) was a Polish doctor and a former Rockefeller Fellow who studied at Johns Hopkins after World War I. On his return to Poland, he was put in charge of the production of sera at the National Institute of Hygiene. He survived internment during the war in a German concentration camp and subsequently emigrated to the United States. (chapters 7, 9)

Mieczysław Centnerszwer (1874–1944) was a professor of chemistry in Latvia before World War II. He taught chemistry in the underground medical school of the Warsaw Ghetto. (chapters 23, 26, 29, 36)

Witold Chodźko (1875–1954) trained as a psychiatrist and was active in assisting children during World War I. Between the wars, he held the positions of Minister of Health, Polish representative to the League of Nations Health Committee, and to the Office International d'Hygiène Publique, he directed the Polish School of Hygiene from 1926 to 1939. (chapters 7, 36)

Arthur F. Coca (1875–1960) was one of the founders of immunology in the United States. He taught at Cornell Medical School from 1911–31 and finished his career at Lederle Laboratories in New Jersey. Coca was founding editor of the *Journal of Immunology* and head of the New York Blood Transfusion Betterment Organization. He met Hirszfeld while both were at Heidelberg in 1908 and remained a lifelong friend. (preface, chapters 2, 9, 13, 21, 39)

Leonard Jan Bruce-Chwatt (1907–89), born in Poland, was studying at the Pasteur Institute when World War II began. He moved to England, where he remained the rest of his life, becoming, after the war, one of the world's leading experts in malaria. (chapters 9, 14)

Adam Czerniaków (1880–1942) was president of the Warsaw Ghetto Jewish Community; he committed suicide in 1943 rather than comply with an order to round up Jews for deportation. (chapters 22, 23, 24, 25, 26, 28)

D

Jan Danysz (1860–1928) was a Polish biologist born in Chylin who spent the better part of his career in Paris at the Pasteur Institute. (chapters 6, 14)

Robert Debré (1882–1978) was a pediatrician and one of the foremost medical reformers in twentieth-century France. He befriended the Hirszfelds, especially Hanna, from the 1920s through the 1950s. (introduction, chapters 9, 13, 16)

F

Abraham Flexner (1866–1959), younger brother of Simon, is best remembered for his work on medical education reform and as an officer of the Rockefeller Foundation. After leaving the foundation in 1928, Flexner became the first director of the Institute for Advanced Study in Princeton. (chapters 14, 39)

Simon Flexner (1863–1946) studied under William Henry Welch and was a professor of bacteriology at the University of Pennsylvania when he was named the first director of the Rockefeller Institute for Medical Research (1901–35). He was a founding trustee of the Rockefeller Foundation. (chapters 8, 14)

G

Tadeusz Ganc (n.d.) was a physician and former major in the Polish Army. A converted Jew, he was appointed the administrative head of the Judenrat (Jewish Council) Health Division when typhus began to rage in the ghetto. (chapters 24, 25)

Abraham Gepner (1872–1943) was a businessman, well-known philanthropist, and member of the Warsaw City Council before World War II. He headed the ghetto's Supply Department, assisting Hirszfeld with equipment as well as Janusz Korczak's orphanage. (chapters 23, 25)

Marceli Godlewski (n.d.) was a Polish priest who had made anti-Semitic speeches before World War II, but from all accounts was quickly converted to sympathy and support for the Jews when faced with the conditions of the Warsaw Ghetto. He furnished fake birth certificates, smuggled children out of the "district," and arranged for hiding with parishioners outside the ghetto walls. He eventually was captured by the Nazis and sent to Auschwitz. (chapter 26)

Samuel Goldflam (1852–1932) was a Polish neurologist best known for his early work on myasthenia gravis (Erb-Goldflam syndrome) and the link between malnourishment and disease during World War I. (chapters 23, 25)

Adam Grabkowski (n.d.) provided hiding, assisted by his wife, for the Hirszfelds in a peasant cottage on the Nida River after they fled from the Warsaw Ghetto. (chapters 30, 33)

H

Wilhelm Hagen (1893–1982) was the second chief physician (*Amtartz*) of Warsaw, after Kurt Schrempf (see below) during World War II. He survived the war and enjoyed a respectable career but has a mixed reputation among historians and survivors of the ghetto. (chapters 23, 24, 26, 36)

Wanda Halber (n.d.–1938) was an assistant to Ludwik Hirszfeld and frequent author on blood groups and immunology. (chapters 8, 9, 14)

Ludwik Maurycy Hirszfeld (1816–76) was an anatomist of the nervous system, educated in Paris. (chapters 1, 14, 23)

K

Marcin Kacprzak (1889–1968) was a doctor and epidemiologist who studied in the United States thanks to a Rockefeller fellowship. He lectured at the National Institute of Hygiene in Warsaw between the wars and was a leader in public health in Poland after World War II. (chapters 7, 20, 21)

Stanisław Kiełbasiński (1883–1955) was Hanna Hirszfeld's brother-in-law (husband of sister Izabela). (introduction, chapters 7, 29, 36)

Wilhelm Kolle (1868–1935) was Paul Ehrlich's successor as director of the Institute for Experimental Therapy. (chapters 7, 10)

Jan Kołodziejski (1880?–1940), a close childhood friend of Rajchman, was the Polish delegate from the city of Warsaw to the First International Blood Transfusion Congress in Rome in 1935. He was killed by the Soviet Army at Starobielsk. (chapters 9,12)

Janusz Korczak (1879–1942), whose given name was Henryk Goldszmit, was a famous pediatrician and director of a well-known orphanage in Warsaw before World War II. In the ghetto, he set up an orphanage for two hundred children and voluntarily went with them to Treblinka. In 1990, his story was the basis of a film, *Korczak*, directed by Andrzej Wajda. (chapters 23, 25)

Nicholas Kossovitch (1884–1948) trained with Hirszfeld in Salonica and went to France after the war where he became one of the world's leading anthropological serologists at the Pasteur Institute. (chapters 5, 15)

Robert Kudicke (1876–1961) was one of Robert Koch's last surviving students, who served as dean of the medical school at Sun Yat-sen University in Canton in the interwar period. He directed the Institute of Hygiene in Warsaw from 1940 to 1944. (chapters 2, 25, 27, 36, 39)

L

Arnold Lambrecht (1902–n.d.) was a German district physician and a Nazi party member as of 1931 who became chief of public health in Warsaw in August 1940. According to a friend, Lambrecht "prided himself with having created the first ghetto in Piotrków in 1940 and then in Warsaw . . . and he [claimed] he had done so to contain typhus." (chapters 21, 22, 25)

Leone Lattes (1887–1954) authored the first Italian study of blood groups and serology in 1915. After service in World War I, he was appointed professor of Legal Medicine first at the University of Messina and eventually at the University of Pavia in 1933. Because of his Jewish ancestry, Lattes left Italy during the war but later returned to his post at Pavia. He was chairman of the Presidium of the First International Blood Transfusion Congress in Rome, 1935. (introduction, chapters 12, 13)

Edmond Lévy-Solal (1882–1971) was a French obstetrician and one of the cofounders in 1926 of the first city-wide transfusion service in Paris. He attended the First International Blood Transfusion Conference in Rome, 1935. (chapters 12, 16)

Sinclair Lewis (1885–1951) was an American novelist and playwright whose book, *Arrowsmith* (1926), criticized the institute model of medical research. The book was partly based on information from Paul de Kruif, his collaborator who previously had worked at the Rockefeller Institute for Medical Research. In 1931 Lewis became the first American to win the Nobel Prize for Literature. (chapters 8, 14)

Wilhelm Liebknecht (1826–1900) was a pacifist and leader of the German Social Democratic party. (chapters 4, 13)

M

Elie Metchnikoff (1845–1916) was a Russian bacteriologist and immunologist recruited by Pasteur himself in 1887. He won the Nobel Prize in Medicine and Physiology in 1908. (chapters 14, 18)

Mieczysław Michałowicz (1876–1965) was rector of the University of Warsaw between the wars and chairman of the Polish Pediatric Society, as well as an important politician in the Socialist (and later Democratic) Party. (chapters 10, 13)

Izrael Milejkowski (1887–1943) was a dermatologist by training who studied the effects of famine in the Warsaw Ghetto. He was head of the Health Council, and a Zionist; he considered Ludwik Hirszfeld a "turncoat." Milejkowski committed suicide on the train to Treblinka in January 1943. (introduction, chapters 22, 23, 24, 25, 26, 34)

Stefan Mutermilch (n.d.) worked at the Pasteur Institute and was a member of the League of Nations Standardisation Commission. (chapters 6, 14)

N

Gabriel Narutowicz (1865–1922) was a hydraulic engineer and a professor in Zurich who returned to Poland after World War I where he became minister of public works and then foreign minister. In December 1922, he was elected the first president of Poland, but five days after his inauguration he was assassinated by Right-wing extremists. (chapters 6, 7)

Ernst Georg Nauck (1897–1967) began working at the Hamburg Institute for Tropical Medicine in 1923 and was named director in 1945. He retained his position until his retirement in 1963, but he was subsequently accused of participating in war crimes. (chapters 21, 27)

Charles Nicolle (1866–1936) was the director of the Pasteur Institute of Tunis and author of the acclaimed *Le destin des maladies infectieuses* (1933) to which Hirszfeld wrote the introduction of the Polish translation. In 1928, Nicolle received the Nobel Prize in Medicine for his work on typhus. (introduction, chapters 4, 7, 9, 31, 39)

O

Wilhelm Ostwald (1853–1932), was a German Nobel laureate in chemistry in 1909. He often wrote about nonscientific matters, including, *Grosse Männer* (1909), which investigated the psychological causes of scientific creativity. (chapters 5, 18)

P

Kazimierz Pelczar (1894–1943) was a physician and a cancer researcher. Between the wars he rose to department chair and dean of the Medical School at the Polish University of Wilno (today Vilnius). Like Hirszfeld, he refused offers to leave the country before the outbreak of World War II. (chapters 14, 36)

Edouard Pożerski de Pomiane (1875–1964) was the son of Polish *émigrés* who fled to France after the thwarted 1863 uprising. He worked at the Pasteur Institute but was also well known for his gastronomic cookbooks. (chapter 6)

Jacek Prentki (1920–2009) was the son of a school friend of Hanna Hirszfeld. Although rarely mentioned in Hirszfeld's memoirs, Jacek Prentki grew up with Marysia Hirszfeld. He worked for a time in the ghetto but lived on the "Aryan" side. He survived the war, emigrated to Switzerland, and became a prominent physicist. (introduction, chapters 14, 17, 20)

Feliks Przesmycki (1892–1974) studied medicine in Kiev and pursued post-doctoral work at the Pasteur Institute in Paris (working with Calmette), the Harvard School of Public Health (working with Zinsser), and the Middlesex Hospital in London (working with McIntosh). He was one of the "pillars" of the National Institute of Hygiene, of which he became the director in 1945, after Hirszfeld had declined the job. (introduction, chapters 7, 8, 39)

R

Ludwik Rajchman (1881–1965), Hirszfeld's first cousin, is best known as the creator of UNICEF. He was also founder of the National Institute of Hygiene in 1918 and the director of the League of Nations Health Organization from 1921–38. (introduction, chapters 6, 7, 10)

Gaston Ramon (1886–1963) was associate director of the Pasteur Institute from 1934–40, then director from 1940–44. (chapters 15, 16)

Otto Reche (1879–66) was a German professor of anthropology who in partnership with Paul Steffan tried to make a science of utilizing blood group distribution as a marker of race, based on the Hirszfelds's 1919 article. He and Steffan were coeditors of the *Zeitschrift für Rassenphysiologie*, which elicited considerable initial interest, but eventually needed Nazi government subsidies to keep the journal going from the 1930s to the end of the war. A Nazi Party member, Reche was an advisor on population resettlement in Poland during the war but resumed his professorship at the University of Leipzig after the war. (chapters 9, 10)

Ernest Rénan (1823–92) was a French historian and philologist most famous for his *Life of Jesus*, but also as an early expert in the study of nationalism. The talk he gave at the Cercle Saint-Simon, on January 27, 1883, was published as *Le Judaïsme comme race et comme religion*. (chapters 23, 27)

Pietro Rondoni (1882–1956) was one of Italy's most highly respected oncologists who directed the National Institute for Cancer Research in Milan from 1935 until his death in 1956. (chapters 14, 15)

Max Rubner (1854–1932) was a distinguished German physiologist who founded the Kaiser-Wilhem Institut für Arbeitsphysiologie in Berlin in 1913. (chapters 1, 5)

Edward Rydz-Śmigły (1886–1941) served under Piłsudski during World War I and the war against Soviet Russia. After Piłsudski's death in 1935, Rydz-Śmigły succeeded him as inspector general of the army and de facto ruler of the country without opposition. When Germany and the Soviet Union invaded Poland, he fled to Romania. (chapters 10, 18, 20, 37)

S

Hans Sachs (1877–1945) was appointed to Ehrlich's Frankfurt Institute for Experimental Therapy, rising to deputy director in 1915. In 1920, he moved to Heidelberg to direct the Institute for Cancer Research, where Hirszfeld had studied before the war. The Nazis forced his resignation in 1935 and he emigrated to the United States. (chapters 3, 6, 7, 10)

Ferdinand Sauerbruch (1875–1951) was a pioneer in thoracic surgery and chair of surgery at Berlin's Charité Hospital from 1927 to the end of World War II. He expressed support as well as some criticism of the Nazi regime, and his legacy has been the subject of much debate. (chapters 10, 25)

Fritz Richard Schaudinn (1871–1906) was a German microbiologist who, along with Erich Hoffman, discovered the syphilis spirochete in 1905. (chapters 1, 25)

Kurt Schrempf (n.d.) was a German physician who was the first chief physician of Warsaw until early 1941 when he was replaced by Wilhelm Hagen. Before the war he had been the city health officer of Frankfurt-am-Main. (chapters 21, 24)

Julia Seydel (1884–1954) came from a polonized German family and studied biology in Liège, zoology at the Sorbonne, and bacteriology at the Pasteur Institute. (chapters 8, 29)

William Silberschmidt (1869–1947) was a professor of hygiene at the University of Zurich. (chapters 3, 9, 10, 19, 21)

T. Sokołowski (n.d.) was the delegate from the Polish Army to the First International Blood Transfusion Congress in Rome, 1935. (chapters 12, 13, 28, 39)

Zygmunt Srebrny (n.d.) was a well-known Polish otolaryngologist. Before the war he established the Society of Preventive Medicine, and he was editor of the *Warsaw Medical Journal.* Hirszfeld describes him as a "fervent Polish patriot." (chapters 21, 23)

Andrija Štampar (1888–1958), an international health leader, was the first minister of public health in Yugoslavia and representative to the League of Nations Health Committee. (chapters 5, 11)

Szymon Starkiewicz (1877–1962) was a physician who established a sanitarium in southern Poland for tubercular children in 1926. The Germans took it over and converted it to care for injured soldiers. (chapters 29, 30)

Paul Steffan (See Reche, above) (chapters 9, 10, 38)

Janusz Supniewski (1899–1964) was a Polish pharmacologist who studied in the United States between the wars as a Rockefeller Fellow. He chaired the Pharmacology Department at Jagiellonian University (Cracow) but was arrested and sent to Sachsenhausen camp in 1940. After the war he returned and served as dean of the Pharmacology School in Cracow. (chapters 8, 36)

Zofja Syrkin-Binstein was a doctor in the Warsaw Ghetto who worked on epidemic prevention and helped to organize the underground medical school. Hagen (unwarrantedly, as Hirszfeld states) fired her as the chief of the Health Service when the typhus epidemic became uncontrollable in the ghetto. She committed suicide—after first assisting others—in a railroad car on the way to a death camp. (chapters 24, 34)

Józef Szeryński (n.d.–1943) had been a colonel in the Polish police before heading the so-called Jewish police, created by the Germans even before the Warsaw Ghetto was established. There were many attempts on his life, and he committed suicide in January 1943. (chapters 23, 28)

Aleksander Szniolis (1891–1963) was a sanitary engineer at the State Institute of Hygiene between the wars who worked on water safety. After World War II he became dean of the Technical University of Wrocław. (chapters 7, 25, 39)

Gustaw Szulc (n.d.) succeeded Ludwik Rajchman as director of the National Institute of Hygiene in 1932. (chapters 7, 10, 18, 21)

Zygmunt Szymanowski (1873–n.d.) was one of Odo Bujwid's students and a prominent figure who assisted Jews during World War II. Several of his students taught in the underground medical school of the Warsaw Ghetto. (chapters 7, 21, 23, 29)

T

Arnault Tzanck (1886–1954) was the foremost pioneer of blood transfusion in France. French representative to the First International Blood Transfusion Congress in Rome, 1935, he hosted the Second Congress in 1937 in Paris. Of Jewish origin, Tzanck fled from occupied France to South America. When he returned after the war, he created the French National Blood Transfusion Service. (chapters 12, 13)

V

Fritz Verzár (1886–1979) was a well-known Hungarian physiologist who did early studies of blood group distribution after World War I. He went on to do his most important scientific work in gerontology and aging. (chapters 9, 11)

Emil von Behring (1854–1917) was a collaborator of Koch and winner of the first Nobel Prize in medicine for his discovery of a diphtheria therapy in 1904. He created the Behringwerke in Marburg, Germany, to produce sera and vaccines against infectious diseases. The Behringwerke operates today under the name of ZLB Behring. (chapters 3, 4, 21, 25)

Emil von Dungern (1867–1961) was director of the Cancer Research Institute in Heidelberg where Hirszfeld trained after medical school. He coauthored the groundbreaking discovery of blood group inheritance in 1910/1911 and remained in touch with Hirszfeld to the end of his life. (introduction, chapters 2, 3, 5, 7, 18, 21, 36)

Max von Pettenkofer (1818–1901) was a chemist by training who helped establish the new field of hygiene (public health) in the second half of the nineteenth century. He established an Institute of Hygiene in 1879 (now named after him) at Ludwig-Maximilians-University in Munich and is perhaps best known for his resistance to the germ theory of Koch. (chapters 6, 18)

W

Rudolf Weigl (1883–1957), of Austrian descent, was born in Moravia and moved for family reasons to Polish Lwów where he was raised and studied biology. He is best known for developing an early anti-typhus vaccine and was able to continue work during both the Soviet and German occupation of the city in World War II. He sent large quantities of vaccine to ghettos and concentration camps. (chapters 23)

Edmund Weil (1880–1922) was an Austrian physician who with Arthur Felix (1887–1956) developed the Weil-Felix test while working at the Koch Institute during World War I. (chapters 13, 23)

Michel Weinberg (1868–1940) emigrated from Russia to Paris where he studied at the Faculty of Medicine and at the Pasteur Institute. He worked initially with Metchnikoff, becoming *chef de laboratoire* in 1910, and finished his career at the institute in the late 1930s. (chapters 14, 16)

Rudolf Wohlrab (1909–n.d.) was German deputy director of the Institute of Hygiene in Warsaw, 1940–44. (Introduction, chapter 39)

Ambassador **Alfred Wysocki** (1873–1959) was a longtime member of the Polish diplomatic corps. (chapters 11, 15)

Z

Hans Zinsser (1878–1940) was a very influential American bacteriologist who made major contributions to the study of typhus and rickettsias. He taught at Stanford, Columbia, and Harvard but is most famous for his popularization of bacteriology in *Rats, Lice and History*. (introduction, chapters 1, 4, 5, 24, 31, 39)

Juliusz Zweibaum (1887–1959) was a Polish biologist who taught at the University of Warsaw between the wars. He was the main organizer of the German-approved "Sanitary Courses for Fighting Epidemics," the underground medical school in the Warsaw Ghetto. After the war he reorganized and directed the Department of Histology at the University of Warsaw until 1956. (chapters 23, 26, 29)

Notes

Preface

1. Marek Jaworski, *Ludwik Hirszfeld* (Warsaw: Wydawnictwo Interpress, 1977). The German translation is *Ludwik Hirszfeld: sein Beitrag zu Serologie und Immunologie* (Leipzig: BSB B. G. Teubner, 1980). There is also a medical thesis, Jakob Wolf Gilsohn, "Prof. Dr. Ludwig Hirszfeld" (Medical dissertation, München; Munich: A. Schubert, 1965).
2. Jaworski, *Ludwik Hirszfeld*, 105.
3. Waldomar Kozuschek, *Ludwik Hirszfeld (1884–1954) Rys życia i dzialalnosc naukowa* (Wrocław: Wydawnictwo Uniwersytetu Wrocławskiego, 2005).

Introduction

1. L. Hirszfeld to "Sein Magnifizenz, den Rektor der Universität Zürich," 9 September 1952, Archives of the Polish Academy of Sciences (hereafter cited as APAN), III/157 Ludwik Hirszfeld Papers (hereafter cited as LHP), folder 105. [Translated from the German by M. Balińska.]
2. The first version of *The Story of One Life*, typed by Hanna Hirszfeld, was later typed over by her sister, Izabela Belin Kiełbasińska, on carbon paper so as to have several copies. All the copies were lost, save that which Hirszfeld had himself buried under the floorboards in Miłosna. (Letter from Joanna Kiełbasińska-Belin, daughter of Izabela, to Marta A. Balińska, 14 April 2006).
3. Ludwik Hirszfeld, *Historia jednego życia* [The Story of One Life] (Warsaw: Czytelnik, 1946); L. Hirszfeld to Bermann Fischer Verlag, 26 October 1948, APAN III/157 LHP, folder 88.
4. L. Hirszfeld to [Robert Debré], 24 November 1947, APAN III/157 LHP, folder 88. Hirszfeld converted to Christianity when he returned to Poland after World War I. See below for more on the consequences, including criticisms he received.
5. L. Hirszfeld to Dr. Guterman [n.d., perhaps 1946], APAN III/157 LHP, folder 88. On the visit by Hirszfeld and his wife to the United States, see Ludwik Hirszfeld, "Some Fragments from the Polish Report about the Voyage to the United States and Canada," [n.d., perhaps March 1947], Rockefeller Archive Center, Sleepy Hollow, NY, RF RG 1.1 ser. 789, box 3, folder 26. He had planned to visit New York to attend the Third International Congress of Microbiology scheduled in September 1939, but his trip was canceled because of the outbreak of World War II.
6. James Oliver Brown to L. Hirszfeld, 3 November 1947, APAN III/157 LHP, folder 88.

7. L. Hirszfeld to Dr. Guterman, 15 November 1947, APAN III/157 LHP, folder 88.

8. Ludwik Hirszfeld to Robert Debré, 28 August 1947, APAN III/157 LHP, folder 104.

9. For background, see Paul K. Hoch, "Whose Scientific Internationalism?" *British Journal for the History of Science* 27 (1994): 345–49. On German science, see Alan Beyerchen, "On the Stimulation of Excellence in Wilhelmian Science," in *Another Germany: A Reconsideration of the Imperial Era*, ed. Jack R. Dukes and Joachim Remak, 139–69 (Boulder, CO: Westview Press, 1988); and Jonathan Harwood, *Styles of Scientific Thought: The German Genetics Community, 1900–1933* (Chicago: University of Chicago Press, 1993).

10. Examples in English of firsthand accounts of the Austro-Serbian front during the war include Gordon Gordon-Smith, *Through the Serbian Campaign: The Retreat of the Serbian Army* (London: Hutchinson, 1916); Fortier Jones, *With Serbia into Exile: An American's Adventures with the Army That Cannot Die* (New York, Century, 1916).

11. On the public health service in Poland, see two articles by Marta A. Balińska: "The National Institute of Hygiene and Public Health in Poland, 1918–1939," *Social History of Medicine* 9, no. 3 (1996): 427–45 and "The Rockefeller Foundation and the National Institute of Hygiene, Poland, 1918–1945," *Studies in History and Philosophy of Biological and Biomedical Sciences* 31C, no. 3 (2000): 419–32. Paul Weindling examines parallel Eastern and Central European developments in "Public Health and Political Stabilisation: The Rockefeller Foundation in Central and Eastern Europe between the Two World Wars," *Minerva* 31 (1993): 253–67. For historical background, see Richard Watt, *Bitter Glory: Poland and Its Fate, 1918–1939* (New York: Simon and Schuster, 1979); and Edward D. Wynot Jr, *Warsaw between the World Wars* (New York: Columbia University Press, 1983).

12. For examples of two Italian scientists under Mussolini, see Spencer M. DiScala, "Science and Fascism: The Case of Enrico Fermi," *Totalitarian Movements and Political Religions* 6, no. 2 (2005): 199–211; and Sandrine Bertaux, "Demographie, Statistique et Fascisme: Corrado Gini et l'ISTAT, entre Science et Ideologie (1926–1932)," *Roma Moderna e Contemporanea* 7, no. 3 (1999): 571–59. On German scientists during World War I, see Klaus Schwabe, "Ursprung und Verbreitung des alldeutschen Annexationismus in der Deutschen Professorenschaft im Ersten Weltkrieg," *Vierteljahrshefte für Zeitgeschichte* 14, no. 2 (1966): 105–38.

13. On the establishment of public health in the new Eastern European country of Yugoslavia, see H. Z. van Hyde, "A Tribute to Andrija Štampar, 1888–1958," *American Journal of Public Health and the Nation's Health* 48, no. 12 (December 1958): 1578–82.

14. The Holocaust accounts by Anne Frank and Elie Wiesel are the best known, but this literature has become quite extensive. For an example of analysis of the genre, see David Patterson, *Sun Turned to Darkness: Memory and Recovery in the Holocaust Memoir* (Syracuse, NY: Syracuse University Press, 1998); and for general historiography, see Raul Hilberg, *Sources of Holocaust Research* (Chicago: Ivan R. Dee, 2001). For an extensive analysis focused on the Holocaust autobiographies by historians, but which contains excellent background on the genre of autobiography itself, see Jeremy D. Popkin, "Holocaust Memories, Historians' Memoirs: First-person Narrative and the Memory of the Holocaust," *History and Memory* 15 (2003): 49–84. For a volume focused on Poland, see Joshua D. Zimmerman, ed., *Contested Memories:*

Poles and Jews during the Holocaust and Its Aftermath (New Brunswick, NJ: Rutgers University Press, 2003).

15. This role in the 1863 insurrection is related by Ludwik Rajchman, in Marta A. Balińska, *Ludwik Rajchman: Medical Statesman* (Budapest: Central European University Press, 1998), 2. For background on Jews in the Polish part of Russia, see Stephen D. Corrsin, *Warsaw before the First World War: Poles and Jews in the Third City of the Russian Empire, 1880–1914* (Boulder, CO: East European Monographs, 1989); and Władysław T. Bartoszewski and Antony Polonsky, *The Jews in Warsaw: A History* (London: Blackwell, 1991).

16. Interview with Felix Milgrom, Buffalo, NY, March 2002. For more on Milgrom, who left Poland and emigrated to the United States following Hirszfeld's death, see Stanisław Dubiski, "The Fifth Basic Sciences Symposium of the Transplantation Society—A Festschrift honoring Felix Milgrom," *Transplantation Proceedings* 31 (1999): 1456–57; and Felix T. Rapaport, "Felix Milgrom, Immunologist and Microbiologist par Excellence," *Transplantation Proceedings* 31 (1999): 1452–55.

17. According to Joanna Kiełbasińska-Belin, the niece of Hanna Hirszfeld.

18. Balińska, *Ludwik Rajchman*.

19. Irena Lepalczyk, *Helena Radlińska: życie i tworczosc´* [Helena Radlinska: life and works] (Torun: Adam Marszałek, 2001).

20. Ludwik Krzywicki, *Wspomnienia* [Memoirs], vol. II (Warsaw: Czytelnik, 1956), 234–35.

21. Andrzej Mencwel, *Etos lewicy* [Ethos of the left] (Warsaw: Państwowy Instytut Wydawniczy, 1990); Zofia Podgorska-Klawe, ed., *Słownik biograficzny polskich nauk medycznych XX wieku*, t1, z1 (Warsaw: IHNOiT, 1991), 37–40.

22. According to Joanna Kiełbasińska-Belin, personal communication to Marta A. Balińska, 2005.

23. Ludwik Hirszfeld, "Untersuchingen über die Hämagglutination und ihre physikalischen Grundlagen" (Munich: R. Oldenbourg, 1907); also published in the *Archiv für Hygiene* 63 (1907): 237–86.

24. The best history of typhus is Paul J. Weindling, *Epidemics and Genocide in Eastern Europe, 1890–1945* (New York: Oxford University Press, 2000). There are numerous firsthand accounts of medical assistance for the epidemic in Serbia, including Richard P. Strong, *Typhus Fever with Particular Reference to the Serbian Epidemic* (Cambridge, MA: Harvard University Press, 1920); and Elsie Corbett, *Red Cross in Serbia, 1915–1919* (Banbury: Cheney and Sons, 1964). For historical assessments, see also Régis Forissier, "L'aide medicale humanitaire apportée a la Serbie par la France et ses Alliés au cours de la Prémière guerre mondiale," *Revue Historique des Armées* 2 (1996): 9–26, as well as Weindling, *Epidemics and Genocide*.

25. In addition to the accounts detailed in note 10 above, for historical background, see Charles Fryer, *The Destruction of Serbia in 1915* (New York: Distributed by Columbia University Press, 1997).

26. For the official British history, see William Grant MacPherson, *History of the Great War: Medical Services General History*, Vol. IV, *Medical Services during the Operations on the Gallipoli Peninsula; in Macedonia . . .* (London: HMSO, 1924). For military background, see George B. Leon, *Greece and the Great Powers, 1914–1917* (Thessaloniki: Institution for Balkan Study, 1974), 256–69.

27. On the work of the institute, in addition to Balińska, "The National Institute of Hygiene and Public Health in Poland, 1918–1939," see Paul Weindling, ed., *International Health Organisations and Movements, 1918–1939* (New York: Cambridge

University Press, 1995). On Hirszfeld's specific research, see the next section of the introduction.

28. Joanna Kiełbasińska-Belin to M. A. Balińska, personal communication, 14 April 2006. On the Home Army, see Michael Alfred Peszke and Piotr Stefan, *The Polish Underground Army, the Western Allies, and the Failure of Strategic Unity in World War II* (London: McFarland, 2005).

29. See chapter 39, which was first published in the Polish edition of the book in 2000.

30. For a brief summary, see T. Heimrath, "Ludwik Hirszfeld (1884–1954) the Wrocław Years, 1945–1954" [In German], *Arch Hist Filoz Med* 63 (2000): 19–23.

31. Interviews by Marta A. Balińska and William H. Schneider with Felix Milgrom and Stanisław Dubiski, Buffalo, NY, and Toronto, March 2002.

32. This memoir is in personal papers of the family. [Translated from the Polish by M. Balińska.]

33. Interview with Dubiski, Toronto, March 2002. For an account of another communist satellite country subjected to the influences of Lysenko, see R. Hagemann, "How Did East German Genetics Avoid Lysenkoism?" *Trends in Genetics* 18, no. 6 (June 2002): 320–24.

34. According to Joanna Kiełbasińska-Belin, personal communication. For background on the Doctors' Plot, see Louis Rapoport, *Stalin's War against the Jews: The Doctors' Plot and the Soviet Solution* (Toronto: Free Press, 1990).

35. Marc Prentki, personal communication to Marta A. Balińska, December 2009.

36. Joanna Kiełbasińska-Belin, "W sprawie Ludwika Hirszfelda," [About Ludwik Hirszfeld] *Gazeta wyborcza* 21 August 2000. [Translated from the Polish by M. Balińska.]

37. This is in the personal papers of the family.

38. Ludwik Hirszfeld to [?], 24 March 1947, APAN III/157 LHP, folder 88. There is a large volume of literature on the subject of Polish-Jewish relations. In addition to Zimmerman, see, for example, M. Magdalena Opalski and Yisrael Bartel, *Poles and Jews: A Failed Brotherhood* (Hanover, NH: University Press of New England, 1992).

39. Ludwik Hirszfeld to Guterman, 2 March 1947, APAN III/157 LHP, folder 88. The chapter is included in this translation.

40. The anorexia was mentioned in separate interviews with Joanna Kiełbasińska-Belin in London during 2000 and Jacek Prentki and Hilary Koprowski during 1997 and 2002. See also Irena Krzywicka, *Wyznania gorzczycielki* [Confessions of a Sinner] (Warsaw: Czytelnik, 1992).

41. Biography has traditionally been and continues to be a major focus of this research, for example, Gerald L. Geison, *The Private Science of Louis Pasteur* (Princeton, NJ: Princeton University Press, 1995). In fact, it was the focus of strong attacks by new "social" historians of science in the last third of the twentieth century. For a defense, see Thomas L. Hankins, "In Defence of Biography: The Use of Biography in the History of Science," *History of Science* 17 (1979): 1–16; and for other examples, see Michael Shortland and Richard Yeo, eds., *Telling Lives in Science: Essays on Scientific Biography* (Cambridge: Cambridge University Press, 1996). For a recent assessment of the balance, see the essays in the special focus, "Biography in the History of Science," *Isis* 97 (2006): 302–32. For medicine, there is a whole journal devoted to the genre, *Medical Biography*.

The extent and usefulness of autobiography, however, has varied greatly. Broader scholarly analysis of autobiography is limited, with the exception of some indivi duals and disciplines, for example, Alain de Mijolla, "Freud, Biography, His Autobiography, and His Biographers," *Psychoanalysis and History* 1, no. 1 (1998): 4–27. Some examples are Pamela Moss, ed., *Placing Autobiography in Geography* (Syracuse, NY: Syracuse University Press, 2001); and Marten Hutt, "Medical Biography and Autobiography in Britain, c. 1780–1920" (PhD dissertation, University of Oxford, 1995).

42. See, for example, Luis Alvarez, *Adventures of a Physicist* (New York: Basic Books, 1987); Emilio Segré, *A Mind always in Motion: The Autobiography of Emilio Segrè* (Berkeley: University of California Press, 1993); and Edward Teller, *Memoirs: A Twentieth-Century Journey in Science and Politics* (Cambridge, MA: Perseus, 2001).

43. James Watson, *Double Helix* (Cambridge, MA: Harvard University Press, 1962); Robert C. Gallo, *Virus Hunting: AIDS, Cancer, and the Human Retrovirus: A Story of Scientific Discovery* (New York: Basic Books, 1991); Luc Montagnier, *Des virus et des hommes* (Paris: Odile Jacob, 1994).

44. See Larry R. Squire, ed., *The History of Neuroscience in Autobiography*, 4 vols. (San Diego: Academic Press, 2004); or the most recent volume in a series that began in 1930, Gardner Lindzey, *A History of Psychology in Autobiography*, 9 vols. (Washington, DC: American Psychological Association, 2007).

45. Hans Zinsser, *As I Remember Him; the Biography of R. S.* (Boston: Little, Brown, 1940).

46. Letter from Joanna Kiełbasińska-Belin to M. A. Balińska, 14 April 2006.

47. Hirszfeld has an entry in the *Dictionary of Scientific Biography* and a number of his contributions have been duly acknowledged in histories of twentieth-century immunology and microbiology. See, for example, Anne-Marie Moulin, *Le dernier langage de la médecine: Histoire de l'immunologie de Pasteur au Sida* (Paris: Presses universitaires de France, 1991); Arthur Silverstein, *A History of Immunology* (New York: Academic Press, 1989); William H. Schneider, "Chance and Social Setting in the Application of the Discovery of Blood Groups," *Bulletin of the History of Medicine* 57, no. 4 (1983): 545–62.

48. Felix Milgrom, "Fundamental Discoveries in Immunohematology and Immunogenetics by Ludwik Hirszfeld," *Vox Sang* 52 (1987): 149–51.

49. E. Witebsky to Nobel Committee for Physiology and Nutrition, 9 January 1950, APAN III/157 LHP, folder 113.

50. Maxwell M. Wintrobe, ed., *Blood Pure and Eloquent* (New York: McGraw-Hill, 1980); Douglas Starr, *Blood: An Epic History of Medicine and Commerce* (New York: Knopf, 1998); William H. Schneider, ed., "The First Genetic Marker: Blood Group Research, Race and Disease, 1900–1950," *History and Philosophy of the Life Sciences* 18, special issue (1996): 273–303.

51. Leone Lattes, an Italian counterpart of Hirszfeld, had the insight and good fortune to select this as the title of his first monograph on the blood groups, *Individuality of the Blood in Biology and in Clinical and Forensic Medicine* (London: Oxford University Press, 1932; originally published in Italian, 1922). Landsteiner titled his 1930 Nobel Prize lecture, "On Individual Differences in Human Blood."

52. Karl Landsteiner, "Über Agglutinationserscheinungen normalen menschlichen Blutes," *Wiener klinische Wochenschrift* 14 (1901): 1134; Karl Landsteiner and Max Richter, "Über die Verwertbarkeit individueller Blutdifferenzen für die forenzische Praxis," *Zeitschrift für Medizinbeamten* 16 (1903): 85–89.

53. Emil von Dungern and Ludwik Hirszfeld, "Über Nachweis und Vererbung biochemischer Strukturen. I," "Über Vererbung gruppenspezifischer Strukturen des Blutes. II," "Über gruppenspezifische Strukturen des Blutes. III," in *Zeitschrift für Immunitätsforschung* 4 (1910): 531–46; vol. 6, 284–92; and vol. 8 (1911): 526–62. Translated from the German in *Transfusion* 2 (1962): 70–74.

54. Ludwik Hirszfeld and Hanna Hirszfeld, "Serological Differences between the Blood of Different Races: The Results of Researches on the Macedonian Front," *Lancet* 2 (1919): 675–79.

55. Pauline Mazumdar, "Two Models for Human Genetics: Blood Grouping and Psychiatry in Germany between the Wars," *Bulletin of the History of Medicine* 70 (1996): 609–57; William H. Schneider, "Blood Group Research in Great Britain, France and the United States between the World Wars," *Yearbook of Physical Anthropology* 38 (1995): 77–104.

56. Hirszfeld, *Konstitutionsserologie und Blutgruppenforschung* (Berlin: Springer Verlag, 1928). For background, see Andrew Mendelsohn, "Medicine and the Making of Bodily Inequality in Twentieth Century Europe," in *Heredity and Infection: The History of Disease Transmission*, ed. Jean-Paul Gaudillière and Ilana Löwy, 21–80 (New York: Routledge, 2001); and Sarah W. Tracy, "George Draper and American Constitutional Medicine, 1916–1946: Reinventing the Sick Man," *Bulletin of the History of Medicine* 66, no. 1 (Spring 1992): 53–89.

57. Hanna Hirszfeld, Ludwik Hirszfeld, and H. Brokman, "On the Susceptibility to Diphtheria (Schick Test Positive) with Reference to the Inheritance of Blood Groups," *Journal of Immunology* 9 (1924): 571–91; Ludwik Hirszfeld and H. Zbrowski, "Sur la permeabilité élective du placenta en rapport avec le groupe sanguin de la mère et la foetus," *Comptes rendus de la Société de Biologie* 92 (1925): 1253–55.

58. Balińska, "The National Institute of Hygiene and Public Health in Poland"; Balińska, "The Rockefeller Foundation and the National Institute of Hygiene, Poland."

59. Tadeusz Heimrath, "Ludwik Hirszfeld (1884–1954)."

60. Andrzej Gorski et al., "Bacteriophages in Medicine," in *Bacteriophage: Genetics and Molecular Biology*, ed. Stephen Mcgrath and Douwe Van Sinderen, 125–58 (Norfolk, UK: Caister Academic Press, 2007); Elizabeth Kutter and Alexander Sulakvelidze, "Bacteriophage Therapy in Humans," in *Bacteriophages: Biology and Applications*, ed. Elizabeth Kutter and Alexander Sulakvelidze, 412–13 (Boca Raton, FL: CRC Press, 2004).

61. Like the Holocaust in general (see note 14), there has been a significant growth in scholarship on the Warsaw Ghetto. Some earlier reviews of the literature are still useful as an introduction to the more important personal accounts, such as Barbara Stern Burstin, "The Warsaw Ghetto: A Shattered Window on the Holocaust," *The History Teacher* 13, no. 4 (August 1980): 531–41. More recent attempts to keep up with the literature are now found on websites of the various Holocaust centers, such as Yad Vashem, http://www1.yadvashem.org/about_holocaust/bibliography/home_bibliography.html (accessed January 7, 2010). One important scholarly study that is closely associated with what Hirszfeld observed is Charles G. Roland, *Courage under Siege: Disease, Starvation and Death in the Warsaw Ghetto.* (New York: Oxford University Press, 1992). See also, Dalia Ofer, "Another Glance through the Historian's Lens: Testimonies in the Study of Health and Medicine in the Ghetto," *Poetics Today* 27, no. 2 (2006): 331–51. On the specific history of typhus, see note 67.

62. Charles Roland, *Courage under Siege*, 30–31. The "turncoat" charge was made by Izrael Milejkowski, head of the Warsaw Ghetto Jewish Health Council. For Hirszfeld's acknowledgment of the charge, see chapters 22 and 25.

63. Through 1919, Hirszfeld's publications appeared under his Germanized name, Ludwig Hirschfeld. After 1921, when he returned to Poland, he published under Ludwik Hirszfeld.

64. Professor Schmidt of the Behringwerke. See chapter 21.

65. Balińska, unpublished notes of interview with Rudolf Wohlrab, February 1995.

66. Roland, *Courage under Siege*, 277, note 11, cites this as a reason for why Hirszfeld has sometimes been given credit for organizing the underground medical school in the ghetto, something Hirszfeld himself denies. See chapter 23.

67. Paul J. Weindling, *Epidemics and Genocide*. On typhus, in addition to Roland's *Courage under Siege*, see Jean Lindenmann, "Herman Mooser, Typhus, Warsaw 1941," *Gesnerus* 59, no. 1–2 (2002): 99–113. Among the documents from Hirszfeld's life in the ghetto, see "Sprawozdanie z działalności Rady Zdrowia w okresie of 15 wrzesnia 1941 r. do kwietnia 1942 r," APAN III/157 LHP, folder 68, 1–14.

68. Hans Zinsser's *Rats, Lice and History* (Boston: Little, Brown, 1934) is the classic account. For Poland there is some scholarship besides Weindling and Roland. See, in particular, *Okupacja i medycyna*, [Occupation and Medicine] vols. I–V (Warsaw: Książka i Wiedza, 1971–84), which contains more than 1,500 pages on the subject of the German occupation and medicine in 1939–45. See also Marta A. Balińska, "La Pologne du choléra au typhus," *Bulletin of the Exotic Pathology Society* 92, no. 5 (1999): 349–54.

69. On typhus vaccines, see Lindenmann, "Herman Mooser"; and Weindling, *Epidemics and Genocide*.

70. Żydowski Instytut Historyczny w Polsce, Archiwum. ARI/363: "Walka z tyfusem, Warszawa, 1942." [The Fight against Typhus, Warsaw, 1942; Translated from the Polish by M. Balińska.]

71. See Edmund Russell, "The Strange Career of DDT: Experts, Federal Capacity and Environmentalism in World War II," *Technology and Culture* 40 (1999): 776–81; and Darwin H. Stapleton, "A Lost Chapter in the Early History of DDT: The Development of Anti-Typhus Technologies by the Rockefeller Foundation's Louse Laboratory, 1942–1944," *Technology and Culture* 46, no. 3 (July 2005): 513–40.

72. Marta A. Balińska, "Choroba jako ideologia: tyfus plamisty w okupowanej Polsce, 1939–1944" [Disease as Ideology: Typhus in Occupied Poland, 1939–1944], *Zeszyty historyczne* 126 (1998): 212–20.

73. In addition to Weindling, *Epidemics and Genocide*, see UNRRA, "Health Conditions in Poland," *Operational Analysis Papers* no. 31 (London: European Regional Office, 1947).

74. Jost Walbaum, ed., *Kampf den Seuchen!* (Krakau: Deutsche-Osten, 1941). On the establishment of the ghettos, see Christopher R. Browning, "Before the 'Final Solution': Nazi Ghettoization Policy in Poland (1940–1941)," in *Ghettos 1939–1945: New Research and Perspectives on Definition, Daily Life, and Survival* (Washington, DC: United States Holocaust Memorial Museum, 2005), 1–14. See also his earlier studies: Browning, "Nazi Ghettoization Policy in Poland, 1940–1941," *Central European History* 19 (1986): 343–68; and Browning, "Genocide and Public Health: German Doctors and Polish Jews, 1939–1941," *Holocaust and Genocide Studies* 3 (1988): 21–36.

75. Marta A. Balińska, "La Pologne face à ses crises médicales du XXe siècle," DEA thesis, Institut d'Etudes Politiques de Paris, 1989. For background on these accounts, see Dalia Ofer, "Another Glance Through the Historian's Lens." The most telling element missing in her otherwise excellent study is Hirszfeld's autobiography, which was not examined, presumably, because it was not available in English translation. Hirszfeld says in chapter 24 that he was Chair of the Health Council, although Roland, *Courage under Siege*, 69, adds somewhat circumspectly, "Hirszfeld takes the major credit for creating the Council, and his role has not been challenged by others."

76. L. Hirszfeld, "Ze wspomnienia epidemiologa" [The Reminiscences of an Epidemiologist] [undated], APAN III/157 LHP, folder 6, 1–26. Emphasis in original. [Translated from the Polish by M. Balińska.]

77. Charles G. Roland, "An Underground Medical School in the Warsaw Ghetto, 1941–1942," *Medical History* 33, no. 4 (October 1989): 399–419. See also Marta A. Balińska, "L'école de médecine du ghetto de Varsovie (1941–1942)," *Revue du Praticien 2008* 58 (31): 227–29.

78. Roland, *Courage under Siege*, 194, also acknowledges some criticism by former students but mostly of his personality. Ludwik Hirszfeld devotes chapter 23 to the school and his lectures, and his papers at the Polish Academy of Sciences contain several documents from the school that survived the war. See APAN III/157 LHP, folder 70.

79. Roland, *Courage under Siege*, 70; Hanna Hirszfeld, "Hunger States in Children" (undated), APAN III/157 LHP, folder 117. See also Anigstein below (note 80) and chapters 22 and 23.

80. L. Hirszfeld to Dr. Leake, 8 December 1947, APAN III/157 LHP, folder 104, 46–47 [original in English]. The editor was Ludwik Anigstein, also a Polish Jew who had worked at the Institute of Hygiene Warsaw and emigrated to the United States from Poland in 1939. The publication referred to is Ludwik Anigstein, ed., "A Symposium of Polish Medical Contributions in World War II," *Texas Reports on Biology and Medicine* 5 (1947): 155–87. Roland, *Courage under Siege*, 98–119, devotes a whole chapter to "Nutrition, Malnutrition and Starvation in the Ghetto," citing English, French, German, and Polish publications based on research smuggled out. See also Leonard Tushnet, *The Uses of Adversity: Studies of Starvation in the Warsaw Ghetto* (New York: Yoseloff, 1966). Hirszfeld's reference to a "Medical Society in the Ghetto," cannot be confirmed in other sources.

81. Balińska, unpublished notes of interview with Rudolf Wohlrab, February 1995, testifies to the impact in West Germany.

82. L. Hirszfeld to L. Doerr, 8 October 1947, APAN III/157 LHP, folder 104, 117–18. [Translated from the German by M. Balińska.]

83. L. Hirszfeld to Joseph Needham, 28 August 1947, APAN III/157 LHP, folder 107, 2 [original in English]. Hirszfeld is probably referring to the Fourth International Microbiological Congress held in Copenhagen, 20–26 July 1947.

84. L. Hirszfeld to the Nobel Committee, 17 October 1952, APAN III/157 LHP, folder 105, 80.

85. Originally published in *Archivum Immunologiae et Therapiae Experimentalis* 3 (1955): 17. [Translated from the German by M. Balińska.]

86. Paweł Jasienica, *Opowieści o żywej materii* [Stories of Live Matter], quoted on the back leaflet of the second edition of Hirszfeld's *Historia jednego życia* (Warsaw: PAX, 1956). Subsequent quotes are from this same source. [Translated from the Polish by M. Balińska.]

Chapter One

1. Bolesław Hirszfeld (1849–99), Ludwik's paternal uncle, was a chemist by training and prominent social activist who devoted his life to the struggle for education, civil liberties, and the defense of political prisoners.
2. Hirszfeld grew up first in Warsaw and then in Łódź, both of which were in the Russian partition of Poland. Łódź was an industrial town producing mainly textiles and known in the nineteenth century as "the Polish Manchester."
3. Würzburg was no sleepy university town. Typical of the high level of research activity at German provincial universities, it was here that Virchow did his fundamental research on cellular pathology in the 1850s, and in 1895 Röntgen made his discovery about mysterious "X-rays." Hirszfeld obviously skips over details of his early life. For more, see introduction.
4. Theodor Boveri (1862–1915), German zoologist who studied at Munich but took a position at Würzburg in 1893 where he remained until the end of his career. The standard biography is Fritz Baltzer, *Theodor Boveri: Life and Work of a Great Biologist*, trans. Dorothea Rudnick (Berkeley: University of California Press, 1967).
5. August Rauber (1841–1917) and Friedrich Kopsch, *Anatomie des Menschen*. 4 vols. (Stuttgart: G. Thieme, 1886).
6. Berlin, when Hirszfeld arrived in 1903, was the center of world scientific thought, the most important political and international capital, and enjoyed its greatest influence in its entire history. See Ronald Taylor, *Berlin and Its Culture: A Historical Portrait* (New Haven, CT: Yale University Press, 1997).
7. Georg Simmel (1858–1918) was a major theorist of German philosophy and social science at the turn of the nineteenth to the twentieth century. He taught at the University of Berlin from 1885 to 1914, when he took a Chair at Strasbourg. Somewhat of a showman, according to authorities as varied as Walter Benjamin and George Santayana, Simmel's lectures were attended by the cultural elite of Berlin as well as by students. On his lecture style, see Lewis H. Coser, "Georg Simmel's Style of Work," *American Journal of Sociology* 63 (1958): 635–41.
8. Johannes Orth (1847–1923).
9. Ernst Bumm (1858–1925), gynecologist and professor at Würzburg.
10. Hirszfeld is talking about the Russian Revolution of 1905 in Poland. For background, see Robert E. Blobaum, *Rewolucja: Russian Poland, 1904–1907* (Ithaca, NY: Cornell University Press, 1995).
11. Martin Ficker (1868–1950) was a professor of hygiene at the University of Berlin who worked with Max Rubner (see note 16). He finished his career after World War I in São Paulo, Brazil. See Karl Wilhelm Jotten, "In Memoriam, Otto Spittal und Martin Ficker," *Archiv für Hygiene und Bakteriologie* 134, no. 2 (1951): 83–84.
12. The side-chain theory was the basis of Ehrlich's immunological research. It assumed unique chemical receptors as part of body cells that interacted with toxins and produced antibodies. See John Parascandola, "The Theoretical Basis of Paul Ehrlich's Chemotherapy," *Journal of the History of Medicine and Allied Sciences* 36 (1981): 19–43. Hermann Oppenheim (1858–1919) was a leading German neurologist at the end of the nineteenth century. See J. M. S. Pearce, "Hermann Oppenheim (1858–1919)," *Journal of Neurology, Neurosurgery, and Psychiatry* 74 (2003): 569.
13. The reference is likely to Ludwik Hirszfeld, *Immunologia ogólna* [General immunology] (Warsaw: Czytelnik, 1949).

14. Alfred Wolff-Eisner (1877–1948) is best known for the Calmette and Wolff-Eisner ophthalmo reaction, a particular form of tuberculine reaction provoking a conjunctivitis of the eye. For more, see Peter Voswinckel, "Grabstein vor dem Verfall bewahrt. In memoriam Professor Alfred Wolff-Eisner (1877–1948), jüdischer Pionier der Pollenkrankheit," *Allergologie* 11 (1988): 41–46.

15. Fritz Schaudinn (1871–1906) was co-discoverer (with Hoffmann) of the cause of syphilis in 1905, while at the Charité Hospital in Berlin.

16. Max Rubner (1854–1932) was the successor to Robert Koch as director of the Hygiene Institute at the University of Berlin. For more, see J. C. Knowles, "Max Rubner," *Diabetes* 6 (1957): 369–71.

17. Ulrich Friedemann (1877–1949), bacteriologist, was a professor of hygiene at the University of Berlin with appointments at the Koch Institute and Virchow Hospital. In 1933, he left first for London, then New York, where Hans Zinsser (see chapter 5, note 6) arranged an appointment at the Jewish Hospital in New York. See chapter 38.

18. Julius Morgenroth (1871–1924) worked at the Berlin Institut für Serumforschung und Serumprüfung with Paul Ehrlich.

Chapter Two

1. Hirszfeld was writing toward the end of World War II. In fact, Heidelberg was spared bombing and was used by the Americans as an army base.

2. Hirszfeld is referring to Theodor von Wasilewski (1868–1941?). See Waldemar Kozuschek, *Ludwik Hirszfeld (1884–1954) Rys życia i działalność naukowa* [Ludwick Hirszfeld (1884–1954) A Sketch of His Life and Scientific Activity] (Wrocław: Wydawnictwo Uniwersytetu Wrocławskiego, 2005), 52.

3. The Cancer Research Institute had opened in 1906 as the Institut für Experimentelle Krebsforschung, led by Vincent Czerny. See Ralf Brüer, "100 Jahre organisierte Krebsforschung," *Deutsche medizinische Wochenschrift* 125, no. 15 (2000): 473–74.

4. Jacques Loeb (1859–1924) had trained in Germany but emigrated to the United States in 1892.

5. Fritz Haber (1868–1934) actually received the Nobel Prize for chemistry in 1918 for his work on the fixation of nitrogen from air. He studied in Heidelberg from 1886–91 and had an appointment in nearby Karlsruhe from 1894 until he moved to Berlin in 1911.

6. Emil von Dungern (1867–1961) is described in detail below and in the introduction. He was born in Würzburg and attended the gymnasium in Freiburg. He continued his medical studies there and then pursued them in Munich and Berlin.

7. Wilhelm von Möllendorf (1887–). Hans Reiter (1881–1969) eventually became president of the Reichsgesundheitsamt, the highest German authority on all fundamental questions of health. Although interrogated after the war, he escaped prosecution. Reiter's Syndrome (an arthritis condition) is named after him, but there has recently been a call to rescind the name because of his Nazi service. See Daniel J. Wallace and Michael Weisman, "Should a War Criminal Be Rewarded with Eponymic Distinction?: The Double Life of Hans Reiter (1881–1969)," *Journal of*

Clinical Rheumatology 6, no. 1 (2000): 49–54; and the February 2003 issue of *Seminars in Arthritis and Rheumatism* 32, no. 4 (2003): 207–45 including six articles beginning with Norman L. Gottlieb and Roy Altman, "An Ethical Dilemma in Rheumatology: Should the Eponym Reiter's Syndrome Be Discarded?"

8. Robert Kudicke (1876–1961), one of Robert Koch's last surviving students, was dean of the Sun Yat Sen University in Canton in the interwar period and later the director of the Institute of Hygiene in Warsaw, from 1940–44. See Paul J. Weindling, *Epidemics and Genocide in Eastern Europe 1890–1945* (New York: Oxford University Press, 2000), 275–77 for Kudicke's work in Warsaw.

9. Otto Warburg (1883–1970) obtained his medical degree at Heidelberg in 1911. He received the Nobel Prize for medicine in 1931 for his discoveries regarding the role of enzymes in oxidation and reduction.

10. Arthur F. Coca (1875–1959) was one of the founders of the *Journal of Immunology* and served as its editor for thirty-two years. He remained a lifelong friend of Hirszfeld. For more, see chapters 13 and 39, plus the introduction.

11. Ludwik Hirszfeld married Hanna Kasman (1884–1964), the daughter of a physician from Łódź who had died from a disease caught from a patient. Hanna studied medicine in Móntpellier and subsequently specialized in pediatrics. For a biographical sketch of Hanna Kasman, see introduction.

12. Indeed, the friendship between Hirszfeld and Dungern lasted to the end of their lives, despite the upheavals of the 1930s and 1940s.

13. Hirszfeld is undoubtedly referring to the bacteriologist Napoleon Jan Gąsiorowski (1876–1941), director of the Institute of Hygiene in Lwów between the wars. See *Polski Słownik Biograficzny* [Polish Biographical Dictionary], Vol. 7 (Cracow: Polska Akademia Nauk and Polska Akademia Umiejętnośći, 1948), 353–54.

14. "Lutek": Polish nickname for "Ludwik."

15. Karl Landsteiner (1868–1943) reported his discovery of the human blood groups in 1901. For this and subsequent work in immunology, he received the Nobel Prize for medicine in 1930.

16. Felix Bernstein (1878–1956) was a mathematician in Göttingen when he learned of Hirszfeld's discoveries about the distribution of blood types in different human populations. His statistical calculations, published in 1924, revealed the underlying genotypes of the ABO blood system that corrected von Dungern and Hirszfeld's erroneous suggestion in 1911.

17. Hirszfeld, *Konstitutionsserologie und Blutgruppenforschung* (Berlin: Springer Verlag, 1928).

18. The *Habilitation* was necessary for university teaching. See chapter 3, note 8.

Chapter Three

1. Willi Von Gonzenbach (1880–1955) became professor of hygiene and bacteriology, Eidgenössische Technische Hochschule Zürich from 1922 to 1950.

2. Little can be found about R. Klinger (n.d.) who, together with Hirszfeld, developed a serodiagnostic reaction test for syphilis in 1914, which did not, however, replace the Wassermann test. See Ludwig Hirschfeld [*sic*] and R. Klinger, "Zur Theorie der Serumreaktionen," *Berliner klinische Wochenschrift* (1914): 1173–76. Between 1913 and 1916, Hirszfeld published nineteen articles with Klinger.

3. For a biography of Silberschmidt, see Maria Loretan, *William Silberschmidt, 1869–1947, Hygieniker und Bakteriologe* (Switzerland: Juris Druck and Verlag, 1988).

4. Of course, Hirszfeld is romanticizing here. This was also the period of resistance to Einstein's inclusion at the top ranks of German physicists, and young Hitler was learning his populist anti-Semitism from the mayor of Vienna, Karl Lüger.

5. The Mianowski Fund was established by alumni of the Warsaw Main School shortly after the death of Józef Mianowski (1804–79), social activist, physician, and rector of the school. It was the only Polish higher education institution under Russian rule, and by the end of the nineteenth century it became the main Polish source of support for research, thanks to generous contributions by the general public and successful businessmen. See http://www.mianowski.waw.pl/history.htm (accessed May 30, 2005).

6. T. Dieterle, L. Hirschfeld, and R. Klinger, "Epidemiologische Untersuchungen über den endemischen Kropf," *Archiv für Hygiene* 81 (1913): 128–78.

7. Polish salt mines near Cracow.

8. This qualification exam was for *Habilitation*, a degree after the doctorate required for university teaching and the position of *privatdozent*, which typically required an additional thesis, a defense, and a public lecture.

9. Hirszfeld remained faithful to his method of lecture, as reported by numerous and varied spectators.

10. Paul Ehrlich was born March 14, 1854; Emil Adolf Behring was born March 15, 1854. Behring received the Nobel Prize in medicine in 1901, Ehrlich in 1908.

11. On Robert Doerr (1871–1952) see R. Walter Schlesinger, "Robert Doerr— Prophet of the Nature of Viruses, Founder of the 'Archives of Virology'," *Archives of Virology* 142, no. 4 (1997): 862–73.

12. For more on Hans Sachs (1877–1945), see Horst Dickel, "Hans Sachs," in *German-speaking Exiles in Ireland 1933–1945*, ed. Gisela Holfter, 183–213 (Amsterdam: Rodopi, 2006).

13. Ernst Friedberger (1875–1932) was later named professor of hygiene at Griefswald University in 1915, where one of his students was Fritz Schiff, who became the leading German serologist in the interwar period. In the 1920s, Friedberger was named editor of the *Zeitschrift für Immunitästsforschug*, and in 1926, he moved to Berlin to become director of the Prussian Research Institute for Hygiene and Immunology.

14. On Rudolf Kraus (1868–1932), see J. Teichmann, "On the 100th Birthday of Rudolf Kraus" [In German], *Wiener medizinische Wochenschrift* 118, no. 42 (October 19, 1968): 869–72.

15. Hirszfeld remained in contact with Landsteiner, who was then in Austria but emigrated to the United States right after World War I, achieving further renown at the Rockefeller Institute for Medical Research.

Chapter Four

1. The so-called Manifesto of the 93 German intellectuals to the civilized world, which supported the Kaiser's declaration of war, was signed on October 4, 1914, by almost one hundred of Germany's leading cultural and intellectual lead-

ers, including fifty-six professors and Nobel Prize winners such as Behring, Ehrlich, Haber, Planck, and Röntgen. It was a series of statements beginning "It is not true," followed by various denials, beginning with "that Germany is guilty of having caused this war"; "Neither the people, the Government, nor the Kaiser wanted war," and so forth. See http://www.gwpda.org/1914/93intell.html (accessed May 31, 2005). See also Klaus Schwabe, "Ursprung und Verbreitung des alldeutschen Annexationismus in der Deutschen Professorenschaft im Ersten Weltkrieg," *Vierteljahrshefte für Zeitgeschichte* 14, no. 2 (1966): 105–38.

2. Hirszfeld was technically a citizen of Russia, which ruled Poland when he was born. In fact, he traveled to Serbia with Russian documents.

3. Karl Radek (1885–1939) was a socialist/communist leader, born in Austrian Poland, who became a member of the Social Democratic party in Poland. He spent the war in Switzerland, where he joined the Bolsheviks in exile. After the revolution, he became an important member of the new Bolshevik government of Russia. For more, see Warren Lerner, *Karl Radek* (Stanford: Stanford University Press, 1970).

4. Wilhelm Liebknecht (1826–1900) was a pacifist and leader of the German Social Democratic party. His son, Karl (see chapter 13), was an early leader of the German Communist party.

5. The typhus epidemic in Serbia was the worst ever witnessed in Europe. Typhus fever, transmitted by body lice, broke out among Serbian refugees and Austrian prisoners who moved south after the initial battles of the war. In the first six months of 1915, the epidemic of typhus fever is estimated to have affected 500,000, or one in six inhabitants. For background, see Paul J. Weindling, *Epidemics and Genocide in Eastern Europe, 1890–1945* (New York: Oxford University Press, 2000); and Richard Pearson Strong, *Typhus Fever with Particular Reference to the Serbian Epidemic* (Cambridge, MA: Harvard University Press, 1920).

6. Ludwik Klocman was the first husband of Hanna's half sister Izabela Belin.

7. Otto Haab (1850–1931) taught ophthalmology at the University of Zurich from 1886 to 1919.

8. Among the missions sent to Serbia were the Scottish Women's Hospitals, which established a hospital unit led by Dr. Elsie Maud Inglis in Kraguievatz, and the American Red Cross Sanitary Commission supported by the Rockefeller Foundation, in which Hans Zinsser served. For an account by the head of the American Red Cross Commission, see Richard Pearson Strong, *Typhus Fever*; and for France, see Régis Forissier, "L'aide Medicale Humanitaire Apportée a la Serbie par la France et ses Alliés au Cours de la Prémière Guerre Mondiale," *Revue Historique des Armées* 2 (1996): 9–26.

9. Valjevo was the northernmost city controlled by the Serbs, and where a large number of Austrian prisoners were held. According to Strong, *Typhus Fever*, 19–20, 87–93, the epidemic began among the Austrian prisoners held at Valjevo in December 1914. The epidemic reached its height in April 1915, when nine thousand cases per day were reported among the military and civilian population. By August, the epidemic was largely over, according to Strong, thanks primarily to Serbian efforts. Although Strong mentions numerous foreign personnel, Hirszfeld is not among them.

10. Vojvoda [Marshal] Živojin Mišić (1855–1921) was the most successful Serbian military commander during the war. (Note: earlier Western spelling, "Mishitch.") There are a number of firsthand accounts of the Austro-Serbian front during the

war. See, for example, Gordon Gordon-Smith, *Through the Serbian Campaign: The Retreat of the Serbian Army* (London: Hutchinson, 1916); and Fortier Jones, *With Serbia into Exile: An American's Adventures with the Army That Cannot Die* (New York: Century, 1916). For a more scholarly account, see Charles Fryer, *The Destruction of Serbia in 1915* (New York: Distributed by Columbia University Press, 1997).

11. Ernest Conseil (1879–1930) was a French medical doctor and biologist. He worked with Charles Nicolle at the Pasteur Institute in Tunis. See http://www.pasteur.fr/infosci/archives/cns0.html (accessed January 10, 2010). In addition to Forissier, on the work of the French mission, which was dispatched by Nicolle, see Kim Pelis, "Prophet for Profit in French North Africa: Charles Nicolle and the Pasteur Institute of Tunis, 1903–1936," *Bulletin of the History of Medicine* 71 (1997): 599–60.

12. Charles Nicolle (1866–1936) was the director of the Pasteur Institute of Tunis. Hirszfeld wrote the preface for the Polish edition of his best-known work *Le destin des maladies infectieuses* (Paris: Alcan, 1933). In 1928, Nicolle received the Nobel Prize in medicine for his work on typhus.

13. Aleksa (shortened Aca) Savić was a member of the Yugoslav parliament from 1925 and also Health Minister. In that role, among other things, he participated at the opening of the School of the Public Health Building in Zagreb in 1927 that had been financed by the Rockefeller Foundation. He unexpectedly died on January 27, 1928. See Milutin M. Ranković, "Dr. Aleksa Savić," *Srpski arhiv za celokupno lekarstvo* 30/2 (1928):154–60.

14. On the Scottish Women's Hospitals and Elsie Inglis (1864–1917), the so-called Scottish Florence Nightingale who established the Serbian mission, see the account by Inglis' sister, Eva Shaw McLaren, *The History of the Scottish Women's Hospitals* (London: Hodder & Stoughton, 1919); and Leah Leneman, *In the Service of Life: The Story of Elsie Inglis and the Scottish Women's Hospitals* (Edinburgh: Mercat Press, 1994). See also I. Emslie Hutton, *With a Woman's Unit in Serbia, Salonica and Sebastopol* (London: Williams and Norgate, 1928).

15. The terms *ague* and *fever* were used interchangeably well into the twentieth century and most commonly described the fever and chills symptoms of malaria, as Hirszfeld does here.

16. Serfdom was abolished in Russian-dominated Poland after the failed insurrection of 1863.

17. August von Mackensen (1849–1945) conducted a successful campaign against the Russians in 1914 and then was sent to lead the Austrians against the Serbs. Joined by the Bulgarians in October 1915, his forces quickly pushed the Serbs and their French and British allies back to Salonica.

18. Nikola Pasić (1845–1926) became prime minister of Serbia in 1904 and led the country during World War I. See Gerald G. Govorchin, "Emergence of the Radical Party in Serbian Politics," *American Slavic and East European Review* 15 (1956): 511–26; and Natasa Milan Matic, "Nikola P. Pasić and the Radical Party, 1845–1941," (dissertation, Boston University, 2000).

19. The distance is about 680 kilometers (420 miles).

20. The Ottoman victory in the Battle of Kosovo Field, June 15, 1389, resulted in Turkish rule over the Serbs for the next five hundred years.

21. Peter I Karadjordjević (1844–1921) had been Serbia's king since 1903. His son, Alexander (1888–1934), who succeeded him in 1921, was assassinated in France in 1934.

22. King Alexander I of Yugoslavia (1888–1934) lived a turbulent life as crown prince, regent, then commander-in-chief and finally king of Serbia (and, latterly, Yugoslavia). See http://www.firstworldwar.com/bio/alexander_serbia.htm.

23. Hirszfeld is referring to Ralph Spencer Paget (1864–1940), a British career diplomat who had served as ambassador to Serbia from 1910–13, and who was sent back to Serbia at the outbreak of war as a Commissioner of the British Red Cross. For more on Paget, see Čedomir Antić, *Ralph Paget: A Diplomat in Serbia* (Belgrade: Serbian Academy of Sciences and Arts, Institute for Balkan Studies, special editions 94, 2006). Also available at http://www.balkaninstitut.com/pdf/izdanja/posebno/pedzet.pdf (accessed August 13, 2009).

Chapter Five

1. Presumably the basis for Hirszfeld's article, "Aus meinem Erlebnissen als Hygieniker in Serbien," *Correspondenz-Blatt für Schweizer Ärzte* 46 (1916): 513–31.

2. Hirszfeld is most likely referring to Wilhelm Ostwald (1853–1932), German Nobel laureate in chemistry 1909. He was a colorful character who often expressed his ideas about nonscientific matters. The probable source of Hirszfeld's paraphrase is a book Ostwald wrote after his retirement in 1906, *Grosse Männer* (Leipzig: Akademische Verlagsgesellschaft, m.b.h., 1909), investigating the psychological causes of scientific creativity.

3. As Hirszfeld notes, following the retreat from the von Mackensen offensive, the Serbian Government, along with 200,000 soldiers and civilians were evacuated to the island of Corfu. It was the largest sea evacuation until Dunkirk in 1940.

4. Etienne Burnet (1873–1960); Marcel Lisbonne (1883–1946).

5. There are a number of firsthand accounts of the Salonica story. See, for example, G. Ward Price, *The Story of the Salonica Army* (London: Hodder & Stoughton, 1917); and Arthur James Mann and William T. Wood, *The Salonica Front* (London: A & C Black, 1920). For background, see a recent general history by Mark Mazower, *Salonica, City of Ghosts: Christians, Muslims and Jews, 1430–1950* (New York: Knopf, 2005).

6. Hirszfeld's description upon seeing Salonica is remarkably similar to that in Hans Zinsser's autobiography, *As I Remember Him; the Biography of R. S.* (Boston: Little, Brown, 1940). In his private papers, Hirszfeld acknowledges that the autobiography of Zinsser, who died in 1941, was a major source of inspiration for his own. Zinsser was a renowned bacteriologist at Columbia and then Harvard who had studied in the same laboratory in Heidelberg as Hirszfeld, but just before Hirszfeld arrived. Although they served at the same time in Serbia in 1916, Hirszfeld was in the city of Valjevo in the northwest, while Zinsser spent only two months in Skopje in the south (Macedonia) setting up a bacteriological laboratory. See Hans Zinsser, "Report of Bacteriologist of the American Red Cross Sanitary Commission to Serbia," in *Typhus Fever with Particular Reference to the Serbian Epidemic*, ed. Richard P. Strong, 261–68 (Cambridge, MA: Harvard University Press, 1920). For more on Zinsser, see William C. Summers, "Hans Zinsser: A Tale of Two Cultures," *Yale Journal of Biological Medicine* 72, no. 5 (1999): 341–47.

7. Hirszfeld's references to the numerous hospital personnel are difficult to trace. He obviously used this chapter to record his gratitude to the many individuals with whom he and his wife shared a very intense experience. Examples of these very

colorful characters mentioned in this chapter are Duchess (also called Princess) Narishkina, an influential person in Salonica, rumored to be the mistress of the French Commander, General Sarrail, and later the Serbian Crown Prince. At the end of the war, she established a hospital in Skopje. See Elsie Corbett, *Red Cross in Serbia, 1915–1919* (Banbury: Cheney and Sons, 1964), 119, 151.

Colonel Roman Sondermeyer was a surgeon and chief of the Serb health service (Hutton, *With a Woman's Unit in Serbia*, 99). Dr. Louise McIlroy had trained as an obstetrician but served as a surgeon and directed one of the first Scottish Women's Hospital units in France in May 1915. In the fall of 1915, her unit was assigned to the French *Armée d'Orient*. She later became a professor in London (Hutton, *With a Woman's Unit in Serbia*; Corbett, *Red Cross in Serbia*, 119, 151). See also Corbett, "Louise McIlroy," *British Medical Journal* 1, no. 589 (February 17, 1968): 451.

Dr. Isabel Emslie (full name Isabel Galloway Emslie Hutton) (1887–1960) was a psychiatrist and author of *With a Woman's Unit* . . . Dr. Alice Hutchinson also served in the Scottish Women's hospital in Serbia and remained there as the Austrians advanced. She was taken prisoner, repatriated, and then was sent back to Salonica.

8.   In Islam, *houris* are beautiful, celestial, black-eyed damsels who are the reward of every believer in the Muslim paradise.

9.   Ague, or malaria, was the biggest health problem on the Salonica front. Unrestricted submarine warfare made it impossible to relieve troops, and this resulted in a high disease rate, primarily from malaria, for northern European troops.

10.   Colonel Roman Sondermeyer was surgeon Chief Inspector of military hospitals. See note 7 above.

11.   Adam Mickiewicz (1798–1855) and Juliusz Słowacki (1809–49) were famous Polish Romantic poets.

12.   Mihailo Milovanović (1879–1941) was a war painter of the Serbian Army's Supreme Command in World War I. His exhibitions became very popular between the wars.

13.   On the British medical services, see William Grant MacPherson, *History of the Great War: Medical Services General History*, Vol. IV. *Medical Services during the Operations on the Gallipoli Peninsula; in Macedonia* . . . (London: HMSO, 1924), 62–162.

14.   Hirszfeld, "A New Germ of Paratyphoid," *Lancet* (February 22, 1919): 296–97. Typhoid, or enteric fever, is caused by *Salmonella paratyphi* and by *S. paratyphi* A, B, and C strains. In 1880, Karl Joseph Eberth coined the term "B typhosus." Four years later Georg Gaffky was able to culture *Salmonella typhi*. In 1896, Emile Charles Achard and Raoul Bensaude isolated *S. paratyphi* B and used the term "paratyphoid fever" to describe it.

15.   Injections of quinine were commonly used in the Salonica hospital, according to Price, *The Story of the Salonica Army*.

16.   This is Ehrlich's phrase to describe the "magic bullet" that would be a "therapy of complete sterilization" to eliminate the invading pathogen without harming the host organism.

17.   A viral disease transmitted by the bite of a sand flea.

18.   On Kosta Todorović (1887–1975), see Živadin Perisić, "Akademik Prof. Kosta Todorović (5.VII 1887–22.IX 1975)" [In Memory of Academician Prof. Kosta Todorović (5 July 1887–22 November 1975)], *Srpski Arhiv za Celokupno Lekarstvo* 103, no. 10 (October 1975): 911–14.

19.   Nicholas Kossovitch (1884–1948) went to France after the war and became one of the world's leading anthropological serologists, thanks to his training with

Hirszfeld in Salonica. See Hirszfeld's elaboration on sero-anthropology in this chapter and Pierre O. Lépine, "Necrologie: Nicholas Kossovitch (1884–1948)," *Annales de l'Institut Pasteur* 75 (1948): 62–64.

20. Hirszfeld's discovery was published as "A New Germ of Paratyphoid."

21. On Andrija Štampar (1888–1958) and public health in the new Yugoslavia, see Linda Killen, "The Rockefeller Foundation in the First Yugoslavia," *East European Quarterly* 24, no. 3 (1990): 349–72; and Patrick Zylberman, "Fewer Parallels than Antitheses: René Sand and Andrija Štampar on Social Medicine, 1919–1955," *Social History of Medicine* 17 (2004): 77–92.

22. Edmund Faustyn Biernacki (1866–1912). Hirszfeld's article was published as "Über ein neues Blutsymptom bei Malariakrankheit," *Correspondenz-Blat für Schweizer Ärzte* 47 (1917): 1007–12.

23. For historical background, see William H. Schneider, "The History of Research on Blood Group Genetics: Initial Discovery and Diffusion," *History and Philosophy of the Life Sciences* 18 (1996): 273–303.

24. This talk was published as "Essai d'application des methodes sérologiques au problème des races," *Anthropologie* 29 (1919): 505–37.

25. Carl Voit (1831–1908) and Max Rubner (1854–1932) were distinguished German physiologists. This discovery also led to Rubner's "law of surface area," which found a correlation between metabolism and the size of the animal.

26. The landmark article, authored by Hirszfeld and his wife, was published as "Serological differences between the blood of different races," *Lancet* 2 (1919): 675–79.

27. Anders Retzius (1796–1860) was a Swedish anatomist who devised a "cephalic index" to measure human skulls for anthropological racial classification.

Chapter Six

1. Richard Paltauf (1858–1924) was an Austrian pathologist and bacteriologist. He helped establish Vienna as a center for pathological anatomy research and founded the state Institute of Sero-therapy in 1905. Arnold Pick (1851–1924) was a Czech neurologist and psychiatrist who was head of the psychiatric clinic at the German university in Prague.

2. Hans Sachs and Walter Georgi, "Zur Kritik des serologischen Luesnachweises mittels Ausflockung," *Münchener medizinische Wochenschrift* 66 (1919): 440–42. This is now known as the Sachs-Georgi reaction. Complement is a series of proteins that acts to cause the precipitation reaction used as the sero-diagnostic method in syphilis.

3. Max von Pettenkofer (1818–1901) is best known for his rejection of the germ theory of Koch, and his demonstration of the fact by drinking water contaminated with cholera.

4. Gabriel Narutowicz (1865–1922) was a hydraulic engineer who had been a professor in Zurich and returned to Poland where he became minister of public works and foreign minister. In December 1922, he was elected the first president of Poland. Five days after his inauguration he was assassinated by right-wing extremists.

5. Konstanty Janicki (1876–1932) was a famous parasitologist who had been educated in Germany and held positions in Italy and Switzerland before returning to

Poland in 1919 to become chair of zoology at Warsaw University. On Rajchman (1881–1965), see introduction and chapter 7. Jan Danysz (1860–1928), biologist; Joseph Jules François Félix Babiński (1857–1932), neurologist, was the son of Polish *émigrés* to France; Edouard Pożerski de Pomiane (1875–1964) was also the son of *émigrés* from the 1863 Polish revolution who went to France. On the brothers Minkowski, Hirszfeld is undoubtedly referring to Eugène Minkowski (1884–1972) who became a prominent French psychiatrist and Paul Minkowski (1888–1947), also a well-known neuropsychiatrist. Stefan Mutermilch (n.d.) worked at the Pasteur Institute and was a member of the League of Nations Biological Standardization Commission.

Chapter Seven

1. On Rajchman, see Marta A. Balińska, *Ludwik Rajchman: Medical Statesman* (Budapest: Central European University Press, 1998).

2. Czesław Wroczyński (1889–1940) was a close associate of Ludwik Rajchman in the early days of the Institute of Hygiene (Warsaw), serving as deputy director and subsequently, director of the Polish Health Service. He was shot by the Red Army in Katyń in 1940. See Jan Bohdan Gliński, *Słownik biograficzny lekarzy i farmaceutów ofiar drugiej wojny światowej* [Biographical Dictionary of Physician and Pharmacist Victims of the Second World War] (Wrocław: Wydawnictwo Medyczne Urban & Partner, 1997), 472–73.

3. Docent (or *dozent*) is the equivalent of associate professor, immediately below the rank of professor.

4. Witold Chodźko (1875–1954) was trained as a psychiatrist and was active in assisting children during World War I. Between the wars, he held the positions of minister of health, Polish representative to the League of Nations Health Committee, and the Office International d'Hygiène Publique, as well as director of the School of Hygiene (1926–39). See Zenon Maćkowiak, "Remarks on Life and Activities of Witold Chodźko (1875–1954)" [In Polish], *Przegląd Lekarski* 37, no. 4 (1980): 437–40.

5. Ignacy Jan Paderewski (1860–1941), world-famous Polish pianist, enlisted his fame in the cause of Polish independence and especially, assistance for Polish refugees during World War I. He signed the Versailles Treaty as Polish Foreign Minister and was Poland's representative to the League of Nations.

6. See Marta A. Balińska, "Assistance and Not Mere Relief: The Epidemic Commission of the League of Nations," in *International Health Organisations and Movements, 1918–1939*, ed. Paul J. Weindling, 81–108 (Cambridge: Cambridge University Press, 1995).

7. Karl Flügge (1847–1923), a prominent German hygienist.

8. This was placed under Thorvald Madsden (1870–1957), director of the Statens Seruminstitut, Copenhagen, who became a close colleague of Hirszfeld. See chapter 10 and Joseph Parnas, "Thorvald Madsen 1870–1957. Leader in International Public Health," *Danish Medical Bulletin* 28, no. 2 (April 1981): 82–86.

9. Wilhelm Kolle (1868–1935) was Paul Ehrlich's successor as director of the Institute for Experimental Therapy.

10. English derivation of French word *charivari*, or mock celebration.

11. Peter Rona (1871–1944) was a pioneer in the field of biochemistry who taught no less than three Nobel laureates: Ernst Chain, Hans Kreb, and Fritz Lippman. Born in Budapest, he headed the Institute for Pathology at the University of Berlin, until he was dismissed.

12. Zygmunt Szymanowski (1873–n.d.) was one of Odo Bujwid's students. See Zygmunt Szymanowski, "Thirtieth Anniversary of Founding of the State Institute of Hygiene" [In Polish], *Polski Tygodnik Lekarski* (Warsaw) 6, no. 35, supplement (August 27–September 2, 1951): 173–74.

13. Jean Cantacuzène (1863–1934) was a prominent Romanian bacteriologist and representative to the League of Nations Health Committee. See Uta Deichmann, "The Expulsion of Jewish Chemists and Biochemists from Academia in Nazi Germany," *Perspectives in Science* 7 (1999): 38; and V. M. Kushnarev, "Jon Cantacuzene (1863–1934)" [In Polish], *Zhurnal Mikrobiologii, Epidemiologii, Immunobiologii* 30 (December 1959): 128–30.

14. Helena Sparrow (1891–1970) was a typhus specialist and close associate of Charles Nicolle whom she joined at the Tunis Pasteur Institute in the 1930s. See Robert Debré, "Helena Sparrow—a Polish and French Scientist," *Materia medica Polona* 11, no. 1 (January–March 1979): 79–82.

15. Rudolf Abel (1868–1942) was a bacteriologist who served as a medical officer in Hamburg and Berlin before being appointed a professor of hygiene in Jena in 1915. He wrote on hunger during World War I, "Von Hungernot und Seuchen in Russland," *MMW* 70 (1923): 485–87, 633–35, as cited in Weindling, *Epidemics and Genocide in Eastern Europe 1890–1945*, 173–74.

16. For background, see Marta A. Balińska, "Assistance and Not Mere Relief." The Serbian epidemic provided valuable experience. For background, see Alfred E. Cornebise, *Typhus and Doughboys: The American Polish Typhus Relief Expedition, 1919–1921* (Newark: Delaware University Press, 1982).

17. Dr. Stanisław Sierakowski (1888–1949) worked in the Division of Vaccines and was a former Rockefeller Fellow who trained at the Harvard School of Public Health and the Institute for Medical Research in London. Among other things, he was the first in Poland to begin research on the microbiology of meteorites. See *Polski Słownik Biograficzny* [Polish Biographical Dictionary] Vol. 37 (Cracow: Polska Akademie Nauk and Polska Akadmie Umiejętnośći, 1999), 305–6. Dr. Józef Celarek (1888–1960?) was also a former Rockefeller Fellow who studied at Johns Hopkins. Upon his return to Poland, he was put in charge of the production of sera at the institute. He survived the war in a German concentration camp and subsequently emigrated to the United States.

18. For background and overview of the institute, see Marta A. Balińska, "The National Institute of Hygiene and Public Health in Poland, 1918–1939," *Social History of Medicine* 9, no. 3 (1996): 427–45.

19. On the Epidemics Commission of the League of Nations, see Balińska, "Assistance and Not Mere Relief," 93–96.

20. Aldo Castellani (1877–1971) was co-discoverer with Bruce of sleeping sickness and founder of the Royal Institute of Tropical Diseases in Rome. He had earlier visited Poland as part of a medical commission of the League of Red Cross Societies. Richard Otto (1872–1952) later was director of the Frankfurt Institut für Serumforschung from 1935 through the war. See Richard Prigger, "Richard Otto," *Archiv für Hygiene und Bakteriologie* 136, no. 8 (1952): 557–58.

21. The school was established with the support of the Rockefeller Foundation. See Marta A. Balińska, "The Rockefeller Foundation and the National Institute of Hygiene," *Studies in History and Philosophy of Biological and Biomedical Sciences* 31C, no. 3 (2000): 419–32.

22. On the American Jewish Joint Distribution Committee, see Zosia Szajkowski, "Disunity in the Distribution of American Jewish Overseas Relief, 1919–1939," *American Jewish Historical Quarterly* 58, no. 3 (1969): 376–407, 484–506. On Rockefeller fellowships, see *Rockefeller Foundation Directory of Fellowship Awards, 1917–1950* (New York: Rockefeller Foundation, 1951); and William H. Schneider, "The Men Who Followed Flexner: Richard Pearce, Alan Gregg, and the Rockefeller Foundation Medical Divisions, 1919–1951," in *Rockefeller Philanthropy and Modern Biomedicine* (Bloomington: Indiana University Press, 2002), 7–60.

23. Henry Hallett Dale (1875–1968) won the 1936 Nobel Prize in medicine for his research on the chemical transmission of nerve impulses. For more on the work of the league, see chapter 10 and W. Charles Cockburn, "The International Contribution to the Standardization of Biological Substances. I. Biological Standards and the League of Nations 1921–1946," *Biologicals* 19 (1991): 161–69; and Pauline Mazumdar, "Serology, Standardisation and Collective Security: The Standardisation Commission of the League of Nations, 1920–1939," in *Proceedings of the European Academy for Standardization Workshop, Paris 2004* (Hamburg: European Academy for Standardization, 2004), 41–52.

24. For background on Rockefeller support for public health in Poland and other East European countries between the wars, see Paul Weindling, "Public Health and Political Stabilisation: The Rockefeller Foundation in Central and Eastern Europe between the Two World Wars," *Minerva* 31 (1993): 253–67.

25. Ludwik Anigstein (1891–1975) studied philosophy in Heidelberg and medicine in Dorpat and became Poland's leading parasitologist between the wars, working with the institute from 1919 to 1939. He also served as a League of Nations expert in Thailand and Liberia, among other places. After his dismissal from the Institute of Hygiene in 1939, he emigrated to the United States. According to his son (M. A. Balińska interview with Robert Anigstein, NYC, 1994), he was dismissed when the authorities discovered that he belonged to B'nai Brith, a Jewish organization reputed to be "masonic," which published its membership list in the newspapers. In the United States, he became professor at the University of Texas Medical Branch in Galveston.

Feliks Przesmycki (1892–1974) studied medicine in Kiev, with additional study at the Pasteur Institute in Paris (working with Calmette), the Harvard School of Public Health (working with Zinsser), and the Middlesex Hospital in London (working with McIntosh). He was one of the "pillars" of the institute and became director in 1945. His unpublished memoirs, *Wspomnienia*, can be found in the Library of the Institute of Hygiene (Warsaw). On Feliks Przesmycki (1892–1974) see Edmund Wojciechowski, "In Memoriam: Feliks Przesmycki (1892–1974)" [In Polish], *Przeglad Epidemiologiczny* 29, no. 3 (1975): 386–89; and Izabela Szenkowa, "80-lecie profesora F. Przesmyckiego" [On the 80th Birthday of Prof. F. Przesmycki], *Służba zdrowia* 7 (1972): 9.

On Marcin Kacprzak (1889–1968), a specialist in vital statistics and former Rockefeller Fellow, see H. Kirschner, "Marcin Kacprzak and the development of social medicine in Poland" [In Polish], *Archiwum Historii i Filozofii Medycyny* 51, no. 1 (1988): 89–100. Brunon Nowakowski (1890–1966), a former Rockefeller fellow, was head of the Division of Epidemiology at the Polish School of Hygiene. During World

War II, he was dean of the Division of Hygiene at the Polish Medical Faculty in Edinburgh (created in 1941). He returned to communist Poland and became director of the School of Occupational Medicine in Katowice (Silesia). See Alfred Puxia, "Brunon Nowakowski, wybitny przedstawiciel higieny i medycyny pracy," [Brunon Nowakowski, Outstanding Representative of Hygiene and Labor Medicine] *Archiwum Historii i Filozofii Medycyny* 51, no. 3 (1988): 359–72.

26. Stanisława Adamowicz (1888–1965), born in Latvia, studied medicine and statistics in St. Petersburg and Geneva. She organized the regional office of the institute in the industrial city of Łódź, working closely with Marcin Kacprzak. She wrote extensively on public health issues for the wider public.

27. Roman Nitsch (1873–1957) was a student of Odo Bujwid. See J. Kostrzewski, "Remembrance of Roman Nitsch in Connection with His Views on the Nature of Fixed Virus" [In Polish], *Przegląd Epidemiologiczny* 11, no. 2 (1957): 195–97. On the Polish school of microbiology, see Władysław Kunicki-Goldfinger, "Migawki z dziejów mikrobiologii polskiej," [Snapshots from the History of Polish Microbiology] *Postępy mikrobiologii* 26, no. 3 (1987): 135–54.

28. Stanisław Grabski (1871–1949) was the most prominent minister of the economy between the wars and introduced major economic reform.

29. On Józef Stanisław Hornowski (1874–1923), see "Historia patomorfologii WSTEP," http://www.cmkp.edu.pl/wstep.htm#HORNOWSKI%20J%D3ZEF%20 STANIS%A3AW (accessed January 11, 2010).

30. Róża Amzel (n.d.) was a frequent coauthor with Hirszfeld on blood group research between the wars. She did not survive World War II. See http://warszawa.getto.pl (accessed January 11, 2010).

31. Hirszfeld cites no specific source in original text. It is likely, "Obsługa bakteriologiczna Państwa. Doniesienie II, » [Le service bactériologique de l'Etat, IIᵉ Communication], *Lekarz Polski*, 12 (1936): 106–13.

32. Filip Eisenberg (1876–1942) was a student of Odo Bujwid and directed the Institute of Hygiene at Cracow. See also, on the Polish school of microbiology, Władysław Kunicki-Goldfinger, "Migawki z dziejów mikrobiologii polskiej," [Moments from the History of Polish Microbiology] *Postępy mikrobiologii* 26, no. 3 (1987): 135–54."

33. Odo Feliks Bujwid (1857–1942) is considered the father of Polish microbiology. He studied with both Robert Koch and Louis Pasteur and set up one of the first rabies vaccine productions outside of France. He was also a prominent member of the Polish Socialist Party and activist for women's education. Eugeniusz J. Kucharz, Marc A. Shampo, and Robert A. Kyle, "Odo Bujwid—Pioneer in Microbiology," *Mayo Clinic Proceedings* 65, no. 2 (February 1990): 286.

34. Brigadier-General Stanisław Rouppert (1887–1945) was head of the Medical Department of the Ministry of War from 1926–39.

35. A. Chrzanowski, [A page from the history of Polish oncology. The foundation of Count Jakub Potocki], *Nowotwory*. 27(3), (1977 Jul–Sep):255–59. [Article in Polish].

36. See Wiesław Magdzik, "Results of the Work of Sanitary-Epidemiologic Service in Poland during the Last Eighty-five Years and Perspectives for the Future" [In Polish], *Przegląd Epidemiologiczny* 58, no. 4 (2004): 569–81.

37. Tadeusz Chrapowicki, "Mikołaj Łącki, MD" [In Polish], *Pediatria Polska* 44, no. 8 (August 1969): 929–31.

38. Hirszfeld is talking about the Mendelian inheritance of blood groups. Accordingly, if the blood type of the mother and child is known, men of certain

blood types could be excluded as the father. For example, a child with type O blood could not have a father of type AB. For background, see William H. Schneider, "Chance and Social Setting in the Application of the Discovery of Blood Groups." *Bulletin of the History of Medicine* 57, no. 4 (1983): 552–55.

39. Hirszfeld is talking about his book, later translated into French under the title, *Les groupes sanguins, leur application à la biologie, à la médecine et au droit* (Paris: Masson & Cie, 1938).

40. Irena Krzywicka (1899–1994), prominent Polish feminist, writer, journalist and translator.

41. "Medics Circles" were associations of medical students. On blood transfusion, see S. Hornowski, "Remembrances on the Beginnings of Blood Donation in Warsaw. The Foundation of the Academic Blood Donation Centre in Warsaw" [In Polish], *Archiwum Historii i Filozofii Medycyny / Polskie Towarzystwo Historii Medycyny i Farmacji* 62, no. 3 (1999): 229–32.

42. Gabriel Narutowicz (1865–1922). See chapter 6.

43. Felicjan Sławoj-Składkowski (1885–1962) was a Polish general who held a number of cabinet ministries in the interwar period, including prime minister.

44. Stanisław Wyspiański (1869–1907). Polish painter and playwright of the *Młoda Polska* movement.

Chapter Eight

1. On Janusz Supniewski (1899–1964), see Ryszard W. Gryglewski, "Prof. Janusz Supniewski (1899–1964)" [In Polish], *Archiwum Historii Medycyny* 37, no. 3 (1974): 383–85.

2. Original appeared as "Wanda Halberówna," *Warszawskie Czasopismo Lekarskie*, 15, (1938):118.

3. On Henryk Brokman (1886–1976), see Radosław Owczuk, J. Ulewicz-Filipowicz, and Anna Balcerska, "Professor Henryk Brokman—Scientist and Paediatrician" [In Polish], *Archiwum Historii i Filozofii Medycyny* 63, no. 1 (2000): 35–43.

4. Jerzy Morzycki (1905–54) worked on typhus vaccine during the war, reactivated the Marine and Tropical Medicine Institute in Gdańsk, and was a cofounder of the Polish Parasitological Society, as well as its first president. See Stefan Krzyński, "In Memoriam Prof. Dr. Jerzy Morzycki, 1905–1954" [In Polish], *Biuletyn Państwawego Instytutu Medycyny Morskiei i Tropikulnej j W Gdansku* 6 (1955): 9–13.

5. Julia Seydel (1884–1954) came from a polonized German family [of German origin but choose Polish culture and identity] and had studied biology in Liège, zoology at the Sorbonne, and bacteriology at the Pasteur Institute. See Teresa Ostrowska, "Julia Seydel," *Służba zdrowia* 9 (March 1969).

6. On Tadeusz Sprożyński (1903–74), see Edmund Wojciechowski, "In Memoriam: Tadeusz Sprożyński (1903–74)," [In Polish] *Medycyny Doświadczalna i Mikrobiologia* 27, no. 1 (1975): 97–99.

7. Hirszfeld is probably referring to Simon Flexner (1863–1946), director of the Rockefeller Institute for Medical Research from 1920–35.

8. Alexis Carrel (1873–1944), who won a Nobel Prize in 1912, worked at the Rockefeller Institute for most of his career (1906–39). He wrote *Man, the Unknown*

after he retired, which became a best seller both in France and in the United States. But Carrel received a great deal of criticism for his work after he came out of retirement, following the French military defeat in 1940, to establish a research institute in Paris under Nazi occupation. For two recent biographies, see Alain Drouard, *Alexis Carrel (1873–1944): De la mémoire à l'histoire* (Paris: L'Harmattan, 1995); and Andres Horacio Reggiani, *God's Eugenicist: Alexis Carrel and the Sociobiology of Decline* (New York: Berghahn Books, 2006).

9.   In his novel *Arrowsmith*, author Sinclair Lewis criticized the institute model, based on information from Paul de Kruif, his collaborator on the novel, who previously had worked at the Rockefeller Institute for Medical Research.

10.   Michael Heidelberger (1888–1991), immunochemist, worked first at the Rockefeller Institute until he joined the Columbia Medical faculty in 1928, where he worked on bacterial polysaccharides, especially of the *Pneumococci* (with O. T. Avery) and on the immunochemistry of protein, antibodies, and antigens. See Julius M. Cruse, "A Centenary Tribute: Michael Heidelberger and the Metamorphosis of Immunologic Science," *Journal of Immunology* 140, no. 9 (1988): 2861–63.

Chapter Nine

1.   A *voivodship* is an administrative region and local government unit in Poland equivalent to *states* in the United States, *Länder* in Germany, or *départements* in France.

2.   Hirszfeld is of course quoting himself.

3.   For background, see Pauline M. H. Mazumdar, "'In the Silence of the Laboratory:' The League of Nations Standardizes Syphilis Tests," *Social History of Medicine* 16, no. 3 (2003): 437–59.

4.   Hanna Hirszfeldowa, Ludwik Hirszfeld, and H. Brokman, "Wrażliwość na błonicę w świetle badań nad konstytutucją i dziedzicznością" [Sensitivity to Diphtheria in Light of Research on Constitution and Heredity], *Medycyna Doświadczalna i Społeczna* 2 (1924): 125–42; and Ludwik Hirszfeld, "Les vaccinations antidiphthériques en Pologne," *Bulletin de l'Académie de Médecine* 121 (1939): 712–22.

5.   Spała is a forest resort in Poland.

6.   Hirszfeld is using Landsteiner's original terminology for what has now come to be called the O group. Verzár and Coca's publications were Fritz Verzár, "Rassenbiologische Untersuchungen mittels Isohamagglutinen," *Biochemische Zeitschrift* 126 (1922): 33–39; Arthur F. Coca and Olin Diebert, "A Study of the Occurrence of the Blood Groups among American Indians," *Journal of Immunology* 8 (1923): 487–91. For more on Verzár, see chapter 11. For more on Coca, see introduction.

7.   Felix Bernstein, "Ergebnisse einer biostatistischen zusammenfassenden Betrachtung uber die erblichen Strukturen des Menschen," *Klinische Wochenschrift* (1924): 1495–97.

8.   Wanda Halber and Jan Mydlarski, "Untersuchung uber die Blutgruppen in Polen," *Zeitschrift für Immunitätsforschung* 43 (1925): 470–84.

9.   See, for example, the early article by the German naval doctor Paul Steffan, "Bedeutung der Blutuntersuchung fur die Bluttransfusion und die Rassenforschung," *Archiv für Rassen- und Gesellschafts-Biologie* 15 (1923): 137–50. In the following years, Steffan and his anthropologist colleague Otto Reche tried to make a science of utiliz-

ing blood group distribution as a marker of race, extending Hirszfeld's suggestion of the evolution of blood types as proof of aboriginal Aryan (blood type A) and non-Aryan (blood type B) races. For more on Steffan and Reche, see chapters 10 and 37; also Schneider, "Chance and Social Setting in the Application of the Discovery of Blood Groups," *Bulletin of Historical Medicine* 57, no. 4 (1983): 545–46; and Pauline Mazumdar "Blood and Soil: The Serology of the Aryan Racial State," *Bulletin of Historical Medicine* 64 (1990): 187–219.

10. This was published in translation as *Konstitutions serologie und Blutgruppenforschung* (Berlin: Springer Verlag, 1928).

11. Erik Bay-Schmith, "Rassenbiologische Untersuchungen auf Gronland," *Acta Pathologica et microbiologica Scandinavica* 4 (1927): 310–40.

12. Robert Debré (1882–1978), pediatrician and one of the foremost medical reformers in France in the twentieth century, befriended the Hirszfelds, especially Hanna, from the 1920s through the 1950s.

13. Hanna Hirszfeldowa, *Rôle de la constitution dans les maladies infectieuses des enfants* (Paris: Masson, 1939).

14. See Ludwik Hirszfeld and Z. Kostuch, "Untersuchungen über die Untergruppen und ihre Vererbung," *Schweizerische Zeitschrift für Allgemeine Pathologie und Bakteriologie* (1938): 407–20. The first confirmation was by C. J. Guthrie and John G. Huck, "On the Existence of More Than Four Isoagglutinin Groups in Human Blood," *Bulletin of the Johns Hopkins Hospital* 34 (1923): 37–48.

15. Hirszfeld's idea of the pleiades assumed that blood types and subgroups evolved from archaic types; hence O preceded other ABO types that evolved from it, as did subgroups from A. See I. Lille-Szyszkowicz, "Development of Studies on Pleiades of Blood Groups" [In Polish], *Post ępy higieny i medycyny doświadczalnej* 11, no. 3 (1957): 229–33; and Ernest Witebsky and Lillian M. Engasser, "Blood Groups and Subgroups of the Newborn: I. The A Factor of the Newborn," *Journal of Immunology* 61 (1949): 171–78.

16. Max Gundel (1901–n.d.) had earlier studied blood groups among prisoners in the 1920s in an attempt to find a link to criminality. He worked on diphtheria at the Robert Koch Institute and then headed the Institute of Hygiene of the Ruhr Area in Gelsenkirchen before being appointed head of health affairs for the city of Vienna in March 1939. See Helmut Groger and G. Stacher, "The Medical Profession in Vienna and the Nazi Regime," *Digestive Diseases* 17 (1999): 286–90.

17. The proceedings of the congress were published in *IIe Congrès international de lutte scientifique et sociale contre le cancer, Bruxelles, 20–26 septembre 1936: Travaux scientifiques*, ed. Marta Fraenkel (Bruxelles: Ligue nationale belge contre le cancer, 1936–37).

18. The Polish eugenics movement has recently become the subject of scholarly research. For example, see Krzysztof Kawalec, "The Dispute over Eugenics in Interwar Poland" [In Polish], *Medycyna Nowożytna* 7, no. 2 (2000): 87–102; and Magdalena Gawin, "Progressivism and Eugenic Thinking in Poland, 1905–1939," in *Blood and Homeland: Eugenics and Racial Nationalism in Central and Southeast Europe, 1900–1940*, ed. Marius Turda and Paul J. Weindling, 167–83 (New York: Central European University Press, 2006).

19. Stefan Zweig, *Die Heilung durch den Geist. Mesmer. Mary Baker-Eddy. Freud.* (Leipzig: Insel-verlag, 1931).

20. The year was 1931. Ludwik Hirszfeld, "Prologomena zur Immunitätslehre," *Klinische Wochenschrift* 10 (1931): 2153–59.

21. On Leonard Jan Bruce-Chwatt (1907–89), see Petrus Gustavus Janssens, "In Memoriam: Leonard Jan Bruce-Chwatt (6/9/1907–3/17/1989)," *Annales de la Société Belge de Medicine Tropicale* 71, no. 2 (June 1991): 163–66. He was studying at the Pasteur Institute when World War II began, but escaped to England, where he remained the rest of his life. After the war, he became one of the world's leading experts in malaria. Born Leonard Chwatt, he added his wife's maiden name to his own when they were married in 1948.

22. Hirszfeld also wrote an obituary of Nicolle in *Warszawskie Czasopismo Lekarskie* 13 (1936): 10–11; and in *Maroc médical* 17 (1937):67–70.

23. Ludwik Hirszfeld, "Die Seuchengesetze in naturgeschichtlicher Betrachtung," *Wiener klinische Wochenschrift* 51 (1938): 732–37.

24. "I am a human being, nothing that is human can be foreign to me." The quote is from the play, *Heauton timoroumenos* by Terence, a Roman playwright during the second century BC.

25. The only publication of Hirszfeld found on the reform of teaching hygiene is, "Kilka uwag w sprawie reformy nauczania higjeny na uniwersytetach [Some remarks on the reform of teaching of hygiene at universities]," *Polska Gazeta Lekarska*, 8 (1929): 839–41.

Chapter Ten

1. Edward Rydz-Śmigly (1886–1941) served under Piłsudski in World War I and the war with Soviet Russia. After Piłsudski's death, Rydz-Śmigły succeeded him (1935) as inspector general of the army and de facto ruler of the country without opposition. When Germany and the Soviet Union invaded Poland, he fled to Romania.

2. Gustaw Szulc succeeded L. Rajchman as director of the Institute of Hygiene in 1932.

3. Piotrowska Street is the "Main Street" in Łódź.

4. The Russian *gimnazjium* or high school was an institution famed for its draconian methods and anti-Polish stance. Students were forbidden even from owning books written in Polish. For a fuller description, see Marta A. Balińska, *Ludwik Rajchman: Medical Statesman* (Budapest: Central European University Press, 1988), 10–11.

5. On the Standardization Commission of the League of Nations, see chapter 7.

6. No record has been found concerning this indirect comment about Madsen's lack of help during the war.

7. On Doerr, see chapter 3.

8. For a description of this meeting, see Balińska, *Ludwik Rajchman.*

9. Fred Neufeld (1869–1945) made fundamental discoveries in pneumococcal biology, including the pneumococcal types, and was director of the Robert Koch Institute until dismissed in 1933. Although not Jewish, the Nazis considered him an opponent.

10. August von Wassermann (1866–1925) is best known for the test or reaction he developed with Neisser to determine syphilis.

11. In fact, Sachs had been dismissed from his post as director of the Heidelberg Cancer Research Institute in December 1935. He did not emigrate to Oxford until

1938. See special exhibit "Juden an der Universität Heidelberg," http://www.tphys.uni-heidelberg.de/Ausstellung/show.cgi?de&D&23&164 (accessed November 22, 2005).

12. Julius Bauer (1887–1979), an Austrian physician, emigrated to the United States in 1938.

13. Gustav von Bergmann (1878–1955) was author of *Handbuch der inneren Medizin* (Springer, 1934). See Gerhard Katsch, "Nachruf für Gustav von Bergmann," *Münchener medizinische Wochenschrift* 97, no. 42 (October 21, 1955): 1398–1400.

14. The "serological constitution" question was an attempt to define races by blood type distribution, based on the Hirszfelds' pioneer work in Salonica.

15. Otto Reche (1879–1966) was the anthropologist who along with Paul Steffan (see chapter 9) were coeditors of the *Zeitschrift für Rassenphysiologie*, published by the Society of Blood Group Investigation. Although developing a substantial interest initially, the subscribers to the journal diminished and it was only kept alive by Nazi Government subsidies at the end of the 1930s and during the war. In addition to Mazumdar, "Blood and Soil," and Schneider, "Chance and Social Setting," cited in chapter 9, see Katja Geisenhainer, *"Rasse ist Schicksal"—Otto Reche (1879–1966), ein Leben als Anthropologe und Völkerkundler* (Leipzig: Evangelische Verlagsanstalt, 2002).

16. Hirszfeld reflected a more open disdain for Steffan's and Reche's society in a letter to Felix Bernstein the year following an international anthropological conference in Amsterdam described in the next chapter. He stated that they must be on guard "to insure the true international and scientific nature of research" on blood groups. As an encouraging example he noted, "it is a sign of healthy instincts that the laughable efforts of the German Society for Blood Group Research are not taken seriously." Hirszfeld letter to Bernstein, 18 November 1927, Bernstein Archives, Niedersächsische Landes- und Universitätsbibliothek.

17. Ferdinand Sauerbruch (1875–1951) was a pioneer in thoracic surgery and one of Germany's top surgeons in the first half of the twentieth century. Despite being the most famous doctor who collaborated with the Nazi regime, he was rehabilitated after the war and served as chief of the Berlin Health Department. For a recent reassessment, see Marc Dewey et al., "Ernst Ferdinand Sauerbruch and His Ambiguous Role in the Period of National Socialism," *Annals of Surgery* 244, no. 2 (2006): 315–21.

18. The German Polish Non-Aggression Pact, signed January 26, 1934 was a major turning point in Polish foreign policy. See Zygmunt J. Gasiorowski, "The German-Polish Non-Aggression Pact of 1934," *Journal of Cultural European Affairs*, 15 (1955):3–29.

Chapter Eleven

1. This was the Third Congress of the International Institute of Anthropology, which was founded in 1919 in the aftermath of World War I. Ten papers were presented on blood group research, including Hirszfeld's on "Les groupes sanguins dans la biologie et la médecine" (Blood Groups in Biology and Medicine).

2. On Fritz Verzár (1886–1979), see below, note 6.

3. Marianne van Herwerden (1874–1934) was trained as an embryologist with an appointment in physiology at the University of Utrecht. She came to direct the Dutch blood group study from an interest she developed in eugenics and population genetics following a visit to the United States in 1920. For more, see C. A. B. van

Herwerden, *Marianne van Herwerden 16 Februari 1874-26 Januari 1934* (Rotterdam: W. L. Brusse, 1948).

4. The published papers on blood group research were listed in a section entitled "Eugenics and Heredity."

5. This paragraph about Hirszfeld's views on race crossing is awkward in the original.

6. Verzár had published a study on blood groups early in his career, during World War I. Studying the differences in blood type distribution among peoples living in Hungary, he independently and on a smaller scale discovered what the Hirszfelds found in Macedonia. Verzár, however, shifted his research focus and became a pioneer in the field of experimental gerontology. See "Present State and Future Developments of Experimental Gerontology: A Memorial to Fritz Verzár (1886–1979)," *Experientia* 37, no. 10 (October 15, 1981): 1039–59.

7. For background, see Paul Weindling, "Public Health and Political Stabilization: The Rockefeller Foundation in Central and Eastern Europe between the Two World Wars," *Minerva* 31 (1993): 253–67.

8. Miklós Horthy de Nagybánya (1868–1957) defeated the revolution of Bela Kun and ruled Hungary from 1920 to 1944.

9. On Štampar, see chapter 5. See also Theodore M. Brown and Elizabeth Fee, "Andrija Štampar: Charismatic Leader of Social Medicine and International Health," *American Journal of Public Health* 96, no. 8 (August 2006): 1382–85.

10. On Aca Savić, see chapter 4.

Chapter Twelve

1. The Italians launched their offensive on October 3, 1935, as Hirszfeld indicates below, only four days after the Congress. It would be difficult to exaggerate the symbolic significance of the Italian invasion of Ethiopia, particularly for those who were internationally minded and for whom this flagrant violation of a nation's rights was to be seen as the "first nail" in the coffin of the League of Nations, on which all hopes were placed.

2. The proceedings were published as *Atti del Primo Congresso Internazionale della Transfusione del Sangue, Roma, 26–29 Settembre 1935* (Milan: A. Colombo, 1935).

3. This is probably [Jan] Kołodziejski, a close childhood friend of L. Rajchman, who was killed by the Soviet Army at Starobielsk in 1940.

4. Eugenio Morelli (1881–1960) was an Italian specialist in tuberculosis who held the title Secretary of the National Fascist Syndicate of Doctors. He created the Carlo Forlanini Institute in Rome in 1934, named after his mentor mentioned below and served as its director until 1945. See A. Omodei Zorini, "Eugenio Morelli," *Lotta contro la tubercolosi* 30 (1960): 665–75.

5. Leone Lattes (1887–1954) was the pioneer of blood groups and serology in Italy. His first publication on the subject appeared in 1915 and included the prescient suggestion that the composition of the blood was such that no two humans were alike. This was expanded into a book in 1923, the English version of which was *Individuality of the Blood in Biology and in Clinical and Forensic Medicine* (London: Oxford University Press, 1932). After service in World War I, Lattes was appointed professor of legal medicine at the University of Messina, then Modena, and finally

the University of Pavia in 1933. As Hirszfeld points out later in this chapter, because of his Jewish ancestry, Lattes left Italy during the war but later returned to his post at the University of Pavia.

6. "Science reflects not only intellectual progress but also a nation's moral values. To give blood is to show compassion, is to imagine and suffer the suffering of another. That is why one can measure not only the cultural value but also the moral force of a country by the importance given to the organization and the question of blood donation. And it is in our field that a unique organization has been created, an organization where anonymous donors give their blood to anonymous sufferers." In addition to Hirszfeld's "Inaugural address," 13–15, he also presented a paper on "Die Serodiagnostik im Dienste der Bluttransfusion," in *Atti del Primo Congresso Internazionale*, 27–38.

7. Carlo Forlanini (1847–1918) first demonstrated this procedure to collapse the lung and allow it to heal in 1882. Mussolini made the hospital named after him a showplace of Fascism.

8. Arnault Tzanck (1886–1954) was the best-recognized pioneer of blood transfusion in France. He hosted the Second Congress in 1937.

9. Alfred Wysocki (1873–1959) was a longtime member of the Polish diplomatic corps.

10. See Alexander A. Bogomoletz, "Les phénomènes de l'autocatalyse et la transfusion du sang," *Atti del Primo Congresso Internazionale della Transfusione del Sangue, Roma, 26–29 Settembre 1935* (Milan: A. Colombo, 1935), 87–104. There had been considerable research on transfusion in the Soviet Union, largely in isolation from the West. This congress was one of the first opportunities to share their results, although it did not contain reports on the use of cadaver blood, which made quite an impression, not entirely favorable, in the West, after it was reported in 1937.

11. Name given to Italian guides for visiting amateur eighteenth-century antiquarians; by extension, generally, any learned guide.

12. Edmond Lévy-Solal (1882–1971) was one of the cofounders in 1926, along with Arnault Tzanck, of the largest transfusion service in France, Transfusion Sanguine d'Urgence.

13. Mussolini's syphilis is taken seriously by historians. See Dennis Mack Smith, *Mussolini* (New York: Knopf, 1982).

14. Hirszfeld does not mention it, but Tzanck also later emigrated from occupied France to South America because he was Jewish.

Chapter Thirteen

1. This was the Second International Congress of Blood Transfusion, the proceedings of which were published as *IIe Congrès international de la transfusion sanguine. 29 septembre–2 octobre 1937. Paris* (Paris: J.-B. Ballière, 1939). This was a much more ambitious undertaking with far larger attendance than the previous meeting in Rome. The World's Fair (international exposition), only less well remembered because it was eclipsed by the one in New York two years later, was held from May to November 1937 and left such memorable monuments as the Trocadero Palace in Paris.

2. Raymond Grasset (1892–1968) became minister of health under the Vichy regime. For more on Tzanck, see William H. Schneider, "Arnault Tzanck, MD (1886–1954)," *Transfusion Medicine Reviews*, 24 (2010):147–50.

3. Arthur Coca was head of the New York Blood Transfusion Betterment Organization, which was not, in fact, the first donor organization. That distinction belongs to a local London Red Cross chapter. But New York quickly rivaled both London and Paris as the biggest transfusion service in the world. Hirszfeld's paper at the congress was published as "Les groupes sanguins à la lumière de la science contemporaine," 2, 7–11. There were fourteen other papers presented in this section.

4. Zakopane is a renowned mountain resort in the Polish Tatras. At the end of the nineteenth century and the beginning of the twentieth, it attracted many artists and prominent members of the Polish intelligentsia.

5. Karl Liebknecht (1871–1919) was the son of a founder of the German Socialist Party who followed his father into socialist politics and was one of the minority of Reichstag deputies who voted against German war credits in 1914. He was killed along with Rosa Luxemburg in the so-called Spartacist Uprising in 1919, following the German defeat in World War I.

6. Alexander A. Bogomoletz (1881–1946), a physiologist based in Ukraine, studied prolongation of life toward the end of his career. He had presented Soviet research at the First Congress in Rome.

7. Hirszfeld was actually eleven years younger than Prosper Émile-Weil (1873–1963), a leading serologist in France and cofounder of the journal *Sang* in 1927. See "Prosper-Emile Weil (1873–1963)," *Bulletins et memoires de la Société Médicale des Hôpitaux de Paris* 114 (December 13, 1963): 1346–51.

8. M. Miserachs Rigalt, "La transfusion sanguine dans les hôpitaux militaires de première ligne pendant la guerre," *IIe Congrès international de la transfusion sanguine. 29 septembre–2 octobre 1937. Paris* (Paris: J.-B. Ballière, 1939), 462–66. There was no published report on the Russian cadaver blood in the proceedings.

9. As mentioned earlier, Ludwik Hirszfeld's book appeared in 1938; Hanna's in 1939.

10. Hirszfeld was not the only one to notice this. The following year, the French established the CNRS (Centre national de la recherche scientifique) under the direction of Henri Laugier. See Willaim H. Schneider, "Henri Laugier: The Science of Work and the Workings of Science in France, 1920–1940," *Cahiers pour l'histoire du CNRS* 5 (1989): 7–34.

11. "Not all of me shall die." This quote from Horace is the epitaph on Hirszfeld's tomb.

12. Paul Chevallier (1884–1960), French hematologist, was a member of the organizing committee for the Congress.

13. Hirszfeld is likely referring to René Fabre (1889–1966), toxicologist and dean of the Paris Faculty of Pharmacy.

14. Robert Debré: see chapter 9.

Chapter Fourteen

1. Hirszfeld has this date wrong. The year was 1936. The proceedings were published as *Travaux scientifiques: IIe congrès international de lutte scientifique et sociale contre le cancer: Bruxelles 20–26 septembre 1936* (Bruxelles: Ligue Nationale Belge Contre le Cancer, 1936).

2. Hirszfeld actually began this research in 1929, with the publication of articles first in Polish and then in German, "Uber die serologische Spezifität der Krebszellen," *Klinische Wochenschrift* 8 (1929): 1563–66.

3. Pietro Rondoni (1882–1956) was one of Italy's most highly respected oncologists who directed the National Institute for Cancer Research in Milan from 1935 until his death in 1956. See Romolo Dietto, "Pietro Rondoni (1882–1956)," *Tumori* 42 (1956): 805–9.

4. Emile Vandervelde (1866–1938) was a longtime Socialist leader in Belgium and one of the founders of European social democracy. The aging leader of the Socialists had been appointed minister of a new public health ministry established only three months earlier in June 1936.

5. Hirszfeld's arrangement to have Polish research appear in the *Comptes Rendus de la Société de Biologie*, published in French under the rubric of the Polish Society of Biology, is evidence of his practice of this admonition.

6. On Danysz (1860–1928), see chapter 6; Ludwik Maurycy Hirszfeld (1816–76) was an anatomist of the nervous system; Joseph Babiński (1857–1932) fled Warsaw with his parents after the Revolution of 1848. He studied with Charcot and became one of France's best-known neurologists. Edouard Pożerski (1875–1964) was born in France to parents who left Poland after the Revolution of 1863. His scientific work was done at the Pasteur Institute, although he is as well known for his gastronomic interests. Skłodowska is, of course, better known by her married name, Marie Curie (1867–1934).

7. For Eugène Minkowski and Mutermilch, see chapter 6. For Chwatt, see chapter 9.

8. Elie Metchnikoff (1845–1916) was a Russian bacteriologist and immunologist recruited to Paris by Pasteur himself in 1887. He won the Nobel Prize in medicine and physiology in 1908. Alexandre Besredka (1870–1940) succeeded Metchnikoff at the Pasteur Institute and continued his work in immunology. Michel Weinberg (1868–1940) also worked initially with Metchnikoff in microbiology. Constantin Levaditi (1874–1953) was a serologist who was born and educated in Romania and came to Paris to work in Metchnokoff's laboratory after spending a year at Ehrlich's serotherapy institute.

9. On *Arrowsmith*. See Chapter 8.

10. Although Hirszfeld may be referring again to Simon Flexner, director of the Rockefeller Institute (see chapter 8), more likely he is pointing to Simon's brother, Abraham Flexner (1866–1959), who is best remembered for his work on medical education reform and as an officer of the Rockefeller Foundation which financed many of them. After leaving Rockefeller, Flexner was the first director of the Institute for Advanced Study at Princeton. The book referred to is likely Abraham Flexner, *Medical Education: A Comparative Study* (New York: Macmillan, 1925).

11. Axel Munthe (1857–1949) was a Swedish physician, psychiatrist, and writer, best known for his autobiography *The Story of San Michele* (London: John Murray, 1929), which was an account of his experience as a doctor in Paris and Rome, and in semi-retirement at the villa of San Michele on the island of Capri (Italy).

12. Ferdinand Blumenthal (1870–1941) directed the Berlin Cancer Research Institute from 1914 to 1933, when the Nazis removed him.

13. Hirszfeld's talk was published in the proceedings of the congress as "Uber die theoretischen Grundlagen der Serodiagnostik des Krebses," 227–39.

Chapter Fifteen

1.   Hirszfeld's paper was published as "Über serologische Mutationen bei Menschen," *Relazioni del IV Congresso Internationale di Patologia Comparata, Roma, 15–20 maggio, 1939* (Milan: Istituto Sieroterapico Milanese, 1939) 1:227–45.

2.   The first Italian anti-Semitic legislation was promulgated in the fall of 1938, forbidding marriage between Jews and non-Jews and removing Jewish teachers from the schools. Many at the time mistakenly attributed this to German pressure and did not anticipate strict enforcement. For background, see Joshua D. Zimmerman, ed., *The Jews of Italy under Fascist and Nazi Rule, 1922–1945* (Cambridge: Cambridge University Press, 2005).

3.   Despite Hirszfeld's criticism of Rondoni's accommodation of the Fascist regime, Rondoni was rehabilitated after the war and enjoyed great prestige. See chapter 14.

4.   Gaston Ramon (1886–1963) was associate director of the Pasteur Institute from 1934–40, then director from 1940–44.

5.   Nikolai Timoféef-Ressovsky (1900–81) was a leading Russian geneticist who had left the country after Stalin's purge of scientists and spent the 1930s through 1945 at the Institute of Brain Research in Berlin. See Yakov G. Rokityanskij, "N. V. Timofeeff-Ressovsky in Germany (July 1925–September 1945)," *Journal of Biosciences* 30, no. 5 (2005): 573–80; and Diane B. Paul and C. B. Krimbas, "Nikolai V. Timoféef-Ressovsky," *Scientific American* 266, no. 2 (February 1992): 86–92.

6.   The malaria program was housed in the Institute of Public Health in Rome, both of which were heavily supported by the Rockefeller Foundation International Health Division between the wars. See Darwin Stapleton, "Internationalism and Nationalism: The Rockefeller Foundation, Public Health, and Malaria in Italy, 1923–1951," *Parassitologia* 42, no. 1–2 (June 2000): 127–34.

7.   Mussolini's slogan was actually *credere, obbedire, combattere*—"believe, obey, fight."

8.   Hirszfeld here means the *Archiv für Rassen- und Gesellschaftsbiologie*, founded by German eugenicists in an effort to lend scientific authority to matters of race and genetics. *Das Schwarze Korps* was the weekly newspaper of the SS (Nazi *Schutzstaffel*) and one of the more radical of the Nazi publications. On *La Razza*, see Sandro Servi, "Building a Racial State: Images of the Jew in the Illustrated Fascist Magazine, *La Difesa della Razza*, 1938–1943," in *The Jews of Italy under Fascist and Nazi Rule, 1922–1945*, ed. Joshua D. Zimmerman, 114–57 (Cambridge: Cambridge University Press, 2005); and for Italian anti-Semitism, see Giorgio Israel, "Science and the Jewish Question in the Twentieth Century: The Case of Italy and What It Shows," *Aleph* 4 (2004): 191–261.

9.   Nicola Pende (1880–1970) was an Italian pathologist and disciple of the Italian criminologist Cesare Lombroso, who is generally highly regarded. See M. Bufano, "Critical Review of the Clinical and Scientific Work of Nicola Pende" [In Italian], *Policlinico* [Med] 78, no. 1 (January–February 1971): 1–18. Nonetheless, he signed the "Manifesto of Racist Scientists" (*Manifesto Degli Scienziati Razzisti*) in July 1938 defending the concept of race and declaring the Italians were a "pure race." See Aaron Gillette, "The Origins of the 'Manifesto of Racial Scientists,'" *Journal of Modern Italian Studies* 6 (2001): 305–23.

10.   Bolesław Długoszowski-Wieniawa (1881–1942) interrupted his military career only from 1938–39 to serve as ambassador to Italy.

11.   Kmicic is a noble and the hero in *The Deluge (Potop)*, an historical novel by Henryk Sienkiewicz.

Chapter Sixteen

1. Emil du Bois-Reymond (1818–96) was a German physiologist who is best known for discovering the electrical nature of nerve signals. In 1850, he visited Paris to present his findings to the Academy of Sciences, not Medicine, as Hirszfeld states. See Gabriel Finkelstein, "M. du Bois-Reymond Goes to Paris," *British Journal for the History of Science* 36, no. 3 (2003): 261–300. Jean-Martin Charcot (1825–93) was a French neurologist who discovered a number of neurological disorders, but is perhaps best known for the students he trained, including Freud. He often spoke at the Academy of Medicine. See Christopher G. Goetz, Michel Bonduelle, and Toby Gelfand, *Charcot: Constructing Neurology* (New York: Oxford University Press, 1995).

2. This is the first time Hirszfeld mentions his daughter at length.

3. Lucien Panisset (1880–1940) was a French veterinarian who worked on diseases transmissible to man, such as tuberculosis.

4. André Boivin (1895–1949) was a biochemist at the Pasteur Institute from 1936 to 1947.

Chapter Seventeen

1. Hirszfeld is referring to Jacek Prentki, the son of a school friend of Hanna Hirszfeld. He survived the war and eventually emigrated to Switzerland where he died in 2009.

Chapter Eighteen

1. On Ostwald, see chapter 5.

2. On Pettenkofer, see chapter 6. He disagreed with the germ theory of Koch, but for a more subtle explanation of his suicide, see David L. Trout, "Max Josef von Pettenkofer (1818–1901)—A Biographical Sketch," *Journal of Nutrition* 107, no. 9 (1977): 1567–74.

3. Franz Werfel (1890–1945) was a Czech-born writer, best known for his novel *Forty Days of Musa Dagh*, about the persecution of Armenians, and *Song of Bernadette*. He escaped the Nazis and settled in the United States.

4. OZON was the government party, Camp of National Union, led by Marshal Rydz-Śmigły.

5. Poland formally adopted a *numerus clausus* (quota) for minority students in 1923, and *numerus nullus* in 1935. See Szymon Rudnicki "From 'Numerus Clausus' to 'Numerus Nullus'," *Polin* 2 (1987): 246–68.

6. Colonel Józef Beck (1894–1944) had been appointed foreign minister by Piłsudski in 1932. He tried, and ultimately failed, to steer a course between Nazi Germany and the Soviet Union, while maintaining alliances with Britain and France.

Chapter Nineteen

1.   This was in conjunction with the Munich Crisis, when Europe came very close to engaging in war. Although he does not give the exact dates, Hirszfeld is probably talking about the month of September, during which the crisis built, with war fears only subsiding on September 29, 1938, with the signing of the agreement by Germany, Italy, France, and Britain. Despite Neville Chamberlain's claim that his policy of appeasement had brought "peace with honor," many doubted the prospects for long-term success. It lasted only until Hitler broke the agreement when he occupied the rest of Czechoslovakia in March 1939.

2.   Bronisława (Bronka) Langrod (1904–72), a relative (by marriage) of the Hirszfelds later became a prominent figure of the Polish resistance (Polish Socialist Party [PPS]), in which capacity she "protected," among other people, Jan Karski. She later joined Żegota (the Polish organization of aid to the Jews), was arrested but survived the war, and emigrated to New York City.

3.   The Polish name for Aachen (Aix-la-Chapelle in French), located just inside the border of Germany.

4.   Hirszfeld is referring to a political slogan of the times.

Chapter Twenty

1.   Reference is made to the "Polish corridor," part of the Versailles Treaty after World War I, which connected the area around the free city of Danzig to Poland, thus cutting off East Prussia from Germany. It was a bone of contention for German nationalists and the immediate pretext for Hitler's invasion of Poland.

2.   Hirszfeld is presumably referring to the much-publicized "Pact of Steel," signed between Hitler and Mussolini in May 1939, committing both countries to support the other if either became engaged in war. When Hitler attacked Poland in September 1939, Mussolini realized his country was not ready for war and did not honor the pact. Only in June 1940, after the defeat of France was assured, did Italy join Germany in the war.

3.   For background on the military campaign, see Steven J. Zaloga, *Poland 1939: The Birth of Blitzkrieg* (Westport, CT: Praeger, 2002).

4.   This occurred on September 17, 1939.

5.   Roman Umiastowski (1893–1982) was an army officer and chief of Propaganda Bureau of the Commander-in-Chief.

6.   "Everyone waits their turn."

7.   A derogatory expression used by other countries to describe interwar Poland, in particular the country's indecisiveness and frequent changes of government.

8.   The defense of Warsaw in 1920 not only turned the tide in the Polish-Soviet War, it made the reputation of Piłsudski. This Battle of Warsaw in 1939 lasted from September 8 to 28, a surprisingly long time, given the disparity of equipment and troops, not to mention the advance of the Soviets from the east.

9.   For more on Róża Amzel, see chapter 7.

10. For more on Stanisława Adamowicz, see chapter 7.

11. Władysław Melanowski (1888–1974) was a leading Polish ophthalmologist.

12. Hirszfeld is referring to Hanna Hirszfeld's friend and her son who had come to live with them in 1928.

13. Kazimierz Przerwa-Tetmajer (1865–1940).

14. Stefan Starzyński (1893–perhaps 1943) was elected president (mayor) of Warsaw in 1934 and achieved popularity for his programs to combat unemployment. During the siege, his speeches made him the symbol of the city's defiance. After surrender, the Germans kept him temporarily as administrator of the city but he soon disappeared and is assumed to have been killed in one of the German prison camps. In 1957, his symbolic grave was dedicated in Warsaw; a street and several schools are named after him; and in 1978, a film was made about his life, *Gdziekolwiek jesteś, panie prezydencie* [Wherever You Are, Mister President], directed by Andrzej Trzos-Rastawiecki.

Chapter Twenty-One

1. In 1904, Emil von Behring (1854–1917), collaborator of Koch and winner of the first Nobel Prize in medicine for his discovery of a diphtheria therapy, created Behringwerke in Marburg, Germany, to produce sera and vaccines to cure infectious diseases. It continued through both world wars and merged with another company in 1995, operating today under the name of ZLB Behring. For more on its role during World War II, see Paul Weindling, *Epidemics and Genocide in Eastern Europe, 1890–1945* (Oxford: Oxford University Press, 2000), 244–51.

2. Arthur Greiser (1897–1946) was one of the more brutal racist regional governors of German-occupied Poland. He was captured after the war, tried by the Poles, and executed.

3. One article, coauthored with Amzel, was published in Switzerland as "Über die Übergangsformen der Blutgruppen," *Schweizerische medizinische Wochenschrift* 70 (1940): 801. The French articles were published in two installments, "Sur les pléiades 'isozériques' du sang," *Annales de l'Institut Pasteur* 65 (1940): 251–78; 386–414.

4. Jerzy Morawiecki (1910–97) worked under Hirszfeld in the Hygiene Institute beginning in 1937 and published one article during the occupation, most likely around 1941. He survived the war and built an important career as an immunologist and oculist. See Krystyna Raczyńska and Barbara Iwaszkiewicz-Bilikiewicz, "Prof. Morawiecki: The Oculist and Outstanding Expert in Immunology, Immunopathology and Serology" [In Polish], *Archiwum Historii i Filozofii Medycyny* 63, no. 3–4 (2000): 68–70.

5. Ujazdowskie Avenue was considered the most elegant avenue in Warsaw.

6. On the contrary, Ernst Georg Nauck (1897–1967) became director of the Hamburg Institute for Tropical Medicine in 1943, surviving the war and remaining in this position until his retirement in 1963. He has nonetheless been subsequently accused of participating in war crimes. See, for example, Ludger Wess, "Menschenversuche und Seuchenpolitik—Zwei unbekannte Kapitel aus der Gesichte der deutschen Tropenmedizin," *Zeitschrift für sozialgeschichte des 20 un 21 Jahrhunderts* (1999), 10–50; and Stefan Wulf, *Das Hamburger Tropeninstitut 1919 bis 1945* (Hamburg: Dietrig Reimer Verlag, 1994).

7. The Holy Ghost Hospital lay in what was to become the ghetto, and although in ruins, a part of a wing remained and housed a number of medical facilities, including those moved in 1941 from the Czyste Hospital where most Jews practiced before the war. See Charles G. Roland, *Courage under Siege: Starvation, Disease, and Death in the Warsaw Ghetto* (New York: Oxford University Press, 1992), 28, 81, 112.

8. This famous incident occurred on November 6, 1939, when 183 professors from Cracow were arrested as part of a Nazi plan to eliminate the Polish intellectual class. A detailed account is given in Jochen August, *Sonderaktion Krakau* (Hamburg: Hamburg Edition, 1997).

9. Hirszfeld is mindful of not being able to describe all that he saw, but he also here slips into unattributed descriptions of things, all of which he could not have witnessed. Whereas earlier he takes care to document his sources, in these and following descriptions of Poland under occupation, he does not. He is still surprisingly accurate, as compared with scholarly accounts based on documentary, archival evidence. See, for example, Janusz Gumkowski and Kazimierz Leszczyński, *Poland under Nazi Occupation* (Warsaw: Polonia Publishing House, 1961).

10. For more on Henryk Brokman, see chapter 8.

11. Zygmunt Srebrny was a well-known Polish otolaryngologist.

12. The *Völkischer Beobachter* was a daily newspaper of the Nazi party, going back to 1923.

13. Kurt Schrempf was a German physician who was public health officer for the civilian Warsaw population until early 1941. He was replaced by Wilhelm Hagen (see following chapters). See Roland, *Courage under Siege*, 77, 144–46. See also Wilhelm Hagen, *Das Gesundheitswesen der Stadt Warschau September 1939 bis März 1942* (Warsaw: Archives of the State Institute of Hygiene, n.d.), 118.

14. The full title in German was *Gesundheit und Leben. Amtsblatt der Gesundheitskammer im Generalgouvernement.*

15. The term *General Gouvernement* is short for General Government for the Occupied Polish Areas (in German: *Generalgouvernement für die besetzten polnischen Gebiete*). It refers to the authority given by the German Army to govern Polish territory after military occupation, but the term is used synonymously with the territory administered by the governing authority.

16. The requirement to wear an armband with a blue star of David was part of the anti-Jewish decrees of November 1939. On the question of Jews as carriers of disease, see below. For a broader historical perspective on the question, see Paul J. Weindling, *Epidemics and Genocide in Eastern Europe*. This rationale was more than an excuse for the ghetto: the standard method of combating typhus epidemics during World War I was "delousing" by gas fumigation. This was the supposed purpose for the gas chambers.

17. It is not clear whether this is the same memorandum that Hirszfeld mentions in chapter 22.

18. On establishment of the ghetto, see chapter 22.

19. Hirszfeld is referring to the effort led by his longtime colleague Arthur Coca. See introduction for more on this.

20. Hirszfeld is referring to the husband of Hanna's half sister (Izabela Belin): Stanisław Kiełbasiński (1883–1955).

21. This was Coca. Hirszfeld kept the letter and it is in his papers at the Polish Academy of Sciences.

22. Although Heinz Auerswald (1908–70) was in charge of the Warsaw Ghetto, he was never convicted of war crimes. For background, see Isaiah Trunk, *Judenrat; the Jewish Councils in Eastern Europe under Nazi Occupation* (New York: Macmillan, 1972), 293–98; and Christopher R. Browning, "Nazi Ghettoization Policy in Poland, 1940–1941," *Central European History* 19 (1986): 343–68.

23. The Main Custodial Council (Rada Główna Opiekuńcza) was a Polish charity organization that looked after civilians in both world wars.

24. Hirszfeld is referring to Dr. Arnold Lambrecht (1902–n.d.), a district physician who became chief of public health in Warsaw in August 1940. According to a friend, Lambrecht "prided himself with having created the first ghetto in Lwów in 1940 and then in Warsaw . . . and he [claimed] he had done so to contain typhus." Interview of M. A. Balińska with Rudolf Wohlrab, Hannover, February 21, 1995. See Marta A. Balińska, "Choroba jako ideologia: tyfus plamisty w okupowanej Polsce 1939–1944," [Disease as Ideology: Typhus in Occupied Poland, 1939–1944] *Zeszyty historyczne* 126 (1998): 212–20. See also Roland, *Courage under Siege*, 127, and Christopher Browning, "Genocide and Public Health: German Doctors and Polish Jews, 1939–1941," *Holocaust and Genocide Studies* 3 (1988): 21–36.

Chapter Twenty-Two

1. This story is found, among other places, in Mark Twain's *Innocents Abroad.* But instead of being shipped to an island, the dogs were reported to have always mysteriously been lost overboard before reaching the island. The end result was the same; when word got out, protest stopped the practice.

2. The first large ghetto was created in Lwów (Lemberg) by the German district physician, Dr. Lambrecht, who also "masterminded" the Warsaw Ghetto. (Interview of Rudolf Wohlrab by Marta A. Balińska, 1995. See Marta A. Balińska, "Choroba jako ideologia: tyfus plamisty w okupowanej Polsce 1939–1944," [Disease as Ideology: Typhus in Occupied Poland, 1939–1944] *Zeszyty historyczne* 126 (1998): 212–19. For background on creation of the ghettos, see Christopher R. Browning, "Nazi Ghettoization Policy in Poland, 1940–1941," *Central European History* 19 (1986): 343–68; and Browning, "Before the 'Final Solution': Nazi Ghettoization Policy in Poland (1940–1941)," in *Ghettos, 1939–1945: New Research and Perspectives on Definition, Daily Life, and Survival* (Washington, DC: United States Holocaust Memorial Museum, 2005), 1–14.

3. Hirszfeld wrote two anonymous memoranda in German directed to the occupying health authorities, copies of which were preserved by the Ringelblum contemporary history working group, organized in the ghetto. These archives are now at the Jewish Historical Institute in Warsaw: "Denkschrift über die Ursachen des Flecktyphus in Warschau und Vorschläge zu seiner Bekämpfung" and "Denkschrift über einige prophylaktische-soziale Massnahmen zu Bekämpfung des Felckfiebers in jüdischen Wohenbezirk zu Warschau," Ringelblum Papers, 15 May 1940, Jewish Historical Institute, Warsaw, A. RingI/85.

4. Although the order was issued on October 2, implementation was delayed a month, but by November 15, 1939, the ghetto existed. The Jewish population of Warsaw, according to other estimates, was reported in 1938 as being above 300,000. *The New Concise Pictorial Encyclopedia* (New York: Garden City Publishing, 1938). Other reports are that 80,000 Christian Poles had to move out of the designated area

and 140,000 Jews moved in. According to Charles G. Roland, *Courage under Siege: Starvation, Disease, and Death in Warsaw Ghetto* (New York: Oxford University Press, 1992), 27–30, the initial population of the ghetto was 390,000 and rose to as high as 460,000 in subsequent months as refugees arrived, before death rates brought a rapid decline.

5. The Warsaw Ghetto covered 1,000 acres and was inhabited by 400,000 to 500,000 people; in other words, 30 percent of the Warsaw population had been forced into 5 percent of the city. This had obvious consequences for health and disease. See Roland, *Courage under Siege*, 130.

6. The most infamous German commissar was Heinz Auerswald (1908–70), a lawyer and SS member who held the position from May 1941 to November 1942. See chapter 21, note 22.

7. The "navy blue police" refers to the Polish collaborators.

8. It should be remembered that the official reason for creating the ghettos was to prevent the spread of infectious disease, and typhus in particular. See Jost Walbaum, ed., *Kampf den Seuchen! Deutscher Ärtze-Einsatz im Osten. Die Aufbarheit im Gesundheitwesen der Generalgouvernements* (Krakau: Bucherverlag "Deutscher Osten" G.m.b.H, 1941). On Walbaum, who was head of public health in the whole Generalgouvernement, see Roland, *Courage under Siege*, 126; and Christopher Browning, "Genocide and Public Health: German Doctors and Polish Jews, 1939–1941," *Holocaust and Genocide Studies* 3 (1988): 22–23. "Ghettoization" describes a debate within Nazi circles between "productionists" wanting some work from the ghettos, and "attritionists" who sought merely the death of the Jewish ghetto inhabitants.

9. *Wacha* is a Polish version of the German word *wache*, which means guard post.

10. "War order of the Führer."

11. Hanna Hirszfeld was one of the pediatricians who worked on starvation in the ghetto. Her memo, "Hunger States in Children," can be found in APAN, HHIII 157/117 LHP.

12. "Strength through joy," was a large, state-owned organization established after the Nazis came to power in 1933 to provide subsidized leisure activities for the working class.

13. Maria (called "Marysia") Hirszfeld had returned from studying history at the Sorbonne in Paris during 1938–39, as related in chapters 16 through 19. She was twenty-one years old when the family was forced into the ghetto.

14. In 1941 alone, it was estimated by Jewish doctors that there were some 15,400 reported cases of typhus in the ghetto, but the real number of cases was probably five times as great. Mortality from the disease was estimated at 40 percent. See *Walka z tyfusem* [Fight against Typhus], Ringelblum Papers, A. Ring/363, Archives of the Jewish Historical Institute, Warsaw; and interview with Dr. Stanisław Tomkiewicz by Marta A. Balińska.

15. To understand this statement and what follows, it should be remembered that while Hirszfeld's parents had not converted to Christianity, they belonged to the "assimilated" Jews and identified with Polish culture.

16. Adam Czerniaków was president of the Warsaw Ghetto Jewish Community; he committed suicide in 1943 rather than comply with an order to round up Jews for deportation. See Raul Hilberg et al., eds, *The Warsaw Diary of Adam Czerniakow: Prelude to Doom* (New York: Stein and Day, 1979).

17. Ludwik Hirszfeld was baptized on his return to Poland in 1920, undoubtedly more to symbolize his identification with the newly reborn Poland than as the result of a religious conversion.

18. Maria Hirszfeld had been suffering from anorexia nervosa since the age of fourteen.

19. In fact, March 27, 1941, was the date of a military coup against the Yugoslav ruler right after he agreed to join the Axis. Hitler attacked on April 16, 1941. For background see Jozo Tomasevich, *War and Revolution in Yugoslovia 1941–1945: Occupation and Collaboration* (Stanford: Stanford University Press, 2001).

20. Józef Piłsudski (1867–1935) was a revolutionary, field marshal, chief of state, and finally dictator of Poland.

21. "Cheeky Jew."

22. According to Roland, *Courage under Siege*, 30–31, Dr. Izrael Milejkowski (1887–1943) was an assimilated Jew who

> had a sharp aversion to Prof. Ludwig [*sic*], who was a convert, and once referred to Hirszfeld with disgust as a "turncoat." Yet these were two of the key figures in the medical world of the ghetto, and a good working relationship between them might have benefited many ghetto dwellers. It is easy to oversimplify, and Milejkowski may have had other reasons to dislike Hirszfeld and vice versa. But the religious/cultural separation certainly played a role.

23. Milejkowski committed suicide in 1943 on the train after deportation from the ghetto to Treblinka. Despite their dislike, Roland, *Courage under Siege*, 227, attributes to Milejkowski the quote from Horace that is Hirszfeld's epitaph on his grave: *non omnis moriar*, "I shall not die completely."

24. Moritz Kon was a converted Jew whom Roland, *Courage under Siege*, 31, lists as head of the Health Department.

25. This is, perhaps, Zdzisław Bieliński, member of the Jewish Health Council, mentioned by Roland, *Courage under Siege*, 69.

Chapter Twenty-Three

1. The first course for Jewish "disinfectors" was organized by the German authorities in November 1940. See [Anonymous], *Pierwszy kurs dla dezynfektorow zydowskich XI 1940*, A.R.I/221, Archives of the Jewish Historical Institute, Warsaw.

2. Dr. Wilhelm Hagen (1893–1982), who survived the war and enjoyed a respectable career, has a mixed reputation. See Paul J. Weindling, *Epidemics and Genocide in Eastern Europe, 1890–1945* (New York: Oxford University Press, 2000), 275–77. All accounts regard him as an improvement over Kurt Strempf whom he replaced as chief physician (Amtartz) for the city of Warsaw. According to Janusz Korczak's "Biography" (see http://korczak.com/Biography/kap-1who.htm), he was supposedly known as a "good German." Christopher Browning, "Genocide and Public Health: German Doctors and Polish Jews, 1939–1941," *Holocaust and Genocide Studies* 3 (1988), largely praises him, and in an interview, his friend Rudolf Wohlrabe indicated that Hagen carried out TB screening for Polish and Jewish children "who had long been condemned to death" (interview of Marta A. Balińska with R. Wohlrab, 1995). But according to Josef Wulf, who survived the ghetto, Hagen had advocated shooting Jews found

"wandering around." See Josef Wulf, *Das Dritte Reich und seine Vollstrecker; die Liquidation von 500 000 Juden im Ghetto Warschau* (Berlin-Grunewald: Arani, 1961). See also Wilhelm Hagen, *Das Gesundheitswesen der Stadt Warschau September 1939 bis März 1942* (Warsaw: Archives of the State Institute of Hygiene, n.d.); and Stanisław Kłodziński, "Konferencja Krynicka Lekarzy Hitlerowskich 13–16 October 1941," [The Bad Krynica Congress of Nazi Physicians, 13–16 October 1941] in *Okupacja i medycyna* (Warszawa: Książka i Wiedza, 1971), 103–16.

3.   Charles Roland gives Dr. Juliusz Zweibaum (1887–1959) credit for organizing the medical school, with Hirszfeld only joining it at a later date. Although Hirszfeld's description of his lectures at first appears to have developed independently, below, in this chapter, he gives full credit to Zweibaum for organizing the lectures, courses, and the school. Roland interviewed a number of the survivors of the ghetto who attended the school and attested to Hirszfeld's high reputation and great impact on the students. See Charles G. Roland, "An Underground Medical School in the Warsaw Ghetto, 1941–42," *Medical History* 33, no. 4 (October 1989): 399–419.

4.   Mieczysław Centnerszwer (1874–1944). On his death, see chapter 36.

5.   To our knowledge, at least one of these students did survive: Stanisław Tomkiewicz (1925–2003), one of France's most prominent child psychiatrists, who recalled Hirszfeld and Zweibaum as his greatest mentors in medicine in his autobiographical memoir, *Une adolescence volée* (Paris: Calmann-Lévy, 1999).

6.   Józef Szeryński headed the so-called Jewish police, created by the Germans even before the ghetto. He had formerly been a colonel in the Polish police, and like Hirszfeld, was a convert to Catholicism. For background, see Frank Fox, "The Jewish Ghetto Police: Some Reflexions," *East European Jewish Affairs* 25 (1995): 41–47.

7.   On Juilian Fliederbaum, Emil Apfelbaum, and their work, see Charles G. Roland, *Courage under Siege: Starvation, Disease, and Death in the Warsaw Ghetto* (New York: Oxford University Press, 1992), 114–19. For examples of their findings, see Myron Winick, ed., "Hunger Disease. Studies by the Jewish Physicians in the Warsaw Ghetto, 1979," *Nutrition* 10, no. 4 (July–August 1994): 365–79, a reprint of a volume first published with the same title in 1979; and Shaul G. Massry and Miroslaw J. Smogorzewski, "The Hunger Disease of the Warsaw Ghetto," *American Journal of Nephrology* 22, no. 2–3 (July 2002): 197–201.

8.   Abraham Gepner (1872–1943) was a Warsaw businessman, philanthropist, and member of the city council before the war. He headed the ghetto's Supply Department, assisting Hirszfeld with equipment and Korczak's orphans home. See Stanislaw Adler, *In the Warsaw Ghetto* (Jerusalem: Yad Vashem, 1982), 250–51.

9.   Hirszfeld is referring to the constant preoccupation of earning enough in the day to feed one's self and one's family.

10.   Hirszfeld himself did not publish on the subject, but others, beginning with Fritz Schiff who studied the Jews of Berlin, found no consistent pattern of difference in blood type distribution between the Jews and the rest of the population. Overall, the Jews more closely resembled the local populace than the Jews in other locations. See chapter 27.

11.   Jan Czekanowski (1882–1965) was a Polish anthropologist who began his career on an expedition to Africa in 1907. In the 1930s, he shifted his research to the physical anthropology of Central Europe. Ernst Rénan (1823–92) was the French historian and philologist most famous for his *Life of Jesus*, but also an early expert in the study of nationalism.

12. The German authorities' veritable obsession with regard to typhus meant that they were inclined to allow extraordinary actions labeled as typhus control within, as well as outside, the ghetto. Some Jews went so far as to produce "an artificial false positive Weil-Felix reaction to deceive the occupying Nazi forces into thinking their Polish village was endemic for typhus. This spared the villagers from being deported to slave labor camps." John D. C. Bennett and Lydia Tyszczuk, "Deception by Immunization Revisited," *British Medical Journal* 301 (1990): 1471–72. Numerous accounts of the German's fear of typhus can be found in the several-volume work entitled *Okupacja i medycyna* [Occupation and Medicine] published in the 1970s and 1980s (Warsaw: Książka i Wiedza).

13. Hirszfeld was able to save some of the mimeographed announcements of the medical school lectures, which are part of his papers at the Polish Academy of Sciences.

14. Hirszfeld is talking here historically of Polish Jews in medicine, including the anatomist Ludwik Maurycy Hirszfeld (1814–76), Samuel Goldflam (1852–1932), and Maksymilian Rose (1883–1937).

15. For more on Zygmunt Srebrny, see chapter 21.

16. Zygmunt Szymanowski was a prominent figure who assisted Jews during World War II. See chapter 7.

17. Rudolf Weigl (1883–1957) was born in Austria (of German-speaking parents) but his mother remarried a Pole after the death of his father. Weigl settled in Poland after World War I and developed his first antityphus vaccine at an institute in Lwów. Because of the value of the vaccine, Weigl was able to continue work under Soviet, then German occupation during the war, and finally in the reestablished Communist Poland after the war. Surprisingly, little scholarly work has been written on him outside of Poland. For more, see Paul J. Weindling, *Epidemics and Genocide*; and Wacław Szybalski, "The Genius of Rudolf Stefan Weigl (1883–1957), a Lvovian Microbe Hunter and Breeder—In Memoriam," *International Weigl Conference: Microorganisms in Pathogenesis and Their Drug Resistance—Programme and Abstracts*, ed. R. Stoika et al. (Lviv: September 11–14, 2003), http://www.lwow.com. pl/weigl/in-memoriam.html (accessed May 30, 2007); and Maurice Huet, "L'elevage du pou au laboratoire," *Histoire des Sciences Medicales* 37 (2003): 43–46.

18. The equivalent of the French *baccalauréat*, or end of high school studies examination.

19. Ludwik Hirszfeld and Róża Amzel, "O postaciach przejściowych (podgrupach) w obrębie grupy O. Doniesienie tymczasowe" [On the Intermediate Forms (Sub-groups) of Group O. Preliminary Communication], *Polski Tygodnik Lekarski* 1 (1946): 1525–27.

20. Edmund Weil (1880–1922) an Austrian physician, and Arthur Felix (1887–1956), a bacteriologist born in Silesia, who studied in Vienna and emigrated first to Palestine, then London, between the wars, developed the Weil-Felix test while working at the Koch Institute during World War I. They isolated strains of a bacillus agglutinated by the serum of patients suffering from typhus. Their classic article is Edmund Weil and Arthur Felix, "Untersuchungen über das Wesen der Fleckfieber-Agglutination," *Wiener Klinische Wochenschrift* 10 (1917): 393–99.

21. See chapter 24 and Roland, *Courage under Siege*, 120–53, who devotes a whole chapter to typhus in the ghetto.

22. For more on this complicated story, see Jean Lindenmann, "Hermann Mooser, Typhus, Warsaw 1941," *Gesnerus* 59, no. 1–2 (2002): 99–113.

23. Michał Szejnman also carried out studies of tuberculosis and has a chapter titled "Changes in Peripheral Blood and Bone Marrow in Hunger Disease," in *Hunger Disease*, ed. Winick, 161–96.

Chapter Twenty-Four

1. The classic study remains Hans Zinsser, *Rats, Lice and History* (Boston: Little, Brown, 1934). For historical background, see Paul J. Weindling, *Epidemics and Genocide in Eastern Europe, 1890–1945* (New York: Oxford University Press, 2000).

2. Infestation with head lice.

3. Erich Martini (1880–1960) organized the department of entomology at the Institute for Tropical Medicine in Hamburg.

4. The term *Congress Kingdom* was an unofficial name for the Kingdom of Poland (1815–31) and referred to an area of Poland from Warsaw to Cracow included in the Germans' "Generalgouvernement."

5. For historians' judgments and background for this and chapters that follow, see Charles G. Roland, *Courage under Siege: Starvation, Disease, and Death in the Warsaw Ghetto* (New York: Oxford University Press, 1992); and Jean Lindenmann, "Herman Mooser, typhus, Warsaw 1941," *Gesnerus* 59 no. 1–2 (2002): 99–113.

6. Professor Dr. Heinrich Teitge (1900–n.d.) also SS-Brigadeführer, was Walbaum's successor as head of the Division of Public Health in the *Generalgouvernement*. See Christopher Browning, "Genocide and Public Health: German Doctors and Polish Jews, 1939–1941," *Holocaust and Genocide Studies* 3 (1988): 32.

7. For an anonymous account of the fight against typhus in the Warsaw Ghetto, see *Walka z tyfusem, Warszawa 1942* [Fight against Typhus, Warsaw, 1942]. A.R.I/363, Archives of the Jewish Historical Institute, Warsaw.

8. Hirszfeld moved to the district on February 28, 1941, thus four months after it was established.

9. See chapter 23. Hagen had in fact the title of "Amtsartz," thus chief physician for the city of Warsaw.

10. Hirszfeld is referring to the Polish disinfecting teams who were employed in the ghetto.

11. A normally calm person who suddenly becomes enraged or violent.

12. Tadeusz Ganc was a former major in the Polish Army and a converted Jew who was appointed administrative head of the Judenrat Health Division when typhus continued to rage in the Ghetto. Roland, *Courage under Siege*, 136.

13. Hirszfeld is referring to the American-Jewish Joint Distribution Committee, which served as an intermediary for submitting his memorandum to the German authorities (see chapter 22).

14. See chapter 22.

Chapter Twenty-Five

1. For background on health in the ghettos, see Isaiah Trunk, *Judenrat; the Jewish Councils in Eastern Europe under Nazi Occupation* (New York: Macmillan, 1972), 143–71.

2. Anna Braude-Heller (n.d.–1943) attended medical school in Switzerland before World War I. She had headed the Bersons and Baumans Children's Hospital (now Warsaw Children's Hospital) since 1930. It had been established in 1878 for Jewish children and was within the boundaries of the ghetto. Dr. Braude-Heller died in the Warsaw Ghetto Uprising. For background, see Charles G. Roland, *Courage under Siege*, 94–96; and for a more extended firsthand account, see Adina Blady Szwajger, *I Remember Nothing More: The Warsaw Children's Hospital and the Jewish Resistance* (New York: Pantheon, 1990).

3. On Szniolis, see chapter 7.

4. On Kudicke, see chapter 2, note 8.

5. On Dr. Lambrecht, see chapter 21, note 24.

6. Janusz Korczak (1879–1942), whose given name was Henryk Goldszmidt, had been a successful pediatrician before the war. Like Hirszfeld, he refused offers of rescue, and once in the ghetto, he set up an orphanage for two hundred children. He voluntarily went with the orphans to Treblinka. See Janusz Korczak, *Ghetto Diary* (New Haven, CT: Yale University Press, 2003); and Betty Jean Lifton, *The King of Children: The Life and Death of Janusz Korczak* (New York: St. Martin's Griffin, 1997). In 1990, his story was the basis of a film, *Korczak*, directed by Andrzej Wajda.

7. For examples of similar "euthanasia," see, for example, Julian Aleksandrowicz, *Kartki z dziennika doktora Twardego* [Pages from the Diary of Dr. Twardy] (Cracow: Wydawnictwo Literackie, 1983); and Adina Blady-Swajger, "Krótka historia szpitala Bersonów i Baumanów (1939–43)" [Short History of the Berson and Bauman Hospital], *Zeszyty Niezależnej Myśli Lekarskiej* 10 (December 1986).

8. Some of these records were saved and are part of the Ringelblum collection at the Archives of the Jewish Historical Institute in Warsaw. On Ringelblum, a historian, see Małgorzata Czeley-Wybieralska and Symcha Wajs. "Emanuel Ringelblum (1900–1944) jako historyk medycyny Żydowskiej [Emanual Ringelbaum (1900–44) Historian of JewishMedicine]," *Archiwum historii I filozofii medycyny* 51, no. 2 (1988): 187–210; and Emanuel Ringelblum, *Notes from the Warsaw Ghetto: The Journal of Emanuel Ringelblum* (New York: Schocken, 1974).

9. Samuel Goldflam (1852–1932).

10. On Behring, see chapter 22. Fritz Schaudinn (1871–1906) was a German microbiologist who, along with Erich Hoffman, discovered the syphilis spirochete in 1905.

11. Initially the Tymczasowy Komitet Pomocy Żydom (Provisional Committee for Aid to Jews) was an underground organization for activist Catholics to provide assistance to Jews in Poland. In September 1942, it added Jews and its name was changed to the Council for Aid to Jews, or Żegota. It engaged in clandestine action, such as providing Aryan papers for Jews in hiding, placement of children, and medical care. See Irene Tomaszewski and Tecia Werbowski, *Zegota: The Council for Aid to Jews in Occupied Poland, 1942–1945* (Montreal: Price-Patterson, 1999).

12. Hirszfeld refers to Adam Mickiewicz's poem "Reduta Ordona," which describes the death of General Ordon, who defended Fort Reduta against the Russian Army in 1831. For the poem in Polish, see "Adam Mickiewicz: Wiersze Wybrane," http://univ.gda.pl/~literat/amwiersz/index. htm (accessed January 12, 2010).

13. Hitler's Alpine retreat.

14. Artur Gold (1897–1943) was a popular Warsaw jazz musician forced into the ghetto in 1940 and deported to Treblinka in 1942. There he organized and played in a small camp orchestra before being put to death in 1943.

15. The Dante quote is from *Inferno,* canto V, lines 121–23.

16. Another film clip of Czerniaków, originally shown by Pathé and in the Grinberg Archives, is available at the U.S. Holocaust Museum. See http://www.ushmm.org/wlc/fi.php?lang=en&ModuleId=10005069&MediaId=205 (accessed January 12, 2010).

17. Dr. Ludwik Lindenfeld (n.d.–1942), former Polish judge and convert to Christianity was superintendent of the Jewish detention facility on the ghetto. He was killed in 1942. Raul Hilberg, Stanisław Staroń, and Josef Kermisz, eds., *The Warsaw Diary of Adam Czerniakow* (New York: Stein and Day, 1979), 319.

18. From January 8 to April 14, 1942, a "Swiss Sanitary Mission," under the sponsorship of the Swiss Red Cross, was stationed in Warsaw. This was one of four such missions of doctors, nurses, and technicians arranged by the Swiss ambassador in Berlin meant as a gesture of goodwill to the Germans who literally surrounded the neutral country. See Jean Lindenmann, "Herman Mooser, Typhus, Warsaw 1941," *Gesnerus* 59, no. 1–2 (2002): 108–10. For a firsthand account, see Franz Blättler, *Warschau 1942: Tatsachenbericht eines Motorfahrers der zweiten schweizerischen Aerztemission 1942 in Polen* (Zürich: F. G. Micha, 1945).

19. Stefania Wilczyńska (1886–1942) had worked with Korczak since they met in Berlin in 1909. She died along with Korczak and the orphans at Treblinka. See Janusz Korczak, *The Warsaw Ghetto Memoirs of Janusz Korczak* (Washington, DC: University Press of America, 1979), 5, 115.

20. Dr. Natalia Zand was shot on August 12, 1942, for escaping the ghetto.

21. The so-called January Insurrection, a Polish revolt against Russian rule, took place in 1863.

Chapter Twenty-Six

1. It should be remembered that Hirszfeld was writing for a Polish, postwar, and overwhelmingly Catholic readership. Although he repeatedly stressed that his intent in writing the book was to bridge the gap between Jews, Poles, and other non-Jews, he realized, nonetheless, that this and other chapters could be offensive to "Western" readers. For example, he requested that the projected English translation of his memoirs remove any passages that might be misinterpreted. See L. Hirszfeld to Guterman, 2 March 1947, APAN III/157 LHP, folder 88.

2. Father Marceli Godlewski had made anti-Semitic speeches before the war but from all accounts was quickly converted to sympathy and support for the Jews when faced with the conditions of the ghetto. His church furnished fake birth certificates, smuggled children out of the district, and arranged for hiding with parishioners outside the district. He eventually was captured by the Nazis and sent to Auschwitz. See Irene Tomaszewski and Tecia Werbowski, *Żegota: The Council for Aid to Jews in Occupied Poland, 1942–1945* (Montreal: Price-Patterson, 1999), 36; and Henry Shoskes, *No Traveler Returns* (Garden City, NY: Doubleday, Doran, 1945), 95.

3. Jan Matejko (1838–93) was a prominent nineteenth-century Polish painter who, in 1862, painted a portrait of Piotr Skarga (1536–1612), Polish Jesuit, writer, and theologian, representative of the Polish counterreformation—a painting that later became famous.

4. It is not clear whether Hirszfeld ever practiced religion. According to a close postwar acquaintance, neither he nor Hanna ever attended mass after 1945. In the family archives, there remains a manuscript, obviously written by Hirszfeld, entitled "My Conversation with Christ," which reads like a failed attempt to come to terms with the unspeakable suffering wrought by World War II.

5. General Władysław Sikorski (1881–1943) was prime minister of the Polish government, living in exile in London.

Chapter Twenty-Seven

1. Hirszfeld requested that this entire chapter be removed from the projected United States edition of his memoirs; see chapter 26, note 1.

2. Hirszfeld is referring to his uncle Bolesław Hirszfeld; see introduction.

3. Roman Kramsztyk (1885–1942) was a major Polish artist of Jewish origin who was a leader of the New Group between the wars. He did a pastel portrait of Hanna, dedicated March 1942 (donated by the family to the National Muzeum in Warsaw), and was shot in the street shortly thereafter.

4. Anton van Miller was a pseudonym for the Austrian Jewish lawyer Franz Rudolf Bienenfeld (1886–1961) who authored *Deutsche und Juden* (M.-Ostrau: Soziologische Verlagsanstalt, 1937), a critique of the Nazi anti-Semitic policies.

5. Theodor Fritsch (1852–1934) was a late-nineteenth-century German racist, whose most influential book, *Handbuch der Judenfrage* (Langensalza: Hermann Beyer und Söhne), was published in 1896 and went through dozens of printings. It was translated into English in 1927 under the title *The Riddle of the Jew's Success*.

6. See chapters 21 (Nauck) and 25 (Kudicke).

7. Alfred Rosenberg (1893–1946) was an early member of the Nazi party (and a newspaper editor) who focused on ideological training and, during the war, administered the conquered eastern territories. He was tried and executed at Nuremberg. The standard biography is Robert Cecil, *The Myth of the Master Race: Alfred Rosenberg and Nazi Ideology* (New York: Dodd Mead, 1972).

8. Hans F. K. Günther (1891–1968) and Ludwig Ferdinand Clauss (1892–1974) were racial anthropologists whose ideas greatly influenced the Nazis. On Günther, see Hans-Jurgen Lutzhöft, *Der nordische Gedanke in Deutschland, 1920–1940* (Stuttgart: E. Klett, 1971); and for Clauss, see Peter Weingart, *Doppel-Leben: Ludwig Ferdinand Clauss: zwischen Rassenforschung und Widerstand* (Frankfurt: Campus, 1995). For an overview in English, see Geoffrey G. Field, "Nordic Racism," *Journal of the History of Ideas* 38 (1977): 523–40.

9. For an extensive later analysis of this question based on blood groups, see Arthur E. Mourant, Ada C. Kopeć, and Kazimiera Domaniewska-Sobczak, *The Genetics of the Jews* (Oxford: Clarendon Press, 1978).

10. This was actually a talk by Ernest Rénan (1823–92) at the Cercle Saint-Simon on January 27, 1883, published as *Le Judaïsme comme race et comme religion* (Paris: C. Lévy, 1883).

11. The quote is from one of the works by Jan Czekanowski (1882–1965), a widely read Polish anthropologist who published a number of works on populations in Africa, as well as the Baltic and Central Europe, using sophisticated linguistics and statistical analysis. For more, see A. Jones, ed., *Jan Czekanowski, Africanist Ethnographer and Physical Anthropologist in Early Twentieth-Century Germany and Poland* (Leipzig: Institut für Afrikanistik, 2002).

12. Ernst Kretschmer (1888–1964) was a German psychiatrist best known for his theories about the relationship between body build and personality type. See Gil F. Pedrosa, M. M. Weber, and W. Burgmair, "Ernst Kretschmer (1888–1964)," *American Journal of Psychiatry* 159, no. 7 (July 2002): 1111.

13. "Bartek zwycięźca" (Bartek the Conqueror) was a short story written in 1882 by Henryk Sienkiewicz (1846–1916), Polish novelist and winner of the Nobel Prize for literature in 1905.

Chapter Twenty-Eight

1. The Wannsee meeting on the Final Solution was held January 20, 1942. According to the Wannsee Protocol (see http://www.yale.edu/lawweb/avalon/imt/wannsee.htm [accessed June 21, 2007]), the part of the minutes dealing with the General Government was as follows:

> State Secretary Dr. Buehler stated that the General Government would welcome it if the final solution of this problem could be begun in the General Government, since on the one hand transportation does not play such a large role here nor would problems of labor supply hamper this action. Jews must be removed from the territory of the General Government as quickly as possible, since it is especially here that the Jew as an epidemic carrier represents an extreme danger and on the other hand he is causing permanent chaos in the economic structure of the country through continued black market dealings.

Work on the Treblinka extermination camp began in May 1942, and it was ready to receive transports on July 22, 1942.

2. Pawiak was the prison in Warsaw where atrocities were carried out by the German occupying forces.

3. Hirszfeld is referring to the Polska Partia Socjalistyczna (PPS); the Bund was the Jewish Socialist Party.

4. Professor Franciszek Raszeja (1886–1942) was a famous Polish surgeon who was called to the ghetto to perform an operation. The incident is related in Wladyslaw Szpilman's book, *The Pianist* (New York: Picador, 1998). Dr. Kazimierz Polak was Raszeja's assistant.

5. On July 19, 1942, Himmler issued the following: "It is my order that the evacuation of the entire Jewish population in the *Generalgouvernement* area be carried out and completed by December 31, 1942."

6. Hirszfeld likely has this name incorrect. These deportations were part of *Aktion Reinhard*, whose overall leadership was directed by Odilo Globocnik, an SS and Police Leader. The head of operations in the Warsaw Ghetto was SS-*Hauptsturmfuehrer*

Hans Hofle. See "Aktion Reinhard" http://www1.yadvashem.org/odot_pdf/micro-soft%20word%20-%205724.pdf, accessed, May 13, 2010.

7. Hirszfeld's reference is to the stand of German intellectuals at the beginning of World War I, which he mentions in the opening of chapter 4.

8. Amzel and the following names have been mentioned before. This is Hirszfeld's final farewell.

Chapter Twenty-Nine

1. Mieczysław Centnerszwer (1874–1944) was a professor of chemistry in Latvia before the war. Both he and Hilai Lachs taught in the underground medical school. See Charles G. Roland, *Courage under Siege: Starvation, Disease, and Death in the Warsaw Ghetto* (New York: Oxford University Press, 1992), 194–95. For his death, see chapter 36.

2. The Hirszfelds' escape was, in fact, orchestrated by the Armia Krajowa (Home Army, or Polish Army of the Resistance), and Hirszfeld had written a separate chapter on the subject, which was censored by the postwar régime; see introduction.

3. Hirszfeld is referring to Julia Seydel, who also worked at the Polish National Institute of Hygiene before the war.

4. Dr. Maria Wierzbowska, friend of Hanna, was the first to hide the Hirszfelds.

5. Stanisław Kiełbasiński, close friend and brother-in-law of Hanna Hirszfeld, also gave immediate assistance.

6. According to Felix Milgrom, Hirszfeld himself was circumcised.

7. Szymon Starkiewicz (1877–1962) was a physician who established a sanitarium in southern Poland for tubercular children in 1926. The Germans took it over and converted it to care for injured soldiers.

8. Kamienna, see subsequent chapters (32, 34).

Chapter Thirty

1. The *Junak* were a paramilitary scouting organization for young Polish boys recruited by the occupying forces.

2. See chapter 29, note 7.

3. Doktor Judym is a figure in Stanisław Żeromski's famous novel, *Homeless People* (*Ludzie bezdomnie*) (Warsaw: B. Natanson, 1900). Judym is a lower-class Polish medical student who studies in Paris, then returns to Poland to take a job at a spa where his sympathy for the lower class comes into conflict with patrons. He sacrifices his personal happiness for the good of society.

Chapter Thirty-One

1. Joseph Viktor Widmann (1842–1911). *The Maikäfer-Komödie* (Frauenfeld: Huber) was published in 1897.

2. The verse is the opening of "Hymn" by Juliusz Słowacki (1809–1849), a Polish Romantic poet. http://slowacki.chez.com/slowhymn.htm (accessed 14 May 2010).

3. Ludwik Hirszfeld, *Immunologia ogólna* (Warsaw: Czytelnik, 1949).

4. The quotation from Horace, *Non omnis moriar* (I shall not die completely), is the epitaph on Hirszfeld's tomb in Wrocław. See chapter 22.

5. These were the authors of the standard textbooks of the day. For examples, see Charles Nicolle, *L'experimentation en médecine* (Paris: Alcan, 1934); Philip C. Jeans, *Essentials of Pediatrics*, 4th ed. (Philadelphia: J. B. Lippincott, 1946); Hans Zinsser and F. F. Russell, *A Textbook of Bacteriology*, 5th ed. (New York: D. Appleton, 1922).

Chapter Thirty-Two

1. Teofil Stanisław Zawartka (1869–1949).

Chapter Thirty-Three

1. The reference here is to the spate of books published shortly after World War I began; these were written by the warring countries, with documents to prove the other side had caused the war. In addition to the "German White Book: Germany's Reasons for War with Russia," and the "Serbian Blue Book," there was an "Austro-Hungarian Red Book (1914)," a "Belgian Grey Book: Diplomatic Correspondence Respecting the War (July 24–August 29, 1914)," a "French Yellow Book," and a "Russian Orange Book, 1914: Documents Respecting the Negotiations Preceding the War, Published by the Russian Government."

Chapter Thirty-Four

1. Zofja Syrkin-Binstein was a doctor in the ghetto who worked on epidemic prevention. See Dalia Ofer, "Another Glance through the Historian's Lens: Testimonies in the Study of Health and Medicine in the Ghetto," *Poetics Today* 27, no. 2 (2006): 344.

2. At its height, fifteen thousand Jews worked for the Walter Toebbens Company in the Warsaw Ghetto.

3. The document is all the more important because Majdanek was one of the first camps liberated, so it was relatively intact since the Germans had no time to destroy or cover up their actions. An English version of the report is available at the Hoover Institution (Stanford University) and on the Internet at http://www.jewishgen.org/ForgottenCamps/Camps/MajdanekReport.html (accessed June 30, 2005). It is titled "Communique of the Polish-Soviet Extraordinary Commission for Investigating the Crimes Committed by the Germans in the Majdanek Extermination Camp in Lublin" (Moscow: Foreign Languages Publishing House, 1944).

4. Drs. Blanke and Rindfleisch show up in other testimonies.

Chapter Thirty-Six

1. Hirszfeld alludes to a still-debated facet of the war in the East. When Soviet General Andrei Andreyevich Vlasov either defected or was captured by the Germans in July 1942, he proposed to Hitler that he be given command of the 1.5 million Soviet prisoners held by the Germans to overthrow Stalin. After much delay, in early 1945, he was given 50,000 men, but realizing the futility of his original plan, Vlasov marched toward Prague and liberated the capital from the Nazis, before the Soviet Army arrived. His attempt to surrender to the Allies was unsuccessful, and he was captured and tried as a traitor by the Soviets. The standard work on this incident is Catherine Andreyev, *Vlasov and the Russian Liberation Movement: Soviet Reality and Emigré Theories* (Cambridge: Cambridge University Press, 1989).

2. Stanisław Kiełbasiński, Hanna Hirszfeld's brother-in-law; see chapter 29.

3. Władysław Sterling (1877–1943) was an internationally famous Polish neurologist, married to Hirszfeld's sister.

4. There is a very large literature on this now, including government documentation: *Dokumenty Polskie z Lat II Wojny Światowej* [Facts and Figures on the Polish Effort in World War II], http://www.unpack.republika.pl/index.html (accessed January 13, 2010).

5. Similarly, the Lwów massacre has now been documented. A summary of Zygmunt Albert's book, *Kaźń Profesorów Lwów skich Lipiec 1941* [The Torture of Lwów Professors, July 1941](Wrocław: University of Wrocław Press, 1989), is available in English at http://lwow.home.pl/Lwow_profs.html (accessed July 19, 2007).

6. Professor Janusz Supniewski (1899–1964) was a pharmacologist.

7. Kazimierz Pelczar (1894–1943) was a physician and as Hirszfeld notes, a cancer researcher. Between the wars he rose to department chair and dean of the Medical School at Vilnius University. Like Hirszfeld, he refused offers to leave the country before the outbreak of World War II. See Rimgaudas V. Kazakevičius and Kristyna Rotkevič, "Kazimierz Pelczar (1894–1943), the Prominent Professor of Vilnius Stefan Batory University (to the 110th Anniversary of His Birth)" [In English], *Acta Medica Lituanica* 11 (2004): 65–67.

8. Aleksander Rajchman (1890–1940) was a mathematician and Hirszfeld's first cousin. See Antoni Zygmund, "Aleksander Rajchman (1890–1940)," *Roczniki Polskiego Towarzystwa Matematycznego* 27 (1987): 219–23.

9. Office international d'hygiène publique, founded in Rome in 1907 with its headquarters in Paris. On Chodźko, see chapter 7.

10. The uprising began August 1, 1944.

11. Bernhard Nocht (1857–1945) was director of the Hamburg Institute for Tropical Medicine from 1901 to 1930. The institute now bears his name.

12. This second volume was never written.

Chapter Thirty-Seven

1. For example, Eliza Orzeszkowa (1841–1910) and Aleksander Świętochowski (1849–1938) belonged to the "Progressive Poles," and several of their novels had Jewish heroes. See Czeslaw Miłosz, *A History of Polish Literature* (Berkeley: University of California Press, 1983).

2. The most infamous versions of this were the Nazi Nuremberg Laws of 1935. Accordingly, three or more Jewish grandparents made one a Jew; one or two Jewish grandparents made one a *mischling* (of mixed blood).

Chapter Thirty-Eight

1. Wilhelm Ehmer (1898–1976) was a minor Nazi functionary who wrote popular volumes praising the Third Reich. It was not possible to identify the source of the quote below, but it is typical of Ehmer's writings as found in *Die Kraft der Seele: Gedanken eines Deutschen im Kriege* (Stuttgart: Engelhorn, 1940).

2. Gerhard Heberer (1901–73) was an important figure in German biology who supported Aryan biology. See Uwe Hossfeld, *Gerhard Heberer (1901–1973): Sein Beitrag zur Biologie im 20. Jahrhundert* (Berlin: VWB-Verlag für Wissenschaft und Bildung, 1997).

3. Franz Six (1909–75) had been dean of the faculty of economics at the University of Berlin before the war, then became a Nazi party functionary involved in Eastern Front operations to eliminate Jews. Tried and sentenced to twenty years in prison at Nuremberg, he was released after serving only four years.

4. Ernst Pascual Jordan (1892–1980) was a professor of physics at the University of Rostock who joined the Nazi party in 1933 and wrote popular articles criticizing "Jewish physics" during the 1930s.

5. See Michael Burleigh, *Germany turns Eastwards: A Study of Ostforschung in the Third Reich* (Cambridge: Cambridge University Press, 1988).

6. The book was published by Masson in 1938. Hirszfeld's characterization is not quite accurate. Paul Steffan's first paper on blood groups, which had clear implications for race, was published in 1923. And as just noted, Hirszfeld's book was not published until 1938. It is true, however, that Hirszfeld condemned Steffan's Deutsche Gesellschaft für Blutgruppenforschung (German Society for Blood Group Research) when it was established, but this was in private correspondence with Bernstein. *Les groupes sanguins* was Hirszfeld's first extensive public critique of the use of his discoveries about blood group distribution.

7. Galileo's "But it moves," supposedly uttered after he admitted the earth was the center of the universe.

8. The author is likely Ernst Wachler (1871–1945), a writer and founder of the open-air Harz Mountain Theater. He was a follower of the movement that revived ancient neo-pagan rituals and symbols, among them the swastika, but the followers were eventually banned by the Nazis and because of Wachler's Jewish ancestry, he was sent to Auschwitz, where he died.

9. Robert Ley (1890–1945) was one of Hitler's closest henchmen and head of the German Labor Front, which replaced labor unions from 1933–45. He committed suicide awaiting trial at Nuremberg.

Chapter Thirty-Nine

1. On James Oliver Brown, of Little, Brown and Company, who discussed the publication of the translation of Hirszfeld's book in English, see the introduction.

2. Lublin was the seat of the provisional, communist-inspired Polish Government in the recovered territories. At the time Warsaw lay in ruins.

3. Probably Feliks Przesmycki.

4. Praga is a suburb of Warsaw on the east bank of the Vistula River.

5. Henryk Raabe (1882–1951) was a cell biologist who organized and became first rector of the university.

6. From the context, this would seem to be the equivalent of dean.

7. Equivalent of the French *baccalauréat* or examinations passed after secondary school.

8. October 23, 1944.

9. On August 1, 1944, as Soviet troops approached Warsaw, the Polish Home Army launched an attack on the German garrisons in the city. The Germans sent in reinforcements as the Soviets held off attacking the city. By October 2, 1944, the Germans had eliminated all Polish resistance. For a recent account of the Warsaw Uprising in English, see Norman Davies, *Rising '44: The Battle for Warsaw* (New York: Viking, 2004).

10. The celebration was held in Moscow and Leningrad in June 1945 as a showcase of the achievements of Soviet science.

11. Frederic (1900–58) and Irene (1897–1956) Joliot-Curie were Nobel laureates from France; Julian Huxley (1887–1975) was a leading British biologist; and Albert Szent-Györgyi (1893–1986) was a Hungarian biochemist and Nobel laureate.

12. As part of the redrawing of boundaries after World War II, Poland lost territory in the east, including Lwów and Wilno. Compensation for this was the addition of territory taken from Germany including Silesia and the city of Wrocław (formerly Breslau) to which many of the people and facilities from Lwów were transferred.

13. Stanisław Kulczyński (1895–1975).

14. The United Nations Relief and Rehabilitation Administration was established in November 1943 to provide economic assistance to countries after the war and to repatriate and help refugees.

15. For a description, see Professor Dr. Ludwik Hirszfeld, "Some Fragments from the Polish Report about the Voyage to the United States and Canada," received 24 May 1947, RG 1.1, ser. 189, box 3, folder 26, Rockefeller Archive Center, Sleepy Hollow, New York. Hirszfeld published his impressions in, "Wrażenia z podróży do Stanów Zjednoczonych i Kanady," [Impressions du voyage aux États Unis et au Canada], *Polski Tygodnik Lekarski*, 2 (1947): 124–26; 157–59; 188–91; 220–24; 253–54.

16. The date is actually 1940. Hans Zinsser, *As I Remember Him* (Boston: Little, Brown, 1940), 441.

17. Dr. Guterman completed translation of Hirszfeld's book into English before Brown stopped the project.

18. Robert A. Lambert (1884–1960) was a staff member of the Rockefeller Foundation, which financed Hirszfeld's trip.

19. Ella Grove Coca (1890–1966) was a former student of Arthur Coca who collaborated with him on serological work until she retired shortly after they married in 1930.

20. Hirszfeld refers here to William H. Welch (1850–1934), the first dean of Johns Hopkins Medical School, and Abraham Flexner (1866–1959), a Rockefeller Foundation officer. Flexner's famous report on American medical schools, published before World War I, adopted the Germans as the model to reorganize American

medical education, pointing to Johns Hopkins University as the example to emulate. Hirszfeld's personal experience spans the time during which these great changes occurred. For more, see Kenneth Ludmerer, *Learning to Heal* (Baltimore: Johns Hopkins University Press, 1985); and Thomas Neville Bonner, *Iconoclast: Abraham Flexner and a Life in Learning* (Baltimore: Johns Hopkins University Press, 2002).

21. Philip Levine (1900–87) worked as an assistant with Landsteiner until 1932, collaborating on many discoveries including the MNP blood system. In 1939, working with Stetson, Wiener, and Landsteiner on a case of stillborn childbirth, he discovered the Rh blood system. For more, see Eloise R. Giblett, "Philip Levine: August 10, 1900–October 18, 1987," *Biographical Memoirs of the National Academy of Science* 63 (1994): 323–47.

22. Published as Ludwik Hirszfeld, "Transition Forms of Blood Groups," *Journal of Immunology* 55 (1947): 141–51. On Hirszfeld's Pleiades, see chapter 9.

23. Coca devised a simple test of food allergy, based on the observation that pulse rate increases with stress induced by allergic reaction to food. See Arthur Coca, *The Pulse Test: Easy Allergy Detection* (New York: Lyle Stuart), first published in 1956. For his scientific publication, see Arthur F. Coca, "Familial Nonreaginic Food Allergy (Idioblapsis) Practical Management," *International Archives of Allergy and Applied Immunology* 1, no. 3 (1950): 173–89.

24. In the 1946 letter describing his visit to North America (see note 15), Hirszfeld refers to meeting "the bacteriologist Dr. Craigie," hence he is most likely referring to James Craigie (1899–1978), a virologist at the University of Toronto. See Christopher Andrewes, "James Craigie. 25 June 1899–26 August 1978," *Biographical Memoirs of Fellows of the Royal Society* 25 (1979): 233–40.

25. Ernst Witebsky (1901–79), immunologist, had worked under Sachs at Heidelberg before the Nazis forced him to leave. He settled in the State University of New York at Buffalo where he later welcomed Felix Milgrom, one of Hirszfeld's last students.

26. John E. Gordon (1890–n.d.) was appointed chair of the Epidemiology Department of the Harvard University School of Public Health in 1937, but before he could assume his post, he served in the U.S. Army for six years. He returned and chaired the department until 1958.

27. Hirszfeld refers to Ernest William Goodpasture (1886–1960), who developed a method of culturing viruses that made vaccine production possible, and Wendell M. Stanley (1904–71), a Nobel laureate chemist (1946) who worked on the tobacco mosaic virus.

28. Both were researchers at the Rockefeller Institute. Oswald T. Avery (1877–1955), a Canadian-born doctor, came to the Rockefeller Institute in 1913, where he spent the remainder of his career. He is best remembered for his discovery in 1944 with McCarty that DNA was the likely material of which chromosomes and genes were made. Rene J. Dubos (1901–82), a microbiologist trained in agronomy, emigrated to the United States between the wars, where he joined the Rockefeller Institute. Although a prolific researcher, he is best remembered, as Hirszfeld notes, for his ability to write science for a popular audience.

29. Rene J. Dubos, *The Bacterial Cell in Its Relation to Problems of Virulence, Immunity and Chemotherapy* (Cambridge, MA: Harvard University Press, 1945).

30. On Heidelberger, see chapter 8, note 9.

31. Ulrich Friedemann (1877–1949), a bacteriologist, had been professor of hygiene at the University of Berlin with appointments at the Koch Institute and

Virchow Hospital. In 1933, he left first for London, then New York, where Zinsser arranged an appointment at the Jewish Hospital in New York.

32. Raymond Fosdick, *Rockefeller Foundation Annual Report, 1945* (New York: Rockefeller Foundation, 1946), 7. Much of what follows is loosely paraphrased or quoted from Fosdick.

33. As quoted by Fosdick, *Rockefeller Foundation Annual Report*, 8. The author is C. E. M. Joad.

34. The original from Fosdick, *Rockefeller Foundation Annual Report*, 8, 10, and 12, respectively, reads:

> But how do we stop the scientists? How can we foresee the use to which knowledge will be put? Almost any discovery can be employed for either social or antisocial purposes.

> The towering enemy of man is not his techniques but his irrationality.

> Have we time? Fear and uneasiness will dog the steps of this génération like menacing shadows. There will be no escape from them. . . . But unless we succeed in building a moral basis for such a world, even the spur of fear will not get us very far.

Bibliography

Unpublished Sources

Archives

Feliks Przesmycki (1892–1974), *Wspomnienia* [Memoirs], Library of the Institute of Hygiene (Paristwowy Zakład Higieny), Warsaw.

Felix Bernstein Papers, Bernstein Archives, Niedersächsische Landes- und Universitätsbibliothek, Göttingen, Germany.

Ludwik Hirszfeld Papers (LHP), Archives of the Polish Academy of Sciences (APAN), Warsaw.

Ringelblum Papers, Archives of the Jewish Historical Institute, Warsaw.

Rockefeller Archive Center, Sleepy Hollow, New York.

Interviews

Robert Anigstein, interviewed by M. A. Balińska, 1995, New York City.

Stanisław Dubiski, interviewed by M. A. Balińska and W. H. Schneider, 2002, Toronto.

Joanna Kiełbasińska-Belin, interviewed by M. A. Balińska, 2000, 2005, London.

Hilary Koprowski, interviewed by M. A. Balińska, 1997, 2002, 2004, London.

Felix Milgrom, interviewed by M. A. Balińska and W. H. Schneider, 2002, Buffalo.

Jacek Prentki, interviewed by M. A. Balińska, 1997, 2002, 2004, London.

Stanisław Tomkiewicz, interviewed by M. A. Balińska, 2002, Paris.

Rudolf Wohlrab, interviewed by M. A. Balińska, 1995, Hannover.

Private Papers

Hirszfeld family papers, in the editor's (Balińska's) possession.

Published sources

Abel, Rudolf. "Von Hungernot und Seuchen in Russland." *MMW* 70 (1923): 485–87, 633–35. As cited in Paul J. Weindling. *Epidemics and Genocide in Eastern Europe, 1890–1945*. New York: Oxford University Press, 2000.

Adler, Stanisław. *In the Warsaw Ghetto*. Jerusalem: Yad Vashem, 1982.

Aktion, Reinhard. http://www1.yadvashem.org/odot_pdf/microsoft%20word%20 -%205724.pdf, accessed, May 13, 2010.

Albert, Zygmunt. *Kaźń Profesorów Lwów skich Lipiec, 1941* [The Torture of Lwów Professors, July 1941]. Wrocław: University of Wrocław Press, 1989.

Aleksandrowicz, Julian. *Kartki z dziennika doktora Twardego* [Pages from the Diary of Dr. Twardy]. Cracow: Wydawnictwo Literackie, 1983.

Alvarez, Luis. *Adventures of a Physicist.* New York: Basic Books, 1987.

Andrewes, Christopher. "James Craigie. 25 June 1899–26 August 1978." *Biographical Memoirs of Fellows of the Royal Society* 25 (1979): 233–40.

Andreyev, Catherine. *Vlasov and the Russian Liberation Movement: Soviet Reality and Emigré Theories.* Cambridge: Cambridge University Press, 1989.

Anigstein, Ludwik, ed. "A Symposium of Polish Medical Contributions in World War II." *Texas Reports on Biology and Medicine* 5 (1947): 155–87.

Atti del Primo Congresso Internazionale della Transfusione del Sangue, Roma, 26–29 Settembre 1935. Milan: A. Colombo, 1935.

August, Jochen. *Sonderaktion Krakau.* Hamburg: Hamburg Edition, 1997.

Balińska, Marta A. "Assistance and Not Mere Relief: The Epidemic Commission of the League of Nations." In *International Health Organisations and Movements, 1918–1939,* edited by Paul J. Weindling, 81–108. Cambridge, UK: Cambridge University Press, 1995.

———. "Choroba jako ideologia: tyfus plamisty w okupowanej Polsce, 1939–1944" [Disease as Ideology: Typhus in Occupied Poland, 1939–1944]. *Zeszyty historyczne* 126 (1998): 212–20.

———. "L'école de médecine du ghetto de Varsovie (1941–1942)." *Revue du Praticien* 58, no. 31 (2008): 227–29.

———. *Ludwik Rajchman: Medical Statesman.* Budapest: Central European University Press, 1998.

———. "The National Institute of Hygiene and Public Health in Poland, 1918–1939." *Social History of Medicine* 9, no. 3 (1996): 427–45.

———. "La Pologne du choléra au typhus." *Bulletin of the Exotic Pathology Society* 92, no. 5 (1999): 349–54.

———. "La Pologne face à ses crises médicales du XXe siècle." DEA thesis, Institut d'Etudes Politiques de Paris, 1989.

———. "The Rockefeller Foundation and the National Institute of Hygiene." *Studies in History and Philosophy of Biological and Biomedical Sciences* 31C, no. 3 (2000): 419–32.

Baltzer, Fritz. *Theodor Boveri: Life and Work of a Great Biologist.* Translated by Dorothea Rudnick. Berkeley: University of California Press, 1967.

Bartoszewski, Władysław T., and Antony Polonsky. *The Jews in Warsaw: A History.* London: Blackwell, 1991.

Bay-Schmith, Erik. "Rassenbiologische Untersuchungen auf Gronland." *Acta Pathologica et microbiologica Scandinavica* 4 (1927): 310–40.

Bennett, John D. C., and Lydia Tyszczuk. "Deception by Immunization Revisited." *British Medical Journal* 301 (1990): 1471–72.

Bernstein, Felix. "Ergebnisse einer biostatistischen zusammenfassenden Betrachtung uber die erblichen Strukturen des Menschen." *Klinische Wochenschrift* (1924): 1495–97.

Bertaux, Sandrine. "Demographie, Statistique et Fascisme: Corrado Gini et l'ISTAT, entre Science et Ideologie (1926–1932)." *Roma Moderna e Contemporanea* 7, no. 3 (1999): 571–59.

Beyerchen, Alan. "On the Stimulation of Excellence in Wilhelmian Science." In *Another Germany: A Reconsideration of the Imperial Era.* Edited by Jack R. Dukes and Joachim Remak, 139–68. Boulder: Westview Press, 1988.

Bienenfeld, Franz Rudolf [Anton van Miller, pseud.]. *Deutsche und Juden.* M.-Ostrau: Soziologische Verlagsanstalt, 1937.

"Biography in the History of Science." *Isis* 97 (2006): 302–32.

Blady-Swajger, Adina. "Krótka historia szpitala Bersonów i Baumanów (1939–1943)" [A Short History of the Berson and Bauman Hospital (1939–1943)]. *Zeszyty Niezależnej Myśli Lekarskiej* 10 (December 1986).

Blättler, Franz. *Warschau, 1942: Tatsachenbericht eines Motorfahrers der zweiten schweizerischen Aerztemission, 1942, in Polen.* Zürich: F. G. Micha, 1945.

Blobaum, Robert E. *Rewolucja: Russian Poland, 1904–1907.* Ithaca, NY: Cornell University Press, 1995.

Bogomoletz, Alexander A. "Les phénomènes de l'autocatalyse et la transfusion du sang." In *Atti del Primo Congresso Internazionale della Transfusione del Sangue, Roma, 26–29 Settembre 1935,* 87–104. Milan: A. Colombo, 1935.

Bonner, Thomas Neville. *Iconoclast: Abraham Flexner and a Life in Learning.* Baltimore: Johns Hopkins University Press, 2002.

Brown, Theodore M., and Elizabeth Fee. "Andrija Štampar: Charismatic Leader of Social Medicine and International Health." *American Journal of Public Health* 96, no. 8 (August 2006): 1382–85.

Browning, Christopher. "Before the 'Final Solution': Nazi Ghettoization Policy in Poland (1940–1941)." In *Ghettos 1939–1945: New Research and Perspectives on Definition, Daily Life, and Survival,* edited by Christopher R. Browning et al., 1–14. Washington, DC: United States Holocaust Memorial Museum, 2005.

———. "Genocide and Public Health: German Doctors and Polish Jews, 1939–1941." *Holocaust and Genocide Studies* 3 (1988): 21–36.

———. "Nazi Ghettoization Policy in Poland, 1940–1941." *Central European History* 19 (1986): 343–68.

Brüer, Ralf. "100 Jahre organisierte Krebsforschung." *Deutsche medizinische Wochenschrift* 125, no. 15 (2000): 473–74.

Bufano, M. "Critical Review of the Clinical and Scientific Work of Nicola Pende." [In Italian.] *Policlinico* 78, no. 1 (January–February 1971): 1–18.

Burleigh, Michael. *Germany Turns Eastwards: A Study of Ostforschung in the Third Reich.* Cambridge: Cambridge University Press, 1988.

Carrel, Alexis. *Man, the Unknown.* New York: Harper and Brothers, 1935. Originally published as *L'Homme, cet inconnu* (Paris: Librairie Plan, 1935).

Cecil, Robert. *The Myth of the Master Race: Alfred Rosenberg and Nazi Ideology.* New York: Dodd Mead, 1972.

Chrapowicki, Tadeusz. "Mikołaj Łącki, MD." [In Polish.] *Pediatria Polska* 44, no. 8 (August 1969): 929–31.

Chrzanowski, A. [A page from the history of Polish oncology. The foundation of Count Jakub Potocki]. *Nowotwory.* 27(3), (1977 Jul–Sep):255–59. [Article in Polish].

Coca, Arthur F. "Familial Nonreaginic Food Allergy (Idioblapsis) Practical Management." *International Archives of Allergy and Applied Immunology* 1, no. 3 (1950): 173–89.

————. *The Pulse Test: Easy Allergy Detection.* New York: Lyle Stuart, 1956.

Coca, Arthur F., and Olin Diebert. "A Study of the Occurrence of the Blood Groups among American Indians." *Journal of Immunology* 8 (1923): 487–91.

Cockburn, W. Charles. "The International Contribution to the Standardization of Biological Substances. I. Biological Standards and the League of Nations, 1921–1946." *Biologicals* 19 (1991): 161–69.

Corbett, Elsie. "Louise McIlroy." *British Medical Journal* 1, no. 589 (February 17, 1968): 451.

————. *Red Cross in Serbia, 1915–1919.* Banbury: Cheney and Sons, 1964.

Cornebise, Alfred E. *Typhus and Doughboys: The American Polish Typhus Relief Expedition, 1919–1921.* Newark: Delaware University Press, 1982.

Corrsin, Stephen D. *Warsaw before the First World War. Poles and Jews in the Third City of the Russian Empire, 1880–1914.* Boulder: East European Monographs, 1989.

Coser, Lewis H. "Georg Simmel's Style of Work." *American Journal of Sociology* 63 (1958): 635–41.

Cruse, Julius M. "A Centenary Tribute: Michael Heidelberger and the Metamorphosis of Immunologic Science." *Journal of Immunology* 140, no. 9 (1988): 2861–63.

Czeley-Wybieralska, Małgorzata, and Symcha Wajs. "Emanuel Ringelblum (1900–1944) jako historyk medycyny żydowskiej" [Emanuel Ringelblum (1900–1944), Historian of Jewish Medicine]. *Archiwum Historii i Filozofii Medycyny* 51, no. 2 (1988): 187–210.

Czerniaków, Adam. *The Warsaw Diary of Adam Czerniakow.* Edited by Raul Hilberg, Stanisław Staroń, and Josef Kermisz. New York: Stein and Day, 1979.

Davies, Norman. *Rising '44: The Battle for Warsaw.* New York: Viking, 2004.

Debré, R. "Helena Sparrow—A Polish and French Scientist." *Materia medica Polona* 11, no. 1 (January–March 1979): 79–82.

Deichmann, Uta. "The Expulsion of Jewish Chemists and Biochemists from Academia in Nazi Germany." *Perspectives in Science* 7 (1999): 38.

De Mijolla, Alain. "Freud, Biography, His Autobiography, and His Biographers." *Psychoanalysis and History* 1, no. 1 (1998): 4–27.

Dewey, Marc, Udo Schagen, Wolfgang U. Eckart, and Eve Schönenberger. "Ernst Ferdinand Sauerbruch and His Ambiguous Role in the Period of National Socialism." *Annals of Surgery* 244, no. 2 (2006): 315–21.

Dickel, Horst. "Hans Sachs." In *German-Speaking Exiles in Ireland, 1933–1945*, edited by Gisela Holfter, 183–213. Amsterdam: Rodopi, 2006.

Dieterle, T., L. Hirschfeld, and R. Klinger. "Epidemiologische Untersuchungen über den endemischen Kropf." *Archiv für Hygiene* 81 (1913): 128–78.

Dietto, Romolo. "Pietro Rondoni (1882–1956)." *Tumori* 42 (1956): 805–9.

DiScala, Spencer M. "Science and Fascism: The Case of Enrico Fermi." *Totalitarian Movements and Political Religions* 6, no. 2 (2005): 199–211.

Drouard, Alain. *Alexis Carrel (1873–1944): De la mémoire à l'histoire.* Paris: L'Harmattan, 1995.

Dubiski, Stanisław. "The Fifth Basic Sciences Symposium of the Transplantation Society—A Festschrift Honoring Felix Milgrom." *Transplantation Proceedings* 31 (1999): 1456–57.

Dubos, Rene J. *The Bacterial Cell in Its Relation to Problems of Virulence, Immunity and Chemotherapy.* Cambridge, MA: Harvard University Press, 1945.

Duffy, Michael. "Peter I Karadjordjevic (1844–1921)." First World War—Who's Who. http://www.firstworldwar.com/bio/alexander_serbia.htm (accessed June 2, 2002).

Ehmer, Wilhelm. *Die Kraft der Seele: Gedanken eines Deutschen im Kriege.* Stuttgart: Engelhorn, 1940.

Field, Geoffrey G. "Nordic Racism." *Journal of the History of Ideas* 38 (1977): 523–40.

Finkelstein, Gabriel. "M. du Bois-Reymond goes to Paris." *British Journal for the History of Science* 36, no. 3 (2003): 261–300.

Flexner, Abraham. *Medical Education: A Comparative Study.* New York: Macmillan, 1925.

Forissier, Régis. "L'aide médicale humanitaire apportée a la Serbie par la France et ses Alliés au cours de la Prémière guerre mondiale." *Revue Historique des Armées* 2 (1996): 9–26.

Fosdick, Raymond. *Rockefeller Foundation Annual Report, 1945.* New York: Rockefeller Foundation, 1946.

Fox, Frank. "The Jewish Ghetto Police: Some Reflexions." *East European Jewish Affairs* 25 (1995): 41–47.

Fritsch, Theodor [F. Roderich-Stoltheim, pseud.]. *The Riddle of the Jew's Success.* Translated by Capel Pownall. Leipzig: Hammer-Verlag, 1927. Originally published *Handbuch der Judenfrage* (Langensalza: Hermann Beyer und Söhne, 1896).

Fryer, Charles. *The Destruction of Serbia in 1915.* New York: Columbia University Press, 1997.

Gallo, Robert C. *Virus Hunting: AIDS, Cancer, and the Human Retrovirus: A Story of Scientific Discovery.* New York: Basic Books, 1991.

"Gąsiorowski, Napoleon Jan" in *Polski Słownik Biograficzny* [Polish Biographical Dictionary]. Edited by Instytut Historii im. Tadeusza Manteuffla. Vol. 7. Cracow: Polska Akademia Nauk and Polska Akademia Umiejętnośći, 1948.

Gasiorowski, Zygmunt J. "The German-Polish Non-Aggression Pact of 1934." *Journal of Central European Affairs* 15 (1955): 3–29.

Gawin, Magdalena. "Progressivism and Eugenic Thinking in Poland, 1905–1939." In *Blood and Homeland: Eugenics and Racial Nationalism in Central and Southeast Europe, 1900–1940.* Edited by Marius Turda and Paul J. Weindling, 167–83. New York: Central European University Press, 2006.

Geisenhainer, Katja. *"Rasse ist Schicksal"—Otto Reche (1879–1966), ein Leben als Anthropologe und Völkerkundler.* Leipzig: Evangelische Verlagsanstalt, 2002.

Geison, Gerald L. *The Private Science of Louis Pasteur.* Princeton: Princeton University Press, 1995.

Giblett, Eloise R. "Philip Levine: August 10, 1900–October 18, 1987." *Biographical Memoirs of the National Academy of Science* 63 (1994): 323–47.

Gillette, Aaron. "The Origins of the 'Manifesto of Racial Scientists.'" *Journal of Modern Italian Studies* 6 (2001): 305–23.

Gilsohn, Jakob Wolf. "Prof. Dr. Ludwig Hirszfeld." Medical dissertation. Munich: A. Schubert, 1965.

Gliński, Jan Bohdan. *Słownik biograficzny lekarzy i farmaceutów ofiar drugiej wojny światowej* [Biographical Dictionary of Physician and Pharmacist Victims of the Second World War]. Wrocław: Wydawnictwo Medyczne Urban & Partner, 1997.

Goetz, Christopher G., Michel Bonduelle, and Toby Gelfand. *Charcot: Constructing Neurology.* New York: Oxford University Press, 1995.

Gordon-Smith, Gordon. *Through the Serbian Campaign: The Retreat of the Serbian Army.* London: Hutchinson, 1916.

Gorski, Andrzej, et al. "Bacteriophages in Medicine." In *Bacteriophage: Genetics and Molecular Biology,* edited by Stephen Mcgrath and Douwe Van Sinderen, 125–58. Norfolk, UK: Caister Academic Press, 2007.

Gottlieb, Norman L., and Roy Altman. "An Ethical Dilemma in Rheumatology: Should the Eponym Reiter's Syndrome be Discarded?" *Seminars in Arthritis and Rheumatism* 32, no. 4 (2003): 207–45.

Govorchin, Gerald G. "Emergence of the Radical Party in Serbian Politics." *American Slavic and East European Review* 15 (1956): 511–26.

Groger, Helmut, and G. Stacher. "The Medical Profession in Vienna and the Nazi Regime." *Digestive Diseases* 17 (1999): 286–90.

Gryglewski, Ryszard W. "Prof. Janusz Supniewski (1899–1964)." [In Polish.] *Archiwum Historii Medycyny* 37, no. 3 (1974): 383–85.

Gumkowski, Janusz, and Kazimierz Leszczyński. *Poland under Nazi Occupation.* Warsaw: Polonia Publishing House, 1961.

Guthrie, C. J., and John G. Huck. "On the Existence of More Than Four Isoagglutinin Groups in Human Blood." *Bulletin of the Johns Hopkins Hospital* 34 (1923): 37–48.

Hagen, Wilhelm. *Das Gesundheitswesen der Stadt Warschau September 1939 bis März 1942.* Warsaw: Archives of the State Institute of Hygiene, n.d.

Hagemann, Rudolf. "How Did East German Genetics avoid Lysenkoism?" *Trends in Genetics* 18, no. 6 (June 2002): 320–24.

Halber, Wanda, and Jan Mydlarski. "Untersuchung uber die Blutgruppen in Polen." *Zeitschrift für Immunitätsforschung* 43 (1925): 470–84.

Hankins, Thomas L. "In Defence of Biography: The Use of Biography in the History of Science." *History of Science* 17 (1979): 1–16.

Harwood, Jonathan. *Styles of Scientific Thought: The German Genetics Community, 1900–1933.* Chicago: University of Chicago Press, 1993.

Heimrath, Tadeusz. "Ludwik Hirszfeld (1884–1954) the Wrocław Years, 1945–1954." [In Polish.] *Archiwum Historii i Filozofii Medycyny* 63 (2000): 19–23.

Hilberg, Raul. *Sources of Holocaust Research.* Chicago: Ivan R. Dee, 2001.

Hirschfeld, L. [*sic*], and R. Klinger. "Zur Theorie der Serumreaktionen." *Berliner klinische Wochenschrift* (1914): 1173–76.

Hirszfeld, Hanna, Ludwik Hirszfeld, and H. Brokman. "On the Susceptibility to Diphtheria (Schick Test Positive) with Reference to the Inheritance of Blood Groups." *Journal of Immunology* 9 (1924): 571–91.

Hirszfeldowa, Hanna. *Rôle de la constitution dans les maladies infectieuses des enfants.* Paris: Masson, 1939.

Hirszfeldowa, Hanna, Ludwik Hirszfeld, and H. Brokman. "Wrażliwość na błonicę w świetle badań nad konstytutucją i dziedzicznością" [Sensitivity to Diphtheria in Light of Research on Constitution and Heredity] *Medycyna Doświadczalna i Społeczna* 2 (1924): 125–42.

Hirszfeld, Ludwik. "Aus meinem Erlebnissen als Hygieniker in Serbien." *Correspondenz-Blatt für Schweizer Ärzte* 46 (1916): 513–31.

———. "Essai d'application des methodes sérologiques au problème des races." *Anthropologie* 29 (1919): 505–37.

———. *Les groupes sanguins, leur application à la biologie, à la médecine et au droit.* Paris: Masson & Cie, 1938. Originally published as *Grupy krwi w zastosowaniu do biologji, medycyny i prawa.* (Warszawa: Delta, 1934).

———. "Les groupes sanguins à la lumière de la science contemporaine." In *IIe Congrès international de la transfusion sanguine. 29 septembre–2 octobre 1937 Paris* 2: 7–11. Paris: J.-B. Ballière, 1939.

———. *Historia jednego życia* [The Story of One Life]. Warsaw: Czytelnik, 1946.

———. *Immunologia ogólna* [General Immunology]. Warsaw: Czytelnik, 1949.

———. "Inaugural Address." In *Atti del Primo Congresso Internazionale della Transfusione del Sangue, Roma, 26–29 Settembre 1935.* Milan: A. Colombo, 1935.

———. "Kilka uwag w sprawie reformy nauczania higjeny na uniwersytetach [Some remarks on the reform of teaching of hygiene at universities]. *Polska Gazeta Lekarska,* 8 (1929): 839–841.

———. *Konstitutionsserologie und Blutgruppenforschung.* Berlin: Springer Verlag, 1928.

———. "A New Germ of Paratyphoid." *Lancet* (February 22, 1919): 296–97.

———. [Obituary of Charles Nicolle]. *Warszawskie Czasopismo Lekarskie* 13 (1936): 10–11.

———. [Obituary of Charles Nicolle]. *Maroc médical* 17 (1937): 67–70.

———. "Obsługa bakteriologiczna Państwa. Doniesienie II, » [The state bacteriological service, 2nd communication]. *Lekarz Polski,* 12 (1936): 106–113.

———. "Prologomena zur Immunitätslehre." *Klinische Wochenschrift* 10 (1931): 2153–59.

———. "Die Serodiagnostik im Dienste der Bluttransfusion." In *Atti del Primo Congresso Internazionale della Transfusione del Sangue, Roma, 26–29 Settembre 1935,* 27–38. Milan: A. Colombo, 1935.

———. "Die Seuchengesetze in naturgeschichtlicher Betrachtung." *Wiener klinische Wochenschrift* 51 (1938): 732–37.

———. "Transition Forms of Blood Groups." *Journal of Immunology* 55 (1947): 141–51.

———. "Über ein neues Blutsymptom bei Malariakrankheit." *Correspondenz-Blatt für Schweizer Ärzte* 47 (1917): 1007–12.

———. "Über serologische Mutationen bei Menschen." In *Relazioni del IV Congresso Internationale di Patologia Comparata, Roma, 15–20 maggio, 1939.* Milan: Istituto Sieroterapico Milanese (1939) 1: 227–45.

———. "Über die serologische Spezifität der Krebszellen." *Klinische Wochenschrift* 8 (1929): 1563–66.

———. "Über die theoretischen Grundlagen der Serodiagnostik des Krebses." In *IIe congrès international de lutte scientifique et sociale contre le cancer: Bruxelles 20–26 septembre 1936. Travaux scientifiques.* Edited by Marta Fraenkel, 227–39. Bruxelles: Ligue Nationale Belge Contre le Cancer, 1936–37.

———. "Les vaccinations antidiphthériques en Pologne." *Bulletin de l'Académie de Médecine* 121 (1939): 712–22.

———. "Wanda Halberówna." *Warszawskie Czasopismo Lekarskie,* 15, (1938):118.

———. "Wrażenia z podróży do Stanów Zjednoczonych i Kanady" [Impressions of a voyage to the United States and Canada]. *Polski Tygodnik Lekarski,* 2 (1947): 124–26; 157–59; 188–91; 220–24; 253–54.

Hirszfeld, Ludwik, and Róża Amzel. "O postaciach przejściowych (podgrupach) w obrębie grupy O. Doniesienie tymczasowe" [On the Intermediate Forms (Subgroups) of Group O. Preliminary Communication]. *Polski Tygodnik Lekarski* 1 (1946): 1525–27.

————. "Sur les pléiades 'isozériques' du sang." *Annales de l'Institut Pasteur* 65 (1940): 251–78; 386–414.

————. "Über die Übergangsformen der Blutgruppen." *Schweizerische medizinische Wochenschrift* 70 (1940): 801.

Hirszfeld, Ludwik, and Hanna Hirszfeld. "Serological Differences between the Blood of Different Races: The Results of Researches on the Macedonian Front." *Lancet* 2 (1919): 675–79.

Hirszfeld, Ludwik, and Z. Kostuch. "Untersuchungen über die Untergruppen und ihre Vererbung." *Schweizerische Zeitschrift für Allgemeine Pathologie und Bakteriologie* (1938): 407–20.

Hirszfeld, Ludwik, and H. Zbrowski. "Sur la perméabilité élective du placenta en rapport avec le groupe sanguin de la mère et le foetus." *Comptes rendus de la Société de Biologie* 92 (1925): 1253–55.

Hoch, Paul K. "Whose Scientific Internationalism?" *British Journal for the History of Science* 27 (1994): 345–49.

Hornowski, S. "Remembrances on the beginnings of blood donation in Warsaw. The foundation of the academic blood donation centre in Warsaw." [In Polish.] *Archiwum Historii I Filozofii Medycyny / Polskie Towarzystwo Historii Medycyny i Farmacji* 62, no. 3 (1999): 229–32.

Hossfeld, Uwe. *Gerhard Heberer (1901–1973): Sein Beitrag zur Biologie im 20. Jahrhundert.* Berlin: VWB-Verlag für Wissenschaft und Bildung, 1997.

Huet, Maurice. "L'élevage du pou au laboratoire. *Histoire des Sciences Medicales* 37 (2003): 43–46.

Hutt, Marten. "Medical Biography and Autobiography in Britain, c. 1780–1920." PhD diss., University of Oxford, 1995.

Hutton, I. Emslie. *With a Woman's Unit in Serbia, Salonica and Sebastopol.* London: Williams and Norgate, 1928.

"In Memoriam: Prof. Dr. Robert Doerr." *Annals of Allergy* 10, no. 2 (1952): 240.

Israel, Giorgio. "Science and the Jewish Question in the Twentieth Century: The Case of Italy and What It Shows." *Aleph* 4 (2004): 191–261.

Iwanski, Jean. *Dokumenty Polskie z Lat II Wojny Światowej* [Facts and Figures on the Polish Effort in World War II]. http://www.republika.pl/unpack/index.html (accessed January 13, 2010).

Janssens, Petrus Gustavus. "In Memoriam: Leonard Jan Bruce-Chwatt, 6/9/1907–3/17/1989." *Annales de la Société Belge de Médicine Tropicale* 71, no. 2 (June 1991): 163–66.

Jasienica, Paweł. "Opowiesci o zywej materii." [Stories of Live Matter] In *Historia jednego zycia*. 2nd ed. Warsaw: PAX, 1956.

Jaworkski, Marek. *Ludwik Hirszfeld.* Warsaw: Wydawnictwo Interpress, 1977. Published in German as *Ludwik Hirszfeld: sein Beitrag zu Serologie und Immunologie.* Leipzig: BSB B. G. Teubner, 1980.

Jeans, Philip C. *Essentials of Pediatrics.* 4th ed. Philadelphia: J. B. Lippincott, 1946.

Jones, Adam, ed. *Jan Czekanowski, Africanist Ethnographer and Physical Anthropologist in Early Twentieth-Century Germany and Poland.* Leipzig: Institut für Afrikanistik, 2002.

Jones, Fortier. *With Serbia into Exile: An American's Adventures with the Army That Cannot Die.* New York: Century, 1916.

Jötten, Karl Wilhelm. "In Memoriam, Otto Spittal und Martin Ficker." *Archiv für Hygiene und Bakteriologie* 134, no. 2 (1951): 83–84.

Juden an der Universität Heidelberg. Special exhibit, University of Heidelberg Archive. http://www.tphys.uni-heidelberg.de/Ausstellung/show.cgi?de&D&23& 164 (accessed November 22, 2005).

"Julius Bauer." *Lancet* 1, no. 8130 (June 23, 1979): 1359.

Katsch, Gerhard. "Nachruf für Gustav von Bergmann." *Münchener medizinische Wochenschrift* 97, no. 42 (October 21, 1955): 1398–1400.

Kawalec, Krzysztof. "The Dispute over Eugenics in Interwar Poland." [In Polish.] *Medycyna Nowożytna* 7, no. 2 (2000): 87–102.

Kazakevičius, Rimgaudus V., and Kristyna Rotkevič. "Kazimierz Pelczar (1894–1943), the Prominent Professor of Vilnius Stefan Batory University (to the 110th anniversary of his birth)." [In English.] *Acta Medica Lituanica* 11 (2004): 65–67.

Kiełbasińska-Belin, Joanna. "W sprawie Ludwika Hirszfelda" [About Ludwik Hirszfeld]. *Gazeta wyborcza* 21 (August 2000).

Killen, Linda. "The Rockefeller Foundation in the First Yugoslavia." *East European Quarterly* 24, no. 3 (1990): 349–72.

Kirschner, H. "Marcin Kacprzak and the Development of Social Medicine in Poland." [In Polish.] *Archiwum Historii i Filozofii Medycyny* 51, no. 1 (1988): 89–100.

Kłodziński, Stanisław. "Konferencja Krynicka Lekarzy Hitlerowskich 13–16 October 1941" [The Bad Krynica Congress of Nazi Physicians, 13–16 October 1941]. In *Okupacja i medycyna*, 103–16. Warsaw: Książka i Wiedza, 1971.

Knowles, J. C. "Max Rubner." *Diabetes* 6 (1957): 369–71.

Kopsch, Friedrich. *Anatomie des Menschen.* 4 vols. Stuttgart: G. Thieme, 1886.

Korczak, Janusz [Henryk Goldszmidt]. *Ghetto Diary.* New Haven, CT: Yale University Press, 2003.

———. *The Warsaw Ghetto Memoirs of Janusz Korczak.* Washington, DC: University Press of America, 1979.

Kostrzewski, Jan. "Remembrance of Roman Nitsch in Connection with His Views on the Nature of Fixed Virus." [In Polish.] *Przegląd Epidemiologiczny* 11, no. 2 (1957): 195–97.

Kozuschek, Waldemar. *Ludwik Hirszfeld (1884–1954) Rys życia i dzialalnosc naukowa* [Ludwik Hirszfeld (1884–1954): A Sketch of His Life and Scientific Activity]. Wrocław: Wydawnictwo Uniwersytetu Wrocławskiego, 2005.

Krzyński, Stefan. "In Memoriam Prof. Dr. Jerzy Morzycki, 1905–1954." [In Polish.] *Biuletyn Państwowego Instytutu Medycyny Morskiej i Tropikalnej j W Gdańsku* 6 (1955): 9–13.

Krzywicka, Irena. *Wyznania gorzczycielki* [Confessions of a Sinner]. Warsaw: Czytelnik, 1992.

Krzywicki, Ludwik. *Wspomnienia* [Memoirs], vol. 2. Warsaw: Czytelnik, 1956.

Kucharz, Eugeniusz J., Marc A. Shampo, and Robert A. Kyle. "Odo Bujwid—Pioneer in Microbiology." *Mayo Clinic Proceedings* 65, no. 2 (February 1990): 286.

Kunicki-Goldfinger, Władysław. "Migawki z dziejów mikrobiologii polskiej" [Snapshots from the History of Polish Microbiology]. *Postępy mikrobiologii* 26, no. 3 (1987): 135–54.

Kushnarev, V. M. "Jon Cantacuzene (1863–1934)." [In Polish.] *Zhurnal Mikrobiologii Epidemiologii i Immunobiologgii* 30 (December 1959): 128–30.

Kutter, Elizabeth, and Alexander Sulakvelidze. "Bacteriophage Therapy in Humans." In *Bacteriophages: Biology and Applications*, edited by Elizabeth Kutter and Alexander Sulakvelidze, 412–13. Boca Raton, FL: CRC Press, 2004.

Landsteiner, Karl. "On Individual Differences in Human Blood." Nobel Prize lecture, December 11, 1930. http://nobelprize.org/nobel_prizes/medicine/laureates/1930/landsteiner-lecture.html (accessed January 13, 2010).

———. "Über Agglutinationserscheinungen normalen menschlichen Blutes." *Wiener klinische Wochenschrift* 14 (1901): 1134.

Landsteiner, Karl, and Max Richter. "Über die Verwertbarkeit individueller Blutdifferenzen für die forenzische Praxis." *Zeitschrift für Medizinbeamten* 16 (1903): 85–89.

Lattes, Leone. *Individuality of the Blood in Biology and in Clinical and Forensic Medicine.* London: Oxford University Press, 1932.

Leneman, Leah. *In the Service of Life: The Story of Elsie Inglis and the Scottish Women's Hospitals.* Edinburgh: Mercat Press, 1994.

Leon, George B. *Greece and the Great Powers, 1914–1917.* Thessaloniki: Institution for Balkan Study, 1974.

Lepalczyk, Irena. *Helena Radlińska: życie i twórczość* [Helena Radlińska: Her Life and Works]. Torun: Adam Marszałek, 2001.

Lépine, Pierre. "Necrologie: Nicholas Kossovitch (1884–1948)." *Annales de l'Institut Pasteur* 75 (1948): 62–64.

Lerner, Warren. *Karl Radek.* Stanford: Stanford University Press, 1970.

Lewis, Sinclair. *Arrowsmith.* New York: Harcourt Brace, 1925.

Lifton, Betty Jean. *The King of Children: The Life and Death of Janusz Korczak.* New York: St. Martin's Griffin, 1997.

———. "Who Was Janusz Korczak?" In *The King of Children.* Edited by Michael Parciak. Munich: Korczak Communication Center. http://www.thehypertexts.com/janusz_korczak.htm (accessed May 19, 2010).

Lille-Szyszkowicz, I. "Development of Studies on Pleiades of Blood Groups." [In Polish.] *Postępy higieny i medycyny doświadczalnej* 11, no. 3 (1957): 229–33.

Lindenmann, Jean. "Hermann Mooser, Typhus, Warsaw 1941." *Gesnerus* 59, no. 1–2 (2002).

Lindzey, Gardner, ed. *A History of Psychology in Autobiography*, vol. 9. Washington, DC: American Psychological Association, 2007.

Loretan, Maria. *William Silberschmidt, 1869–1947, Hygieniker und Bakteriologe.* Switzerland: Juris Druck + Verlag, 1988.

Ludmerer, Kenneth. *Learning to Heal.* Baltimore: Johns Hopkins University Press, 1985.

Lutzhöft, Hans-Jurgen. *Der nordische Gedanke in Deutschland, 1920–1940.* Stuttgart: E. Klett, 1971.

Maćkowiak, Zenon. "Remarks on Life and Activities of Witold Chodźko (1875–1954)." [In Polish.] *Przegląd Lekarski* 37, no. 4 (1980): 437–40.

MacPherson, William Grant. *History of the Great War: Medical Services General History.* Vol. 4, *Medical Services during the Operations on the Gallipoli Peninsula; in Macedonia . . .* London: HMSO, 1924.

Magdzik, Wiesław. "Results of the Work of Sanitary-Epidemiologic Service in Poland during the Last 85 Years and Perspectives for the Future." [In Polish.] *Przegląd Epidemiologiczny* 58, no. 4 (2004): 569–81.

Mann, Arthur James, and W. T. Wood. *The Salonica Front.* London: A&C Black, 1920.

Massry, S. G., and M. Smogorzewski. "The Hunger Disease of the Warsaw Ghetto." *American Journal of Nephrology* 22, no. 2–3 (July 2002): 197–201.

Mazower, Mark. *Salonica, City of Ghosts: Christians, Muslims and Jews, 1430–1950.* New York: Knopf, 2005.

Mazumdar, Pauline. "Blood and Soil: The Serology of the Aryan Racial State." *Bulletin of the History of Medicine* 64 (1990): 187–219.

———. "'In the Silence of the Laboratory:' The League of Nations Standardizes Syphilis Tests." *Social History of Medicine* 16, no. 3 (2003): 437–59.

———. "Serology, Standardisation and Collective Security: The Standardisation Commission of the League of Nations, 1920–1939." In *Proceedings of the European Academy for Standardization Workshop, Paris, 2004*, 41–52. Hamburg: European Academy for Standardization, 2004.

———. "Two Models for Human Genetics: Blood Grouping and Psychiatry in Germany between the Wars." *Bulletin of the History of Medicine* 70 (1996): 609–57.

McLaren, Eva Shaw. *The History of the Scottish Women's Hospitals.* London: Hodder & Stoughton, 1919.

Mencwel, Andrej. *Etos lewicy* [Ethos of the Left]. Warsaw: Państwowy Instytut Wydawniczy, 1990.

Mendelsohn, Andrew. "Medicine and the Making of Bodily Inequality in Twentieth Century Europe." In *Heredity and Infection: The History of Disease Transmission.* Edited by Jean-Paul Gaudillière and Ilana Löwy, 21–79. New York: Routledge, 2001.

The Mianowski Fund. "History of the Józef Mianowski Fund." http://www.mianowski.waw.pl/history.htm (accessed May 30, 2005).

Mickiewicz, Adam. "Reduta Ordona." See "Adam Mickiewicz: Wiersze Wybrane" [In Polish.]. http://univ.gda.pl/~literat/amwiersz/index. htm (accessed January 12, 2010).

Milgrom, Felix. "Fundamental Discoveries in Immunohematology and Immunogenetics by Ludwik Hirszfeld." *Vox Sang* 52 (1987): 149–51.

Miłosz, Czesław. *A History of Polish Literature.* Berkeley: University of California Press, 1983.

Miserachs Rigalt, M. "La transfusion sanguine dans les hôpitaux militaires de première ligne pendant la guerre." *IIe Congrès international de la transfusion sanguine 29 septembre–2 octobre 1937, Paris.* Paris: J.-B. Ballière, 1939, 462–66.

Montagnier, Luc. *Des virus et des hommes.* Paris: Odile Jacob, 1994.

Moss, Pamela, ed. *Placing Autobiography in Geography.* Syracuse, NY: Syracuse University Press, 2001.

Moulin, Anne-Marie. *Le dernier langage de la médecine: Histoire de l'immunologie de Pasteur au Sida.* Paris: Presses universitaires de France, 1991.

Mourant, Arthur E., Ada C. Kopeć, and Kazimiera Domaniewska-Sobczak. *The Genetics of the Jews.* Oxford: Clarendon Press, 1978.

Munthe, Axel. *The Story of San Michele.* London: John Murray, 1929.

The New Concise Pictorial Encyclopedia. New York: Garden City Publishing, 1938.

Nicolle, Charles. *Le destin des maladies infectieuses.* Paris: Alcan, 1933.

———. *L'experimentation en médecine.* Paris: Alcan, 1934.

Ofer, Dalia. "Another Glance through the Historian's Lens: Testimonies in the Study of Health and Medicine in the Ghetto." *Poetics Today* 27, no. 2 (2006): 344.

Okupacja i medycyna [Occupation and Medicine]. 4 vol. Warsaw: Książka i Wiedza, 1971–84.

Opalski, Magdalena M., and Yisrael Bartel. *Poles and Jews. A Failed Brotherhood.* Hanover, NH: University Press of New England, 1992.

Ostrowska, Teresa. "Julia Seydel." *Służba zdrowia* 9 (March 1969).

Ostwald, Wilhelm. *Grosse Männer.* Leipzig: Akademische verlagsgesellschaft m.b.h., 1909.

Owczuk, Radosław, J. Ulewicz-Filipowicz, and Anna Balcerska. "Professor Henryk Brokman—Scientist and Paediatrician." [In Polish.] *Archiwum Historii i Filozofii Medycyny* 63, no. 1 (2000): 35–43.

Parascandola, John. "The Theoretical Basis of Paul Ehrlich's Chemotherapy." *Journal of the History of Medicine and Allied Sciences* 36 (1981): 19–43.

Parnas, Joseph. "Thorvald Madsen 1870–1957. Leader in International Public Health." *Danish Medical Bulletin* 28, no. 2 (April 1981): 82–86.

Pasteur Institute. *Biography of Ernest Conseil (1879–1930).* [In French.] http://www.pasteur.fr/infosci/archives/cns0.html.

Patterson, David. *Sun Turned to Darkness: Memory and Recovery in the Holocaust Memoir.* Syracuse, NY: Syracuse University Press, 1998.

Paul, Diane B., and C. B. Krimbas. "Nikolai V. Timoféef-Ressovsky." *Scientific American* 266, no. 2 (February 1992): 86–92.

Pearce, J. M. S. "Hermann Oppenheim (1858–1919)." *Journal of Neurology, Neurosurgery, and Psychiatry* 74 (2003): 569.

Pedrosa, Gil F., M. M. Weber, and W. Burgmair. "Ernst Kretschmer (1888–1964)." *American Journal of Psychiatry* 159, no. 7 (July 2002): 1111.

Pelis, Kim. "Prophet for Profit in French North Africa: Charles Nicolle and the Pasteur Institute of Tunis, 1903–1936." *Bulletin of the History of Medicine* 71 (1997): 583–622.

Perišić, Živadin. "Akademik Prof. Kosta Todorović (5.VII 1887–22 IX 1975)" [In Memory of Academician Prof. Kosta Todorovic (5 July 1887–22 November 1975)]. *Srpski Arhiv za Celokupno Lekarstvo* 103, no. 10 (October 1975): 911–14.

Peszke, Michael Alfred, and Piotr Stefan. *The Polish Underground Army, the Western Allies, and the Failure of Strategic Unity in World War II.* London: McFarland, 2005.

Podgórska-Klawe, Zofia, ed. *Słownik biograficzny polskich naukowców medycznych XX wieku,* t1, z1 [Biographical Dictionary of Polish Medical Scientists of the 20th Century]. Warsaw: IHNOiT, 1991.

Popkin, Jeremy D. "Holocaust Memories, Historians' Memoirs: First-Person Narrative and the Memory of the Holocaust." *History and Memory* 15 (2003): 49–84.

"Present State and Future Developments of Experimental Gerontology: A Memorial to Fritz Verzár (1886–1979)." *Experientia* 37, no. 10 (October 15, 1981): 1039–59.

Price, G. Ward. *The Story of the Salonica Army.* London: Hodder & Stoughton, 1917.

Prigger, Richard. "Richard Otto." *Archiv für Hygiene und Bakteriologie* 136, no. 8 (1952): 557–58.

"Prosper-Emile Weil (1873–1963)." *Bulletins et memoires de la Société Médicale des Hôpitaux de Paris* 114 (December 13, 1963): 1346–51.

Puxia, Alfred. "Brunon Nowakowski, wybitny przedstawiciel higieny i medycyny pracy" [Brunon Nowakowski, Outstanding Representative of Hygiene and Labor Medicine]. *Archiwum Historii i Filozofii Medycyny* 51, no. 3 (1988): 359–72.

Raczyńska, Krystyna, and Barbara Iwaszkiewicz-Bilikiewicz. "Prof. Morawiecki: The Oculist and Outstanding Expert in Immunology, Immunopathology and Serology." [In Polish.] *Archiwum Historii i Filozofii Medycyny* 63, no. 3–4 (2000): 68–70.

Ranković, Milutin M. "Dr. Aleksa Savić." *Srpski arhiv za celokupno lekarstvo* 30/2 (1928):154–60.

Rapaport, Felix T. "Felix Milgrom, Immunologist and Microbiologist Par Excellence." *Transplantation Proceedings* 31 (1999): 1452–55.

Rapoport, Louis. *Stalin's War against the Jews: The Doctors' Plot and the Soviet Solution.* Toronto: Free Press, 1990.

Reggiani, Andres Horacio. *God's Eugenicist: Alexis Carrel and the Sociobiology of Decline.* New York: Berghahn Books, 2006.

Rénan, Ernest. *Le Judaïsme comme race et comme religion.* Paris: C. Lévy, 1883.

Ringelblum, Emanuel. *Notes from the Warsaw Ghetto: The Journal of Emanuel Ringelblum.* New York: Schocken, 1974.

Rockefeller Foundation Directory of Fellowship Awards, 1917–1950. New York: Rockefeller Foundation, 1951.

Rokityanskij, Yakov G. "N. V. Timofeeff-Ressovsky in Germany (July, 1925–September, 1945)." *Journal of Biosciences* 30, no. 5 (2005): 573–80.

Roland, Charles G. *Courage under Siege: Starvation, Disease, and Death in the Warsaw Ghetto.* New York: Oxford University Press, 1992.

———. "An Underground Medical School in the Warsaw Ghetto, 1941–42." *Medical History* 33, no. 4 (October 1989): 399–419.

Rudnicki, Szymon. "From 'Numerus Clausus' to 'Numerus Nullus'." *Polin* 2 (1987): 246–68.

Russell, Edmund. "The Strange Career of DDT: Experts, Federal Capacity and Environmentalism in World War II." *Technology and Culture* 40 (1999): 776–81.

Sachs, Hans, and Walter Georgi. "Zur Kritik des serologischen Luesnachweises mittels Ausflockung." *Münchener medizinische Wochenschrift* 66 (1919): 440–42.

Schlesinger, R. Walter. Robert Doer—Prophet of the Nature of Viruses, Founder of the Archives of Virology. *Archives of Virology* 142, no. 4 (1997): 862–73.

Schneider, William H. "Arnault Tzanck, MD (1886–1954)." *Transfusion Medicine Reviews* 24 (2010): 147–50.

———. "Blood Group Research in Great Britain, France and the United States between the World Wars." *Yearbook of Physical Anthropology* 38 (1995): 77–104.

———. "Chance and Social Setting in the Application of the Discovery of Blood Groups." *Bulletin of the History of Medicine* 57, no. 4 (1983): 545–62.

———, ed. "The First Genetic Marker: Blood Group Research, Race and Disease, 1900–1950." *History and Philosophy of the Life Sciences* 18, special issue (1996): 273–303.

———. "Henri Laugier: The Science of Work and the Workings of Science in France, 1920–1940." *Cahiers pour l'histoire du CNRS* 5 (1989): 7–34.

———. "The History of Research on Blood Group Genetics: Initial Discovery and Diffusion." *History and Philosophy of the Life Sciences* 18 (1996): 273–303.

———. "The Men Who Followed Flexner: Richard Pearce, Alan Gregg, and the Rockefeller Foundation Medical Divisions, 1919–1951." In *Rockefeller Philanthropy and Modern Biomedicine.* Bloomington: Indiana University Press, 2002.

Schwabe, Klaus. "Ursprung und Verbreitung des alldeutschen Annexationismus in der Deutschen Professorenschaft im Ersten Weltkrieg". *Vierteljahrshefte für Zeitgeschichte* 14, no. 2 (1966): 105–38.

IIe [Second] *Congrès international de la transfusion sanguine, 29 septembre–2 octobre 1937, Paris.* Paris: J.-B. Ballière, 1939.

Segré, Emilio. *A Mind Always in Motion: The Autobiography of Emilio Segrè.* Berkeley: University of California Press, 1993.

Servi, Sandro. "Building a Racial State: Images of the Jew in the Illustrated Fascist Magazine, *La Difesa della Razza*, 1938–1943." In *The Jews of Italy under Fascist and Nazi Rule, 1922–1945.* Edited by Joshua D. Zimmerman, 114–57. Cambridge: Cambridge University Press, 2005.

Shortland, Michael, and Richard Yeo, eds. *Telling Lives in Science: Essays on Scientific Biography.* Cambridge: Cambridge University Press, 1996.

Shoskes, Henry. *No Traveler Returns.* Garden City, NY: Doubleday, Doran, 1945.

Sienkiewicz, Henryk. "Bartek zwycięzca" [Bartek the Conqueror]. Warsaw: Nasza Ksiegarnia, 1996.

———. *Potop* [*The Deluge*]. Warsaw: Paristwowy Instytut Wydawniczy, 1984.

"Sierakowski, Dr. Stanisław." In *Polski Słownik Biograficzny* [Polish Biographical Dictionary]. Edited by Instytut Historii im. Tadeusza Manteuffla. Vol. 37. Cracow: Polska Akademia Nauk and Polska Akademia Umiejętności, 1948.

Silverstein, Arthur. *A History of Immunology.* New York: Academic Press, 1989.

Słowacki, Juliusz. "Hymn." http://slowacki.chez.com/slowhymn.htm (accessed 14 May 2010).

Słownik biograficzny lekarzy i farmaceutów ofiar drugiej wojny swiatowej [Biographical Dictionary of Physician and Pharmacist Victims of the Second World War]. Wrocław: Wydawnictwo Medyczne Urban & Partner, 1997.

Smith, Dennis Mack. *Mussolini.* New York: Knopf, 1982.

Squire, Larry R., ed. *The History of Neuroscience in Autobiography.* 4 vols. San Diego: Academic Press, 2004.

Stapleton, Darwin. "Internationalism and Nationalism: The Rockefeller Foundation, Public Health, and Malaria in Italy, 1923–1951." *Parassitologia* 42, no. 1–2 (June 2000): 127–34.

———. "A Lost Chapter in the Early History of DDT: The Development of Anti-Typhus Technologies by the Rockefeller Foundation's Louse Laboratory, 1942–1944." *Technology and Culture* 46, no. 3 (July 2005): 513–40.

Starr, Douglas. *Blood: An Epic History of Medicine and Commerce.* New York: Knopf, 1998.

Steffan, Paul. "Bedeutung der Blutuntersuchung fur die Bluttransfusion und die Rassenforschung." *Archiv für Rassen- und Gesellschafts-Biologie* 15 (1923): 137–50.

Stern-Burstin, Barbara. "The Warsaw Ghetto: A Shattered Window on the Holocaust." *The History Teacher* 13, no. 4 (August 1980): 531–41.

Strong, Richard Pearson. *Typhus Fever With Particular Reference to the Serbian Epidemic.* Cambridge, MA: Harvard University Press, 1920.

Summers, William C. "Hans Zinsser: A Tale of Two Cultures." *Yale Journal of Biological Medicine* 72, no. 5 (1999): 341–47.

Szajkowski, Zosia. "Disunity in the Distribution of American Jewish Overseas Relief 1919–1939." *American Jewish Historical Quarterly* 58, no. 3 (1969): 376–407, 484–506.

Szejnman, Michał. "Changes in Peripheral Blood and Bone Marrow in Hunger Disease." In *Hunger Disease.* Edited by Myron Winick, 161–99. New York: Wiley, 1979.

Szenkowa, Izabela. "80-lecie profesora F. Przesmyckiego" [On the 80th Birthday of Prof. F. Przesmycki]. *Służba zdrowia* 7 (1972): 9.

Szpilman, Władysław. *The Pianist.* New York: Picador, 1998. Originally published as *Śmierć miasta* [Death of a City], 1945.

Szwajger, Adina Blady. *I Remember Nothing More: The Warsaw Children's Hospital and the Jewish Resistance.* New York: Pantheon, 1990.

Szybalski, Wacław. "The Genius of Rudolf Stefan Weigl (1883–1957), a Lvovian Microbe Hunter and Breeder—In Memoriam." *International Weigl Conference (Microorganisms in Pathogenesis and Their Drug Resistance—Programme and Abstracts).* September 11–14, 2003. Edited by R. Stoika et al. Lviv: SPOLOM Publishers, 2003, 10–31. http://www.Lwów.com.pl/weigl/in-memoriam.html (accessed May 30, 2007).

Szymanowski, Zygmunt. "Thirtieth Anniversary of Founding of the State Institute of Hygiene." [In Polish.] *Polski Tygodnik Lekarski* (Warsaw) 6, no. 35, supplement (August 27–September 2, 1951): 173–74.

Taylor, Ronald. *Berlin and Its Culture: A Historical Portrait.* New Haven, CT: Yale University Press, 1997.

Teichmann, Jens. "On the 100th Birthday of Rudolf Kraus." [In German.] *Wiener medizinische Wochenschrift* 118, no. 42 (October 19, 1968): 869–72.

Teller, Edward. *Memoirs: A Twentieth-Century Journey in Science and Politics.* Cambridge, MA: Perseus, 2001.

Tomasevich, Jozo. *War and Revolution in Yugoslavia 1941–1945: Occupation and Collaboration.* Stanford: Stanford University Press, 2001.

Tomaszewski, Irene, and Tecia Werbowski. *Żegota: The Council for Aid to Jews in Occupied Poland, 1942–1945.* Montreal: Price-Patterson, 1999.

Tomkiewicz, Stanisław. *Une adolescence volée.* Paris, Calmann-Lévy, 1999.

Tracy, Sarah W. "George Draper and American Constitutional Medicine, 1916–1946: Reinventing the Sick Man." *Bulletin of the History of Medicine* 66, no. 1 (Spring 1992): 53–89.

Trauring, Philip, ed. "Communique of the Polish-Soviet Extraordinary Commission for Investigating the Crimes Committed by the Germans in the Majdanek Extermination Camp in Lublin." In *JewishGen: The Home of Jewish Genealogy.* http://www.jewishgen.org/ForgottenCamps/Camps/MajdanekReport.html (accessed June 30, 2005).

Trout, David L. "Max Josef von Pettenkofer (1818–1901)—a Biographical Sketch." *Journal of Nutrition* 107, no. 9 (1977): 1567–74.

Trunk, Isaiah. *Judenrat; the Jewish Councils in Eastern Europe under Nazi Occupation.* New York: Macmillan, 1972.

Trzos-Rastawiecki, Andrzej, director. *Gdziekolwiek jesteś, panie prezydencie* [Wherever You Are, Mister President]. Motion picture. 1978.

Tushnet, Leonard. *The Uses of Adversity: Studies of Starvation in the Warsaw Ghetto.* New York: Yoseloff, 1966.

[UNRRA] United Nations Relief and Rehabilitation Administration. "Health Conditions in Poland." *Operational Analysis Papers* no. 31. London: European Regional Office, 1947.

Van Herwerden, C. A. B. *Marianne van Herwerden 16 Februari 1874–26 Januari 1934.* Rotterdam: W. L. Brusse, 1948.

Van Hyde, H. Z. "A Tribute to Andrija Štampar, 1888–1958." *American Journal of Public Health and the Nation's Health* 48, no. 12 (December 1958): 1578–82.

Verzár, Fritz. "Rassenbiologische Untersuchungen mittels Isohamagglutinen." *Biochemische Zeitschrift* 126 (1922): 33–39.

Von Behring, Emil, Paul Ehrlich, Fritz Haber et al. "Manifesto of the 93 German Intellectuals to the Civilized World." In *World War I Document Archive.* Edited by Jane Plotke and Richard Hacken. http://www.gwpda.org/1914/93intell.html (accessed May 31, 2005).

Von Dungern, Emil, and Ludwik Hirszfeld. "Über Nachweis und Vererbung biochemischer Strukturen. I." "Über Vererbung gruppenspezifischer Strukturen des Blutes. II." "Über gruppenspezifische Strukturen des Blutes. III." In *Zeitschrift für Immunitätsforschung* 4 (1910): 531–46; 6, 284–92; and 8 (1911): 526–62. Translated from the German in *Transfusion* 2 (1962): 70–74.

Voswinckel, Peter. "Grabstein vor dem Verfall bewahrt. In memoriam Professor Alfred Wolff-Eisner (1877–1948), jüdischer Pionier der Pollenkrankheit." *Allergologie* 11 (1988): 41–46.

Walbaum, Jost, ed. *Kampf den Seuchen! Deutscher Ärtze-Einsatz im Osten. Die Aufbarheit im Gesundheitwesen der Generalgouvernements.* Krakau: Bucherverlag "Deutscher Osten" G.m.b.H, 1941.

Wallace, Daniel J., and Michael Weisman. "Should a War Criminal Be Rewarded with Eponymic Distinction? The Double Life of Hans Reiter (1881–1969)." *Journal of Clinical Rheumatology* 6, no. 1 (2000): 49–54.

Wannsee Protocol. Meeting, Berlin, January 20, 1942. http://www.yale.edu/lawweb/avalon/imt/wannsee.htm (accessed June 21, 2007).

Watson, James. *Double Helix.* Cambridge, MA: Harvard University Press, 1962.

Watt, Richard. *Bitter Glory: Poland and Its Fate, 1918–1939.* New York: Simon and Schuster, 1979.

Weil, Edmund and Arthur Felix. "Untersuchungen über das Wesen der Fleckfieber-Agglutination." *Wiener Klinische Wochenschrift* 10 (1917): 393–99.

Weindling, Paul J. *Epidemics and Genocide in Eastern Europe 1890–1945.* New York: Oxford University Press, 2000.

———, ed. *International Health Organisations and Movements, 1918–1939.* New York: Cambridge University Press, 1995.

———. "Public Health and Political Stabilisation: The Rockefeller Foundation in Central and Eastern Europe between the Two World Wars." *Minerva* 31 (1993): 253–67.

Weingart, Peter. *Doppel-Leben: Ludwig Ferdinand Clauss: zwischen Rassenforschung und Widerstand.* Frankfurt: Campus, 1995.

Werfel, Franz. *Forty Days of Musa Dagh.* New York: Carroll & Graf, 1933.

———. *Song of Bernadette.* American Limited, 1941.

Wess, Ludger. "Menschenversuche und Seuchenpolitik—Zwei unbekannte Kapitel aus der Geschichte der deutschen Tropenmedizin." *Zeitschrift für sozialgeschichte des 20 un 21 Jahrhunderts* (1999), 10–50.

Widmann, Joseph Viktor. *Maikäfer-Komödie.* Frauenfeld: Huber, 1897.

Winick, Myron, ed. "Hunger Disease. Studies by the Jewish Physicians in the Warsaw Ghetto. 1979." *Nutrition* 10 no. 4 (July–August 1994): 365–79. First published with the same title in 1979.

Wintrobe, Maxwell M., ed. *Blood Pure and Eloquent.* New York: McGraw-Hill, 1980.

Witebsky, Ernest, and Lillian M. Engasser. "Blood Groups and Subgroups of the Newborn: I. The A Factor of the Newborn." *Journal of Immunology* 61 (1949): 171–78.

Wojciechowski, Edmund. "In Memoriam: Feliks Przesmycki (1892–1974)." [In Polish.] *Przegląd Epidemiologiczny* 29, no. 3 (1975): 386–89.

———. "In Memoriam: Tadeusz Sprożyński (1903–1974)." [In Polish.] *Medycyna Doświadezalna Mikrobiologia* 27, no. 1 (1975): 97–99.

Wood, William T. *The Salonica Front.* London: A & C Black, 1920.

Wulf, Josef. *Das Dritte Reich und seine Vollstrecker; die Liquidation von 500 000 Juden im Ghetto Warschau.* Berlin-Grunewald: Arani, 1961.

Wulf, Stefan. *Das Hamburger Tropeninstitut 1919 bis 1945.* Hamburg: Dietrig Reimer Verlag, 1994.

Wynot, Edward D., Jr. *Warsaw between the World Wars.* New York: Columbia University Press, 1983.

Yad Vashem Holocaust Research Center. http://www1.yadvashem.org/about_holocaust/bibliography/home_bibliography.html. (accessed May 19, 2010)

Zaloga, Steven J. *Poland 1939: The Birth of Blitzkrieg.* Westport, CT: Praeger, 2002.

Żeromski, Stanisław. *Ludzie bezdomne* [Homeless People]. Warsaw: B. Natanson, 1899.

Zimmerman, J. D., ed. *Contested Memories. Poles and Jews during the Holocaust and Its Aftermath.* New Brunswick, NJ: Rutgers University Press, 2003.

Zimmerman, Joshua D., ed. *The Jews of Italy under Fascist and Nazi Rule, 1922–1945.* Cambridge: Cambridge University Press, 2005.

Zinsser, Hans. *As I Remember Him; the Biography of R. S.* Boston: Little, Brown, 1940.

———. *Rats, Lice and History.* Boston: Little, Brown, 1934.

———. "Report of Bacteriologist of the American Red Cross Sanitary Commission to Serbia." In *Typhus Fever with Particular Reference to the Serbian Epidemic.* Edited by Richard P. Strong, 261–68. Cambridge, MA: Harvard University Press, 1920.

Zinsser, Hans, and F. F. Russell. *A Textbook of Bacteriology.* 5th ed. New York: D. Appleton, 1922.

Zorini, A. Omodei. "Eugenio Morelli." *Lotta contro la tubercolosi* 30 (1960): 665–75.

Zweig, Stefan. *Die Heilung durch den Geist. Mesmer. Mary Baker-Eddy. Freud.* Leipzig: Insel-verlag, 1931.

Zygmund, Antoni. "Aleksander Rajchman (1890–1940)." *Roczniki Polskiego Towarzystwa Matematycznego* 27 (1987): 219–23.

Zylberman, Patrick. "Fewer Parallels Than Antitheses: René Sand and Andrija Štampar on Social Medicine, 1919–1955." *Social History of Medicine* 17 (2004): 77–92.

Index

constitutional serology, 15, 101, 102, 103, 136
Corfu, 46, 51, 389n3
Council for Aid to Jews. See Committee for the Aid to Jews
Cox, H. R. (bacteriologist), 360
Craigie, James, 359, 425n24
Curie, Marie, 65, 98, 140, 153, 404n6
Czarnecki, Antoni, 245, 261
Czerniaków, Adam, 196, 197, 201, 204, 205, 207, 215, 216, 217, 227, 228, 234, 238, 242, 245, 260, 261, 365, 411n16
Czerny, Vincent, 13, 21, 384n3

Dale, Henry Hallett, 71, 394n23
Damond (doctor and hospital director in Salonica), 47, 48, 49, 52
Danysz, Jan, 65, 140, 365, 391–92n5
DDT, xxxi, 360
De Kruif, Paul, 95, 367, 397n9
Debré, Robert, xvii, 103, 137, 150–51, 152, 365, 398n12
Detienne (doctor and researcher at Heidelberg), 16
Dieterle, T. (doctor and researcher at Zürich), 25
Długoszowski-Wieniawa, Bolesław, 148, 405n10
Dmochowski, 106
Dobrzaniecki (professor of medicine at University of Lwów liquidated by Germans), 311
Doerr, Robert, 27, 119, 143
du Bois-Raymond, Emil, 150, 406n1
Dubos, René, 360, 425n28
Dubrovnik, 60–61
Duke of Baden, 14
dysentery, 52, 53, 54, 67, 71, 76, 77, 119, 207, 214, 226, 227, 236, 239

Eckerman (member of Health Council in Warsaw ghetto), 217
Ehmer, Wilhelm, 335, 423n1
Ehrlich, Paul, xi, 7, 9, 18, 27, 64, 68, 98, 251, 366, 370, 383n12, 384n18, 386n10, 386n1, 390n16, 392n9, 404n8
Einstein, Albert, 59, 251, 388n4
Eisenberg, Filip, 79, 395n32

Ekerman, (councilor at cemetery in Warsaw Ghetto) 237
Emslie, Isabel Hutton, 51, 390n7
Endelman, Tadeusz, 261
Engineer D. (helped Hirszfelds escape ghetto), 274, 276
Epstein, Tekla, xxxiii, 207, 265, 270
eugenics, 106, 258, 338, 398n18, 405n8; Eugenics Society, 106
Experimental and Social Medicine journal, 73, 74, 85

Fabre, René, 136, 137, 403n13
Farreus (Swedish typhoid researcher), 57
fascism, xxv, 117, 127, 128, 129, 130, 131, 145, 146, 147, 149, 401n4, 402n7, 405n3
Feer, Emil, 21
Fejgin, Bronisława, 80, 99, 270
Ficker, Martin, 7, 8, 383n11
Final Solution, xxx, 260–72, 419n1
Flaks, Doctor, 138
Flatan, Edward, 206
Flexner, Abraham, 141, 358, 365, 404n10, 424n20
Flexner, Simon, 92, 365, 396n7
Fliederbaum, Julian, 204, 298, 413n7
Floksztrumf, M. (researcher at Hygiene Institute), 105, 106, 138
Flugge, Karl, 68, 392n7
Forlanini, Carl, 401n7, 401n4
Fosdick, Raymond, 361, 426n32, 426n34
Free Polish University (Free University), 72–73, 89, 94, 207, 355
Frenkiel, (chairman of Łódź Medical Society), 310
Friedberg, Michal, 220
Friedberger, Ernst, 27, 386n13
Friedemann, Ulrich, 9, 24, 360, 384n17, 425n31
Funk, Kazimierz, xxix

Galvani, Luigi, 97, 98
Ganc, Tadeusz, 216, 220, 228, 365, 415n12
gas chambers, xxx, 237, 271, 284, 301, 303, 304, 305, 340, 409n16

Syrkin-Binstein, Zofja, 216, 297, 371, 421n1
Szejnman, Michal, 209, 415n23
Szent-Gyorgyi, Albert, 352, 424n11
Szeryński, Jozef, 203, 263, 371, 413n6
Szeynman, Mieczyslaw, 270
Szniolis, Aleksander, 72, 220, 355, 371, 416n3
Szper, Miss (worked at Institute of Hygiene), 99
Szulc, Gustaw, 86, 117, 160, 163, 175, 227, 371, 399n2
Szymanowski, Zygmunt, 69, 74, 75, 177, 207, 274, 364, 371, 393n12, 414n16

Taylor (doctor with Scottish Women's Hospital at Salonica), 51
Teitge, Heinrich, 212, 415n6
Teplitz, Henryk, 261
Timofeeff-Ressovsky, Nikolai, 145, 405n5
Todorović, Kosta, 55, 56, 57, 390n18
Toebbens (shop in Warsaw Ghetto), 267, 299, 421n2
Toporol. See Society for Supporting Agriculture
Treaty of Versailles, 68, 327, 392n5, 407n1
Treblinka, 284, 300–301, 312, 316, 368, 412n23, 416n6, 417n14, 417n19, 419n1
Tubbet, Miss (befriended Hirszfelds at Salonica), 54
tuberculosis, 8, 18, 57, 94, 98, 105–6, 129, 182, 220, 221, 223–27, 229, 239, 271, 401n4, 406n3, 412n2, 415n23
typhoid, 51, 52, 54, 55, 57, 76, 99, 110, 222, 226, 359, 360, 390n14. See also paratyphoid bacillus
typhus, xxii, xxvi, xxxi, 51, 53, 67, 69–70, 76, 182–83, 266–69, 236, 239, 241–42, 249, 251, 270, 360, 367, 368, 373, 377n24, 381n67, 388n12, 393n14, 396n4, 409n16, 410n24, 411n8, 414n12, 414n20; in Serbia, xviii, xxii, xxix, 30–34, 36, 357, 387n5; in the Warsaw ghetto, xvi,

xxx, xxxii, xxxiii, 190, 195, 197, 199, 205–9, 210–18, 220–23, 226–29, 236, 239, 241–42, 249, 270, 365, 371, 410n3, 411n14, 413n12, 414n21, 415n7, 415n12
Tzanck, Arnault, 130, 131, 133, 372, 402n8, 402n12, 402n14

Ujazdowski Hospital, 169, 178
Umiastowski, Roman, 167, 174, 407n5
UNICEF, xx, 369
United Nations Relief and Rehabilitation Administration, xxxi, 424n14
University of Wroclaw Medical Faculty, 353, 354

Valéry-Radot, Louis Pasteur, 152
Valjevo, xxvii, 31, 32, 34, 36, 49, 56, 62, 387n9, 389n6
Valjevo Infectious Disease Hospital, 31, 32, 34–36, 38
van Herwerden, Marianne, 123, 400n3
van Miller, Anton (Franz Rudolf Bienenfeld), 250, 251, 418n4
Vandervelde, Emile, 139, 404n4
venereal disease, 95, 220, 221, 227. See also syphilis
Verzár, Fritz, 100, 123, 124, 164, 183, 372, 397n6, 400n2, 401n6
Vieweg, Doctor (head of Social Security Health Center of General Government), 181
Virchow, Rudolf, 8, 383n3, 384n17
virology, 159
Voit, Carl, 59, 276, 391n25
voivodship, 97, 181, 277, 326, 397n1
von Behring, Emil, 27, 67, 175, 228, 372, 387n1, 408n1, 416n10
von Bergmann, Gustav, 121, 400n13
von dem Bach, 351
von Dungern, Emil, xi, xxi, xxviii, xxxiv, 12, 13–16, 18, 20–21, 23–24, 27, 58, 59, 82, 86, 104, 158, 176, 313, 372, 380n53, 384n6, 385n12, 385n16
von Gonzenbach, Willi, 22, 385n1
von Mackensen, August, 38, 388n17, 389n3